World Atlas of Seagrasses

Published in association with
UNEP-WCMC by the University of
California Press
University of California Press
Berkeley and Los Angeles, California
University of California Press, Ltd.
London, England

© 2003 UNEP World Conservation Monitoring Centre
UNEP-WCMC
219 Huntingdon Road
Cambridge CB3 0DL, UK
Tel: +44 (0) 1223 277 314
Fax: +44 (0) 1223 277 136
E-mail: info@unep-wcmc.org
Website: www.unep-wcmc.org

No part of this book may be reproduced by any means or transmitted into a machine language without the written permission of the publisher.

The contents of this volume do not necessarily reflect the views or policies of UNEP-WCMC, contributory organizations, editors or publishers. The designations employed and the presentations do not imply the expression of any opinion whatsoever on the part of UNEP-WCMC or contributory organizations, editors or publishers concerning the legal status of any country, territory, city or area or its authority, or concerning the delimitation of its frontiers or boundaries or the designation of its name or allegiances.

Cloth edition ISBN
0-520-24047-2

Cataloging-in-publication data is on file with the Library of Congress

Citation Green E.P. and Short F.T. (2003) *World Atlas of Seagrasses*. Prepared by the UNEP World Conservation Monitoring Centre. University of California Press, Berkeley, USA.

World Atlas of Seagrasses

Edmund P. Green & Frederick T. Short

UNIVERSITY OF CALIFORNIA PRESS
BERKELEY LOS ANGELES LONDON

World Atlas of Seagrasses

Prepared by
UNEP World Conservation
Monitoring Centre
219 Huntingdon Road
Cambridge CB3 0DL, UK
Tel: +44 (0) 1223 277 314
Fax: +44 (0) 1223 277 136
E-mail: info@unep-wcmc.org
Website: www.unep-wcmc.org

Director
Mark Collins

Scientific editors
Edmund P. Green
Frederick T. Short

Assistant scientific editor
Michelle Taylor

Cartographer
Corinna Ravilious

Technical editor
Catherine Short

Layout
Yves Messer

A Banson production
27 Devonshire Road
Cambridge CB1 2BH, UK

Color separations
Swaingrove

Printed in China

Jackson Estuarine Laboratory contribution number 396.

The UNEP World Conservation Monitoring Centre is the biodiversity assessment and policy implementation arm of the United Nations Environment Programme (UNEP), the world's foremost intergovernmental environmental organization. UNEP-WCMC aims to help decision-makers recognize the value of biodiversity to people everywhere, and to apply this knowledge to all that they do. The Centre's challenge is to transform complex data into policy-relevant information, to build tools and systems for analysis and integration, and to support the needs of nations and the international community as they engage in joint programs of action.

UNEP-WCMC provides objective, scientifically rigorous products and services that include ecosystem assessments, support for implementation of environmental agreements, regional and global biodiversity information, research on threats and impacts, and development of future scenarios for the living world.

Illustrations in Appendix 3 and at foot of pages:

Mark Fonseca: *Zostera marina*.

Phillips RC, Meñez EG (1988). *Seagrasses*. Smithsonian Contributions to the Marine Sciences 34. Smithsonian Institution Press, Washington DC: *Zostera asiatica*.

QDPI Northern Fisheries Centre, Cairns: *Halophila australis*, *Halophila capricorni*.

Ron Phillips: All remaining illustrations.

Supporting institutions

 The United Nations Environment Programme is the principal United Nations body in the field of the environment. Its role is to be the leading global environmental authority that sets the global environmental agenda, promotes the coherent implementation of the environmental dimension of sustainable development within the United Nations system and serves as an authoritative advocate for the global environment. http://www.unep.org

 The UK Department for Environment, Food and Rural Affairs is working for sustainable development: a better quality of life for everyone, now and for generations to come. This includes a better environment at home and internationally, and sustainable use of natural resources; economic prosperity through sustainable farming, fishing, food, water and other industries that meet consumers' requirements; thriving economies and communities in rural areas and a countryside for all to enjoy.

 The Department for International Development is the UK Government department working to promote sustainable development and eliminate world poverty. This publication is an output from a research program funded by DFID for the benefit of developing countries. The views expressed are not necessarily those of DFID. http://www.dfid.gov.uk/

The David and Lucile Packard Foundation The David and Lucile Packard Foundation, started in 1964, provides international and national support to non-profit organizations in conservation, science and many other areas. The foundation currently provides funding to SeagrassNet, a global seagrass monitoring program based at the University of New Hampshire. http://www.packard. org/, http://www.seagrassnet.org/

 The University of New Hampshire is a land-grant, sea-grant and space-grant public institution with 10 000 undergraduate and 2 000 graduate students, and a well-established marine program. The Jackson Estuarine Laboratory is the primary marine research organization at UNH and has a strong seagrass research component. http://www.unh.edu/, http://marine.unh.edu/jel/home.html

 The World Seagrass Association is committed to the science, protection and management of the seagrass ecosystem worldwide. The members come from many countries and include leading scientists in marine and seagrass biology. The association supports training and information exchange and raises global awareness of seagrass science and environmental management issues. http://www.worldseagrass.org/

 The Convention on Wetlands, signed in Ramsar, Iran, in 1971, is an intergovernmental treaty which provides the framework for local, regional and national actions for the conservation and wise use of wetlands and their resources. There are presently 135 Contracting Parties to the Convention, with 1 230 wetland sites, totaling 105.9 million hectares, designated for inclusion in the Ramsar List of Wetlands of International Importance. http://ramsar.org/

 The International Coral Reef Action Network is an innovative and dynamic global partnership of many of the world's leading coral reef science and conservation organizations. Established in 1999 to halt and reverse the decline of the health of the world's coral reefs the partnership draws on its partners' investments in reef monitoring and management to create strategically linked actions at local, national and global scales. http://icran.org/

 The Scientific Committee on Oceanic Research (SCOR) was the first interdisciplinary body formed by the International Council for Science. SCOR activities focus on promoting international cooperation in planning and conducting oceanographic research. http://www.jhu.edu/~scor

 The Estuarine Research Federation (ERF) is a private, non-profit organization. The federation was created in 1971 to address broad estuarine and coastal issues; it holds biannual international meetings and supports the scientific publication *Estuaries*. http://www.erf.org/

Acknowledgments

The *World Atlas of Seagrasses* is a product of global collaboration between many different people but a few merit a special mention and deserve extra thanks. First and foremost amongst these are the 58 authors who have given freely and extensively of their time and experience in writing the 25 chapters that constitute this *World Atlas*. Without their attention to detail and efforts in sourcing data outside the mainstream scientific literature this book would not have been possible. The contributions of the assistant scientific editor, Michelle Taylor, the cartographer, Corinna Ravilious, and technical editor, Catherine Short, have been equally invaluable in the synthesis of information from many hundreds of disparate sources and for ensuring consistency throughout every section. The editors have a particular debt of gratitude to all these people.

The origins of the *World Atlas of Seagrasses* go back to late 1997 when the need for a global compendium of information on seagrasses was first acknowledged. Hans de Jong, Eddie Hegerl, Paul Holthus, Richard Luxmore and the participants at the third International Seagrass Biology Workshop, 19-26 April 1998, Manila and Bolinao, Philippines, were particularly helpful in pulling these ideas together. Nicholas Davidson, Salif Diop, Will Rogowski, Ed Urban, Genevieve Verbrugge and Marjo Vierros provided great support during the fundraising, support which was instrumental in making the necessary resources available. We are, of course, notably grateful to our sponsors, listed at the beginning of the *World Atlas*, for investing funds in this work. Special mention is due here to the David and Lucile Packard Foundation and University of New Hampshire for supporting Fred Short's time.

A long period of data collection followed soon after work began and involved much detailed correspondence with very many people. Our thanks go out to everyone who answered our 'phone calls and e-mails, but even more so to those who provided us with seagrass distribution data or maps, especially William Allison, Alex Arrivillaga, Susanne Baden, Seth Barker, David Blackburn, Simon Blyth, Christoffer Boström, Nikki Bramwell, Marnie Campbell, Jacoby Carter, Rob Coles, Helen Cottrell, Lucy Conway, Charlie Costello, Jeffrey Dahlin, Dick de Jong, Karen Eckert, Caroline Erftemeijer, Randolph Ferguson, Mark Finkbeiner, Terence Fong, Mark Fonseca, Sarah Gage, Martin Gullström, Rob Hughes, Herman Hummel, Hitoshi Iizumi, Chung Il Choi, Emma Jackson, Pauline Kamermans, Hilary Kennedy, Ryo Mabuchi, Ian May, Pete McLain, Thomas Meyer, Mark Monaco, Kenji Morita, Ivan Nagelkerken, Brian Pawlak, Karin Pettersson, Ron Phillips, Martin Plus, Chris Pickerell, Jean Pascal Quod, Thorsten Reusch, Ron Rozsa, Jan Steffen, Marieke van Katwijk, Mikael von Numers, Rob Williams, Lisa Wood, Masumi Yamamuro and members of the Wider Caribbean Sea Turtle Conservation Network (WIDECAST): Timothy Austin, Andy Caballero, Didiher Chacon, Juan Manuel Díaz and Alan Mills. Clearly a number of data sharing agreements were necessary as these data were identified and we thank Mary Cordiner for sorting out these institutional complexities.

The coordinators of the Global Seagrass Workshop, Mark Spalding and Michelle Taylor, and all 23 delegates (page 283) are gratefully acknowledged for the time and effort they made to review, amend and correct the seagrass distribution data. The workshop itself would not however have been possible without considerable logistic and organizational support from Janet Barnes, Joy Bartholomew, Jean Finlayson, Anne Giblin, Pam Price, Ed Urban and Susan White, also all the staff at the Tradewinds Hotel, St Petersburg Beach, Florida.

Jamie Adams, Mary Edwards and Sergio Martins have provided additional geographical information systems support at various stages during the preparation of the maps for the *World Atlas* and Elizabeth Allen, Janet Chow, Mary Cordiner and Michael Stone have spent very many long hours formatting and organizing the reference sections for the chapters and on-line bibliography (http://www.unep-wcmc.org/marine/seagrassatlas/references).

Readers will quickly note the wonderful photographs which have been kindly donated. Credit is given next to individual photographs but thanks are also due here to Nancy Diersing, Florence Jean, Karine Magalhaes, Kate Moran and John Ogden all of whom helped us track down owners of pictures which we wanted to use. Most of the drawings of seagrasses in Appendix 3 and illustrated at the foot of the pages were donated by Ron Phillips, whom we especially thank for this contribution. Rob Coles, Mark Fonseca and Mike Fortes provided additional drawings for Appendix 3 and the page corners, and we thank them as well.

When reviewing correspondence and notes spanning the last five years it is all too easy to overlook or forget someone. Sincere apologies to anyone whom we have neglected to mention here. Please be assured that this was simply an oversight brought about by the effort of completing the book and nothing else!

Ed Green and Fred Short

Preface

It is with great pleasure that I introduce this new book from the UNEP World Conservation Monitoring Centre. The World Summit on Sustainable Development adopted, in the area of biodiversity, a commitment to reverse the trend of losses by 2010. To achieve this we need hard facts on which to base decisions. The *World Atlas of Seagrasses* meets that need for a vital marine ecosystem whose importance has largely been overlooked until now.

This book would not have been possible without a remarkable collaboration between the 58 authors from 25 countries. The *World Atlas of Seagrasses* has played a role in fostering international collaboration by gathering information from many different sources all over the world. On behalf of UNEP I would like to express my gratitude to all the authors who have contributed their knowledge.

I would also like to thank the sponsors of the *World Atlas of Seagrasses* including the UK Department for Environment, Food and Rural Affairs, the UK Department for International Development, the Secretariat of the Ramsar Convention on Wetlands, the David and Lucile Packard Foundation, the University of New Hampshire, the World Seagrass Association, the Scientific Committee on Oceanic Research, the Estuarine Research Federation and the International Coral Reef Action Network.

I am confident that this book will help not only UNEP but all interested parties to focus on the implementation of sustainable development in the marine environment worldwide.

Klaus Toepfer
Executive Director, United Nations Environment Programme (UNEP)

Foreword

In describing the complex relationships that exist in the living world we all too often focus on the maestros that take center stage. It is true that rain forests, coral reefs, whales, tigers and the like carry an important representational role as they fill our television screens and become a priority in our conservation programs. But we should not forget the many other ecological players that make up nature's orchestra. The living world is an interactive and integrated continuum that we partition into ecosystems for our own scientific convenience. The less well-known ecosystems often play a distinct and very important part in the overall harmony that we need to maintain, but only poorly understand. One such ecosystem is the beds of seagrass that are found on coastlines around the world.

Seagrass beds are unusual in that they are very widespread, occurring on shallow coastlines in all but the coldest waters of the world. A small group of flowering plants, just 60 among the 270 000 species of fish, plants and other organisms that have colonized the sea, they owe their success to this ability to tolerate a wide range of conditions. So why have they been selected for this global report?

First of all, seagrass beds are an important but under-rated resource for coastal people. Physically they protect coastlines from the erosive impact of waves and tides, chemically they play a key role in nutrient cycles for fisheries and biologically they provide habitat for fish, shellfish and priority ecotourism icons like the dugong, manatee and green turtle. And yet, despite these important attributes, they have been overlooked by conservationists and coastal development planners throughout their range.

This *World Atlas of Seagrasses* is literally putting seagrass beds onto the map, for the first time. It is a groundbreaking synthesis that provides people everywhere with the first world view of where seagrasses occur and what has been happening to them. It is a worrying story. Seagrass beds have been needlessly destroyed for short-term gain without real analysis of the values that the intact ecosystems bring to coastal society. There is no proper strategy for their protection. Their significance is not well appreciated and awareness is very low. This *World Atlas* will go a long way towards reversing these trends.

As ever in the production of an analysis of this kind, our scientists at the UNEP World Conservation Monitoring Centre have been able to achieve their results only by standing on the shoulders of giants. We acknowledge and applaud the dedicated band of seagrass ecologists and taxonomists who have laid the groundwork for this *World Atlas* and prepared much of the text. I hope it will bring well-deserved recognition for them and for their seagrasses, and establish a baseline from which to build a more sustainable future for coastal peoples and the home of the gentle dugong.

Mark Collins
Director, UNEP-WCMC

Contents

Acknowledgments		vi
Preface		vii
Foreword		viii
Introduction to the World Atlas of Seagrasses		1
Key to maps and mapping methods		4

GLOBAL OVERVIEW
The distribution and status of seagrasses — 5

MAPS
- 1 World seagrass distribution — 21
- 2 Global seagrass biodiversity — 22

FIGURES
- 1 Relative size-frequency distribution of 538 seagrass polygons in latitudinal swathe 20-30°S — 14
- 2 Growth of marine protected areas which include seagrass ecosystems, shown both as the number of sites (line) and the total area protected (shaded area) — 20

TABLES
- 1 A list of seagrass species by family — 6
- 2 Major taxonomic groups found in seagrass ecosystems, with brief notes — 11
- 3 Threatened species regularly recorded from seagrass communities worldwide — 12
- 4 Estimates of seagrass coverage for selected areas described in this *World Atlas* — 14
- 5 Functions and values of seagrass from the wider ecosystem perspective — 15
- 6 Summary of the goods and services provided by seagrass ecosystems — 16
- 7 Summary of marine protected areas that contain seagrass ecosystems, from the UNEP-WCMC Protected Areas Database — 19

REGIONAL CHAPTERS

1 THE SEAGRASSES OF Scandinavia and the Baltic Sea — 27

MAP
- 1.1 Scandinavia — 29

FIGURES
- 1.1 Average (±1 SE) above-ground biomass values for eelgrass (*Zostera marina*) along the Baltic Sea coastline — 28
- 1.2 Aerial photographs of two typical exposed eelgrass (*Zostera marina*) sites at the Hanko Peninsula, southwest Finland — 30
- 1.3 Norwegian eelgrass coverage — 31
- 1.4 Map of eelgrass area distribution in Danish coastal waters — 32
- 1.5 Maximum colonization depth of eelgrass patches in Danish estuaries and along open coasts in 1900 and 1996-97 — 33
- 1.6 Secchi depths and maximum colonization depths of eelgrass patches in Danish estuaries and open coasts in 1900 and 1992 — 33
- 1.7 Long-term changes in the distribution of eelgrass (*Zostera marina*) in the southeastern Baltic Sea (Puck Lagoon, Poland) — 33

2 THE SEAGRASSES OF Western Europe — 38

MAPS
- 2.1 Western Europe (north) — 39
- 2.2 Western Europe (south) — 39

CASE STUDIES
- 2.1 The Wadden Sea — 41
- 2.2 Glénan Archipelago — 45

3 THE SEAGRASSES OF The western Mediterranean — 48

MAP
- 3.1 The western Mediterranean — 49

CASE STUDIES
- 3.1 Italy — 50
- 3.2 France — 51
- 3.3 Spain — 54

TABLES
- 3.1 Examples of general features of Mediterranean seagrass meadows — 49
- 3.2 Distribution of seagrasses throughout the western Mediterranean (Italy, France and Spain) — 52

4 THE SEAGRASSES OF The Black, Azov, Caspian and Aral Seas — 59

MAP
- 4.1 The Black, Azov, Caspian and Aral Seas — 61

5 THE SEAGRASSES OF The eastern Mediterranean and the Red Sea — 65

MAPS
- 5.1 The eastern Mediterranean — 67
- 5.2 The Red Sea — 67

CASE STUDY
- 5.1 Israeli coast of the Gulf of Elat — 70

6 THE SEAGRASSES OF The Arabian Gulf and Arabian region		74
MAP		
6.1	The Arabian Gulf and Arabian region	75
CASE STUDIES		
6.1	The Bahrain Conservancy	76
6.2	Rapid assessment technique	77
6.3	Marine turtles and dugongs in the Arabian seagrass pastures	79
TABLE		
6.1	Seagrass species in the Arabian region	75
7 THE SEAGRASSES OF Kenya and Tanzania		82
MAP		
7.1	Kenya and Tanzania	83
CASE STUDIES		
7.1	Gazi Bay, Kenya: Links between seagrasses and adjacent ecosystems	84
7.2	Seagrass beach cast at Mombasa Marine Park, Kenya: A nuisance or a vital link?	88
8 THE SEAGRASSES OF Mozambique and southeastern Africa		93
MAPS		
8.1	Mozambique and southeastern Africa	94
8.2	The Seychelles	94
8.3	Mauritius	94
CASE STUDY		
8.1	Inhaca Island and Maputo Bay area, southern Mozambique	96
FIGURE		
8.1	Digging of *Zostera capensis* meadows at Vila dos Pescadores, near Maputo city	99
TABLES		
8.1	Area cover and location for the seagrass *Zostera capensis* in South Africa	94
8.2	Seagrass cover and area lost in Mozambique	99
9 THE SEAGRASSES OF India		101
MAPS		
9.1	India	103
9.2	Andaman and Nicobar Islands	103
CASE STUDY		
9.1	Kadmat Island	106
	TABLE: Characterization of a seagrass meadow at Kadmat Island, Lakshadweep	
	TABLE: Benthic macrofauna in the seagrass bed at Kadmat Island, Lakshadweep	
FIGURE		
9.1	Abundance of seagrass species at various depths in the Gulf of Mannar (southeast coast)	105
TABLES		
9.1	Quantitative data for major seagrass beds in Indian waters	102
9.2	Occurrence of seagrasses in coastal states of India	104
9.3	Associated biota of seagrass beds of India	105
10 THE SEAGRASSES OF Western Australia		109
MAP		
10.1	Western Australia	111
CASE STUDY		
10.1	Shark Bay, Western Australia: How seagrass shaped an ecosystem	116
TABLES		
10.1	Western Australian endemic seagrass species	110
10.2	Summary of major human-induced declines of seagrass in Western Australia	113
11 THE SEAGRASSES OF Eastern Australia		119
MAP		
11.1	Eastern Australia	121
CASE STUDIES		
11.1	Mapping deepwater (15-60 m) seagrasses and epibenthos in the Great Barrier Reef lagoon	124
	FIGURE: Probability of the occurrence of deepwater seagrasses in the Great Barrier Reef Lagoon	
	FIGURE: Frequency of the probability of occurrence of seagrasses within each depth stratum	
11.2	Westernport Bay	126
	FIGURE: Distribution of estuarine habitats in Westernport Bay, Australia	
11.3	Expansion of Green Island seagrass meadows	128
	FIGURE: Seagrass distribution at Green Island in 1994, 1972, 1959 and 1936	
12 THE SEAGRASSES OF New Zealand		134
MAP		
12.1	New Zealand	135
CASE STUDY		
12.1	A seagrass specialist	140
FIGURE		
12.1	An example of changes in the historical distribution of seagrasses in New Zealand	139
TABLES		
12.1	Area of seagrass in New Zealand estuaries where benthic habitats have been mapped	135
12.2	List of locations where seagrasses have been recorded in New Zealand	136
13 THE SEAGRASSES OF Thailand		144
MAP		
13.1	Thailand	145
CASE STUDY		
13.1	The dugong – a flagship species	147

TABLE
13.1 Occurrence of seagrass species in Thailand — 146

14 THE SEAGRASSES OF Malaysia — 152
MAPS
14.1 Peninsular Malaysia — 153
14.2 Sabah — 153
CASE STUDIES
14.1 The seagrass macroalgae community of Teluk Kemang — 155
14.2 The subtidal shoal seagrass community of Tanjung Adang Laut — 156
14.3 Coastal lagoon seagrass community at Pengkalan Nangka, Kelantan — 158
TABLE
14.1 Estimate of known seagrass areas in Peninsular Malaysia — 157

15 THE SEAGRASSES OF The western Pacific islands — 161
MAPS
15.1 Western Pacific islands (west) — 163
15.2 Western Pacific islands (east) — 163
CASE STUDIES
15.1 Kosrae — 164
MAPS: Lelu Harbour ca 1900 and 1975; Okat Harbour and Reef 1978 and 1988
15.2 SeagrassNet – a western Pacific pilot study — 168

16 THE SEAGRASSES OF Indonesia — 171
MAP
16.1 Indonesia — 173
CASE STUDIES
16.1 Banten Bay, West Java — 176
16.2 Kuta and Gerupuk Bays, Lombok — 177
16.3 Kotania Bay — 178
TABLE: Distribution of seagrass
TABLES
16.1 Average biomass of seagrasses at various locations throughout Indonesia — 172
16.2 Average density of seagrasses at various locations throughout the Indonesian Archipelago — 172
16.3 Average shoot density of seagrass species in mixed and monospecific seagrass meadows in the Flores Sea — 173
16.4 Average growth rates of seagrass leaves using leaf-marking techniques — 174
16.5 Indonesian seagrass-associated flora and fauna: number of species — 175
16.6 Present coverage of seagrasses in Indonesia — 178

The Philippines and Viet Nam — 183

17 THE SEAGRASSES OF Japan — 185
MAP
17.1 Japan — 187
CASE STUDIES
17.1 Akkeshi, eastern Hokkaido — 189
17.2 Rias coast in Iwate Prefecture, northeastern Honshu — 190
TABLES
17.1 Seagrasses recorded in Japan — 186
17.2 Traditional uses of seagrasses in Japan — 188
17.3 Estimates of total areas of algal and seagrass beds in Japan in 1978 and 1991, and the percent area lost — 188

18 THE SEAGRASSES OF The Republic of Korea — 193
MAP
18.1 Republic of Korea — 195
CASE STUDY
18.1 Recent research on seagrasses — 196
TABLES
18.1 Physical characteristics of seagrass beds on the west, south and east coasts of the Republic of Korea — 194
18.2 Seagrass species distributed on the coasts of the Republic of Korea — 194
18.3 Habitat characteristics of seagrass species in the Republic of Korea — 195
18.4 Morphological characteristics of seagrasses distributed in the Republic of Korea — 195
18.5 The estimated areas of seagrasses distributed on the coasts of the Republic of Korea — 197

19 THE SEAGRASSES OF The Pacific coast of North America — 199
MAP
19.1 The Pacific coast of North America — 201
CASE STUDIES
19.1 The link between seagrass and migrating black brant along the Pacific Flyway — 200
19.2 The link between the seagrass *Zostera marina* (ts'áts'ayem) and the Kwakwaka'wakw Nation, Vancouver Island, Canada — 202
19.3 The link between seagrasses and humans in Picnic Cove, Shaw Island, Washington, United States — 203
TABLE
19.1 *Zostera marina* and *Zostera japonica* basal area cover in the Northeast Pacific — 204

20 THE SEAGRASSES OF The western North Atlantic — 207
MAP
20.1 The western North Atlantic — 209
CASE STUDIES
20.1 Portsmouth Harbor, New Hampshire and Maine — 208
FIGURE: Eelgrass distribution by depth in Portsmouth Harbor, Great Bay Estuary, on the border of New Hampshire and Maine, United States

20.2	Ninigret Pond, Rhode Island	211	Case Studies	
	Figure: Eelgrass distribution in Ninigret Pond, Rhode Island (United States) plotted by depth for 1974 and 1992		23.1 Florida's east coast	236
			23.2 Parque Natural Tayrona, Bahía de Chengue, Colombia	239
			23.3 Puerto Morelos Reef National Park	240
	Figure: Change in eelgrass area in Ninigret Pond, Rhode Island (United States) plotted against increasing number of houses in the watershed			

20.3 Maquoit Bay, Maine — 213

TABLE

20.1 The area of eelgrass, *Zostera marina*, in the western North Atlantic — 212

21 THE SEAGRASSES OF The mid-Atlantic coast of the United States — 216

MAP

21.1 The mid-Atlantic coast of the United States — 217

CASE STUDIES

21.1 Seagrasses in Chincoteague Bay: a delicate balance between disease, nutrient loading and fishing gear impacts — 220
  FIGURE: Recovery and recent decline of seagrass (*Zostera marina* and *Ruppia maritima*) distribution in Chincoteague Bay
  FIGURE: Aerial photograph taken in 1998 of a portion of Chincoteague Bay, Virginia, seagrass bed showing damage to the bed from a modified oyster dredge

FIGURES

21.1 Seagrass distribution (mainly *Zostera marina* and *Ruppia maritima*) in Chesapeake Bay — 218

21.2 Changes in seagrass (*Zostera marina* and *Halodule wrightii*) distribution in the Cape Lookout area (southern Core Sound, North Carolina) between 1985 and 1988 — 218

22 THE SEAGRASSES OF The Gulf of Mexico — 224

MAP

22.1 The Gulf of Mexico — 225

CASE STUDIES

22.1 Tampa Bay — 226
22.2 Laguna Madre — 228
  FIGURE: Seagrass cover in the Laguna Madre of Texas
22.3 Laguna de Términos — 231

23 THE SEAGRASSES OF The Caribbean — 234

MAP

23.1 The Caribbean — 235

24 THE SEAGRASSES OF South America: Brazil, Argentina and Chile — 243

MAP

24.1 South America — 245

CASE STUDIES

24.1 Itamaracá Island, northeast Brazil — 244
24.2 Abrolhos Bank, Bahia State, northeast Brazil — 247
24.3 *Ruppia maritima* in the Patos Lagoon system — 248

FIGURE

24.1 Cumulative number of companion species to the Brazilian seagrasses reported since 1960 — 245

APPENDICES

1 Seagrass species, by country or territory — 251

2 Marine protected areas known to include seagrass beds, by country or territory — 256

3 Species range maps — 262

The Global Seagrass Workshop — 287

INDEX — 288

REGIONAL MAPS

Europe	Plate I, facing p 38
Africa, West and South Asia	Plate III, facing p 102
Australasia	Plate V, facing p 118
The Pacific	Plate VII, facing p 166
Asia	Plate IX, facing p 182
North America	Plate XI, facing p 214
The Caribbean	Plate XIII, facing p 230
South America	Plate XIV, facing p 231

COLOR PLATES

The beauty of seagrasses	Plate II, facing p 39
Impacts to seagrass ecosystems	Plate IV, facing p 103
Seagrass ecosystems	Plate VI, facing p 119
Seagrasses and people	Plate VIII, facing p 167
The sex life of seagrasses	Plate X, facing p 183
Diversity of seagrass habitats	Plate XII, facing p 215

Introduction to
THE WORLD ATLAS OF SEAGRASSES

Seagrasses are valuable and overlooked habitats, providing important ecological and economic components of coastal ecosystems worldwide. Although there are extensive seagrass beds on all the world's continents except Antarctica, seagrasses have declined or been totally destroyed in many locations. As the world's human population expands and continues to live disproportionately in coastal areas, a comprehensive overview of coastal resources and critical habitats is more important than ever. The *World Atlas of Seagrasses* documents the current global distribution and status of seagrass habitat.

Seagrasses are a functional group of about 60 species of underwater marine flowering plants. Thousands more associated marine plant and animal species utilize seagrass habitat. Seagrasses range from the strap-like blades of eelgrass (*Zostera caulescens*) in the Sea of Japan, at more than 4 m long, to the tiny, 2-3 cm, rounded leaves of sea vine (e.g. *Halophila decipiens*) in the deep tropical waters of Brazil. Vast underwater meadows of seagrass skirt the coasts of Australia, Alaska, southern Europe, India, east Africa, the islands of the Caribbean and other places around the globe. They provide habitat for fish and shellfish and nursery areas to the larger ocean, and performing important physical functions of filtering coastal waters, dissipating wave energy and anchoring sediments. Seagrasses often occur in proximity to, and are ecologically linked with, coral reefs, mangroves, salt marshes, bivalve reefs and other marine habitats. Seagrasses are the primary food of manatees, dugongs and green sea turtles, all threatened and charismatic species of great public interest.

Seagrasses are subject to many threats, both anthropogenic and natural. Runoff of nutrients and sediments from human activities on land has major impacts in the coastal regions where seagrasses thrive; these indirect human impacts, while difficult to measure, are probably the greatest threat to seagrasses worldwide. Both nutrient and sediment loading affect water clarity; seagrasses' relatively high light requirements make them vulnerable to decreases in light penetration of coastal waters. Direct harm to seagrass beds occurs from boating, land reclamation and other construction in the coastal zone, dredge-and-fill activities and destructive fisheries practices. Human-induced global climate change may well impact seagrass distribution as sea level rises and severe storms occur more frequently. The *World Atlas of Seagrasses* makes it clear that seagrasses receive little protection despite the myriad threats to this habitat.

Most of our understanding of seagrass ecosystems is based on site-specific studies, usually in developed nations. Very little is known about the importance of seagrasses in maintaining regional or global biodiversity, productivity and resources, partly because seagrasses are under-appreciated and their distribution is so poorly documented. As a result, seagrasses are rarely incorporated specifically into coastal management plans and are vulnerable to degradation. Seagrass ecosystems in the Caribbean, Indian Ocean, Southeast Asia and Pacific are especially poorly researched, yet it is in these regions that the direct economic and cultural dependence of coastal communities upon marine resources, including seagrasses, tends to be highest.

The purpose of the *World Atlas of Seagrasses* is to present a global synthesis of the distribution and status of seagrasses. Such syntheses are available for other coastal ecosystems and have been instrumental in creating awareness, driving clearer conservation and management efforts and focusing priorities at the international level. For example, over the last ten years, opinion on the status of coral reefs has changed from a predominant view that the majority of coral reefs were unaffected by human activities, to the present view in which the global decline of coral reefs, and the increasing threats to them, are widely acknowledged. A similar understanding of seagrass ecosystems is needed in

A patch reef in the Philippines surrounded by a luxuriant mixed bed of *Thalassia hemprichii* and *Syringodium isoetifolium*.

order to achieve the visibility and recognition necessary to protect this valuable global resource. Public perception translates into political interest. Perceptions of seagrass ecosystems must achieve comparable status with those of coral reef and mangrove ecosystems, through the creation of global maps, global estimates of loss, knowledge of human impacts to the ecosystem, regular monitoring of ecosystem status and a global plan of action to reverse seagrass ecosystem decline. It is our hope that the *World Atlas of Seagrasses* will contribute to the more widespread recognition, understanding, and protection of seagrass ecosystems worldwide.

ORGANIZATION OF THIS *WORLD ATLAS*

The *World Atlas of Seagrasses* is presented in two sections. The first section comprises a Global Overview of the state of our knowledge of seagrasses. It presents detailed ecosystem distribution and species diversity maps and the most accurate possible estimate of global seagrass area. Appendices supply seagrass species lists for almost 180 countries and territories, a list of marine protected areas known to include seagrasses and a collection of species range maps. The Global Overview was based on a compilation of seagrass literature and a workshop held in Florida with seagrass scientists from around the world contributing their regional knowledge and expertise. The Global Seagrass Workshop, sponsored by UNEP-WCMC, with considerable assistance from the World Seagrass Association, was held in St Petersburg, Florida in November 2001 specifically to begin assembling information on global seagrass distribution for the *World Atlas* (see page 287). Twenty-three delegates from 15 countries participated, and all are represented here as chapter authors. The workshop was a forum for discussion on the organization of an atlas, regional seagrass distribution, and seagrass functions and threats at a global level. Later, additional chapter authors were asked to contribute to represent regions of the world not yet well covered; also, chapter authors invited co-authors to join their effort. The geographical coverage of the *World Atlas* reflects this process.

The second section of the *World Atlas of Seagrasses* consists of 24 regional and national chapters. In each chapter, the authors have synthesized knowledge of seagrasses, the plants' biogeography, ecology and associated species, historical perspectives and threats to the ecosystem as well as management policies pertaining to seagrasses. Wherever possible, the authors have estimated the area of seagrass in their region and summarized its status. Case studies throughout the chapters highlight particularly interesting seagrass habitats and areas where human or natural impacts to seagrasses are of concern.

Dugong feeding on *Halophila ovalis*, Vanuatu, western Pacific islands.

Of course, any comprehensive atlas builds on the work of many scientists beyond the chapter authors. Seagrass science owes much to den Hartog's *Seagrasses of the World* and the many subsequent publications and books that are referenced throughout the *World Atlas*. All of the references used to compile the World Seagrass Distribution Map (reproduced on page 21), as well as the individual chapter references, appear in an online bibliography at http://www.unep-wcmc.org/marine/seagrassatlas/references.

Additionally, the sources of information that contribute to the World Seagrass Distribution Map may be queried online through a GIS database at http://www.stort.unep-wcmc.org/imaps/marine/seagrass. Inevitably, in a complex collaboration of this type, some sources of data are overlooked. Indeed, we have become aware of additional sources of information on the distribution of seagrasses since our printing deadline. Readers with information on seagrass distribution that they would like to add to the database may contact us directly.

The seagrass distributions mapped in this *World Atlas* were derived from scientific journals, books, other publications and reports, reliable websites and personal communications. Where these sources provided maps of actual seagrass beds, that mapped extent of seagrass (polygon) was entered directly onto the World Seagrass Distribution Map. More frequently, publications and other sources simply mention the occurrence of

Eutrophication reduces water clarity and stimulates growth of epiphytic algae, as on this *Zostera marina* in southern Norway.

Women harvesting shellfish, *Pinna muricata*, from an intertidal seagrass flat at low tide, Matibane, Mozambique.

seagrass at a particular location (e.g. a bay, beach, town or known latitude/longitude). In these cases, the seagrass occurrence is shown on the distribution map as a dot, designating the mentioned location. The World Seagrass Distribution Map at the beginning of the *World Atlas* gives the compilation of all the available information on seagrass distribution, as both actual beds and as locations indicated by dots, of all seagrass species combined. Species range maps (in Appendix 3) depict the area where a certain seagrass species may be expected to occur, based on individual species reports collected for the World Seagrass Distribution Map. Using an overlay of all the species range maps, a global map of seagrass species diversity was created (reproduced on page 22). Additionally, regional maps show the same information as the World Seagrass Distribution Map, but at a finer scale and with the locations of the case studies in the region. Finally, each of the chapters has its own map, showing seagrass distribution and important locations discussed in the chapter.

THREATS TO SEAGRASSES

The synthesis represented by the *World Atlas of Seagrasses* confirms that seagrasses are one of the most widespread marine ecosystems, quite possibly the most widespread shallow marine ecosystem, in the world. They cover an area that can only be crudely estimated at present; the area we are able to document in the *World Atlas* is certainly a gross underestimate. The threats to seagrasses worldwide are similar and widespread. Seagrasses everywhere are vulnerable to eutrophication from nutrient over-enrichment of the environment and to turbid conditions caused by upland clearing and disturbance, both leading to reduced light availability. Seagrasses are also subject to total destruction through coastal construction and other direct human impacts. Direct use of seagrass plants by humans is limited, but seagrass beds support important coastal fisheries worldwide, and because they occur in easily accessible, shallow, sheltered areas these are often subsistence fisheries. Seagrasses are an important coastal ecosystem in need of more study, awareness and protection.

Ed Green
Fred Short

Essential information

LEGEND TO MAPS

 Seagrass (location only, extent unknown)

▪ Seagrass area

Number of species (map page 22)

- 1-2
- 3-6
- 7-9
- 10-11
- 12-15

Bathymetry

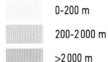

- 0-200 m
- 200-2 000 m
- >2 000 m

Species range maps (Appendix 3)

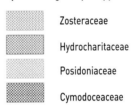

- Zosteraceae
- Hydrocharitaceae
- Posidoniaceae
- Cymodoceaceae

MAPPING METHODS

The seagrass features mapped throughout this *World Atlas* were derived from very many different sources. Selection criteria were used when reviewing thousands of records from hundreds of sources to determine which features would be mapped.

The approach adopted was one that minimized subjectivity. For example a statement such as "...in Gerupuk Bay, southern Lombok, *Halodule uninervis* densities ranged from..." (page 173) would result in a point at that location. At the scale of a global atlas a point in Gerupuk Bay is sufficiently accurate. Statements such as "extensive terracing of these expanses of the intertidal zone [of the Kimberley Coast, Western Australia] often results in seagrass, particularly *Enhalus acoroides*, high in the intertidal just below the mangroves" (page 110) have not been recorded on the maps because no exact locations or extent of seagrass were available. At the scale of a global atlas an assumption that seagrass occurs along large sections of the Kimberley coast would have been too inaccurate without independent reference. Some islands or coastal areas have comprehensive coverage on the maps. These are derived from studies where an entire area has been mapped in great detail, often using aerial photography or satellite remote sensing. Corsica is one example[1] and the data were available for inclusion in the *World Atlas* maps. The decision to construct the maps only on referenced sources (e.g. Corsica) and not extrapolation from rather inexact statements (e.g. the Kimberley Coast) does create some apparent discrepancies but in all cases these are due to this decision. As such the collected total of seagrass features mapped in the *World Atlas* should be regarded as a minimal representation of actual coverage.

Two further rules were applied to the making of the seagrass maps. Firstly, in some cases only crude maps were available, often covering very large areas with swathes simply indicative of the presence of seagrass (e.g. the global National Geographic 2000 Coral World map). They were cut to match shallow bathymetry data to avoid misrepresenting the depths at which seagrasses are found. Secondly, when no specific location was available beyond the name of a very small island a point was placed in the center of that island. Yap, Micronesia, is one example. Seagrass is recorded as occurring all around Yap with no more precise locators so this is recorded as a point centered on the island. Yap is small enough so that, at the scale at which these maps are most useful, a visible point covers the island entirely.

1. Pasqualini V, Pergent-Martini C, Pergent G (1999). Environmental impact identification along the Corsican coast (Mediterranean sea) using image processing. *Aquatic Botany* 65: 311-320.

ABBREVIATIONS USED

m	meter	mg	milligram	psu	practical salinity units (almost equal to parts per thousand)
km	kilometer	g	gram		
ha	hectare	kg	kilogram	UV	ultraviolet
cm/s	centimeters per second	kcal	calorie	°C	degrees Centigrade

Bold type is used to indicate the corresponding author and contact details at the end of each chapter.

Global overview
THE DISTRIBUTION AND STATUS OF SEAGRASSES

M. Spalding
M. Taylor
C. Ravilious
F. Short
E. Green

Seagrasses are a mixed group of flowering plants which grow submerged in shallow marine and estuarine environments worldwide. In many places they cover extensive areas, often referred to as seagrass beds or seagrass meadows. Although there are relatively few species of seagrass, the complex physical structure and high productivity of these ecosystems enable them to support a considerable biomass and diversity of associated species. Seagrasses themselves are a critically important food source for dugong, manatee, sea turtles and waterfowl. Many other species of fish and invertebrates, including sea horses, shrimps and scallops, utilize seagrass for part of their life cycles, often for breeding or as juveniles. Seagrasses are considered to be one of the most important shallow marine ecosystems to humans, playing a significant role in fisheries production as well as binding sediments and providing some protection from coastal erosion.

The overview summarizes the distribution, importance and status of seagrasses worldwide. Firstly we consider the definition of seagrasses, both as species and as habitats, and look at their geographic distribution patterns. Much of this work is the presentation of entirely new datasets that have been developed for this atlas, including a detailed distribution database and digital maps compiled from numerous sources, often generously contributed. Next we consider the importance of seagrasses to humans. Finally we look at human impacts on these ecosystems, including both threats and management measures for the protection of seagrass beds. Much of this chapter has benefited from the specialist input of seagrass experts worldwide, and especially those who are also contributors to this *World Atlas*.

Definitions
Seagrasses are flowering plants which grow fully submerged and rooted in estuarine and marine environments. They are not true grasses. Although they are all monocotyledons, they do not have a single evolutionary origin, but are a polyphyletic group, defined by the particular ecological niche they inhabit. Five particular adaptations to enable survival in this niche have been identified[1]:

o   an ability to grow whilst completely submerged, which presents problems, notably of lowered gas concentrations and rates of diffusion;
o   an adaptation to survive in high, and often varying, salinity;
o   an anchoring system to withstand water movements;
o   a submarine pollination mechanism;
o   an ability to compete with other species in the marine environment.

The adaptations have led to a number of morphological characteristics which are widespread amongst seagrasses, notably: flattened leaves (with the exception of *Syringodium* and some *Phyllospadix* spp.); elongated or strap-like leaves (with the exception of species in the genus *Halophila*); and an extensive system of roots and rhizomes[1].

Considerable arguments remain over the nomenclature and taxonomic relations of the seagrasses, and it is likely that there will be considerable changes to the accepted classification in coming years[2-4] and hence to the number of species considered to be seagrasses. In the present work we have adopted a conservative approach, and consider 59 species, based on species lists used in Hemminga and Duarte[2] and in Short and Coles[5], with further advice from the authors of this *World Atlas*. These species are listed in Table 1. It is important to bear in mind, however, that "the actual number of seagrass species is a matter of debate, depending in part on their proximity to the marine environment and on the level

of discrimination in physical taxonomy and genetics"[5].

Many species of the genus *Ruppia* are accepted as seagrasses, commonly occurring in the marine environment and often intermingled with other seagrass species[5]. Species in the genera *Potamogeton* and *Lepilaena* are occasionally important members of seagrass ecosystems, but are often regarded as seagrass associates or facultative members of the seagrass community. We have included *Ruppia* spp. when they occur in marine and estuarine environments, but these species are less well covered in the

Overview Table 1
A list of seagrass species by family

Genus	Species	Author
Hydrocharitaceae		
Enhalus	acoroides	(L.f.) Royle
Halophila	australis	Doty & Stone
Halophila	baillonii	Ascherson
Halophila	beccarii	Ascherson
Halophila	capricorni	Larkum
Halophila	decipiens	Ostenfeld
Halophila	engelmanni	Ascherson
Halophila	hawaiiana†	Doty & Stone
Halophila	johnsonii†	Eiseman
Halophila	minor†	(Zollinger) den Hartog
Halophila	ovalis	(R. Brown) Hooker f.
Halophila	ovata†	Gaudichaud
Halophila	spinulosa	(R. Brown) Ascherson
Halophila	stipulacea	(Forsskål) Ascherson
Halophila	tricostata	Greenway
Thalassia	hemprichii	(Ehrenberg) Ascherson
Thalassia	testudinum	Banks ex König
Cymodoceaceae		
Amphibolis	antarctica	(Labill.) Sonder et Ascherson
Amphibolis	griffithii	(Black) den Hartog
Cymodocea	angustata	Ostenfeld
Cymodocea	nodosa	(Ucria) Ascherson
Cymodocea	rotundata	Ehrenberg & Hemprich ex Ascherson
Cymodocea	serrulata	(R. Brown) Ascherson
Halodule	beaudettei*	(den Hartog) den Hartog
Halodule	bermudensis*	den Hartog
Halodule	emarginata*	den Hartog
Halodule	pinifolia*	(Miki) den Hartog
Halodule	uninervis	(Forsskål) Ascherson
Halodule	wrightii	Ascherson
Syringodium	filiforme	Kützing
Syringodium	isoetifolium	(Ascherson) Dandy
Thalassodendron	ciliatum	(Forsskål) den Hartog
Thalassodendron	pachyrhizum	den Hartog

Genus	Species	Author
Posidoniaceae		
Posidonia	angustifolia	Cambridge & Kuo
Posidonia	australis	Hooker f.
Posidonia	coriacea*	Cambridge & Kuo
(including the conspecific *Posidonia robertsoniae*[82])		
Posidonia	denhartogii*	Kuo & Cambridge
Posidonia	kirkmanii*	Kuo & Cambridge
Posidonia	oceanica	(L.) Delile
Posidonia	ostenfeldii*	den Hartog
Posidonia	sinuosa	Cambridge & Kuo
Zosteraceae		
Zostera	asiatica	Miki
Zostera	caespitosa	Miki
Zostera	capensis	Setchell
Zostera	capricorni	Ascherson
(including the conspecific *Zostera mucronata*, *Zostera muelleri* and *Zostera novazelandica*[80])		
Zostera	caulescens	Miki
Zostera	japonica	Aschers. & Graebner
Zostera	marina	Linnaeus
Zostera	noltii	Hornemann
Zostera	tasmanica	(Martens ex Aschers.) den Hartog
(formerly *Heterozostera*)		
Phyllospadix	iwatensis	Makino
Phyllospadix	japonicus	Makino
Phyllospadix	scouleri	Hooker
Phyllospadix	serrulatus	Ruprecht ex Aschers.
Phyllospadix	torreyi	S. Watson
Ruppiaceae		
Ruppia	cirrhosa	(Petagna) Grande
(formerly *spiralis*)		
Ruppia	maritima	Linnaeus
Ruppia	megacarpa	Mason
Ruppia	tuberosa	Davis & Tomlinson

Note:
* Species designations that are a matter of debate and currently under genetic and morphometric investigation.
† Species proposed as conspecific with *Halophila ovalis*[83].

literature than many other species and have not been universally accepted as seagrasses.

Typically, seagrasses grow in areas dominated by soft substrates such as sand or mud, but some species can be found growing on more rocky substrates (e.g. *Phyllospadix*). Seagrasses require high levels of light, more than other marine plants, because of their complex below-ground structures which include considerable amounts of non-photosynthetic tissues. Thus, although they have been recorded to 70 m in clear waters[6], they are more generally restricted to shallow waters due to the rapid attenuation of light with depth.

Seagrasses can form extensive monospecific stands or areas of mixed species. Such areas are known as seagrass beds or meadows, and make up a unique marine ecosystem or biotope. Seagrasses can also grow in isolated patches, or as part of a habitat mosaic with other habitats such as corals, mangroves, bivalve reefs, rocky benthos or bare sediments. Generally it is the larger seagrass beds and meadows which have been the subject of intensive study and mapping worldwide. Although typically permanent over periods of decades, seagrass systems can be highly dynamic, moving into new areas and disappearing from others over relatively short timeframes.

DEVELOPING SEAGRASS DISTRIBUTION INFORMATION AND MAPS

In order to develop a clearer picture of the distribution of seagrasses worldwide, a new dataset was developed at UNEP-WCMC, based on literature review and outreach to expert knowledge. An output from this dataset is presented here in the World Seagrass Distribution Map (Map 1, which appears on page 21).

Initial efforts focused on the acquisition of point-source information which was compiled into a spreadsheet with details on species as well as information on location in both descriptive terms and, wherever possible, geographic coordinates. This work continued throughout a second data-gathering phase, during which maps on the distribution of seagrasses were developed on a geographical information system (GIS). The two datasets remained closely linked: the point locations from the first phase were linked to the GIS, and the GIS layer also allowed for the incorporation of boundary information delimiting particular seagrass areas (polygons). A third phase involved the presentation of the initial maps prepared by UNEP-WCMC to the Global Seagrass Workshop in Florida, 2001, where they were thoroughly checked by regional and national seagrass experts. As a result, new data points were added, new datasets and references were provided, and incorrectly located or spurious data points were removed.

At the conclusion of this effort, over 520 major

Quadrat sampling in an intertidal *Zostera marina* bed, Maine, USA.

sources had been used in developing seagrass distribution data (see the online bibliography at http://www.unep-wcmc.org/marine/seagrassatlas/references). These sources provide information on seagrasses in more than 120 countries and territories worldwide, and the majority include information on specific species. All data sources were documented and can be queried online through the GIS (go to http://stort.unep-wcmc.org/imaps/marine/seagrass).

Despite the broad range of sources, the geographic information can be seen largely to fall into three categories, as discussed below.

Direct habitat maps

Direct habitat maps are high-resolution maps, typically prepared from remotely sensed data but in some cases mapped entirely from field observations; they represent the polygons showing the true spatial extent of seagrass distribution. They provide the most accurate data available for habitat distribution, but are available for only a very limited area worldwide. In some cases they do not provide species-specific distribution information. Sources included some broader maps showing seagrasses over several kilometers or tens of kilometers of coastline, but also many maps prepared and presented for individual study sites in expert publications.

Expert interpolations

In some cases, maps have been based on the interpolation of ground-based knowledge and observation – seagrasses may be known from a series

of point locations, and with an accurate benthic chart it is possible to interpolate between these points to generate an outline of assumed seagrass area. Clearly the accuracy of such maps is highly variable, but can be relatively reliable with sufficient background information and cautious interpretation. These maps were utilized with caution and included in the GIS only if better data were unavailable and the source was considered to be reliable.

Point-based samples

For wide areas of the globe, maps of any sort were unavailable; however, it was possible to gather accurate point locations of seagrass beds from a large number of site-based seagrass publications, herbarium records and national species inventories. Clearly, as points, these give no indication of actual seagrass area, but they are very useful in a broader mapping context where no further information is available.

Developing the distribution map

The source maps used for producing the World Seagrass Distribution Map were created using many mapping techniques, and with various goals. There are also differences in resolution, which will clearly influence the area of seagrass portrayed on a map. With remote sensing, accuracy is limited by the resolution and bandwidths utilized by the sensor, the degree of ground-truthing and sensitivity of the interpretation, as well as by the depth of the water column, the clarity of the water and other attributes of the benthos. Some remotely sensed images will pick up only shallow (<10 m) seagrass beds with a high shoot density, while large pixel size will fail to capture small or highly patchy seagrass areas. Error also plays a part, and some mapping systems may incorporate non-seagrass species, notably macroalgae. Although typically more accurate, direct sampling can have many similar problems, particularly associated with water depth and clarity.

Combining data from multiple sources, as undertaken here, exacerbates these problems, as there are always differences in both quality and definition between studies. Seagrass shoot density varies considerably and, while some studies will consider only seagrass ecosystems where seagrass shoots are continuous at high densities (such a definition may in fact be forced by the mapping techniques), others may include all areas of even very sparse seagrass growth. Differences in scale between studies introduce further variance: lower resolution maps may tend to ignore minor breaks in seagrass beds, while finer resolution maps will pick up even small breaks which, it could be argued, are still a part of the seagrass habitat. Further problems may be associated with time. Seagrass systems are highly variable through time, with some showing seasonal variations and others showing dramatic interannual variation. Finally, it is important on a composite map, such as that presented here, to be aware that gaps where there are no data cannot be distinguished from gaps where seagrasses do not occur.

The results of this data gathering have been used to show the distribution of individual species, and to show the overall distribution of seagrass habitat. The World Seagrass Distribution Map includes all the species-specific information as well as additional points and areas where species were not specified.

Interpolation of the species distributions was used to generate species range maps (see Appendix 3). The known occurrences of each species were used to set the limits to a generalized outline of the range of that species. Like the raw datasets, preliminary range maps were reviewed at the Global Seagrass Workshop in Florida. It should be noted that they do not indicate definite occurrence of a seagrass species, but rather show where a species might be expected to occur should environmental conditions be suitable. Such maps are useful in biogeographic studies and comparisons between species, and also for predicting possible species occurrence in areas which have not been previously investigated.

The data that constitute the World Seagrass Distribution Map were also used to make a preliminary calculation of seagrass area at global and regional levels. Such work has been done in more detail for other nearshore marine habitats[7,8]; however the weaknesses and gaps in the seagrass dataset mean that initial area calculations, presented below, are only broadly indicative.

SPECIES DISTRIBUTION

From the world seagrass distribution datasets described above, we assembled species records for more than 120 countries and territories. The datasets include some records for countries where point locations were unavailable (i.e. only species lists were available), and hence these are not shown on the maps. All the datasets were used to generate species lists by country, presented in Appendix 1. The species lists show that the countries with greatest seagrass diversity are countries which extend into both tropical and temperate climates, including Australia (29 species), the United States (23 species including all overseas territories) and Japan (16 species). The greatest seagrass species diversity in single-climate countries occurs in the tropics. Tropical countries with the highest seagrass species diversity include India and the Philippines (both with 14 species) and Papua New Guinea (12 species). The Philippines and Papua New Guinea, together with

Indonesia (12 species), are considered to be the center of global seagrass biodiversity.

The geographic data from the same seagrass distribution datasets were used to generate the species range maps presented in Appendix 3. (Range maps were not prepared for *Ruppia* species as the existing data were deemed insufficient.) The species range maps update earlier work by den Hartog[9] and by Phillips and Meñez[1]. They show areas where the species may be expected to occur, but they may leave out some areas where seagrass information is not available.

By amalgamating the species range maps, a global map of seagrass biodiversity was created (Map 2, page 22). The biodiversity map indicates the number of seagrass species in various parts of the globe; a previous effort is provided in Hemminga and Duarte[2]. Map 2 is modeled on similar maps compiled for corals[10] and for mangroves[11].

Biogeographic patterns

Map 2 shows the three clear centers of high diversity, all of which occur in the eastern hemisphere. The first and largest of these lies over insular Southeast Asia. The other two centers are adjacent to this region but remain distinctive, being Japan/Republic of Korea and southwestern Australia. Other areas of significant diversity include southern India and eastern Africa. Looking at diversity patterns in more detail, and also at the individual species ranges that underpin them, it is possible to distinguish general regions of seagrass occurrence, each with distinctive floral characteristics[2, 12]. The following list of seagrass regions is largely based on Short et al.[12].

1. Tropical Indo-Pacific (IX in Short et al.[12]). Mirroring the biodiversity found in coral reefs and mangrove forests, this is a region dominated by tropical seagrass species, with a great focus of diversity in insular Southeast Asia and northern Australia, continued high diversity across the Indian Ocean and up the Red Sea, but relatively rapid attenuation of biodiversity across the Pacific islands. Key genera include *Cymodocea, Enhalus, Halodule, Halophila, Syringodium, Thalassia* and *Thalassodendron*.

2. Southern Australia (X). A highly diverse region, dominated by temperate species. The particular center of diversity occurs in southwestern Australia (with species in the genera of *Amphibolis, Halophila, Posidonia* and *Zostera*).

3. Northwestern Pacific (I). The third-highest diversity region which, although connected to insular Southeast Asia, is dominated by temperate species (notably species of *Zostera* and *Phyllospadix*). The genus *Phyllospadix* is

Halophila capricorni female flower, Lizard Island, Queensland, Australia.

unique to the North Pacific, occurring in both the east and the west.

4. Northeastern Pacific (I): A lower-diversity temperate area, dominated by *Zostera* and *Phyllospadix* species. This region is closely linked to the more diverse western North Pacific but also includes three endemic species, *Phyllospadix scouleri, Phyllospadix serrulatus* and *Phyllospadix torreyi*.

5. North Atlantic (III). A low-diversity temperate area, dominated by *Zostera* and *Ruppia* species, with *Halodule* reaching its northern limit at 35°N in North Carolina, USA. Europe is distinguished by having a second species of the *Zostera* genera, *Zostera noltii. Zostera marina* is the main species of the region.

6. Wider Caribbean (IV). A tropical area with moderate seagrass diversity, including species of *Halodule, Halophila, Syringodium* and *Thalassia*. Although the tropical communities of Brazil are geographically isolated they are not sufficiently distinct to merit consideration as a separate flora (limited to species of *Halodule, Halophila* and *Ruppia*).

7. Mediterranean (VI). An area of relatively diverse temperate and tropical seagrass flora, which includes seagrass communities just outside the Mediterranean in northwest Africa as well as communities in the Black Sea Basin and the Caspian and Aral Seas. Species of *Cymodocea, Posidonia* and *Zostera* are common; *Ruppia* also plays an important role in the region, particularly in the Black, Caspian and Aral Seas.

8   South Africa (VIII). The region has both temperate and tropical species from the genera *Halodule, Halophila, Ruppia, Syringodium, Thalassodendron* and *Zostera*.

In addition to these floristically distinct regions there are three other geographically distinct seagrass areas which are of biogeographic interest, but which are poorly known and lack a distinctive floral characteristic, being largely depauperate.

9   Chile (II). One species, *Zostera tasmanica* (formerly *Heterozostera*), has been found along this coast.
10 Southwest Atlantic (V). Along the coast of Argentina and southern Chile there are extensive communities of *Ruppia*.
11 West Africa (VII). Only one species, *Halodule wrightii*, has been recorded; the distribution is poorly known.

Considerable further work is required in order to understand fully the distribution patterns of seagrasses; to determine the patterns of evolution and migration of species; and to uncover the interconnections between these regions. Some of the patterns observed in the tropical floras mirror the patterns observed in corals and mangroves. The Southeast Asian center of diversity is a particular feature of several marine biodiversity maps produced to date, including mangroves[11] and several major groups of coral reef taxa[13]. It is important to distinguish this Southeast Asian region from the separate centers of diversity seen in southwestern Australia and Japan, as these two areas have larger ranges of climate from temperate to tropical (Map 2).

Theories for the development of the Southeast Asian center of diversity have been advanced for a number of species groups. It has been variously suggested that this region may have been a center for species accumulation linked to favorable ocean currents ("the vortex model of coral reef biogeography"[14]); a location where high diversity was maintained thanks to benign climatic conditions during recent ice ages[15]; or a center for species evolution with the combination of benign conditions and changing sea levels ("eustatic diversity pump model"[16]).

The high diversity of temperate species in Japan and southwestern Australia is also of considerable evolutionary interest, but its cause remains a matter of speculation. There is evidence that the southwestern Australian flora may contain important relict elements[17] but more recent events associated with the dramatic changes during and following the last ice age must also be considered.

It is important to consider the evolutionary origin of seagrasses. The relatively low number of seagrass species could lead to the inference of a recent evolutionary history; however den Hartog[9] reports evidence for the existence of marine angiosperms as long ago as 100 million years, and there are clear examples of seagrass fossils from the Cretaceous. Further studies have failed to produce evidence of any massive diversification or of major extinction events, and so it may be that seagrasses have simply followed a relatively conservative evolutionary pathway. More work is required in this field[2, 17].

ASSOCIATED SPECIES AND HABITATS

Seagrasses do not grow in isolation but form an integral and often defining part of highly complex ecosystems. The seagrasses themselves are an important standing stock of organic matter, which is relatively stable in the tropics and has broad intra-annual variation in temperate regions. The productivity of these ecosystems is usually enhanced by other primary producers, including macroalgae and epiphytic algae. The abundant plant material of seagrass beds forms an integral part of many food chains. Additionally, the complex three-dimensional structure of the seagrass bed is important, providing shelter and cover, binding sediments and, at fine scales, even altering the patterns and strength of currents in the water. The complex, modified seagrass environment provides a great variety of niche spaces on and within the sediments, on the plant surfaces and within the water column.

Thus, despite the relatively small number of seagrass species, a vast array of other species can be found within seagrass ecosystems. Many are obligate members of the seagrass ecosystem, found nowhere else. Others may be restricted to seagrass areas for shorter periods of their life histories, using them as breeding or nursery areas, or settling there for their adult lives. Many more are found across a broad range of marine habitats, but regularly inhabit seagrass areas. Table 2 provides a list of some of the major taxonomic groups typically associated with seagrass ecosystems.

Seagrass ecosystems often play an important role in the functioning of a wider suite of coastal and marine ecosystems, including coral reefs and mangroves in the tropics, but also soft muddy bottoms, intertidal flats, salt marshes, oyster reefs and even pelagic ecosystems.

Levels of species diversity in seagrass ecosystems can be very high indeed. Humm[19] listed 113 species of algal epiphytes from *Thalassia testudinum* beds in Florida. Using this, combined with lists from 26 other publications worldwide, Harlin[20] produced a list of some 450 algal species that are epiphytic

Overview Table 2
Major taxonomic groups found in seagrass ecosystems, with brief notes

Taxonomic group	Notes
Bacteria	
Fungi	Including *Plasmodiophora*
Diatoms (Bacillariophyta)	
Blue-green algae (Cyanophyta)	
Red algae (Rhodophyta)	Including calcareous species
Brown algae (Phyaeophyta)	Including *Padina*
Green algae (Chlorophyta)	Notably *Ulva, Halimeda* and *Caulerpa*
Protozoa	Includes the slime molds *Labyrinthula* spp., and Foraminifera
Sponges	Includes epiphytic and free-standing species
Cnidarians	Includes epiphytic hydrozoans, sea anemones, solitary corals and Scleractinia such as *Pavona, Psammacora, Porites, Pocillopora, Siderastrea*
Polychaetes	Including rag-worms (nereids)
Ribbon worms	
Sipunculid worms	
Flatworms	
Crustaceans	Includes amphipods, and many decapod crustaceans including crabs, stomatopods and commercially important shrimp and lobster
Bivalve mollusks	Some oysters and scallops, also many boring species
Gastropod mollusks	A broad range including *Conus, Cypraea* and commercially important species of *Strombus*
Cephalopod mollusks	Squid and cuttlefish often found over seagrass areas
Bryozoans	Epiphytic on seagrass and rocks
Echinoderms	A range of commercially important holothurian species, ophiroids are widespread, but also asteroids and echinoids
Tunicates	Ascideans
Fish	All groups, but including the commercially important Haemulidae (grunts), Siganidae (rabbitfish), Lethrinidae (emperors), Lutjanidae (snappers), Bothidae (left-eye flounders), Syngnathidae (pipefishes and sea horses); many of the latter, which are used in the aquarium trade and Chinese medicine trade, are considered threatened
Reptiles	Notably the green turtle *Chelonia mydas*
Birds	Notably brant (geese) and other migrating waterfowl and wading birds
Mammals	Notably the sirenian species dugong *Dugong dugon* and manatee *Trichechus manatus, Trichechus senegalensis*

Source: Key references for this table include various chapters in Phillips and McRoy[18], and review comments by the contributors to this *World Atlas*.

on seagrasses, still probably an underestimate. Hutchings[21] listed some 248 arthropods, 197 mollusks, 171 polychaetes and 15 echinoderm species from Jervis Bay in New South Wales, Australia. In Florida, Roblee *et al.*[22] noted 100 species of fish and 30 species of crustaceans in seagrass beds.

A number of studies have compared diversity in seagrass beds with that observed in adjacent ecosystems. Seagrasses consistently have higher levels of diversity than adjacent non-vegetated surfaces; however, if other vegetated surfaces, or coral reefs, are compared these often have similar to significantly higher levels of diversity[2].

Despite this high diversity and the importance of associated species, there is no detailed database of species associated with seagrass beds. Many of the species that have been recorded are also found in other ecosystems, although some appear to be restricted to seagrass ecosystems or dependent on them for at least a part of their life cycles. Such seagrass-dependent species range from particular epiphytic algae[20] to the large seagrass-grazing manatee and dugong. Most of the comprehensive faunal assessments have been undertaken in temperate waters, or the relatively low-diversity waters of the Caribbean, and it seems likely that further work in the Indo-Pacific in particular will

lead to large increases in the recorded numbers of seagrass associates.

Threatened and restricted range species

Within the wider conservation arena, species with restricted distributions, together with threatened species, are often singled out for attention. Apart from concerns over these individual species, they are often used as "flagship species" to draw attention to particular areas and issues. Amongst the seagrasses, however, the problems of taxonomic uncertainty undermine the determination of both threat and restricted range.

Two species of seagrass have been listed as threatened by IUCN–The World Conservation Union[23] (see Table 3): *Halophila johnsonii* and *Phyllospadix serrulatus*. A number of countries harbor the sole populations of a seagrass species (national endemics), most notable of which is Australia, with 13 species found nowhere else in the world. Such national endemism has no inherent ecological significance, although it can be used as a basis to support conservation actions. Using the species range maps, it is possible to calculate the total area of each species range. For such calculations it was necessary to modify the broad range maps and not to include areas outside the continental shelf in the calculations. These range-area statistics are provided next to the species range maps in Appendix 3. From this work we can see that only a small number of species have truly restricted ranges, notably: *Halodule bermudensis* (1 000 km^2), *Halophila hawaiiana* (7 000 km^2), *Halophila johnsonii* (12 000 km^2), *Posidonia ostenfeldii* (66 000 km^2), *Posidonia kirkmanii* (66 000 km^2) and *Halodule beaudettei* (74 000 km^2). However, all these six species are in the process of taxonomic review and their individual species designations are presently in question.

Given the problems of taxonomy, and the low threat to the existence of individual seagrass species, measures of restricted range, endemism or threat of extinction are probably of little value in seagrass conservation efforts. Similar arguments are not true for seagrass-associated animals, although here lack of knowledge hampers a true assessment of the full

Overview Table 3
Threatened species regularly recorded from seagrass communities worldwide

Species	Common name	Status	Species	Common name	Status
Halophila johnsonii	Johnson's seagrass	Vu	*Hippocampus kuda*	Spotted or yellow sea horse	Vu
Phyllospadix serrulatus	Surf grass	R	*Hippocampus reidi*	Slender sea horse	Vu
Carcinoscorpius rotundicauda	Horseshoe crab	DD	*Hippocampus whitei*	White's sea horse	Vu
Tachypleus tridentatus	Horseshoe crab	DD	*Hippocampus zosterae*	Dwarf sea horse	Vu
Hippocampus abdominalis	Big-bellied sea horse	Vu	*Epinephelus striatus**	Nassau grouper	En
Hippocampus borboniensis	Sea horse	Vu	*Mycteroperca cidi**	Venezuelan grouper	Vu
Hippocampus breviceps	Short-headed sea horse	DD	*Mycteroperca microlepis**	Gag grouper	Vu
Hippocampus erectus	Lined sea horse	Vu	*Chelonia mydas*	Green turtle	En
Hippocampus fuscus	Sea pony	Vu	*Dugong dugon*	Dugong	Vu
Hippocampus histrix	Spiny or thorny sea horse	Vu	*Trichechus manatus*	West Indian manatee	Vu
Hippocampus jayakari	Sea horse	Vu	*Trichechus senegalensis*	West African manatee	Vu

Notes:
* Juveniles regularly observed in seagrass beds.
This list includes only species which are partially or wholly dependent on seagrasses and may be incomplete.
DD – Data Deficient: A taxon is Data Deficient when there is inadequate information to make a direct, or indirect, assessment of its risk of extinction based on its distribution and/or population status. A taxon in this category may be well studied, and its biology well known, but appropriate data on abundance and/or distribution is lacking.
R – Rare: Taxa with small world populations that are not at present Endangered or Vulnerable but are at risk. These taxa are usually localized within restricted geographic areas or habitats or are thinly scattered over a more extensive range.
Vu – Vulnerable: A taxon is Vulnerable when it is not Critically Endangered or Endangered but is facing a high risk of extinction in the wild in the medium-term future.
En – Endangered: A taxon is Endangered when it is not Critically Endangered but facing a very high risk of extinction in the wild in the near future.
Critically Endangered: A taxon is Critically Endangered when it is facing an extremely high risk of extinction in the wild in the immediate future.

Source: Walter and Gillett[23]; IUCN[24].

threats facing many species. Table 3 provides a list of some of the known seagrass species and seagrass associates listed as threatened by IUCN[23,24]. The clear focus of this list towards a few groups is probably indicative of the general lack of knowledge of the status of many seagrass associates. This problem has also been more widely recognized by IUCN[24] which acknowledges that "there has been no systematic assessment" apart from some limited groups. Of the species which have been listed, most remain poorly known or are ranked at a relatively low level of threat such as "Vulnerable".

DISTRIBUTION OF SEAGRASS HABITAT

The known locations of seagrass ecosystems, based on the mapping efforts described above, are presented in the World Seagrass Distribution Map (Map 1) and in the maps which appear in Chapters 1-24. In some parts of the world, notably the western North Atlantic, the Gulf of Mexico, Queensland (Australia), Western Australia and some parts of the Mediterranean, the maps are based on fairly comprehensive information on seagrass distribution. Elsewhere, available information is more sporadic, restricted to individual sites, bays or national coverages for smaller countries, though there may be some documentation of broader distribution patterns. Typically this is the case for areas such as the western Pacific, the Indian Ocean and the Caribbean. Over a few large stretches of the world's coasts, there exists almost no information on whether or not seagrasses occur, let alone their density, extent or species composition. This is notably the case for West Africa, South America, Greenland, northern China and the Siberian coast, and parts of Southeast Asia and the Pacific islands.

The World Seagrass Distribution Map shows the broad distribution of seagrasses in most of the world's oceans and seas, including the Black, Caspian and Aral Seas, and further shows the considerable latitudinal range of seagrasses. The most northerly locations for seagrasses are for *Zostera marina* which is recorded at Veranger fjord in Norway at 70°30'N, Chëshskaya Guba in Russia (67°30'N) and in Alaska (at 66°33'N). The most southerly locations are for *Zostera capricorni* in New Zealand, with the southernmost record being at 46°55'S on Stewart Island, and *Ruppia maritima* in the Straits of Magellan (54°S).

A limitation of these distribution maps is that they provide no information on the extent of coastlines surveyed without finding seagrass and hence do not distinguish between "no seagrass" and "no information". Gaps in the distribution maps may result from the lack of available data for certain parts of the world, but in other areas they reflect knowledge that no seagrass exists. Thus the western coastlines

A sea horse, *Hippocampus kuda*, among *Enhalus acoroides*, southern Peninsular Malaysia.

of South America and of much of West Africa may indeed have more seagrass communities than are reflected here.

CALCULATING GLOBAL SEAGRASS AREA

The calculation of a global seagrass habitat area is very important and useful for an assessment of the role of seagrasses in global processes, particularly in global carbon budgets, and also in assessing historical and future loss of seagrass and in priority setting and management of natural resources for activities such as fisheries and conservation.

To date the only global area estimate for seagrasses has been one of some 600 000 km^2 [25,26] reportedly derived from Charpy-Roubaud and Sournia[27]. The latter paper, however, does not provide an area estimate directly, and it would appear that the figure of 600 000 km^2 is derived from a global estimate of seagrass productivity[28,29] and typical seagrass productivity figures taken from an unspecified source. This estimate[28] seems too large, as the original source of global productivity was itself based on an area estimate of only 350 000 km^2 for seagrasses, salt marshes and mangrove communities combined.

The calculation of global and regional habitat areas for the marine environment can be done using two broad approaches. The first is to estimate or model probable habitat area utilizing known and mapped parameters, such as bathymetry, coastal features or existing biogeographic knowledge. The second involves

Overview Table 4
Estimates of seagrass coverage for selected areas described in this *World Atlas*

Chapter	Location	Area (km²)
1	Scandinavia	1 850
2	Western Europe	338
3	Western Mediterranean	4 152
4	Euro-Asian Seas	2 600
6	Saudi Arabia	370
8	Mozambique	439
9	India	39
10	Western Australia	25 000
11	Eastern Australia	71 371
12	New Zealand	44
13	Thailand	94
14	Peninsular Malaysia	3
15	Kosrae, Federated States of Micronesia	4
16	Indonesia	30 000
16	Philippines	978
16	Viet Nam	440
17	Japan	495
18	Korea, Republic of	70
19	Pacific coast of North America	1 000
20	Western North Atlantic coast of USA	374
21	Mid-Atlantic coast of USA	292
22	Gulf of Mexico	19 349
22	East coast of Florida	2 800
23	Mexico	500
23	Belize	1 500
23	Curaçao	8
23	Bonaire	2
23	Tobago	1
23	Martinique	41
23	Guadeloupe	82
23	Grand Cayman	25
24	Brazil	200
24	Chile	2
24	Argentina	1

Note: Almost certainly an underestimate in most cases.

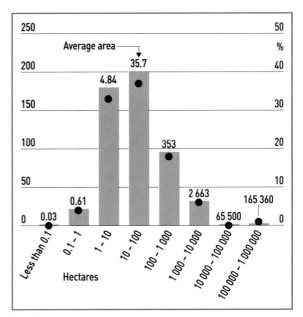

Overview Figure 1
Relative size-frequency distribution of 538 seagrass polygons in latitudinal swathe 20-30°S

Notes: The number of polygons is plotted on the primary y-axis (bars) against a logarithmic scale of area. The percentage frequency of each size class is plotted on the secondary y-axis (dots) and the mean area of all polygons in each size category is stated at the top of the columns. In this swathe there are 180 seagrass polygons of 1-10 ha in area. In other words 33 percent of the polygons in this swathe have an average area of 4.84 ha. In the area calculation it was therefore assumed that a third of all points at these latitudes were each representative of a seagrass area 4.84 ha in size, that 37 percent of points were representative of areas 35.7 ha in size, etc.

the use of mapped data to develop a more direct calculation. In many studies, elements of both approaches have been combined.

Using a simple modeling approach, the total area of continental shelf (coastal waters to a depth of 200 m) worldwide has been estimated at almost 25 million km²[30]. Assuming a constant slope, this estimate would imply an area of approximately 5 million km² of benthos within the depth range of most seagrasses, although for large parts of the globe turbidity, substrate characteristics and other factors reduce this area of potential seagrass. In reality, seagrasses occupy only a fraction of the world's nearshore waters. If the total area of seagrasses is less than 10 percent of the shallow water area of the world's continental shelves, then the maximum area would be 500 000 km². This upper limit incorporates many assumptions and is likely to be an overestimate.

Many of the authors of the subregional and national chapters of this *World Atlas of Seagrasses* have either summarized the existing seagrass maps for their area or consulted expert opinion to produce estimates of seagrass coverage. Further details are provided in the relevant chapters but these totals are summarized in Table 4.

These chapters document some 164 000 km² of seagrass but as these cover a limited geographic area and a subset of known locations they cannot be used to generate a global area.

The World Seagrass Distribution Map, developed

on a GIS, is now the most comprehensive map of global seagrass occurrence in existence. Using this we have begun to explore the direct calculation of global seagrass area.

The World Seagrass Distribution Map dataset includes more than 37000 polygons and some 8800 points. A total area of 124000 km² is clearly defined by the polygons but these provide only partial geographic coverage from a few areas which tend to be well known. Point data represent seagrass areas where habitat maps are not available. Though more poorly known than mapped areas, these locations are likely to have large and important seagrass meadows and should be factored into any calculation of area. We have experimented with methods of using the polygon data to estimate the seagrass area of these points by calculating logarithmic size-frequency distributions of polygon data in 10-degree latitudinal swathes. The distribution was then applied to the points within the swathe, generating an estimate for total seagrass area (Figure 1). Very small polygons, from data derived from remote sensing (these small polygons tend to be single or clusters of few pixels), and very large polygons, derived from sketch maps covering

Overview Table 5
Functions and values of seagrass from the wider ecosystem perspective

Function	Ecosystem values
Primary production – including benthic and epibenthic production	Seagrasses are highly productive, and play a critical role as food for many herbivores (manatee, dugong, turtles, fish, waterfowl, etc.). This productivity lies at the base of the food chain and is also exported to adjacent ecosytems.
Canopy structure	The growing structures of seagrasses provide a complex three-dimensional environment, used as a habitat, refuge and nursery for numerous species, including commercially important fish and shellfish.
Epiphyte and epifaunal substratum	The large surface area of seagrass above-ground biomass provides additional space for epiphytes and epifauna, supporting high secondary productivity.
Nutrient and contaminant filtration	Seagrasses help to both settle and remove contaminants from the water column and sediments, improving water quality in the immediate environment and adjacent habitats.
Sediment filtration and trapping	The canopy of seagrasses helps to encourage settlement of sediments and prevent resuspension, while the root systems help to bind sediments over the longer term, improving water quality and in some places helping to counter sea-level rise.
Creating below-ground structure	The complex and often deep structures of the seagrass roots and rhizomes support overall productivity and play a critical role in binding sediments.
Oxygen production	The oxygen released from photosynthesis helps improve water quality and support faunal communities in seagrasses and adjacent habitats.
Organic production and export	Many seagrass ecosystems are net exporters of organic materials, supporting estuarine and offshore productivity.
Nutrient regeneration and recycling	Seagrasses hold nutrients in a relatively stable environment, and nutrient recycling can be relatively efficient, supporting overall ecosystem productivity.
Organic matter accumulation	Along with sediments the organic matter of roots, rhizomes and even leaves can remain bound within the sediment matrix, or accumulate on adjacent coastlines or other habitats, building up the level of the benthos and supporting other food webs.
Wave and current energy dampening	By holding and binding sediments, and by preventing the scouring action of waves directly on the benthos, seagrasses dampen the effects of wave and current energy, reduce processes of erosion, reduce turbidity and increase sedimentation.
Seed production/vegetative expansion	Seagrasses are capable of both self-maintenance and spreading to new areas via sexual and asexual reproduction. Recovery following storms, disease or human-induced damage can be relatively rapid.
Self-sustaining ecosystem	The complex community of the seagrass ecosystem supports important biodiversity and provides trophic interactions with other important ecosystems such as coral reefs, mangroves, salt marshes and shellfish reefs.
Carbon sequestration	As perennial structures, seagrasses are one of the few marine ecosystems which store carbon for relatively long periods. In a few places such carbon may be bound into sediments or transported into the deeper oceans and thus play an important role in long-term carbon sequestration.

Source: Derived from Short et al.[31] and Global Seagrass Workshop recommendations.

Overview Table 6
Summary of the goods and services provided by seagrass ecosystems

Commercial and artisanal fisheries[32]
Finfish (snappers, emperors, rabbitfish, surgeonfish, flounder)[33]
Mollusks (conch, oysters, mussels, scallops, clams)[34]
Crustacea (shrimp, lobster, crab)[35]
Mammals and reptiles (dugongs, manatee, green turtle)[36, 37]

Nursery habitat for offshore fisheries[2, 38]

Food
Seeds of *Zostera marina* used to make flour by Seri Indians[39]
Rhizomes of *Enhalus* used as food in Lamu, Kenya[40]

Fodder or bedding for animals[41, 42]

Fiber
Used in mat weaving, Lamu, Kenya[40]
Basket making, thatch, stuffing mattresses, upholstery[43]
Insulation[44]

Packing material[45]

Fertilizer and mulch[46, 47, 48]

Building dikes[49]

Coastal protection from erosion[50-52]

Water purification
Reducing eutrophication and phytoplankton blooms[53]
Removing toxic organic compounds from water column and sediment[54, 55]

Interaction with adjacent ecosystems[25]
Nutrient export[56]
Source of food or shelter, as a nursery, resting ground or feeding ground[2]
Water column filtration[53]

Maintenance of biodiversity and threatened species[2]
Dugongs, manatee, green turtle[57, 58]

Carbon dioxide sink[26]

Cultural, esthetic and intrinsic values[44]
Places of natural beauty
Recreational value
Educational value

Stabilizing sediments
Binding function of roots[53]
Role of shoots in reducing surface flow and encouraging settlement[59, 60]

Source: Various sources – see references by entries.

enormous areas (e.g. the global National Geographic "Coral World" map), were excluded from this analysis to avoid serious under- and overestimates respectively.

When combined with polygon data this method generates an estimate for the global coverage of seagrass of 177 000 km^2 (using median polygon areas reduced the estimate by 4 percent). It is based on the most comprehensive dataset on seagrass distribution to date. However it is necessarily and unavoidably based upon a number of crude assumptions and is intended to be no more than indicative of the global extent of seagrass. In any event, even the 177 000 km^2 is an underestimate of the actual global seagrass area, since for many areas seagrasses have not been documented. Until our knowledge of seagrasses in large areas such as insular Southeast Asia, the east coast of South America and the west coast of Africa improves, it is unlikely that a better estimate can be generated.

THE VALUE OF SEAGRASSES

Seagrasses are a critical ecosystem: their role in fisheries production, and in sediment accumulation and stabilization, is well documented, but there are many other important roles, both in terms of their place in the ecosystem and their value to humanity. Table 5 lists a number of the functions of seagrasses from a wider ecosystem perspective.

Seagrasses have a relatively low biomass compared with terrestrial ecosystems, but have a very high biomass in relation to planktonic-based marine communities. Figures for average biomass vary considerably between seagrass species and between studies; communities of *Amphibolis*, *Phyllospadix* and *Posidonia* in particular are noted for their high biomass, the last's enhanced by extensive stem and root systems. In contrast, species of *Halophila*, with their small petiolate leaves and high turnover rates, rarely achieve high biomass.

Duarte and Chiscano[26], in a literature review, calculated from nearly 400 samples an average biomass for different seagrass species, and by averaging these values derived an average biomass for seagrass of 460 g dry weight/m^2 (above- and below-ground biomass combined). As an estimate of global seagrass biomass, such estimates are biased towards large seagrass species. Taking these factors into account, the median biomass statistic of 205 g dry weight/m^2, also from data in Duarte and Chiscano, may be a more accurate reflection of the typical biomass for seagrass communities worldwide.

In terms of productivity, Duarte and Chiscano[26] estimated an average net primary production of about 1 012 g dry weight/m^2/year. Even allowing for overestimation, such figures are very high for marine communities, with the same source citing productivity figures for macroalgal communities of 1 g dry weight/m²/day and of phytoplankton of 0.35 g dry weight/m²/day.

The high productivity and biomass of seagrasses are an integral part of many of their uses and values from a human perspective. A broad sample of the goods and services provided by seagrasses is shown in Table 6, while further information on a number of these is given in the text, both here and in many of the regional and national chapters.

Fisheries

Seagrass ecosystems are highly productive and also have a relatively complex physical structure, thus providing a combination of food and shelter that enables a high biomass and productivity of commercially important fish species to be maintained[2, 61]. Seagrasses also provide an important nursery area for many species utilized in offshore fisheries and in adjacent habitats such as coral reefs and mangrove forests. In most cases, the association between commercially important species and seagrasses is not obligatory; the same species are found in other shallow marine habitats. There are, however, a number of studies which clearly show the higher biomass of such species associated with seagrasses as compared with adjacent unvegetated areas[2].

Sediment stabilization and coastal protection

Seagrasses are the only submerged marine photo-trophs with an underground root and rhizome system. This below-ground biomass is often equal to that of the above-ground biomass, and can be considerably more e.g. *Posidonia*[26]. The role of these roots and rhizomes in binding sediments is highly important, as has been illustrated in a number of studies that have compared erosion on vegetated versus non-vegetated areas during storm events. The role of seagrass shoots in this process is also important, as these provide a stable surface layer above the benthos, baffling currents and therefore encouraging the settlement of sediments and inhibiting their resuspension[62].

Water purification and nutrient cycling

By enhancing processes of sedimentation, and through the relatively rapid uptake of nutrients both by seagrasses and their epiphytes, seagrass ecosystems remove nutrients from the water column. Once removed these nutrients can be released only slowly through a process of decomposition and consumption, quite different from the rapid turnover observed in phytoplankton-dominated systems. In this way seagrasses can reduce problems of eutrophication and bind organic pollutants[2].

Mitigating climate change

The role of the world's oceans in removing carbon dioxide from the atmosphere is still being investigated and remains poorly understood. It appears that biological processes in the surface layers of the world's oceans are one of the few mechanisms actively removing carbon dioxide from the global carbon cycle[63]. Within these processes, seagrasses clearly have a minor role to play, although their high productivity gives them a disproportionate influence on primary productivity in the global oceans on a unit area basis, and they typically produce considerably more organic carbon than the seagrass ecosystem requires[25, 64]. Any removal of carbon either through binding of organic material into the sediments or export into the deep waters off the continental shelf represents effective removal of carbon dioxide from the ocean-atmosphere system which could play some role in the amelioration of climate change impacts.

Maintaining biodiversity and threatened species

The concept of seagrasses as high-diversity marine ecosystems has often been overlooked, but this role has already been briefly outlined above. Seagrasses also play a role in safeguarding a number of threatened species, including those such as sirenians, turtles and sea horses, which are widely perceived to have very high cultural, esthetic or intrinsic values by particular groups. The wider functions of biodiversity include the maintenance of genetic variability, with potential biochemical utility, and a possible, though poorly understood, role in supporting ecosystem function and resilience.

Economic valuation

There have been very few studies of the direct economic value of seagrasses. In Monroe County, Florida, the value of commercial fisheries for five species which depend on seagrasses was estimated at US$48.7 million per year, whilst recreational fisheries, as well as the diving and snorkeling industry in that county, contribute large sums to the economy and are also indirectly dependent on seagrasses[65].

Costanza *et al.*[66] calculated a global value of annual ecosystem services for "seagrass/algae beds" of US$19 004 per hectare per year. With their estimated total area for these combined ecosystems of 2 000 000 km² they calculated a global annual value of US$3 801 000 000 000 (i.e. US$3.8 trillion), based almost entirely on their role in "nutrient cycling", which is only one of many values of the ecosystem. The same source gives no value to seagrass/algae beds for food production.

Further information is needed to demonstrate the full economic value of seagrass ecosystems worldwide.

Damage to seagrass beds caused by yachts in Jersey, Channel Islands, UK.

Photo: Emma Jackson/Paul Tucker

It will be important not only to measure direct value from activities such as fisheries but also indirect values associated with various functions (Table 5) including maintenance of water quality and protecting coastlines. In many ways dollar values provide only a part of the true picture of the value of an ecosystem, and it is important to consider other possible means to quantify value, including employment, protein supply or even quality of life as alternative measures which address value from a human perspective. It should be noted that even when dollar values are estimated they do not represent the entire worth of the ecosystem and in no way constitute a purchase value.

THREATS TO SEAGRASSES

The global threats to seagrasses have received considerable attention from a number of authors (e.g. Short et al.[12], Phillips and Durako[62], Short and Wyllie-Echeverria[67], Hemminga and Duarte[2]) and their efforts are only summarized here. In many cases it seems likely that declines in seagrass areas have been the result not of individual threats but a combination of impacts. Typical combined impacts may include increased turbidity, increased nutrient loads and direct mechanical damage. Seagrasses exist at the land-sea margin and are highly vulnerable to the world's human populations which live disproportionately along the coasts. Such conditions threaten seagrass ecosystems and have resulted in substantial loss of many seagrass areas in the more populated parts of the world, as well as degradation of much wider areas over the last 100 years.

A number of natural threats to seagrasses have been recorded. Geological impacts may include coastal uplift or subsidence, raising or lowering beds to less than ideal growing conditions. Meteorological impacts can also affect seagrasses: major storm events in particular may remove surface biomass and even uproot and erode wide areas of shallow water. Finally there are biological impacts. Typically these are part of the ongoing processes in seagrass ecosystems, such as grazing by fish, sea urchins, sirenians, geese or turtles; they also include disruption to the sediments by burrowing animals or foraging species such as rays. It is rare that such activities should disrupt seagrass beds over large areas. Diseases, however, represent an important biological impact which can have very widespread effects. The eelgrass wasting disease recorded from the North Atlantic in the 1930s[68, 69] was caused by the slime mold *Labyrinthula zosterae*[70, 71]. This wasting disease continues to occur and remains a threat to eelgrass in the North Atlantic[70]. Similarly in Florida Bay, disease caused by *Labyrinthula* sp. has been implicated in an extensive seagrass die-off[81].

Human threats to seagrasses are now widespread. Many result in direct destruction of these habitats. Dredging to develop or widen shipping lanes and open new ports and harbors, and certain types of fishery such as benthic trawling, have led to losses of wide areas of seagrass. Boating activities frequently lead to propeller damage, groundings or anchor damage, often increasing sediment resuspension or creating holes and initiating "blow-out" areas in seagrass beds. Construction activities within coastal waters have sometimes led to losses: land reclamation is a clear example, as is the construction of aquaculture ponds in some areas. Even the construction of docks and piers can lead to some direct losses, and to further losses arising from shading or fragmentation of seagrass beds. The alteration of the hydrological regime as a result of coastal development and the building of sea defenses can also impact seagrasses. There are examples of direct and deliberate removal of seagrasses, for example to "clean" tourist beaches or to maintain navigation channels.

In addition, many seagrass beds have been affected by the indirect impacts of human activities. Land-based threats include increases of sediment loads: higher turbidity reduces light levels, while very high sedimentation smothers entire seagrass beds.

Similarly, while seagrasses can assimilate certain levels of nutrient and toxic pollutants, high levels of increased nutrients from sewage disposal, overland runoff and enriched groundwater discharge can reduce seagrass photosynthesis by excess epiphytic overgrowth, planktonic blooms or competition from macroalgae. Toxins can poison and kill seagrasses rapidly. Another indirect threat comes from the introduction of alien or exotic species. The alga *Caulerpa taxifolia*, released into the Mediterranean in the 1980s, has smothered and killed wide areas of seagrass beds. In 1999, the same species was first observed off the coast of California and could have the same impact there[72].

Climate change represents a relatively new threat, the impacts of which on seagrasses are largely undetermined[64]. Potential threats from climate change may come from rising sea levels, changing tidal

Overview Table 7
Summary of marine protected areas (MPAs) that contain seagrass ecosystems, from the UNEP-WCMC Protected Areas Database

Country or territory	Number of sites	Country or territory	Number of sites
Anguilla	2	Monaco	1
Antigua and Barbuda	1	Mozambique	8
Australia	11	Netherlands Antilles	2
Bahamas	1	Nicaragua	1
Bahrain	1	Palau	1
Belize	4	Panama	1
Brazil	3	Papua New Guinea	6
British Indian Ocean Territory	1	Philippines	2
Cambodia	1	Puerto Rico	6
Canada	1	Réunion	5
Cayman Islands	4	Russian Federation	3
China	1	Saint Lucia	3
Colombia	3	Saint Vincent and the Grenadines	1
Costa Rica	2	Saudi Arabia	2
Croatia	1	Seychelles	3
Cuba	1	Singapore	1
Cyprus	1	Slovenia	1
Dominica	1	South Africa	4
Dominican Republic	4	Spain	3
France	3	Tanzania	10
French Polynesia	1	Thailand	2
Germany	2	Tonga	2
Guadeloupe	1	Trinidad and Tobago	1
Guam	1	Tunisia	1
Guatemala	1	Turks and Caicos Islands	1
Honduras	3	Ukraine	6
India	5	United Kingdom	4
Indonesia	7	United States	31
Israel	1	United States minor outlying island	1
Italy	5	Venezuela	9
Jamaica	6	Viet Nam	1
Kenya	5	Virgin Islands (British)	5
Korea, Republic of	1	Virgin Islands (US)	5
Madagascar	2		
Malaysia	13		
Martinique	1		
Mauritania	1		
Mauritius	5		
Mexico	6		

Note: Few of these sites are managed directly to support seagrass protection, and in many cases they do not protect the most important areas of seagrass in a region.

regimes, localized decreases in salinity, damage from ultraviolet radiation, and unpredictable impacts from changes in the distribution and intensity of extreme events. In contrast there could be increases in productivity resulting from higher carbon dioxide concentrations[64].

Various studies have attempted to quantify the decline of seagrasses, although it must be accepted that seagrasses have been degraded or lost over vast areas without any knowledge of their existence. Short and Wyllie-Echeverria[67,73] provide an analysis of seagrass losses from reports worldwide. They found that a loss of 2 900 km² of seagrass was documented between the mid-1980s and the mid-1990s, and they extrapolated likely seagrass losses over that time period alone of up to 12 000 km² worldwide.

PROTECTING SEAGRASSES

The dramatic and accelerating declines in seagrass areas worldwide are mirrored in other coastal ecosystems such as mangroves and coral reefs[30]. Concerns about these declines have prompted some increase in efforts to protect these ecosystems. Perhaps the most valuable protection measure is the wholesale reduction of the full suite of anthropogenic impacts via legislation and enforcement at local and regional scales. Unfortunately the cost is high and rates of improvement are low.

More practical protection, although only localized in effect, is the establishment of marine protected areas (MPAs), legally gazetted sites where certain (but by no means all) human activities are controlled or prohibited in order to provide some protection of marine resources or to promote sustainable fisheries. Whether out of direct interest or as an indirect beneficiary, seagrass habitat is present in an increasing number of sites in the expanding MPA network. The total number of MPAs has increased dramatically in recent years, from less than 500 MPAs worldwide in 1960 to more than 4 000 by 2001 (UNEP-WCMC data, but note that this figure includes intertidal as well as subtidal sites). No MPAs have been designated solely for the protection of seagrasses; however seagrasses are often one of a list of key habitats singled out when sites are recommended for protection (e.g. the Great Barrier Reef Marine Park in Australia). Many other sites include seagrasses even when the key natural resource behind their protection may be something else, such as a coral reef. In the majority of MPAs, seagrasses are not acknowledged or directly protected. With increased awareness, MPA boundaries and protection could be expanded to incorporate adjacent seagrass habitats (e.g. Florida Bay adjacent to the Everglades).

UNEP-WCMC maintains a global database on MPAs on behalf of the IUCN World Commission on Protected Areas. Linked to the current work, a list of the areas which are known to contain seagrass habitat has been prepared and is presented in Appendix 2. A summary of this information is provided in Table 7.

Worldwide there are some 247 MPAs known to include seagrasses. These are located in 72 countries and territories. These numbers are likely to be conservative: seagrasses may well occur at a site but not be recorded, or not be listed in literature which has been used to develop this database. Even so, it seems likely that this list is far smaller than the equivalent network for coral reefs (more than 660[8]) and mangrove forests (over 1 800, unpublished data 2000) and clearly does not present any form of global network. Added to this must be the recognition that the vast majority of these sites do not provide any clear protection for seagrasses – their inclusion within MPAs is largely fortuitous.

Figure 2 shows the increase in seagrass MPAs over the past century. It should be noted that the area figures (shaded area) are a measure of the total area covered by these MPAs. At the present time it is impossible to determine the area of seagrasses within these sites, although it is likely to be only a very small fraction of the total area. It should further be noted that

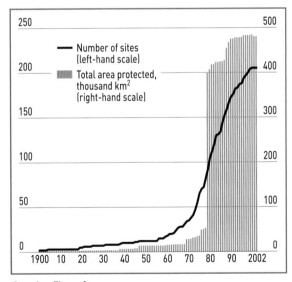

Overview Figure 2
Growth of marine protected areas which include seagrass ecosystems, shown both as the number of sites (line) and the total area protected (shaded area)

Notes: The total area statistics are for the entire MPAs; there is no information on the area of seagrass within these sites but it is likely to be only a small fraction of the total area. Figure 2 covers only those sites for which a date of designation has been recorded. In addition to the 205 sites shown here there are a further 42 with a total area of some 3 500 km² whose year of designation is not known.

The distribution and status of seagrasses

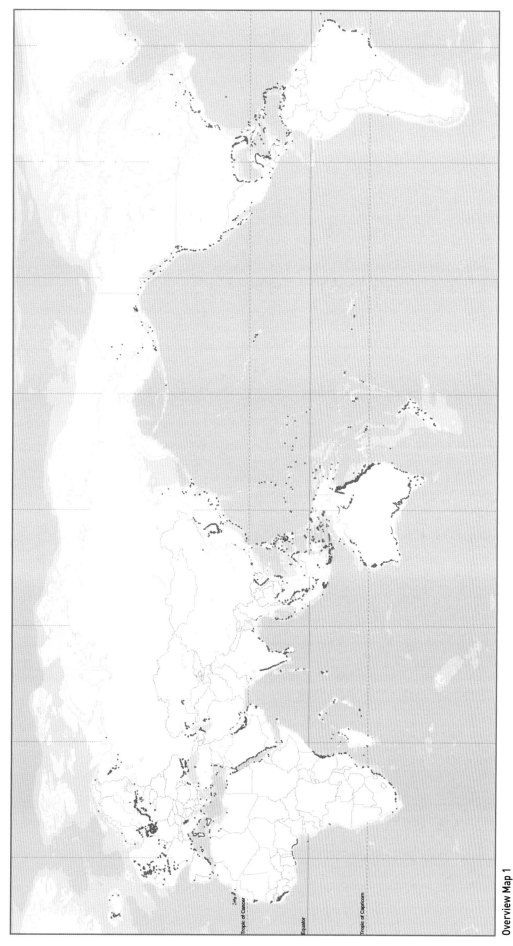

Overview Map 1
World seagrass distribution

Source: Data are from highly varied source material as described in the text (and see http://www.unep-wcmc.org/marine/seagrassatlas/references). In order to make all data visible on a map at this scale, all data points and polygons have been exaggerated, and hence this map is indicative of general distribution rather than areal extent.

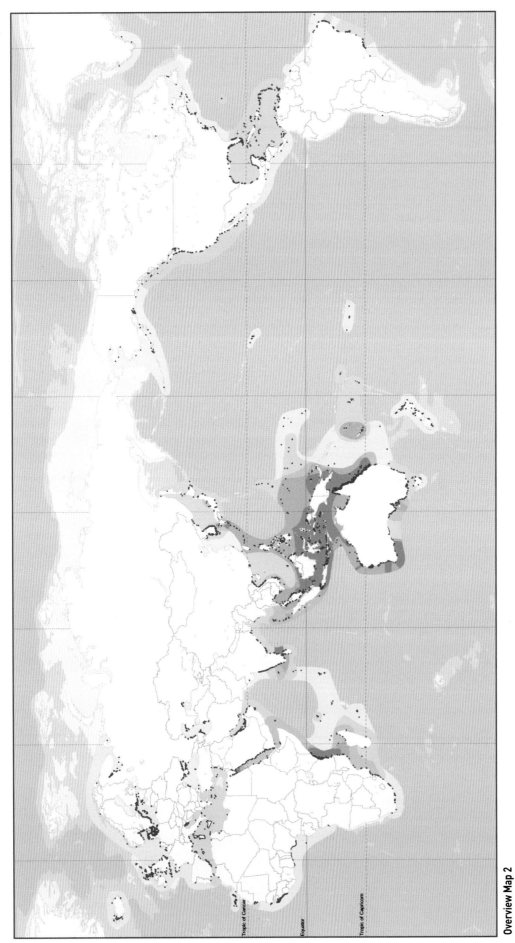

Overview Map 2
Global seagrass biodiversity

Diversity of seagrass species, derived from a combination of the known ranges of 59 individual species (not including *Ruppia* spp.)

designation as "protected" covers a broad range of types of protection, both in terms of legal status and practical application of that status. Some sites, such as the National Estuarine Research Reserves in the United States, do not provide any direct habitat protection under their supporting legislation. In many other cases, even where the legislation may provide a formal safeguard, management may be inadequate. The world's largest MPA, Australia's Great Barrier Reef Marine Park, has made some efforts to prevent trawling in seagrass areas, but this entire park, like most others worldwide, is still subject to influences from beyond the park boundaries.

In a recent analysis by regional experts, management effectiveness was considered for some 342 MPAs in Southeast Asia and was rated as "good" for only 46 sites (14 percent)[74]. Finally, many of the threats facing seagrasses come from remote sources, notably terrestrial runoff. Few protected areas currently manage entire watersheds and the legal framework is typically powerless to control nutrient and toxic pollution and sedimentation arising outside an MPA.

In addition to the designation of MPAs, other legal measures have proved beneficial to seagrasses in some places, although seagrasses themselves are rarely singled out as the object of protection. Such legislation includes restrictions on particular activities such as trawling, dredging or the release of land-based sources of degradation such as sediments and pollutants. For example, in Queensland waters (Australia) all seagrasses and other marine plants are specifically protected under the Fisheries Act of 1994, for the protection of commercial and recreational fishing activities. In South Australia seagrass is protected under the Native Vegetation Act 1992. In the United States, seagrass habitats are protected under Section 404(c) of the Clean Water Act from direct dredge and fill activities without a permit[75]. Although clearly important, such legislation is rare, and still insignificant at the global level.

In addition to legal protection, public education can play an important role in safeguarding seagrasses, notably via the protection of charismatic seagrass associates such as turtles and dugongs, but also in the directing of activities which could impact seagrasses. Coles and Fortes[76] provide a valuable review of methods for direct and indirect protection of seagrasses.

Seagrass restoration may include both the improvement of overall conditions for seagrass growth in an area, such as an improvement in water clarity resulting from decreased runoff or nutrient inputs, as well as direct transplanting or seeding of seagrasses[77]. Sometimes transplanting is mandated as mitigation for unavoidable damage to seagrasses incurred in coastal

The shallow seagrass beds of Montepuez Bay, Mozambique, at low tide.

development. Transplanting cannot be successful unless the conditions for seagrass to thrive pre-exist[76]. Although widely undertaken in some areas, many transplantation efforts have had low success rates and transplanting can be quite labor intensive and expensive[77]. Technologies for more uniformly successful and less expensive seagrass transplanting are evolving, and include developing models for site selection[78], advanced methods for transplanting[75] and seeding[79], and scientific success criteria[31]. Restoration of seagrasses is now at the stage where technologies are available, but overcoming insufficient water quality conditions remains the greatest obstacle to seagrass restoration worldwide.

CONCLUSIONS

We know a substantial amount about seagrasses in many parts of the world, but there remain considerable gaps in our knowledge. As the taxonomies of various species are revised, even our understanding of how many species of seagrass there are will be subject to debate and change.

The range of individual seagrass species is presented in a new series of maps. By combining these range maps we are also able to look at biodiversity patterns in seagrasses as a whole. The primary centers of seagrass biodiversity are identified here as insular Southeast Asia, Japan and southwest Australia, with additional areas in southern India and eastern Africa. While there are some important parallels between seagrasses and the two other major tropical coastal ecosystems of coral reefs and mangroves, there are also important divergences, notably with the seagrass centers of diversity in Japan and in southwestern Australia but also with the occurrence of seagrasses in high latitudes as well as the tropics.

It is clear that, despite the relative paucity of seagrass species, as a habitat these communities are in fact highly diverse. There are many thousands of species recorded living in association with seagrass communities, although only a small proportion of these are strictly confined to seagrass ecosystems. There is an urgent need to develop a more comprehensive understanding of the full range and diversity of life in seagrasses.

The work presented here includes a detailed map of the known locations of seagrass habitats around the world. Once again we are made aware of considerable gaps in our knowledge. There is an urgent need for clearer documentation of the existence and location of seagrass ecosystems in western South America and in West Africa, for example. Even within areas of high seagrass biodiversity, in many cases little is known about the actual distribution of seagrasses. Much of our data for the World Seagrass Distribution Map is based on individual points of occurrence and not on area of coverage. The importance of the high levels of primary productivity in seagrasses is well known, and these are clearly disproportionate to the total area covered by these habitats. It would be invaluable to develop an accurate estimate of the total area of seagrasses worldwide in order to better analyze the role that these may play in global and regional fisheries, and in climatic and oceanic carbon cycles. In the absence of any better data we have undertaken an analysis of seagrass area and suggest a conservative estimate of 177 000 km^2.

There can be no doubt of the value of seagrasses, although such values are often overlooked. For fish, many species are not obligatory users of seagrass ecosystems, but appear to benefit from their presence. Many others use seagrass ecosystems for a short (but often critical) part of their life histories, and seagrasses are rarely considered in assessing these fisheries. Economic evaluations are often constrained by analytical procedures and many fail to calculate the total economic value of an ecosystem. The critical role of seagrasses in stabilizing sediments, reducing erosion and even cleaning coastal waters is rarely accounted for in such analyses. In addition, other measures, which include social welfare, health and well-being, are difficult to measure.

The threats to seagrasses have been widely considered by other authors and include natural and anthropogenic causes. The latter appear to have increased dramatically in recent years, and include direct physical destruction and a range of indirect threats, the most critical being decreases in water clarity resulting from nutrient and sediment inputs but also including climate change. In many cases, seagrass declines have been linked to multiple stresses, acting together. In only a few places around the world are measures being taken to address these threats. In the present work we have assembled an assessment of marine protected areas with seagrasses worldwide. Some 247 sites are known to include seagrass ecosystems. This is a far lower figure than for other shallow marine ecosystems, while further concern must be expressed about the effectiveness of these sites in protecting seagrasses, both from direct impacts and from the indirect impacts such as pollution and sedimentation which may be carried into the seagrass areas from beyond the reserve boundaries.

The chapters which make up the bulk of this work provide a more detailed examination of seagrass distribution and of the various themes considered here. They provide detailed examples of seagrass communities around the world, and illustrate issues relating to distribution, status and management of these beautiful and critically important ecosystems.

ACKNOWLEDGMENTS

This chapter would not have been possible without the help and input from all the participants at the Global Seagrass Workshop which was held at the Estuarine Research Federation meeting in St Petersburg, Florida in October 2001. The chapter's authors have also drawn heavily on all the chapters written by the regional authors (see Table of Contents). The assistance provided by Sergio Martins and Mary Edwards is gratefully acknowledged. Jackson Estuarine Laboratory contribution number 397.

AUTHORS

M. Spalding, UNEP World Conservation Monitoring Centre, 219 Huntingdon Road, Cambridge, CB3 0DL, UK. **Contact address:** 17 The Green, Ashley, Newmarket, Suffolk, CB8 9EB, UK. **Tel:** +44 (0)1638 730760. **E-mail:** mark@mdspalding.co.uk

M. Taylor, C. Ravilious, E. Green, UNEP World Conservation Monitoring Centre, 219 Huntingdon Road, Cambridge, CB3 0DL, UK.

F. Short, University of New Hampshire, Jackson Estuarine Laboratory, 85 Adams Point Road, Durham, NH 03824, USA.

REFERENCES

1. Phillips RC, Meñez EG [1988]. *Seagrasses*. Smithsonian Contributions to the Marine Sciences 34. Smithsonian Institution Press, Washington DC.
2. Hemminga MA, Duarte CM [2000]. *Seagrass Ecology*. Cambridge University Press, Cambridge.
3. Kuo J, den Hartog C [2001]. Seagrass taxonomy and identification key. In: Short FT, Coles RG (eds) *Global Seagrass Research Methods*. Elsevier Science, Amsterdam. pp 31-58.
4. Kuo J, McComb AJ [1989]. Seagrass taxonomy, structure and development. In: Larkum AWD, McComb AJ, Shepherd SA (eds) *Biology of Seagrasses – A Treatise on the Biology of Seagrasses with Special Reference to the Australian Region*. Elsevier, New York. pp 6-73.
5. Short FT, Coles RG (eds) [2001]. *Global Seagrass Research Methods*. Elsevier Science, Amsterdam.
6. Lipkin Y [1979]. Quantitative aspects of seagrass communities, particularly of those dominated by *Halophila stipulacea*, in Sinai (Northern Red Sea). *Aquatic Botany* 7: 119-128.
7. Spalding MD, Blasco F, Field CD (eds) [1997]. *World Mangrove Atlas*. International Society for Mangrove Ecosystems, Okinawa.
8. Spalding MD, Ravilious C, Green EP [2001]. *World Atlas of Coral Reefs*. University of California Press, Berkeley, California.
9. den Hartog C [1970]. The seagrasses of the world. *Verhandelingen der Koninklijke Nederlandse Akademie van Wetenschappen Afdeling Natuurkunde* 59: 1-275.
10. Veron JEN [2000]. *Corals of the World*. Australian Institute of Marine Science, Townsville.
11. Groombridge B, Jenkins MD (eds) [2000]. *Global Biodiversity: Earth's Living Resources in the 21st Century*. World Conservation Press, Cambridge.
12. Short FT, Coles RG, Pergent-Martini C [2001]. Global seagrass distribution. In: Short FT, Coles RG (eds) *Global Seagrass Research Methods*. Elsevier Science, Amsterdam.
13. Roberts CM, Mclean CJ, Allen GR, Hawkins JP, McAllister DE, Mittermeier C, Schueler F, Spalding M, Veron JEN, Wells F, Vynne C, Werner T [2002]. Marine biodiversity hotspots and conservation priorities for tropical reefs. *Science* 295: 1280-1284.
14. Jokiel PL, Martinelli FJ [1992]. The vortex model of coral reef biogeography. *J Biogeog* 19: 449-458.
15. McCoy ED, Heck KL [1976]. Biogeography of corals, seagrasses and mangroves: An alternative to the center of origin concept. *Sys Zool* 25: 201-210.
16. Rosen BR [1984]. Reef coral biogeography and climate through the late Cainozoic: Just islands in the sun or a pattern of islands? In: Brenchley P (ed) *Fossils and Climate*. John Wiley and Sons, London.
17. Larkum AWD, den Hartog C [1989]. Evolution and biogeography of seagrasses. In: Larkum AWD, McComb AJ, Shepherd SA (eds) *Biology of Seagrasses – A Treatise on the Biology of Seagrasses with Special Reference to the Australian Region*. Elsevier, New York. pp 112-56.
18. Phillips RC, McRoy PC [1980]. *Handbook of Seagrass Biology: An Ecosystem Perspective*. Garland STPM Press, New York.
19. Humm HJ [1964]. Epiphytes of the seagrasses, *Thalassia testudinum*, in Florida. *Bull Mar Sci Gulf Caribbean* 14: 306-341.
20. Harlin MM [1980]. Seagrass epiphytes. In: Phillips RC, McRoy PC (eds) *Handbook of Seagrass Biology: An Ecosystem Perspective*. Garland STPM Press, New York. pp 117-131.
21. Hutchings P [1994]. Jervis Bay Infauna Data from Seagrass Beds 1988-1991, CSIRO Division of Fisheries Jervis Bay Baseline Studies, Final Report, May 1994, Vols 1-3. CSIRO Division of Fisheries, North Beach, WA.
22. Roblee MB, Barber TR, Carlson PR, Durako MJ, Fourqurean JW, Muehlstein LK, Porter D, Yarbro LA, Ziemen RT, Zieman JC [1991]. Mass mortality of tropical seagrass *Thalassia testudinum* in Florida Bay (USA). *Marine Ecology Progress Series* 71: 297-299.
23. Walter KS, Gillett HJ (eds) [1998]. *1997 IUCN Red List of Threatened Plants*. Compiler: World Conservation Monitoring Centre. IUCN–The World Conservation Union, Gland, Switzerland, and Cambridge.
24. IUCN [2000]. *The 2000 IUCN Red List of Threatened Species*. IUCN, Gland, Switzerland.
25. Duarte CM, Cebrian J [1996]. The fate of marine autotrophic production. *Limnology and Oceanography* 41: 1758-1766.
26. Duarte CM, Chiscano CL [1999]. Seagrass biomass and production: A reassessment. *Aquatic Botany* 65: 159-174.
27. Charpy-Roubaud C, Sournia A [1990]. The comparative estimation of photoplanktonic and microphytobenthic production in the oceans. *Mar Microbial Food Webs* 4: 31-57.
28. Woodwell GM, Rich PH, Hall CA [1973]. Carbon in estuaries. In: Woodwell GM, Pecan EV (eds) *Carbon and the Biosphere*. AEC Symposium Series 20. NTIS US Dept of Commerce, Springfield, Virginia. pp 221-239 (cited in de Vooys[29]).
29. de Vooys CGN [1979]. Primary production in aquatic environments. In: Bolin B, Degens ET, Kempe S, Ketner P (eds) *SCOPE, 13: The Global Carbon Cycle*. John Wiley and Sons, Chichester. pp 259-262.
30. Burke L, Kura Y, Kassem K, Revenga C, Spalding M, McAllister D [2001]. *Pilot Analysis of Global Ecosystems: Coastal Ecosystems*. World Resources Institute, Washington, DC.
31. Short FT, Burdick DM, Short CA, Davis RC, Morgan PA [2000]. Developing success criteria for restored eelgrass, salt marsh and mud flat habitats. *Ecological Engineering* 15: 239-252.
32. Bell JD, Pollard DA [1989]. Ecology of fish assemblages associated with seagrasses. In: Larkum AWD, McComb AJ, Shepherd SA (eds) *Biology of Seagrasses – A Treatise on the Biology of Seagrasses with Special Reference to the Australian Region*. Elsevier, New York.
33. Walker DI, Lukatelich RJ, Bastyan G, McComb AJ [1989]. Effect of boat moorings on seagrass beds near Perth, Western Australia. *Aquatic Botany* 36: 69-77.
34. Heck KLJr, Able KW, Fahay MP, Roman CT [1989]. Fishes and decapod crustaceans of Cape Cod eelgrass meadows: Species composition, seasonal abundance patterns and comparison with unvegetated substrates. *Estuaries* 12: 59-65.
35. Thayer GW, Kenworthy WJ, Fonseca MS [1984]. *The Ecology of Eelgrass Meadows of the Atlantic Coast: A Community Profile*. US Fish and Wildlife Service, FWS/OBS0-84/02.
36. Short FT, Short CA [2000]. Identifying seagrass growth forms for leaf and rhizome marking applications. *Biologia Marina Mediterranea* 7: 131-134.
37. This volume, Chapter 11.
38. Smith KA, Suthers IM [2000]. Consistent timing of juvenile fish recruitment to seagrass beds within two Sydney estuaries. *Marine and Freshwater Research* 51: 765-776.
39. Felger RS, Moser MB [1973]. Eelgrass (*Zostera marina* L.) in the Gulf of California: Discovery of its nutritional value by Seri Indians. *Science* 81: 355-356.
40. Crafter SA, Njuguna SG, Howard GW [1992]. *Wetlands of Kenya: Proceedings of the KWWG Seminar on Wetlands of Kenya, National Museums of Kenya, Nairobi, Kenya, 3-5 July 1991*. p 122.
41. Loo MGK, Tun KPP, Low JKY, Chou LM [1994]. A review of seagrass communities in Singapore. In: Wilkinson CR, Sudara S, Ming CL (eds) *Proceedings, Third ASEAN-Australia Symposium on Living Coastal Resources*. Vol 1. pp 311-316.
42. UNEP [1986]. Environmental Problems of the Marine and Coastal Area of Sri Lanka: National Report. UNEP Regional Seas Report and Studies No. 74. UNEP, Nairobi.

43. Wyllie-Echeverria S, Cox PA [1999]. The seagrass (*Zostera marina* [ZOSTERACEAE]) industry of Nova Scotia (1907-1960) *Economic Botany* 53: 419-426.
44. Wyllie-Echeverria S, Arzel P, Cox PA [2000]. Seagrass conservation: Lessons from ethnobotany. *Pac Cons Biol* 5: 329-335.
45. Hurley LM [1990]. *US Fish and Wildlife Service Field Guide to the Submerged Aquatic Vegetation of Chesapeake Bay*. Chesapeake Bay Estuary Program, Annapolis, Maryland.
46. Capps PG [1977]. Use of Seaweed as Hydromulch for Revegetation. South Australian Coast Protection Board, Amdel Report No. 1182. Australian Mineral Development Laboratories, Frewville, South Australia.
47. Stewart CM, Mills JA [1975]. Some notes on the chemistry and utilization of *Posidonia australis*. Notes compiled by CM Stewart and JA Mills. CSIRO Division of Chemical Technology, South Melbourne.
48. Walker DJ [1977]. Report of the Seaweed Problem on Taperoo Beach. South Australian Coast Protection Board Report.
49. van Katwijk MM [2000]. *Zostera marina* and the Wadden Sea. In: Sheppard, C (ed) *Seas at the Millennium: An Environmental Evaluation*, Vol 3. Elsevier Science, Amsterdam. pp 6-7.
50. Patriquin DG [1975]. Migration of blowouts in seagrass beds at Barbados and Carriacou West Indies and its ecological and geological implications. *Aquatic Botany* 1: 163-189.
51. Talbot MMB, Knoop WT, Bate GC [1990]. The dynamics of estuarine macrophytes in relation to flood/siltation cycles. *Botanica Marina* 33: 159-164.
52. Lee Long W, Thom RM [2001]. Improving seagrass habitat quality. In: Short FT, Coles RG (eds) *Global Seagrass Research Methods*. Elsevier Science, Amsterdam. pp 407-424.
53. Short FT, Short CA [1984]. The seagrass filter: Purification of estuarine and coastal waters. In: Kennedy V (ed) *The Estuary as a Filter*. Academic Press. pp 395-413.
54. Ward TJ [1987]. Temporal variation of metals in the seagrass (*Posidonia australis*) and its potential as a sentinel accumulator near a lead smelter. *Marine Biology* 95: 315-321.
55. Hoven HM, Gaudette HE, Short FT [1999]. Isotope ratios of 206Pb/207Pb in eelgrass, *Zostera marina*, indicate sources of Pb in an estuary. *Mar Env Res* 48: 377-387.
56. Klumpp DW, Howard RK, Pollard DA [1989]. Trophodynamics and nutritional ecology of seagrass communities. In: Larkum AWD, McComb AJ, Shepherd SA (eds) *Biology of Seagrasses – A Treatise on the Biology of Seagrasses with Special Reference to the Australian Region*. Elsevier, New York. pp 394-457.
57. Lanyon J, Limpus CJ, Marsh H [1989]. Dugongs and turtles: Grazers in the seagrass system. In: Larkum AWD, McComb AJ, Shepherd SA (eds) *Biology of Seagrasses – A Treatise on the Biology of Seagrasses with Special Reference to the Australian Region*. Elsevier, New York. pp 610-634.
58. Gell FR, Whittington MW [2002]. Diversity of fishes in seagrass beds in the Quirimba Archipelago, northern Mozambique. *Marine and Freshwater Research* 53: 115-121.
59. Gambi MC, Nowell ARM, Jumars PA [1990]. Flume observations on flow dynamics in *Zostera marina* (eelgrass) beds. *Marine Ecology Progress Series* 61: 159-169.
60. Koch EW, Verduin JJ [2001]. Measurements of physical parameters in seagrass habitat. In: Short FT, Coles RG (eds) *Global Seagrass Research Methods*. Elsevier Science, Amsterdam. pp 325-344.
61. Orth RJ, Heck KLJr, van Montfrans J [1984]. Faunal communities in seagrass beds: A review of the influence of plant structure and prey characteristics on predator-prey relationships. *Estuaries* 7: 339-350.
62. Phillips RC, Durako MJ [2000]. Global Status of Seagrasses. In: Sheppard C (ed) *Seas at the Millennium: An Environmental Evaluation*, Vol 3. Elsevier Science, Amsterdam. pp 1-16.
63. Baliño BM, Fasham MJR, Bowles MC (eds) [2001]. *IGBP Science, 2: Ocean Biogeochemistry and Global Change*. International Geosphere-Biosphere Programme, Stockholm.
64. Short FT, Neckles AH [1999]. The effects of global climate change on seagrasses. *Aquatic Botany* 63: 169-196.
65. Heck C [2001]. The economics of seagrass. Viewed online June 2002 at http://www.floridabay.org/pub/newspaper/economics.shtml Community Outreach, Florida Keys National Marine Sanctuary.
66. Costanza R, Arge R d', Groot R de, Farber S, Grasso M, Hannon B, Limburg K, Naeem S, O'Neill RV, Paruelo J, Raskin RG, Sutton P, Belt M van den [1997]. The value of the world's ecosystem services and natural capital. *Nature* 387: 253-260.
67. Short FT, Wyllie-Echeverria S [2000]. Global seagrass declines and effects of climate change. In: Sheppard C (ed) *Seas at the Millennium: An Environmental Evaluation*, Vol. 3. Elsevier Science, Amsterdam. pp 10-11.
68. Milne LJ, Milne MJ [1951]. The eelgrass catastrophe. *Scientific American* 184: 52-55.
69. Rasmussen E [1977]. The wasting disease of eelgrass (*Zostera marina*) and its effects on environmental factors and fauna. In: McRoy CP, Helfferich C (eds) *Seagrass Ecosystems: A Scientific Perspective*. Marcel Dekker, New York. pp 1-52.
70. Short FT, Muehlstein LK, Porter D [1987]. Eelgrass wasting disease: Cause and recurrence of a marine epidemic. *Biological Bulletin* 173: 557-562.
71. Muehlstein LK, Porter D, Short FT [1991]. *Labyrinthula zosterae* sp. Nov., the causative agent of wasting disease of eelgrass, *Zostera marina*. *Mycologia* 83(2): 180-191.
72. Goldschmid A, Yip M [2001]. Essay about *Caulerpa taxifolia*. Salzburg, 6 May 1999 (revised in April 2001). Viewed online May 2002 at http://www.sbg.ac.at/ipk/avstudio/pierofun/ct/caulerpa.htm
73. Short FT, Wyllie-Echeverria S [1996]. Natural and human-induced disturbance of seagrasses. *Environmental Conservation* 23: 17-27.
74. Burke L, Selig L, Spalding M [2002]. *Reefs at Risk in Southeast Asia*. World Resources Institute, Washington, DC.
75. Davis RC, Short FT [1997]. Restoring eelgrass, *Zostera marina* L., habitat using a new transplanting technique: The horizontal rhizome method. *Aquatic Botany* 59: 1-15.
76. Coles R, Fortes M [2001]. Protecting seagrasses – approaches and methods. In: Short FT, Coles RG (eds) *Global Seagrass Research Methods*. Elsevier Science, Amsterdam. pp 445-463.
77. Calumpong H, Fonseca M [2001]. Seagrass transplantation and other seagrass restoration methods. In: Short FT, Coles RG (eds) *Global Seagrass Research Methods*. Elsevier Science, Amsterdam. pp 425-444.
78. Short FT, Davis RC, Kopp BS, Short CA, Burdick DM [2002]. Site selection model for optimal restoration of eelgrass, *Zostera marina* L. *Marine Ecology Progress Series* 227: 253-267.
79. Granger SL, Traver MS, Nixon SW [2000]. Propagation of *Zostera marina* L. from seed. Ch 107. In: Sheppard CRC (ed) *Seas at the Millennium: An Environmental Evaluation*. Vol. III, Global Issues and Processes. Elsevier Science, Amsterdam. pp 4-5.
80. Les DH, Moody ML, Jacobs SWL, Bayer RJ [2002]. Systematics of seagrasses (Zosteraceae) in Australia and New Zealand. *J Sys Botany* 27: 468-484.
81. Hall MO, Durako MD, Fourqurean JW, Zieman JC [1999]. Decadal scale changes in seagrass distribution and abundance in Florida Bay. *Estuaries* 22(2B): 445-459.
82. Campey ML, Waycott M, Kendrick GA [2000]. Re-evaluating species boundaries among members of the *Posidonia ostenfeldii* species complex (Posidoniaceae) – morphological and genetic variation. *Aquatic Botany* 66: 41-56.
83. Waycott M, Freshwater DW, York RA, Calladine A and Kenworthy WJ [2002]. Evolutionary trends in the seagrass genus *Halophila* (Thouars): insights from molecular phylogeny. *Bulletin of Marine Science*.

1 The seagrasses of
SCANDINAVIA AND THE BALTIC SEA

C. Boström
S.P. Baden
D. Krause-Jensen

Scandinavia supports only a small fraction of the global seagrass resource; however, the first reports on the importance of seagrass meadows for coastal ecosystems derive from this area, from Denmark[1-4]. This chapter summarizes the distribution and importance of eelgrass, *Zostera marina*, in Scandinavian and Baltic coastal waters. Although most of the quantitative information is based on research carried out in non-tidal areas of Denmark, Sweden and Finland, the approach is holistic, and includes distribution maps and anecdotal information on eelgrass from Iceland, Norway and the coastal areas of the Baltic Sea, including Germany, Poland, Lithuania, Latvia and Estonia (see Map 1.1).

DISTRIBUTION PATTERNS
Norway
In the north Atlantic, eelgrass is found around Iceland, where about 30 sites have been identified since the 1950s[5]. Eelgrass forms isolated populations on shallow exposed and sheltered sandy bottoms along the entire Norwegian coast[6] and extends into the White Sea. The only Norwegian seagrass paper reports eelgrass densities between 50 and 160 shoots/m^2 and canopy heights generally between 15 and 60 cm, although in extreme cases the length of an individual plant may exceed 180 cm[6]. Areas of low density have the highest canopies. The average biomass (April-November) at the two sites studied was 20 and 40 g dry weight/m^2, respectively (range: 12-60 g dry weight/m^2, Figure 1.1) The associated fauna is rich (265 taxa, including mobile macrofauna and epiphytes) and ranges between 5 000 and 10 000 individuals/m^2. The crustacean species assemblage is dominated by six or seven families of amphipods, while the epiphytic community is characterized by hydroids, bryozoans and crustose and upright algae[6]. Consequently, these shallow, vegetated sites are of great importance for young year classes of fish in the Skagerrak area[7, 8].

The Swedish west coast and Denmark
On the Swedish west coast, as well as in Danish waters, eelgrass is the most widely distributed seagrass, and dominates sandy and muddy sediments in coastal areas of low to moderate wave exposure. In Denmark, very exposed areas facing the North Sea are devoid of eelgrass. Along moderately exposed Danish and Swedish coasts eelgrass forms extended belts interrupted by sandbars, while protected eelgrass populations generally form more coherent patches. Due to its wide salinity tolerance (5-35 psu)[9], eelgrass grows in the inner parts of brackish estuaries and sheltered bays and in fully marine waters. In areas of low salinity, *Ruppia* spp. and *Zostera noltii* can co-occur at the inner edges (0.5-1.5 m depth) of eelgrass.

Eelgrass occurs from shallow (0.5-1 m) water down to maximum colonization depths that often match the Secchi depth. In the inner parts of estuaries, the maximum colonization depth is about 3 m, in outer parts 4 m and along open coasts about 5 m[10, 11]. In rare cases of very clear waters, eelgrass penetrates to 10 m mean sea level (tidal range ±0.1 to 0.4 m). Eelgrass displays a bell-shaped distribution pattern along the depth gradient, with maximum abundance at intermediate depths and lower abundances in shallow and deep water[12, 13]. The biomass of Danish and Swedish eelgrass populations peaks in late summer at levels reaching above 250 g dry weight/m^2. Maximum shoot densities range between 1 000 and 2 500 shoots/m^2 [12, 14-19]. Exposure, desiccation and ice scour may reduce seagrass abundance in shallow water, while reductions in seagrass abundance towards the lower depth limit correlate with light attenuation along the depth gradient[13, 18, 20].

In southernmost Sweden, eelgrass meadows flourish on stony and sandy bottoms at 2-4 m depth, and may reach densities and standing crops corresponding to 3 600 shoots/m^2 and 470 g dry weight/m^2, respectively (site 11 in Figure 1.1)[21]. The

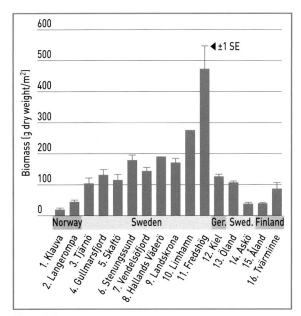

Figure 1.1
Average (±1 SE) above-ground biomass values (g dry weight/m²) for eelgrass (*Zostera marina*) along the Baltic Sea coastline (>1 500 km)

Source: Various sources[6, 19, 21].

Öresund area between Denmark and Sweden (sites 8-10 in Figure 1.1) also supports well-developed eelgrass meadows at 1.5-6 m depth[18]. In September 2000, four eelgrass sites along this 100-km coastline showed the following features: coverage: 20-80 percent; density: 293-1 573 shoots/m²; above-ground biomass: 69-193 g dry weight/m²; shoot length: 25-125 cm; and shoot width: 0.2-0.5 cm[22].

There are qualitative and quantitative data on the leaf fauna (defined as the sessile and motile fauna living on the leaves), mobile epifauna (intermediate predator invertebrates and fish) and piscivore fish (secondary predators) from the Swedish west coast[12, 23]. Data on infauna are more scarce[24, 25]. Due to the high organic content of most western Swedish seagrass beds (2-24 percent ash-free dry weight), the infauna (40-130 000 individuals/m²) is dominated by polychaetes and nematodes. The leaf fauna is dominated by tube-building amphipods, mainly detritivores and suspension feeders (80-250 000 individuals/m²), whereas the abundance of herbivores is low. Shrimps and crabs make up 90 percent of the mobile epifauna, and fishes constitute only about 10 percent of the intermediate predator abundance (30-160 individuals/m², with maximum abundances in late summer). The piscivore fishes (eelpout, cod and salmon) are few during daytime[12]. Faunal communities of Danish eelgrass beds are similarly rich, but have received little attention since the 1960s[26] and 1970s[27].

Western Baltic Sea and Germany

The western Baltic Sea, composed of the Kiel and the Mecklenburger Bights, is a transition zone between marine (North Sea) and brackish (Baltic proper) water and shows fluctuations in salinity (generally 10-18 psu but occasionally 8-28 psu[28, 29]). In this region eelgrass is found both along exposed sandy shores and in long, inner bays ("Förden") and shallow lagoons ("Bodden", "Haffs") with reduced water exchange and muddy substrate[30, 31]. Along exposed shores, the upper limit of distribution is set by wave-induced disturbance. Typically, continuous beds are found from 2.5 m depth and deeper. Additionally, patchy beds are found between the sand reefs and the shore at depths of 1-2 m. In sandy areas, eelgrass grows down to a depth of 8 m, and there is an almost continuous belt of eelgrass all along the shoreline, although on gravel- and stone-dominated substrates plants are rare. Extended populations are found in Orth Bay, in Kiel Fjord (Falkenstein) between Travemünde and Klützhöved, in the Wismar Bay and north of Zingst Peninsula[31]. *Zostera noltii* has been reported from Schleimünde, Heiligenhafen, Wismar Bay and Greifswald Lagoon[31].

In the Kiel area (Belt Sea) the eelgrass growing period is approximately 210 days, and growth is initiated in June, peaks in August-September and stops in March. Shoot lengths range between 20 and 140 cm[28, 29]. In Kiel Fjord (Friedrichsort and Moeltenort), eelgrass density is 600-1 600 shoots/m²[29]. The biomass range in Kiel Bight is 450-600 and 200-800 g dry weight/m² on mud and sand, respectively, and the daily production is 1.5-2.2 g carbon/m²[28]. In the 1970s, the mean annual eelgrass standing stock for two sites in Schleswig-Holstein (Kiel Bight) was 42.5 metric tons/ha[28].

A typical feature of shallow (depths of 1-3 m) eelgrass beds is their co-occurrence with blue mussels (*Mytilus edulis*), which represents a facultative mutualism[32]. Isopods (*Idotea* spp.) and snails (*Hydrobia* spp., *Littorina* spp.) are abundant grazers, and remove eelgrass biomass and epiphytes, respectively, highlighting the importance of biological interactions, which may locally override the negative symptoms of eutrophication[33, 34]. In shallow lagoons (e.g. Schlei Estuary), eelgrass is also consumed by birds, especially mute swans (*Cygnus olor*).

The Swedish east coast

Eelgrass penetrates into the brackish (0-12 psu) Baltic Sea, and is common in most coastal areas. The northern and eastern distribution limits of eelgrass correlate with the 5 psu halocline. The usual depth of eelgrass in the Baltic Sea is 2-4 m (range 1-10 m). *Zostera noltii* extends to southern Sweden, and to Lithuania in the eastern Baltic[35]. At present, the northern limit of *Zostera noltii* in

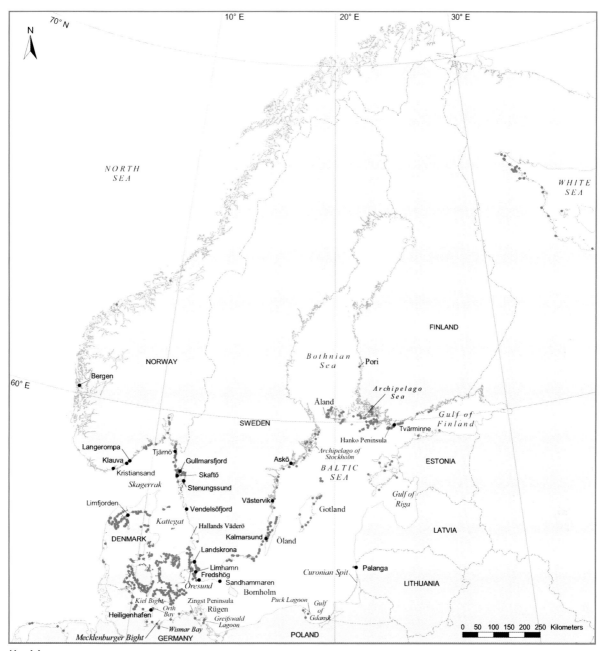

Map 1.1
Scandinavia

the Baltic Sea is unknown. Due to lack of tides, all seagrass beds in the Baltic Sea are permanently submerged, and often mixed with limnic angiosperms (e.g. *Potamogeton* spp. and *Myriphyllum* spp.).

On the brackish (6-8 psu) east coast of Sweden, the most extensive eelgrass meadows are probably found in the sandy Kalmarsund-Öland. Along the southeastern coast of Sweden (Sandhammaren to Västervik), eelgrass is common on sandy bottoms with good water exchange. The demographic information from this area is based on anecdotal evidence, diving observations made during coastal monitoring (University of Kalmar), and unpublished data by S. Tobiasson. Dense shallow stands have short (20 cm), narrow (2-3 mm) leaves and usually grow in mixed stands with *Potamogeton pectinatus*, *Ruppia maritima*, *Zannichellia palustris* and bladderwrack *Fucus vesiculosus*, while the deepest stands are sparse, monospecific and have longer (>80 cm) and broader (5 mm) leaves.

The coverage pattern is usually patchy (patch area 10-50 m^2 with a mean coverage of 50-75 percent, range 5-100 percent). At the main distribution depth, the mean shoot density is 500-600 shoots/m^2, but ranges between 100 and 1 040 shoots/m^2, depending on depth. The

Figure 1.2
Aerial photographs of two typical exposed eelgrass (*Zostera marina*) sites at the Hanko Peninsula, southwest Finland, northern Baltic Sea (adjacent to site 16 in Figure 1.1)

Photos: J. Lindholm

a. Kolaviken (59°49'N, 22°59'E): the high-energy regime at this site is reflected in a complex, patchy bed structure.

b. Ryssholm (59°60'N, 23°05'E): the continuous eelgrass bed is interrupted by sandbars, while circular to highly irregular, elongated patches are found at the outer edge of the bed.

Notes: The areas covered by eelgrass in (a) and (b) are 23 and 6 hectares, respectively. The depth range covered by eelgrass is approximately 2-6 m.

above-ground biomass may exceed 100 g dry weight/m^2 (site 13 in Figure 1.1). The steep, exposed coastal areas of southern Sweden (Skåne) and the east coast of the Öland Island lack eelgrass[36]. The semi-exposed sandy shores of Gotland Island support extensive eelgrass meadows. The northern limit of distribution is in the northern Archipelago of Stockholm[37]. Few studies of the associated fauna have been carried out[38, 39], but these meadows support more than 20 infaunal species and a rich leaf fauna with over 30 species[19].

Finland and Åland Islands
In Finland, eelgrass grows exclusively on exposed or moderately exposed bottoms with sandy sediments. The spatial patterns of eelgrass beds in shallow water are mainly controlled by physical factors (Figure 1.2). In the Archipelago Sea, eelgrass beds are found towards the leeside of islands, while more sheltered, inner bays on the mainland do not support eelgrass beds. Eelgrass sites in Finland vary in terms of patch size (1-75 m^2), shoot density (50-500 shoots/m^2), shoot length (20-100 cm), biomass (10-32.1 g ash-free dry weight/m^2 [40]) and sediment properties (organic content 0.5-1.5 percent, grain size 0.125-0.5 mm). The low shoot densities result in low areal production rates (138-523 mg dry weight/ m^2/day[41]). The associated fauna of Finnish seagrass beds is well described[42]. A rich sedimentary fauna (25 000-50 000 individuals/m^2, 50 species[42-44] and a distinct leaf fauna[19, 43, 45]) contributes significantly to coastal biodiversity in Finland. Northern Baltic seagrass communities lack crabs and echinoderms, and the nursery role for economically important fish species is limited, but seagrass beds serve as feeding grounds for fish.

HUMAN USE OF SEAGRASSES
The direct use and manufacture of eelgrass-based materials has been local and intermittent. In Denmark and other countries dried eelgrass leaves have been used as fuel, packing and upholstery material, insulation and roof material, feeding and bedding for domestic livestock, fertilizer and as a resource to obtain salt[3, 46-49]. In Sweden, dried eelgrass leaves have mainly been used for insulation of houses. Historically, the abundant eelgrass resources of the sheltered lagoons in the western Baltic Sea (Germany) have been utilized for upholstery and insulation. The last eelgrass collector at Maasholm, Germany, retired in the 1960s. In the southeastern Baltic Sea, human communities on the Curonian Spit (Lithuania) used eelgrass as upholstery material before the Second World War, indicating abundant eelgrass meadows in the area before the 1940s[50]. The current appreciation of seagrasses primarily concerns the services that seagrasses provide to the overall functioning of coastal ecosystems in terms of enhancing biodiversity, providing nursery and foraging areas for commercially important species, improving water quality by reducing particle loads and absorbing dissolved nutrients, stabilizing sediments and influencing global carbon and nutrient cycling[51].

HISTORICAL AND PRESENT DISTRIBUTION
Norway
Along the southeastern coast of Norway (between the Norwegian-Swedish border and Kristiansand), almost 100 sites have been monitored since the 1930s in connection with beach seine surveys each autumn (September-October) by the Institute of Marine Research[8]. The presence of vegetation has been estimated by aquascope, and seagrass cover has been

divided into the following categories: 1 = no vegetation, 2 = few plants, 3 = some plants, 4 = many plants, 5 = bottom totally covered. Unfortunately, only a small fraction of this dataset has been compiled and most is unpublished. The general impression, however, is that the coverage of eelgrass increased during the 1930s, and since then it has varied irregularly (Figure 1.3 a, b). Some areas showed signs of reduction in the late 1960s, and apparently there was a reduction probably indirectly related to the great bloom of *Chrysochromulina* in 1988. Now the coverage seems generally to be good[52].

The western and eastern coasts of Sweden

During the 1980s inventories of the shallow coastal areas including eelgrass were carried out along the Swedish west coast as a basis for coastal zone management. In 2000, a revisit and inventory of 20 km^2 of eelgrass meadows in five coastal regions along 200 km of the Skagerrak coast was carried out using the same methods (aquascope) as during the 1980s, but mapping accuracy was improved by using the global positioning system (GPS). This study showed that areal cover had decreased 58 percent (with regional variations) in 10-15 years. In the 1980s, eelgrass covered about 20 km^2 of bottom along this 200-km section of the west coast, while only about 8.4 km^2 was present in 2000[11]. Since 1994, one eelgrass site in southwest Sweden near Trelleborg (site 11 in Figure 1.1) has been included in the local coastal monitoring program. Shoot density and biomass of eelgrass at this site has increased significantly since 1994 (linear regression for biomass: $p<0.001$, $r = 0.81$), and this positive trend seems to be true for many of the eelgrass monitoring sites in the Öresund region (sites 8-11 in Figure 1.1[22]) probably due to greater exposure and/or invertebrate grazing[21]. No estimates of the total area covered by eelgrass along the whole Swedish west coast (>400 km) exist. An estimation of the total eelgrass coverage along the southeastern Swedish coast (including the Öland Island) yields minimum and maximum numbers between 60 and 130 km^2, respectively[53]. Between this region and the northern distribution limit in the Stockholm Archipelago eelgrass is still common, but far less abundant due to lack of suitable substrate[37].

Denmark

In Denmark, records of eelgrass distribution date back to around 1900, and provide a unique opportunity to describe long-term changes. In 1900, eelgrass was widely distributed in Danish coastal waters, and covered approximately 6 726 km^2 or one seventh of all Danish marine waters (Figure 1.4[2, 3]). The standing crop ranged between 270 and 960 g dry weight/m^2, in sparse and dense stands, respectively, and total annual eelgrass production was estimated at 8 million metric tons dry

Figure 1.3
Norwegian eelgrass coverage

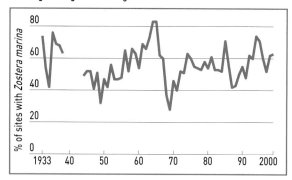

a. Long-term trends in the presence of eelgrass (*Zostera marina*) at shallow, soft-bottom sites assessed by aquascope in southeastern Norway (Kristiansand to the Norwegian-Swedish border) during the period 1933-2000.

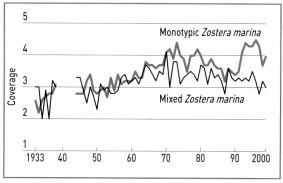

b. Coverage at sites where eelgrass occurs in single stands (green line) and mixed with benthic algae (black line).

Notes: 1 = no coverage, 5 = bottom totally covered. Number of sites sampled each year (38-134, mean 93) vary due to variation in water turbidity. No data obtained during 1940-44.

Source: Norwegian Institute of Marine Research[88].

weight[3]. In the 1930s, wasting disease led to substantial declines in eelgrass populations, especially in northwest Denmark where salinity is highest (Figure 1.4[54]). In 1941, eelgrass covered only 7 percent of the formerly vegetated areas, and occurred only in the southern, most brackish waters and in the low-saline inner parts of Danish estuaries (Figure 1.4[27, 55]). No national monitoring took place between 1941 and 1990, but analyses of aerial photos during the period from 1945 to the 1990s show an initial lag after the wasting disease followed by marked recolonization in the 1960s[56, 57].

Today eelgrass again occurs along most Danish coasts but has not reached the former areal extension[58, 59]. Based on comparisons of eelgrass area distribution in two large regions, Öresund and Limfjorden, in 1900 and in the 1990s, we estimate that the present distribution area of eelgrass in Danish

Figure 1.4
Map of eelgrass area distribution in Danish coastal waters in 1901, 1933, 1941 and 1994

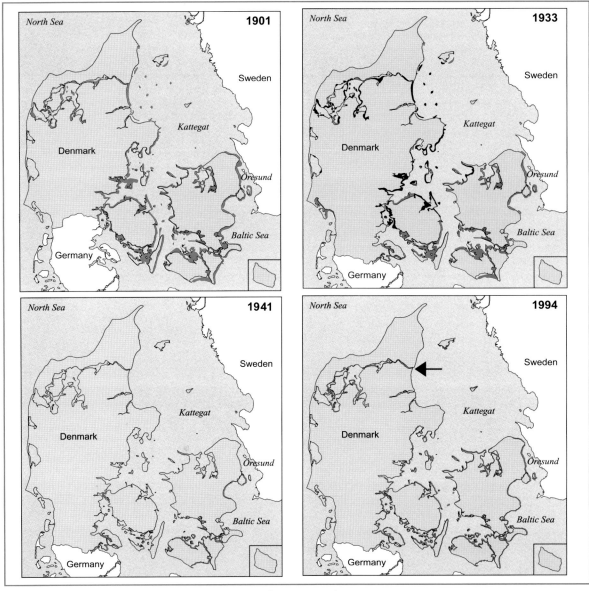

Notes: Dark green areas indicate healthy eelgrass while black areas (on the 1933 map) indicate where eelgrass was affected by the wasting disease but still present in 1933. The arrow shows the location of Limfjorden.

Source: Various sources: 1901 (redrawn[2]), 1933 (redrawn[54]), 1941 (redrawn[55]) and 1994 (coarse map based on visual examination of aerial photos and data from the national Danish monitoring program, produced by Jens Sund Laursen).

coastal waters constitutes approximately 20-25 percent of that in 1900 (Figure 1.4). The area distribution of eelgrass in Limfjorden was thus estimated at 345 km² in 1900[4] and at only 84 km² in 1994 (based on aerial photography data from the Limfjord counties). In Öresund, eelgrass covered about 705 km² in 1900[4] and only about 146 km² in 1996-2000[60]. Differences in methodology influence these comparisons since the distribution maps of eelgrass from the beginning of the last century were based on extrapolation between sites visited in field surveys, while maps from the 1990s were based on image analysis of aerial photography. This large areal reduction is partly attributed to the loss of deep eelgrass populations as a consequence of impoverished light conditions due to eutrophication. In 1900, maximum colonization depths averaged 5-6 m in estuaries and 7-8 m in open waters (Figures 1.5 and 1.6). In the 1990s, colonization depths were reduced by about 50 percent to 2-3 m in estuaries and 4-5 m in open waters.

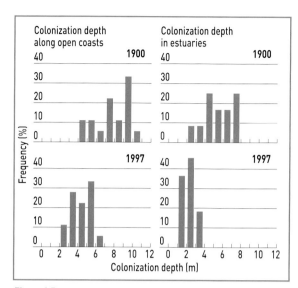

Figure 1.5
Maximum colonization depth of eelgrass patches in Danish estuaries and along open coasts in 1900 and 1996-97

Source: Based on data from 12 sites in estuaries and 18 sites along open coasts investigated by Ostenfeld[4] in 1900 and by the national Danish monitoring program in 1996-97.

Germany, Poland and Lithuania

In Germany (Kiel Bight), eelgrass competes with increasing amounts of filamentous algae, and in some areas the depth distribution of eelgrass decreased from 6 m in the 1960s to less than 2 m at the end of the 1980s[61]. In the Greifswald Lagoon (island of Rügen), the distribution of eelgrass has remained fairly stable, despite the almost total disappearance of red algal belts during the period 1930 to 1988[62]. Nevertheless, eelgrass is by far the most abundant macrophyte on sandy to muddy shores in this area[33].

In Poland (Gulf of Gdansk, Puck Lagoon), abundant eelgrass meadows grew down to a depth of 10 m in the 1950s, but were almost totally replaced by filamentous brown algae and *Zannichellia palustris* during the period 1957-87[63-65] (Figure 1.7). The change from dense seagrass beds to algal-dominated assemblages has caused a shift in the commercially important fish communities. Hence, eel (*Anguilla anguilla*) and pike (*Esox lucius*) have decreased in abundance and have been partly replaced by roach (*Rutilus rutilus*)[66, 67]. In addition, eelgrass suffers from heavy metal contamination[68]. Transplantation of eelgrass has been tested in the Puck Lagoon[66]. Recently natural recolonization has taken place in some areas of this lagoon[69].

Along Lithuanian coasts in the southeastern Baltic Sea, eelgrass had virtually disappeared before any scientific evaluation was made. Eelgrass most likely occurred along the 90-km-long sea side of the Curonian Spit, covering thousands of hectares[50]. In

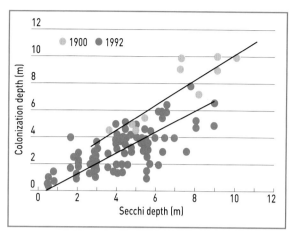

Figure 1.6
Secchi depths and maximum colonization depths of eelgrass patches in Danish estuaries and open coasts in 1900 and 1992

Source: Measured by Ostenfeld[4] in 1900 and by the national Danish monitoring program in 1992.

1998, filamentous green algae (*Cladophora glomerata*) dominated along the coast, and eelgrass was considered rare and endangered; no eelgrass was found during underwater surveys during 1993-97[31].

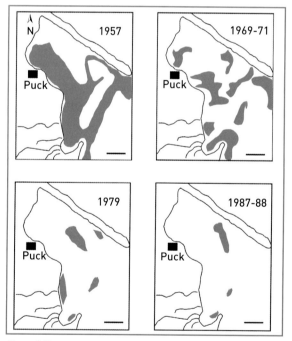

Figure 1.7
Long-term changes in the distribution of eelgrass (*Zostera marina*) in the southeastern Baltic Sea (Puck Lagoon, Poland)

Notes: Scale bar in lower right corner corresponds to approximately 5 km. Green areas indicate eelgrass cover.

Source: Modified after Kruk-Dowgiallo[63].

One northern site (Palanga) supported eelgrass, indicating that eelgrass was probably present formerly along the whole Lithuanian coast[31, 35]. The seagrass literature from Latvia and Estonia is scarce, but eelgrass has been reported to occur sparsely among algal-dominated assemblages in the Gulf of Riga[70].

Finland

The only long-term analysis of an eelgrass site in southwest Finland recorded no change in density and standing stock between 1968 and 1993[71]. In 1993, eelgrass biomass was about 85 g dry weight/m^2 and corresponded well with the yearly means for 1968-70 in terms of ash-free dry weight (20 g/m^2). By contrast, the associated eelgrass fauna showed marked signs of eutrophication. Total abundance of infauna had increased almost fivefold, and the total animal biomass had more than doubled over 25 years. The number of taxa showed minor changes over time. These faunal changes indicate increased food availability, due to eutrophication. Unfortunately, no long-term data from other Finnish eelgrass sites exist to verify this result. Genetic analysis of Finnish eelgrass meadows suggests an age of these plant ecosystems between 800 and 1 600 years[72, 73], indicating that eelgrass colonization must have taken place at present salinities. Those eelgrass populations near their limit of distribution in terms of salinity were not affected by the wasting disease in the 1930s[27] and have also persisted through severe anthropogenic stress and long-term physical stress in terms of landlift, wind disturbance, sedimentation and fluctuations in temperature and ice cover. Based on very crude areal estimates, and extrapolations from the number of known eelgrass sites verified by diving (totally about 50 sites), our guess is that the total coverage of eelgrass in Finland is probably less than 10 km^2.

THREATS
Kattegat and Skagerrak

Since the lower depth limit of eelgrass is determined by water transparency, eutrophication is a main threat to especially deep eelgrass populations. Maximum Secchi depths and colonization depths approached 12 m in open Danish waters in 1900 but rarely exceeded 6 m in the 1990s (Figure 1.6). The maximum colonization depth is also correlated to the concentration of water column nitrogen, which is the main determinant of phytoplankton biomass in Danish coastal waters[20]. Eutrophication-gained filamentous algae (mainly ephemeral) may shade seagrasses, hamper water exchange and cause a decline in associated faunal communities, e.g. shrimps and crabs[11, 19, 74, 75]. In shallow stagnant waters with limited oxygen pools, as well as in deeper stratified waters, the oxygen-consuming decomposition of ephemeral algae and detritus may lead to anoxia. High water temperature also stimulates microbial decomposition rates and thereby further increases the risk of anoxia. Oxygen deficiency in the meristematic region of eelgrass is a likely key factor explaining events of mass mortality in eelgrass beds[11, 76], possibly in combination with sulfide exposure[77]. Shallow eelgrass populations often show large and rapid fluctuations, suggesting that stochastic interactions between water temperature, light, nutrients and physical disturbance like strong wave action and ice scouring play important regulating roles, and that recolonization may also happen relatively fast in deeper water if conditions improve[56, 57].

Other threats include siltation and mechanical damage. For example, the construction in 1995-2000 of the Öresund bridge between Denmark and Sweden, almost 8 km long and one of the most massive marine constructions in Scandinavia, was likely to affect the large eelgrass populations in Öresund. However, strict regulations on dredged quantities and spillage during the construction works prevented detectable negative impacts on eelgrass[60, 78, 79].

In some Danish estuaries where eelgrass and blue mussel occur in mixed populations, mussel fishery may constitute a threat to eelgrass populations[80]. In Sweden the increasing leisure boat harbors with uncontrolled anchoring, dredging and water currents from propellers are the main physical threats to seagrass meadows.

The Baltic Sea

As in Denmark and Sweden, the drifting and sessile forms of fast-growing, filamentous algae constitute a serious threat to seagrasses in other areas of the Baltic[61, 71, 81], which will probably have negative effects on the whole eelgrass community[82]. During the past ten years, increasing amounts of ephemeral, filamentous algal mats have been observed at shallow localities in the northern Baltic Sea[43, 83], with profound negative effects on the benthic communities[81]. In 1968-71 filamentous algal mats were already common at eelgrass sites, but their biomass was less than 5 g ash-free dry weight/m^2 [40]. Today, the biomass of drifting algae in Finland commonly exceeds 1 000 g dry weight/m^2 [71, 83] and subsequent periodic anoxia is also common in shallow areas. It is clear that these algae are a major threat to the Baltic Sea seagrass ecosystems. In the heavy traffic coastal areas of the Baltic Sea, oil spill accidents could be detrimental to seagrass vegetation. Other threats include sand suction and construction.

NATIONAL AND SCANDINAVIAN POLICY

Several political initiatives affect Scandinavian seagrass populations. In 1987, the Danish Government passed an Action Plan on the Aquatic Environment including

measures on wastewater treatment, the storage of animal manure and reductions of agricultural nitrogen and phosphorus. The aim was to reduce annual total nitrogen discharge by 50 percent, and that of phosphorus by 80 percent, within five years. A second action plan containing further measures was passed in 1998 to ensure that the planned reductions of nitrogen and phosphorus discharges will be in effect before 2003. In addition there are several directives concerning point sources and protection of groundwater. An announcement on mussel fishery in Denmark prohibits fishery at water depths shallower than 3 m in order to protect eelgrass beds. A nationwide Danish monitoring program was established in 1988 to demonstrate the effects of the Action Plan (for latest adjustments, see Environmental Protection Agency[84]). Large construction works typically have associated monitoring programs, as was the case for the fixed link across Öresund.

As in Denmark, a series of action plans aiming to reduce nutrient discharge have been agreed in Sweden since the late 1980s, but not fulfilled. The latest action plan against coastal nutrient pollution is part of Swedish national environmental goals (Governmental Proposition 2000), and specifically says that total nitrogen discharge with anthropogenic origin from land should be reduced by 30 percent from 1995 not later than 2010, whereas phosphorous should decrease continuously from 1995 to 2010 with no specific aim. However, not only nutrient pollution but also overfishing might be part of the decreasing extension of seagrass through a possible, but still unverified, top-down control mechanism[85]. In the Baltic, as well as in the Kattegat and Skagerrak, most fish stocks are overfished to levels below biological safe limits. This is a much-debated topic, but has so far not been the subject of serious action plans.

Finland is not committed to monitor seagrass meadows. However, Finland follows political agreements, which are carried out by national (e.g. Water Protection Targets for 2005, Renewed Nature Conservation Act (1996), Renewed Water Act and EIA (Environment Impact Assessment) procedures) and international (Habitat Directive and Natura 2000) environmental programs. Thus, seagrasses in Finland are only indirectly protected through limitations on nutrient discharges. A new governmental program initiated in June 2001 aims at reducing nutrient discharges to the Baltic Sea and protecting and monitoring marine coastal biodiversity.

At the international level, seagrasses are listed in the Rio Declaration (1992/93:13) as diverse habitats in need of protection and monitoring (Chapter 17 part D 17.86 d). Further, the European Water Framework Directive, the Habitat Directive, the Helsinki Convention (HELCOM), the Oslo-Paris Convention (OSPAR) and the Convention on Biodiversity also place demands for monitoring seagrasses in Scandinavia[86]. More details on political initiatives on nutrient reductions and monitoring are summarized in Lääne et al.[86] (Chapter 6.2). On the initiative of the Helsinki Commission, a Red List of marine biotopes in the Baltic Sea[31] serves as an instrument in conservation, management and policy-making. In the Red List "sublittoral sandy bottoms dominated by macrophytes" and "sand banks of the sublittoral photic zone with or without macrophyte vegetation" are classified as "heavily endangered" and "endangered", respectively[31]. Accordingly, during the implementation of the European Union Water Framework Directive, eelgrass should be included as an indicator species. In future years, the coverage, depth range and biodiversity of eelgrass beds may potentially be used for ecological classification of Baltic coastal waters. Guidelines for monitoring eelgrass and other key macrophytes are included in the HELCOM COMBINE program[87].

However, classification of Baltic Sea seagrass meadows as threatened is only a first step obligating regular quantitative estimates of the distribution patterns, dynamics and diversity of seagrass meadows. Consequently, these parameters should be obtained and evaluated within standardized, national monitoring programs. Presently, only a fraction of the Baltic Sea seagrass resources undergo regular monitoring. In future, such measures are crucial in order to understand and sustain these important ecosystems.

ACKNOWLEDGMENTS

The authors would like to thank the following persons for data, comments or logistical support during the preparation of the manuscript: Penina Blankett, Hartvig Christie, Jan Ekebom, Stein Fredriksen, Jakob Gjøsæter, Frida Hellblom, Agnar Ingolfsson, Hans Kautsky, Jonne Kotta, Hordur Kristinsson, Jouni Leinikki, Mikael von Numers, Sergej Olenin, Panu Oulasvirta, Lars-Eric Persson, Eeva-Liisa Poutanen, Thorsten Reusch, Aadne Sollie, Stefan Tobiasson and Jan Marcin Weslawski. Susanne P. Baden gained financial support from WWF (World Wildlife Fund) and the County of Västra Götaland, and Dorte Krause-Jensen from the European Union (#EVK3-CT-2001-00065 "CHARM" and EVK3-CT-2000-00044 "M&MS").

AUTHORS

Christoffer Boström, Åbo Akademi University, Department of Biology, Environmental and Marine Biology, Akademigatan 1, FIN-20500 Åbo, Finland. **Tel:** +358 (0)2 2154052, 4631045. **Fax:** +358 (0)2 2153428. **E-mail:** christoffer.bostrom@abo.fi

Susanne P. Baden, Göteborg University, Department of Marine Ecology, Kristineberg Marine Research Station, S-45034 Fiskebäckskil, Sweden.

Dorte Krause-Jensen, National Environmental Research Institute, Department of Marine Ecology, Vejlsøvej 25, 8600 Silkeborg, Denmark.

REFERENCES

1. Petersen CGJ [1891]. Fiskens biologiske forhold i Holbæk Fjord. *Report of the Danish Biological Station* 1: 1-63.
2. Petersen CGJ [1901]. Fortegnelse over ålerusestader i Danmark optaget i årene 1899 og 1900 med bemærkninger om ruseålens vandringer etc. In: *Beretning til Landbrugsministeriet fra den danske biologiske station*. 1900 og 1901. København, Centraltrykkeriet. X pp 3-28.
3. Petersen JCG [1914]. Om bændeltangens (*Zostera marina*) aars-produktion i de danske farvande. Kap. X i Jungersen, HFE og Warming, Eug. Mindeskrift i anledning af hundredåret for Japetus Steenstrups fødsel. I Kommission hos GEC Gad, København. Bianco Lunos Bogtrykkeri.
4. Ostenfeld CH [1908]. Ålegræssets (*Zostera marina's*) udbredelse i vore farvande. In: Petersen CGJ *Beretning til Landbrugsministeriet fra den danske biologiske station*. København, Centraltrykkeriet. XVI pp 1-61.
5. Ingolfsson A, Kristinsson H. Personal communication.
6. Fredriksen S, Christie H [2002, in press]. *Zostera marina* (Angiospermae) and *Fucus serratus* (Phaeophyceae) as habitat for flora and fauna – seasonal and local variation. *Proceedings from the 17th International Seaweed Symposium*.
7. Fjøsne K, Gjøsæter J [1996]. Dietary composition and the potential food competition between 0-group cod (*Gadus morhua* L.) and some other fish species in the littoral zone. *ICES Journal of Marine Science* 53: 757-770.
8. Johannessen T, Sollie A [1994]. Overvåkning av gruntvannsfauna på Skagerrakkysten – historiske forandringer i fiskefauna 1919-1993 og ettervirkninger av den giftige algeoppblomstringen i mai 1988. Havsforskningsinstitutet/Institute of Marine Research, *Fisken og havet* nr. 10.
9. Pinnerup SP [1980]. Leaf production of *Zostera marina* L. at different salinities. *Ophelia* 1: 219-224.
10. Henriksen P et al. [2000]. Marine områder 2000. Miljøtilstand og udvikling. NOVA 2003. National Environmental Research Institute Technical Report No. 375. p 110. http://www.dmu.dk/1_viden/2_Publikationer/3_fagrapporter/rapporter/FR375.pdf (accessed July 2002).
11. Baden S, Gullström M, Lundén B, Pihl L, Rosenberg R [2003, in press]. Vanishing seagrass (*Zostera marina* L.) in Swedish coastal waters. *Ambio* 32.
12. Baden SP, Pihl L [1984]. Abundance, biomass and production of mobile epibenthic fauna in *Zostera marina* (L.) meadows, western Sweden. *Ophelia* 23: 65-90.
13. Krause-Jensen D, Pedersen MF, Jensen C [Accepted manuscript]. Regulation of *Zostera marina* cover in Danish coastal waters. *Estuaries*.
14. Sand-Jensen K [1975]. Biomass, net production and growth dynamics in a eelgrass (*Zostera marina* L.) population in Vellerup Vig, Denmark. *Ophelia* 14: 185-201.
15. Wium-Andersen S, Borum J [1984]. Biomass variation and autotrophic production of an epiphyte-macrophyte community in a coastal Danish area: I. Eelgrass (*Zostera marina* L.) biomass and net production. *Ophelia* 23: 33-46.
16. Pedersen MF, Borum J [1993]. An annual nitrogen budget for a seagrass *Zostera marina* population. *Marine Ecology Progress Series* 101: 169-177.
17. Olesen B, Sand-Jensen K [1994]. Biomass-density patterns in the temperate seagrass *Zostera marina*. *Marine Ecology Progress Series* 109: 283-291.
18. Krause-Jensen D, Middelboe AL, Sand-Jensen K, Christensen PB [2000]. Eelgrass, *Zostera marina*, growth along depth gradients: Upper boundaries of the variation as a powerful predictive tool. *Oikos* 91: 233-244.
19. Baden S, Boström C [2001]. The leaf canopy of *Zostera marina* meadows – faunal community structure and function in marine and brackish waters. In: Reise K (ed) *Ecological Comparisons of Sedimentary Shores*. Springer Verlag, Berlin. pp 213-236.
20. Nielsen SL, Sand-Jensen K, Borum J, Geertz-Hansen O [2002]. Depth colonization of eelgrass (*Zostera marina*) and macroalgae as determined by water transparency in Danish coastal waters. *Estuaries* 25: 1025-1032.
21. TOXICON AB [2000]. Årsrapport, Sydkustens vattenvårdförbund.
22. Undersökningar i Öresund, Öresunds vattenvårdförbund. *ÖVF Rapport* 2001: 1.
23. Baden SP [1990]. The cryptofauna of *Zostera marina* (L.): Abundance, biomass and population dynamics. *Netherlands Journal of Sea Research* 27: 81-92.
24. Möller P, Pihl L, Rosenberg R [1985]. Benthic faunal energy flow and biological interaction in some shallow marine soft bottom habitats. *Marine Ecology Progress Series* 27: 109-121.
25. Berg T [2000]. Quantitative Studies of Infauna in *Zostera marina* Meadows on the Swedish West Coast. Biomass and Abundance Correlated to Organic Content of the Sediment. MSc thesis, Department of Marine Ecology, Göteborg University. 29 pp.
26. Muus B [1967]. The fauna of Danish estuaries lagoons. Distribution and ecology of dominating species in the shallow reaches of the mesohaline zone. *Medd Danm Fisk- og Havsunders* NS 5: 1-316.
27. Rasmussen E [1973]. Systematics and ecology of the Isefjord marine fauna (Denmark) with a survey of the eelgrass (*Zostera*) vegetation and its communities. *Ophelia* 11: 1-507.
28. Feldner J [1977]. Ökologische und Produktionsbiologische Untersuchungen am Seegrass *Zostera marina* L. in der Kieler Bucht (Westliche Ostsee). Reports Sonderforschungsbereich 95: Wechselwirkung Meer-Meeresboden, Kiel University. 170 pp (in German with English abstract).
29. Reusch TBH [1994]. Factors Structuring the *Mytilus*- and *Zostera* Community in the Western Baltic: An Experimental Approach. PhD thesis, Christian-Albrechts-Universität, Kiel. 162 pp (in English).
30. Lotze H [1998]. Population Dynamics and Species Interactions in Macroalgal Blooms: Abiotic versus Biotic Control at Different Life-cycle Stages. Berichte aus dem Inst für Meereskunde. PhD thesis, Christian Albrechts-Universität, Kiel. 134 pp (in English).
31. HELCOM [1998]. Red list of marine coastal biotopes and biotope complexes of the Baltic Sea, Belt Sea and Kattegat. *Baltic Sea Environment Proceedings* 75. 128 pp.
32. Reusch TBH, Chapman ARO, Gröger JP [1994]. Blue mussels (*Mytilus edulis*) do not interfere with eelgrass (*Zostera marina*) but fertilize shoot growth through biodeposition. *Marine Ecology Progress Series* 108: 265-282.
33. Reusch T. Personal communication.
34. Boström C. Unpublished data.
35. Labanauskas V [2000]. Baltijos juros lietuvos priekrantés bentoso makrofitu bendijos, *Botanica Lithuania* 6: 401-413.
36. Persson LE, Tobiasson S. Personal communication.
37. Kautsky H. Personal communication.
38. Göthberg A, Röndell B [1973]. Ekologiska studier i *Zostera*-samhället i norra Östersjön. *Inf Sötvattenslab Drottningholm* No. 11.
39. Kautsky H, van der Maarel E [1990]. Multivariate approaches to the variation in phytobenthic communities and environmental vectors in the Baltic Sea. *Marine Ecology Progress Series* 60: 169-184.
40. Lappalainen A, Hällfors Kangas GP [1977]. Littoral benthos of the Northern Baltic Sea. IV. Pattern and dynamics of macrobenthos in a sandy-bottom *Zostera marina* community in Tvärminne. *Internationale Revue der gesamten Hydrobiologie* 62: 465-503.
41. Roos C [2000]. A Seasonal Study of the Production of Eelgrass (*Zostera marina* L.) at Two Sites in the Northern Baltic Sea. MSc thesis, Åbo Akademi University. 56 pp (in Swedish with English abstract).
42. Boström C [2001]. Ecology of Seagrass Meadows in the Baltic Sea. PhD thesis, Åbo Akademi University. 47 pp + 6 appendices.
43. Boström C, Bonsdorff E [1997]. Community structure and spatial variation of benthic invertebrates associated with *Zostera marina* (L.) beds in the northern Baltic Sea. *Journal of Sea Research* 37: 153-166.
44. Boström C, Bonsdorff E [2000]. Zoobenthic community establishment and habitat complexity – the importance of seagrass shoot density, morphology and physical disturbance for faunal recruitment. *Marine Ecology Progress Series* 205: 123-138.
45. Boström C, Mattila J [1999]. The relative importance of food and shelter for

seagrass-associated invertebrates: A latitudinal comparison of habitat choice by isopod grazers. *Oecologia* 120: 162-170.

46 Lehmann MCG [1814]. Der endeckte Nutzen des Seegrases zum Füllen der Kissen und Polster. Kobenhagen bey Schubothe.

47 Cottam C, Munro DA [1954]. Eelgrass status and environmental relations. *Journal of Wildlife Management* 18: 449-460.

48 Thayer GW, Wolfe DA, Williams RB [1975]. The impact of man on seagrass systems. *American Scientist* 63: 288-29.

49 Brøndegaard VJ [1987]. *Folk og Flora: Dansk etnobotanik.* Bind 1. 2nd edn. Rosenkilde og Bagger, København.

50 Olenin S. Personal communication.

51 Hemminga MA, Duarte CM [2000]. *Seagrass Ecology.* Cambridge University Press, Cambridge.

52 Gjøsæter J, Sollie A. Personal communication.

53 Tobiasson S. Personal communication.

54 Blegvad H [1935]. En epidemisk sygdom i bændeltangen (*Zostera marina* L.).. In: Blegvad H (ed) *Beretning til Ministreriet for Søfart og Fiskeri fra Den Danske Biologiske Station 1934.* CA Reitzels Forlag, København. XXXIX pp 1-8.

55 Lund S [1941]. Tangforekomsterne i de danske farvande og mulighederne for deres udnyttelse. *Dansk Tidssk Farm* 15(6): 158-174.

56 Frederiksen M, Krause-Jensen D, Laursen JS [Submitted manuscript, a]. Long-term changes in area distribution of eelgrass (*Zostera marina*) in Danish coastal waters.

57 Frederiksen M, Krause-Jensen D, Holmer M [Submitted manuscript b]. Long-term changes in eelgrass (*Zostera marina*) landscapes: Influence of physical setting on spatial distribution.

58 Olesen B [1993]. Population Dynamics of Eelgrass. PhD thesis, Department of Plant Ecology, University of Aarhus.

59 Kaas H, Møhlenberg F, Josefson A, Rasmussen B, Krause-Jensen D, Jensen HS, Svendsen LM, Windolf J, Middelboe AL, Sand-Jensen K, Pedersen MF [1996]. Marine Areas. Danish Inlets – State of the Environment, Trends and Causal Relations. The Monitoring Programme under the Action Plan for the Aquatic Environment 1995. Ministry of Environment and Energy. National Environmental Research Institute Technical Report No. 179. 205 pp. http://www.dmu.dk/1_viden/2_Publikationer/3_fagrapporter/rapporter/FR179.pdf (accessed July 2002).

60 Krause-Jensen D, Middelboe AL, Christensen PB, Rasmussen MB, Hollebeek P [2001]. Benthic Vegetation *Zostera marina, Ruppia* spp., and *Laminaria saccharina.* The Authorities' Control and Monitoring Programme for the fixed link across Øresund. Benthic vegetation. Status report 2000. SEMAC JV on behalf of the Danish and Swedish Environmental Authorities. 115 pp.

61 Schramm W [1996]. The Baltic Sea and its transitions zones. In: Schramm, W, Nienhuis, PH (eds) *Marine Benthic Vegetation – Recent Changes and the Effects of Eutrophication.* Springer-Verlag, Berlin. pp 131-163.

62 Messner U, von Oertzen JA [1991]. Long-term changes in the vertical distribution of macrophytobenthic communities in the Greifswalder Bodden. *Acta Ichthyologica et Piscatoria* 22: 135-143.

63 Kruk-Dowgiallo L [1991]. Long-term changes in the underwater meadows of the Puck Lagoon. *Acta Ichthyologica et Piscatoria* 22: 77-84.

64 Kruk-Dowgiallo L [1994]. Distribution and biomass of phytobenthos of the inner Puck Bay in the summer of 1987. In: *The Puck Bay – Possibilities of Restoration.* IOS, Warszawa. pp 109-122 (in Polish with English abstract).

65 Kruk-Dowgiallo L [1996]. The role of filamentous brown algae in the degradation of the underwater meadows in the Gulf of Gdansk. *Oceanological Studies* 1-2: 125-135.

66 Ciszewski P, Kruk-Dowgiallo L, Zmudzinski L [1992]. Deterioration of the Puck Bay and biotechnical approaches to its reclamation. In: Bjørnestad E, Hagerman L, Jensen K (eds) *Proceedings of the 12th Baltic Marine Biologists Symposium.* Olsen & Olsen, Fredensborg. pp 43-46.

67 Ciszewski P, Ciszewska L, Kruk-Dowgiallo L, Osowiecki A, Rybicka D, Wiktor J, Wolska-Pys M, Zmudzinsky L, Trokowicz, D [1992]. Trends of long-term alterations of the Puck Bay ecosystem. Studia i materialy oceanologiczne nr 60. *Marine Biology* 8: 33-84.

68 Kruk-Dowgiallo L, Pempkowiak J [1997]. Macrophytes as indicators of heavy metal contamination in the Puck Lagoon (southern Baltic). In: Ojaveer E (ed) *Proceedings from the 14th Baltic Marine Biologists Symposium, Tallinn 1997.* pp 86-100.

69 Weslawski JM. Personal communication.

70 Orav H, Kotta J, Martin G [2000]. Factors affecting the distribution of benthic invertebrates in the phytal zone of the north-eastern Baltic Sea. *Proceedings Estonian Academy of Sciences Biology Ecology* 49:253-269.

71 Boström C, Bonsdorff E, Kangas P, Norkko A [2002]. Long term changes in a brackish water *Zostera marina* community indicate effects of eutrophication. *Estuarine Coastal Shelf Science* 55: 795-804.

72 Reusch TBH, Boström C, Stam WT, Olsen JL [1999]. An ancient eelgrass clone in the Baltic. *Marine Ecology Progress Series* 183: 301-304.

73 Fries M [1959]. En *Zostera marina*-förekomst I Stockholms norra skärgård. *Svensk Botanisk Tidskrift* 53: 469-474.

74 Borum J [1985]. Development of epiphytic communities on eelgrass (*Zostera marina*) along a nutrient gradient in a Danish estuary. *Marine Biology* 87: 211-218.

75 Isaksson I, Pihl L [1992]. Structural changes in benthic macrovegetation and associated epibenthic faunal communities. *Netherlands Journal of Sea Research.* 30:131-140.

76 Greve T, Borum J, Pedersen O [in press, 2003]. Meristematic oxygen dynamics in eelgrass (*Zostera marina*). *Limnology and Oceanography* 48.

77 Holmer M, Bondgaard EJ [2001]. Photosynthetic and growth response of eelgrass to low oxygen and high sulfide concentrations during hypoxic events. *Aquatic Botany* 70: 29-38.

78 Danish Ministry of the Environment and Energy, Danish Ministry of Transport, Swedish Control and Steering Group for the Öresund Fixed Link [2001]. Final Report on the Environment and the Öresund Fixed Link's Coast-to-coast Installation. 11th semi-annual report. 22 pp.

79 Feedback Monitoring Centre [2001]. Annual Compliance Test of Eelgrass Variables in 2000. 42 pp + appendices.

80 Limfjordsovervågningen [1996]. Vegetationsforhold i Limfjorden i områder lukket for muslingefiskeri i perioden 1988-1995. Limfjordsovervågningen ved Ringkøbing Amtskommune, Viborg Amt og Nordjyllands Amt.

81 Norkko J, Bonsdorff E, Norkko A [2000]. Drifting algal mats as an alternative habitat for benthic invertebrates: Species specific responses to a transient resource. *Journal of Experimental Marine Biology and Ecology* 248: 79-104.

82 Norkko A, Bonsdorff E [1996]. Population responses of coastal zoobenthos to stress induced by drifting algae. *Marine Ecology Progress Series* 140: 141-151.

83 Vahteri P, Mäkinen, A., Salovius S, Vuorinen I [2000]. Are drifting algal mats conquering the bottom of the Archipelago Sea, SW Finland? *Ambio* 29: 338-343.

84 Environmental Protection Agency [2000]. NOVA-2003 Programbeskrivelse for det nationale program for overvågning af vandmiljøet 1998-2003. Redegørelse fra Miljøstyrelsen no. 1, Ministry of Environment and Energy. 397 pp.

85 Heck Jr KL, Pennock JR, Valentine JF, Coen LD, Sklenar SA [2000]. Effects of nutrient enrichment and large predator removal on seagrass nursery habitats: An experimental assessment. *Limnology and Oceanography* 45: 1041-1057.

86 Lääne A, Pitkänen H, Arheimer B, Behrendt H, Jarosinski W, Lucane S, Pachel K, Räike A, Shekhovtsov A, Svendsen L, Valatka S [2002]. Evaluation of the Implementation of the 1988 Ministerial Declaration regarding Nutrient Load Reductions in the Baltic Sea Catchment Area. Finnish Environment Institute, Helsinki.

87 Bäck, S [1999]. Guidelines for Monitoring of Phytobenthic Plant and Animal Communities in the Baltic Sea. Annex for HELCOM COMBINE programme. 12 pp.

88 Institute of Marine Research, Flødevigen Marine Research Station. Unpublished data.

2 The seagrasses of WESTERN EUROPE

C. Hily

M.M. van Katwijk

C. den Hartog

Western Europe is considered here as the coasts of the North Sea, the Channel and Irish Sea as well as the Atlantic coasts of the British Isles, France, Spain and Portugal. Two seagrass species are found in coastal and estuarine areas: *Zostera marina* and *Zostera noltii*. A third species, *Cymodocea nodosa*, occurs less abundantly in the southern part of the area (Portugal). The widgeon grasses, *Ruppia maritima* and *Ruppia cirrhosa*, sometimes considered to be seagrasses, occur in brackish water sites[1], such as low-salinity ponds and mesohaline to polyhaline coastal lagoons; occurrence under marine conditions is very rare in western Europe and generally ephemeral.

The seagrasses are found on soft sediments to a maximum depth of about 10 m. They occupy a large variety of marine and estuarine habitats. They often grow in dense beds and extensive meadows creating a productive and diverse habitat used as shelter, nursery, spawning or food area by a large variety of animal species. Among these, several are of commercial interest or cultural value. Therefore, seagrass beds are recognized as an important reservoir of coastal biodiversity; they shelter in the same habitat endofaunal and epifaunal species in the sediment, and creeping and walking species on the leaves, as well as swimming species[2, 3]. Seagrasses are consequently of considerable economic and conservation importance. The dense root network of the seagrasses is able to stabilize the underlying sediment and to increase the sedimentation fluxes by reducing the hydrodynamic forces. Their essential ecological role in terms of primary production at the scale of the coastal ecosystem is mainly recognized in the areas where hard bottom surfaces with macroalgae cover are scarce. The importance of the beds was higher at the beginning of the 20th century, before the "wasting disease" struck. The proliferation of the pathogenic slime mold (*Labyrinthula zosterae*) in the leaves of *Zostera marina*, considered to be the consequence of weakening of the plants under continuous unfavorable environmental conditions, resulted in the 1930s in the loss of almost 90 percent of the *Zostera marina* populations of western Europe[4, 5]. After this period, many beds progressively recovered but the area covered remained low in most areas compared with previous distribution. *Zostera marina* lives mainly in the infralittoral (or sublittoral) zone but can develop occasionally in the lower and middle part of the mediolittoral (or eulittoral) zone. There the species develops a morphological variety with narrow and short leaves previously considered as a separate species, named *Zostera angustifolia*, and which in many areas behaves as an annual. It is noteworthy that in the United Kingdom a specific distinction between *Zostera angustifolia* and *Zostera marina* is still made[6]. *Zostera noltii* lives higher on the shore and occurs in the middle and upper parts of the mediolittoral belt. The species can also live under permanent subtidal conditions in small brackish streams and coastal lagoons with euhaline conditions.

In the United Kingdom, *Zostera marina* is the more common species; it is widely, but patchily, distributed around the coasts of England, Scotland and Wales; the main concentrations occur along the west coast of Scotland including the Hebrides and in southwest England including Devon and Cornwall, as well as the Scilly Isles and Channel Islands. The intertidal form of *Zostera marina* is also widely distributed, but less abundant; sites with major concentrations occur in the Exe Estuary, in Hampshire, the Thames Estuary, and the Moray and Cromarty Firths in Scotland. *Zostera noltii* has a predominantly eastern distribution in the United Kingdom, more or less coinciding with the distribution of the intertidal *Zostera marina* form[6, 7].

In Ireland, Whelan[8] carried out an extensive survey of *Zostera* spp.; *Zostera marina* is frequently found along the coasts under subtidal conditions (Ventry Bay,

Regional map: Europe

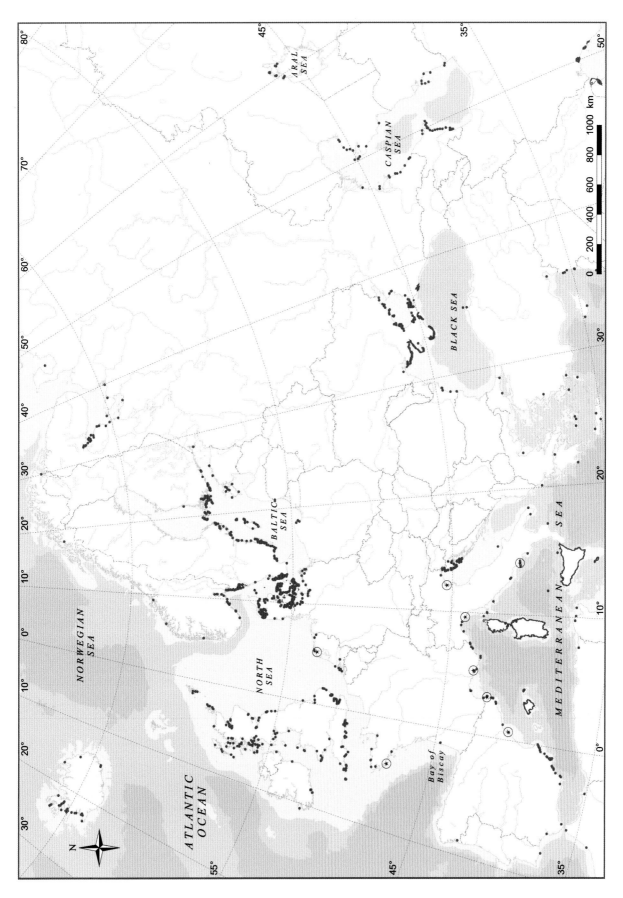

THE BEAUTY OF SEAGRASSES

Thalassia testudinum meadow in Florida, USA.

Fish sheltering in a cluster of sea urchins in an *Enhalus acoroides* bed, Komodo National Park, Indonesia.

Spotted eagle ray feeding on benthic invertebrates in a bed of *Thalassia testudinum* and *Syringodium filiforme*, Turks and Caicos Islands, Caribbean.

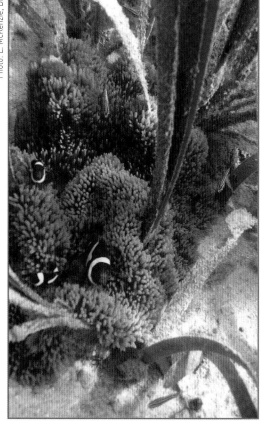

Clownfish and anemone in *Enhalus acoroides* and *Thalassia hemprichii* meadow in Kavieng, Papua New Guinea.

Western Europe

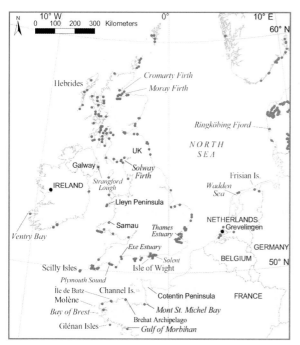

Map 2.1
Western Europe (north)

Map 2.2
Western Europe (south)

Galway sites, West Cork sites). Wasting disease symptoms were observed in this country in the 1930s[9].

The sandy, surf-exposed North Sea coasts of Denmark, Germany, the Netherlands, Belgium and France are devoid of seagrass. Seagrasses are restricted to the Wadden Sea area, which is protected from the full hydrodynamic forces of the ocean by the Frisian Islands. Along the Danish west coast, *Zostera marina* occurs in the enclosed Ringkøbing Fjord. In the southern part of the Netherlands some intertidal populations of the two *Zostera* species occur in the estuarine branches of the mouths of the rivers Rhine and Meuse. Some of these branches were diked in the second part of the 20th century, but still contain some submerged beds of *Zostera marina* (Lake Grevelingen, Lake Veere).

In France, the two *Zostera* seagrass species are widely distributed. The brackish water species *Ruppia maritima* and *Ruppia cirrhosa* are uncommon and mostly encountered in brackish ponds along the Channel coast, and the Guérande north of the Loire Estuary; their distribution is insufficiently known. Many small *Zostera marina* beds occur along the Channel coasts from the west of Normandy to the west of Brittany, mainly on the sandy bottom under both intertidal and subtidal marine conditions[10]. From the west of Brittany to the south of the Bay of Biscay, the sites are either subtidal around islands (Molène and Glénan Archipelagos) or in very sheltered bays (Bay of Brest, Gulf of Morbihan, Arcachon Basin) in which they can occupy large areas.

In Spain and Portugal, seagrass beds are localized in drowned river mouths (*rías*) and protected bays (e.g. Vigo Bay), with a zoned occurrence of the two species of *Zostera*. *Zostera noltii* occurs on the large muddy mediolittoral flats and *Zostera marina* (accompanied sometimes by *Cymodocea nodosa* (e.g. Ría Formosa)) in the upper part of the infralittoral. Many of these beds are concentrated in the numerous Galician *rías* in northwestern Spain[11]. In Europe, *Zostera marina* reaches the southern limit of its Atlantic distribution in southern Spain near Gibraltar.

BIOGEOGRAPHY

Zostera marina occurs under a large range of environmental conditions which can be identified as the following three main biotopes:

o Sheltered habitats in enclosed and semi-enclosed bays, estuaries and *rías*, with turbid low-salinity waters and muddy sediments. Eelgrass beds are limited to a narrow depth range (<2 m) and because of the high turbidity do not extend much below mean sea level. These beds often appear as long narrow (<30 m wide) ribbons along the small subtidal channels which groove the muddy intertidal flats (many North Sea sites, Galicia, Arcachon Basin). In some sites (United Kingdom, Brittany) the intertidal *Zostera marina* form ("*Zostera angustifolia*") can extend across large muddy areas in the mediolittoral belt, mainly on poorly drained sediments with a thin water layer remaining at the sediment surface during the low-tide period.

o Semi-open habitats under marine conditions (salinity of 32-36 psu). These beds occur on sandy and locally even on coarse sediments from a depth of +2 m mean sea level to –3 m. Their spatial extension depends on the rocky platforms, small islands and hard substrate structures which protect the beds from the most extreme hydrodynamic forces (strong currents, swell and waves). This type is common along the western coasts of the Channel.

o Open habitats under fully marine and subtidal conditions (–2 to –10 m mean sea level) mainly around islands in very clear waters. Swell is probably the main factor limiting the extension of these beds to the intertidal zone. *Zostera marina* can occasionally be observed in artificial lagoons, brackish pools and abandoned salt production areas on the French Atlantic coast[12] with a morphology close to its intertidal morph.

Zostera noltii is very often found in estuarine and sheltered environments as described above but occupies higher levels on the shore. The species mainly occurs in muddy and sandy sediments and can form extensive beds on tidal flat areas (Wadden Sea, United Kingdom, Ireland, Gulf of Morbihan and Arcachon Basin). The species is never found below the low-tide mark.

PRODUCTIVITY, BIOMASS AND ROLE IN NUTRIENT CYCLES

The primary production of *Zostera marina* meadows is the highest of the coastal sedimentary environments of the region. The associated organisms supported by eelgrass production are numerous and diverse. The beds are used as refuge and nursery areas by many species, including commercial fish and invertebrates. At this latitude, growth of the perennial morph is continuous throughout the year, although limited in winter, so there is a permanent flux of seagrass tissues inducing a detritus-based food chain. The detritus often becomes accumulated by waves and tidal currents outside the beds which thus spatially extend their functional role in the marine ecosystem[4]. It has been calculated that 1 g (dry weight) of seagrass detritus supports on average 9 mg of bacteria and protists[6]. The living leaves are used as a substrate by diatoms, bacteria and heterotrophic protists and many macro-epiphytes (algae and invertebrates).

The total surface area available for the superficial biofilm and epiphytes, calculated by adding the surface area of all the leaves of the shoots, can reach 6 to 8 m^2 in 1 m^2 of sediment (leaf area index). Whelan and Cullinane[13] identified 60 algal species in Ventry Bay (Ireland) and Connan and Hily[14] found 82 epiphytic algal species in Brittany beds including 60 species in only one bed in the Bay of Brest. Some of these species are found only on *Zostera* leaves, or have their most luxuriant development on these, such as the small Phaeophytes *Ascocyclus magnusii*, *Myriotrichia clavaeformis*, *Cladosiphon zosterae* and *Punctaria tenuissima*, and the Rhodophytes *Fosliella lejolisii*, *Erythrotrichia bertholdii*, *Erythrotrichia boryana* and *Rhodophysema georgii*. This epiphytic community, described as *Fosliello-Myriotrichietum clavaeformis*, occurs only along the oceanic coasts and is very sensitive to pollution.

In western Europe, many commercial fish and shellfish species use eelgrass meadows as a habitat. Some fish predators occupy the beds during tidal and nocturnal migrations (Labridae, *Morone labrax* and flatfish). Others use the beds as spawning sites and nursery areas (*Mullus surmuletus*). The juveniles of the crab *Maia squinado*, an important commercial species in France, hibernate in the sediments of the subtidal beds[14]. The beds are actively exploited by handnet fishermen for the shrimp *Leander serratus*. Many commercial bivalves such the clams *Venerupis pullastra*, *Venus verrucosa*, razor shells, *Lutraria lutraria*, and pectinids, *Chlamys opercularis* and *Pecten maximus*, are especially abundant in the *Zostera marina* beds and are heavily exploited for recreational fishing.

Some rare and endangered species like the sea horse (*Hippocampus* sp.) still occur in *Zostera marina* beds in the area. A few invertebrates directly consume the eelgrass leaves, e.g. the sea urchin *Sphaerechinus granularis* and the sea rabbit *Aplysia punctata*. Brent geese (*Branta bernicla*) used to be strongly dependent on the eelgrass meadows (*Zostera noltii* and *Zostera marina*) which are generally found in their main migration sites. At present they have found alternative food sources following the loss of eelgrass. Other birds such as teal, widgeon, pintail, mallard, shoveler, pochard, mute swan and coot are also consumers of eelgrass.

HISTORICAL PERSPECTIVES

Before the outbreak of the wasting disease in the 1930s, eelgrass beds were very common along the European coasts. The beds were locally harvested for different uses (soil improvement (Galicia, Spain), embankment or dikes around fields on small islands (Brittany, France), sea walls (the Netherlands), filling of mattresses and cushions (Normandy, France, the Netherlands),

packaging, roofing and insulation material); therefore eelgrass was historically of economic importance. As an example, about 150 km² were covered by eelgrass in the western Wadden Sea[15]. Despite this abundance, already by the 18th century, Martinet[16] was urging development of a method to multiply eelgrass, because "one cannot have too much of it"[17]. Though little documentation is available, it seems that *Zostera marina* was more abundant than *Zostera noltii*. Most of the subtidal *Zostera marina* beds did not recover from the wasting disease. From the 1960s onwards the eulittoral beds of *Zostera marina* and *Zostera noltii* declined, probably as a consequence of increased turbidity[18]. In one site the increased turbidity was found to be more related to increased sediment particles, dredging and filling activities than to increased phytoplankton.

AN ESTIMATE OF HISTORICAL LOSSES

Without doubt, the losses of areas occupied by eelgrass have been very great since the beginning of the 20th

---

Case Study 2.1
THE WADDEN SEA

The Wadden Sea is one of the world's largest international marine wetland reserves. Before the 1930s it contained large beds of subtidal and low-intertidal *Zostera marina*, whereas many mid-intertidal flats were covered with a mixed bed of *Zostera marina* and *Zostera noltii*. After the wasting disease in the 1930s, seagrasses survived only in the mid-intertidal zone (a narrow zone around 0 m mean sea level). Here, new losses occurred from the 1970s onwards. Increased turbidity, increased shell-fisheries, increased construction activities and increased nutrient loads are the main factors that have caused the losses and lack of recovery, although the causes of the wasting disease losses during the 1930s are still open to dispute[20, 37].

Currently, Dutch seagrass beds cover 2 km², German seagrass beds 170 km² and Danish seagrass beds some 30 km². In the Netherlands, an intensive monitoring program has revealed large fluctuations in cover of *Zostera marina* particularly: for example a sixfold increase in area was observed within two years (followed by some decrease), whereas at another area an 80 percent decrease was observed between two different years (followed by some increase). The *Zostera noltii* bed cover fluctuates less than twofold, which may be ascribed to some habitat characteristics, including firm clay banks, and the plants' perennial reproductive strategy. The *Zostera marina* beds in the Wadden Sea are mainly (but not totally) annual. These fluctuations in cover make the populations vulnerable to local and temporal disturbances, caused by human actions or by ice scour or gales, particularly when the area becomes small, and the habitat offers no local refugia.

Since 1987, the University of Nijmegen, assigned by and in cooperation with the Dutch Government, has investigated the possibilities for restoration of *Zostera marina* in the western Wadden Sea. Water clarity of the Dutch Wadden Sea has improved and shellfisheries have been locally prohibited. Experiments in the field, in outdoor mesocosms and in the laboratory, as well as literature, long-term environmental data and global information system (GIS) studies, all provided knowledge of suitable donor populations, habitat requirements and potential habitats[37, 38, 39]. In 2002, transplanting began in the western Wadden Sea. Risks will be spread in space and time, the transplants will be protected in the field during the first years (to prevent seed-bearing shoots drifting to open sea) and protective mussel ridges will be constructed to provide refugia for the transplants.

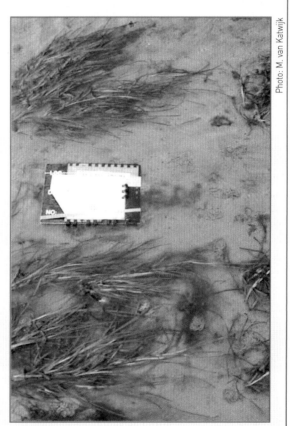

Transplanted *Zostera marina* in the Wadden Sea, Netherlands.

century in western Europe. After the wasting disease in the 1930s which destroyed most of the *Zostera marina* beds, the recovery was very slow and at many sites eelgrass did not recover at all. From the 150 km² in the western Wadden Sea estimated to be covered by seagrass in 1919 by van Goor[15], the area estimate in 1971 was reduced to 5 km² only in intertidal areas[19], and the beds were estimated to cover approximately 2 km² in 1994[20], mainly consisting of *Zostera noltii*[21]. Giesen[5] considered that in 1990 *Zostera marina* had declined to the point of virtual disappearance in the Dutch Wadden Sea; at present only two locations of this species are still known in the area.

In a few cases, anthropogenic shoreline modifications have facilitated the growth of eelgrass beds. In the second part of the 20th century several large constructions along the coast of the Netherlands modified the sites colonized by eelgrass. The construction of a dam in 1964, 25 km upstream of the Grevelingen Estuary, isolated the ecosystem from the freshwater influence; at that period the eelgrass beds covered about 12 km² in the intertidal belt. Then, in 1971, a second dam at the mouth of the estuary isolated the system from the sea's influence. The Grevelingen Estuary was transformed into a stagnant salt water lake. The new conditions favored the extension of the *Zostera marina* beds, which became permanently submerged, and occupied about 34 km² in the period 1971-85[22]. *Zostera noltii*, which was the most common seagrass before this human intervention, declined to almost complete disappearance; in the early 1990s a small, completely submerged stand of a few square meters was found in very shallow water. The extension of the *Zostera marina* beds soon came to a halt. A large-scale die-off started around 1986-87, and has so far not been explained in a convincing way[22, 23].

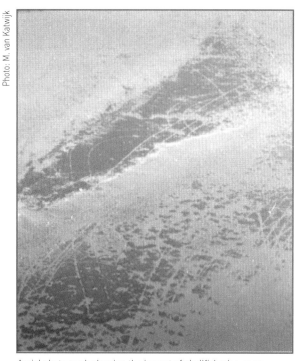

Aerial photograph showing the impact of shellfisheries on an intertidal *Zostera noltii* bed.

Most of the recent observations underline, however, the gradual regression of the eelgrass bed areas under anthropogenic influence. Human impact may be exercised directly by dredging, filling and marina development, aquaculture of mollusks (Ostreidae, Mytilidae, Veneridae) and fish farms, anchoring and other boat activities, and directly and indirectly by the effects of eutrophication, such as increasing turbidity, development of invasive macroalgae and floating blankets of macroalgae which may suffocate the seagrasses, and development of high biomass of epiphytic microalgae and macroalgae on the leaves. In the geographic area considered, the intensity of these perturbations varies from one region to the other.

Along the Channel coasts the natural harbors are used as semi-permanent anchoring sites for pleasure boats as here the optimal conditions for this activity, including tidal level considerations, protection from swell and currents, and distance from the shore are met. Unfortunately, these are exactly the sites where *Zostera marina* has its ecological optimum. This activity is an important cause of the erosion of eelgrass beds and it is increasing very rapidly everywhere. In the same way, recreational fishing is increasing; the digging of mollusks during low tide at spring tides by very destructive tools induces a rapid regression of the intertidal parts of *Zostera marina* beds. Numerous eelgrass sites of *Zostera noltii* and *Zostera marina* are progressively disappearing with the rapid extension of aquaculture, in France and Spain, on intertidal sites. In Cornwall and Devon (southern England) many losses were pointed out by Giesen[5] by comparing the results given by Covey and Hocking[24], and Holme and Turk[25]; no explanations were found for these losses. The less-threatened beds are probably the deeper subtidal *Zostera marina* meadows under semi-exposed conditions particularly around small islands, but the continuously increasing turbidity of the coastal waters in western Europe, generally recognized but not really quantified, is probably a factor in the variations of the lower limits observed in many beds. As an example, the lower limit of *Zostera marina* in Ventry Bay (Ireland) was 13 m in 1977-78, 10 m in 1980 and continued to fall after 1980[8].

The loss of eelgrass has not been quantified for the whole region, but probably more than 50 percent of the beds are subject to one or other of the types of perturbations mentioned, and are threatened by total or partial destruction over the next ten years. A review of the abundant literature concerning eelgrass in western Europe suggests that the general trend of recovery after the almost complete disappearance of the sublittoral beds in the 1930s is largely being reversed by the diverse, and generally adverse, local and regional anthropogenic impacts.

AN ESTIMATE OF PRESENT COVERAGE

In France, along the Channel and Atlantic coasts, most of the eelgrass sites are known. Along the western coast of the Cotentin Peninsula, beds of *Zostera marina* occur near Granville, and the most eastern beds of *Zostera noltii* are in Baie des Veys near Isigny on the eastern side of the peninsula[26]. Eelgrass beds in Brittany were located and mapped in 1999 by Hily *et al.*[10], although the exact area of each bed has not been determined. This study identified more than 70 sites from the Mont St Michel Bay in the north to the Loire Estuary in the south of Brittany. Most of them are small beds between 1 and 5 ha, but there are at least ten large beds covering 10 to more than 100 ha (including the Gulf of Morbihan, Glénan Archipelago, Île de Batz, Bay of Brest, Brehat Archipelago, Abers Estuaries and Etel Ría). To the south, the Marennes-Oléron Basin is a large site of *Zostera noltii*. Further to the south the Arcachon Basin is the largest site of *Zostera noltii* (70 km^2 in 1984) in Europe, and also a large site of *Zostera marina* (4 km^2 in 1984)[27].

In the United Kingdom, most of the beds are mapped and consist of about 140 sites of *Zostera marina* (including the intertidal sites) and about 70 sites of *Zostera noltii*[7]. Some of them extend over a considerable area, such as the *Zostera marina* bed in the Cromarty Firth, Scotland, which covers 12 km$^{2\,[28]}$ and is considered the largest bed in the United Kingdom. In the cross-border sites of Scotland and England, the Solway Firth and the northern Northumberland coasts have coverage respectively of 2 and 9 km$^{2\,[29,\,30]}$. Along the coast of England, the seagrass coverage of some large sites has been documented: Essex estuaries (8.44 km^2), North Thames Estuary (3.25 km^2), Solent and Isle of Wight (4.40 km^2), Plymouth Sound and estuaries (6.50 km^2). Some smaller beds occur in Devon and Cornwall. In Wales, the main sites are in the Lleyn Peninsula and the Sarnau, while in Northern Ireland the beds in the Strangford Lough cover 6.30 km$^{2\,[6]}$. *Zostera marina* beds are also common on the semi-sheltered sediments of the Channel Islands.

Along the southeastern coasts of the North Sea, seagrass beds are restricted to the sheltered Wadden Sea and southwest Netherlands and cover a total of 200 km^2.

In Spain, the actual seagrass coverage is not known, but many beds are recorded from the numerous *rías* of the Galicia region (*Zostera noltii* covers approximately 20 km^2 between the French and Portuguese coasts[31]). In Portugal, the Ría Formosa is recorded as a site for intertidal *Zostera noltii* beds and subtidal beds of *Zostera marina* and *Cymodocea nodosa*.

It is at present not possible to measure the potential seagrass habitat in the whole region. However, it can be estimated that it would be more than three times the actual coverage both for *Zostera marina* and *Zostera noltii*. An estimate for Brittany is planned in 2003 with a long-term survey of the coastal benthic communities (REBENT network survey) including *Zostera* beds. In the United Kingdom the Habitat Action Plan for seagrass beds developed by the UK Biodiversity Steering Group may result in quite an accurate estimate of the potential habitat in this area.

PRESENT THREATS

Direct destruction of beds

As a result of the rapid development of pleasure fishing and sailing over the last 20 years, filling and dredging for extension or creation of harbors have destroyed many eelgrass beds. As a consequence of economic and environmental arguments such developments are becoming less harmful nowadays, but the damage has been done.

Oyster and mussel aquaculture on littoral sediments has been the cause of the destruction of many eelgrass beds because the optimal conditions for the culture of these animals correspond with the optimal conditions for the beds. This activity is still expanding, and will probably be one of the main threats to the beds in the future.

Anchoring and mooring outside harbors is damaging. Anchoring causes the formation of deep holes which in their turn may become points of impact for the eroding forces, while the chains dragging across the bottom destroy the surrounding biocenoses including the seagrass communities.

Hand fishing for clams using rakes, forks and hoes to catch the endofaunal bivalves at low tide, as well as within the seagrass beds, results in whole plants with their rhizomes being pulled out of the sediment. This causes considerable damage to the seagrass beds, because it is generally followed by erosion. Collecting clams in this way is becoming very popular, so this type of perturbation is increasing in western Europe[32]. The same kind of perturbation is caused in the eulittoral seagrass beds by digging for polychaetes such as *Arenicola* and *Nereis* to be used as bait.

Professional fishermen on boats dredge on the limits of the beds to catch bivalves. The natural

acclimation of the Japanese clam *Venerupis philippinarum* in the south of the area (from south Brittany to Spain) increases the direct impact of this activity on the seagrass beds because this species develops dense populations in and around the eelgrass beds of the sheltered bays and estuaries[14]. In the Netherlands, much damage has been done to the few still existing eelgrass beds by professional cockle fisheries with their modern, effective, but environmentally unfriendly equipment.

Indirect destruction of beds

Eutrophication is the main cause of indirect destruction of seagrass beds. The increase of organic matter and nutrients from terrestrial effluents favors phytoplankton, causing blooms which in their turn decrease light availability. Moreover, plankton production increases not only in terms of instantaneous biomass but the period of production also becomes longer and longer, and can be observed all year in some areas with numerous successive small blooms[33]. As a consequence of these plankton blooms, the water transparency decreases, limiting the light available for the growth of *Zostera*.

Apart from this shading effect by plankton, a further reduction of light is brought about by increased epiphyte cover on *Zostera* leaves in which both diatoms and macroalgae participate. The specific community of small epiphytes mentioned above is, however, the first element to disappear from the seagrass bed in the case of eutrophication. Eutrophication also increases the production of green macroalgae (*Enteromorpha*, *Ulva*); particularly in semi-enclosed, sheltered bays the green algae can form thick blankets which float around and can be deposited on the *Zostera* beds of the sandy and muddy intertidal flats during periods of very calm weather. Under such conditions the seagrass beds become smothered and suffocated, leading to complete die-off within a very short time[34]. When these blankets are deposited on the bare surface of the intertidal flats, the spatial competition favors the green algae which prevent the extension of the seagrass beds, and reduce the growth of shoots by shading and suffocation effects in the areas where they border the seagrass beds. In these conditions the beds decrease progressively, and may completely disappear in a few years.

The increase of turbidity is not only associated with eutrophication, but can also result from an increasing input of terrigenous particles by river effluents as a consequence of large-scale changes in agricultural practices; modern practices encourage the leaching of soil in winter. Extraction of calcareous sediments and calcareous macroalgae (*Lithophyllum* sp.) from sublittoral beds induces high turbidity; the high levels of sediment in suspension in the water reduce light and cover the leaves of seagrass during resedimentation. Dredging for harbor and channel maintenance and releasing the dredged sediments on the seafloor also increases turbidity and lowered light levels.

Spatial competition with invasive species may also limit the extension of seagrass beds in western Europe. The brown algae *Sargassum muticum* is able to develop in the eelgrass beds where the sediment floor is coarse or includes gravel, stones and/or shells. In these beds *Sargassum* gradually takes over and prevents the rejuvenation of eelgrass[35].

Most of these threats concern the eelgrass *Zostera marina*. The intertidal species *Zostera noltii* has been assumed to be threatened by a combination of various factors including turbidity, eutrophication and associated epiphyte cover, the decrease of mud snail populations (*Hydrobia ulvae*) which graze on the epiphytes, and also as a result of bioturbation by the lugworm *Arenicola marina*. These processes have been well studied in the Dutch Wadden Sea by Philippart[20].

Finally, a very important potential threat is shipping. The Channel and the southern North Sea are among the world's busiest shipping routes and the chance that accidents will occur cannot be excluded (adverse weather conditions; human error). Notorious disasters were those with the tankers *Torrey Canyon* and *Amoco Cadiz* in the western Channel and recently *Erika* in North Biscay. The impact of these oil spills on the whole coastal ecosystem has been disastrous.

POLICY RESPONSES

The European Union (EU) elaborated a Habitats Directive for both terrestrial and marine habitats which identifies the main natural habitats and their cultural value for further consideration in terms of protection and conservation. In this context eelgrass beds are identified as particular ecological units of several marine habitats: sandy shore, mud flats and coastal subtidal sandy sediments. These initiatives have led to eelgrass habitats being specifically targeted for conservation and restoration[36]. But although they are considered as biotopes of special interest, they are not considered as "endangered" and so not considered for immediate and strong protection. In France *Zostera marina* is listed in the Red Book of threatened species but is not in the list of protected species. *Zostera noltii* is not considered. Additionally, very locally and in few localities, some *Zostera* beds are protected by municipal authorities. In March 2002, *Zostera marina* and *Zostera noltii* were both incorporated in the Dutch Red List of threatened plants.

In the United Kingdom, the eelgrass beds have been considered for many years as targets for conservation and a habitat action plan for seagrass

Case Study 2.2
GLÉNAN ARCHIPELAGO

In the northern part of the Bay of Biscay, the Glénan Isles are located 9 miles off the continental coast of south Brittany, France. The area is characterized by ten small islands and numerous rocky islets surrounding an enclosed, shallow (< 5m deep), sandy area well protected from the oceanic swell. Aerial photographs are available from the year 1932 and allowed estimates of the long-term development of the areas covered by eelgrass[40].

This is an interesting experimental site because the continental influence (eutrophication and associated consequences) is minimized which allows the observation of the natural dynamics of the beds under climatic factors, but also because human activities (anchoring, fishing) induce local perturbations in the eelgrass beds. So it is possible to separate the role of each of the factors that control the dynamics.

Based on the cover in 1932, it can be considered that a surface of 10 km² is suitable for eelgrass beds, but in 1990 only 25 percent of this area was colonized by eelgrass. In 2000, this percentage increased to around 40 percent as a result of positive climatic conditions since 1995; this tendency is also observed in many beds of the Brittany coasts[14]. However this evolution is moderated by the negative impacts of numerous human activities[14]:

o dredging for clams by professional fishing boats prevents recolonization in the opened central subtidal part of the area;
o anchoring by numerous pleasure boats throughout the year induces fragmentation of the beds in five main sheltered subtidal sites;
o recreational fishing for clams induces fragmentation of the intertidal beds;
o extraction of calcareous sediments (maerl beds) 1.5 miles off the archipelago induces heavy turbidity in the northern waters of the archipelago which may limit the extension of the beds in depth (a decrease of the deeper limits of the laminarians close to the beds was recently demonstrated[41]).

This example underlines the complexity of the dynamics of the eelgrass beds which are under the influence of factors working at various spatial and temporal scales: here the positive climatic factors working at the global scale compensate for the negative impacts of the perturbations induced by human activities at the local scale.

This example also underlines the difficulties of seagrass conservation: it is hard to explain to the authorities and users alike that human activities must be moderated in the beds because of their impacts while the spatial cover is actually increasing. It is necessary to explain that under adverse climatic conditions (which are expected in the future) the cumulative effect, with human impacts, would induce dramatic and rapid loss of the beds, and consequently preventive action should be planned.

Fortunately, the management authorities at this site are working with the scientific teams on a sustainable development plan to preserve the image of high environmental quality in this tourist area.

Anchoring on a *Zostera marina* bed in Glénan Archipelago.

Photo: C. Hily

Zostera marina on a maerl bed in the Bay of Brest.

beds was prepared by the UK Biodiversity Steering Group. In a complementary way, the South West Regional Biodiversity Habitat Action Plan has also been developed. These initiatives are integrated in the EU Habitats Directive which requires the identification of European marine sites in a network called "Natura 2000": sites which should be managed in order to maintain or restore the favorable conservation status of their habitats and species. Each state of the EU has the statutory responsibility, via the conservation agencies, for developing conservation objectives in each site, defined as a statement of the nature conservation aspirations for a site. In the United Kingdom these sites are called SACs (special areas of conservation), in France they are called "sites Natura 2000". The regulations suggest that relevant authorities from the various sites should work together within a management group. In most countries, the presence of *Zostera* beds has been a criterion (but not the only one) to retain a site as a SAC. When a bed is included in a SAC, specific management is required for the bed. This procedure is to be applied independently by each country, and has not yet been achieved. Some sites derive their conservation status from a combination of several different directives, and this can reinforce the conservation of *Zostera* beds. For example, some sites are also RAMSAR sites and/or sites of the EU Birds Directive, which reinforces the international recognition of the site's importance and requires the government to strongly protect the site. However, the sites indicated according to these directives are far from covering all the eelgrass beds in Europe. It therefore remains very important to give global consideration to eelgrass habitats on a wide scale and it remains necessary to define specific conservation regulations at the level of the species or genus and/or the habitat.

ACKNOWLEDGMENTS

Dr R.M. Asmus kindly provided *Zostera marina* cover percentages for the German Wadden Sea. Thanks to Ingrid Peuziat who provided data on the Glénan Archipelago.

AUTHORS

Christian Hily, Institut Universitaire Européen de la Mer (University of western Brittany), Technopole Brest Iroise, 29280, Plouzané, France. **Tel:** +33 (0)2 98 49 86 40. **Fax:** +33 (0)2 98 49 86 45. **E-mail:** christian.hily@univ-brest.fr

Marieke M. van Katwijk, Department of Environmental Studies, University of Nijmegen, P.O. Box 9010, 6500GL Nijmegen, Netherlands.

Cornelius den Hartog, Department of Aquatic Ecology and Environmental Biology, University of Nijmegen, P.O. Box 9010, 6500GL Nijmegen, Netherlands.

REFERENCES

1. den Hartog C [1981]. Aquatic plant communities of poikilosaline waters. *Hydrobiologia* 81: 15-22.
2. den Hartog C [1983]. Structural uniformity and diversity in *Zostera*-dominated communities in western Europe. *Marine Technology Society Journal* 17: 105-117.
3. Hily C, Bouteille M [1999]. Modifications of the specific diversity and feeding guilds in an intertidal sediment colonized by an eelgrass meadow (*Zostera marina*) (Brittany, France). *Comptes Rendus de l'Académie des Sciences Serie III, Sciences de la Vie/Life Sciences* 322: 1121-1131.
4. den Hartog C [1987]. "Wasting disease" and other dynamic phenomena in *Zostera* beds. *Aquatic Botany* 27: 3-13.
5. Giesen W [1990]. "Wasting Disease" and Present Eelgrass Condition. Report to Dutch Ministry of Transport and Waterways. University of Nijmegen, Netherlands. 138 pp.
6. Davison DM, Hughes DJ [1998]. *Zostera* Biotopes: An Overview of Dynamics and Sensitivity Characteristics for Conservation Management of Marine SACs. Reports UK Marine SACs Project, Task Manager, AMMW Wilson, SAMS. 95 pp.
7. Stewart A, Pearman DA, Preston CD [1994]. *Scarce Plants in Britain*. JNCC, Peterborough.
8. Whelan PM [1986]. The Genus *Zostera* in Ireland. PhD thesis, University College Cork, Ireland. 215 pp.
9. Whelan PM, Cullinane JP [1987]. The occurrence of "wasting disease" of *Zostera marina* in Ireland in the 1930s. *Aquatic Botany* 27: 285-289.
10. Hily C, Raffin C, Connan C [1999]. Les herbiers de zostères en Bretagne: Inventaire des sites, faune et flore. Rapport Diren Région Bretagne, Université de Bretagne Occidentale, Brest. 57 pp.
11. Curras A, Sanchez-Mata A, Mora J [1993]. Estudio comparativo de la macrofauna bentonica de un fondo de *Zostera marina* y un fondo

arenoso libre de cubierta vegetal. *Cahiers de Biologie Marine* 35: 91-112.
12 Gruet Y [1976]. Répartition des herbiers de *Zostera* (Monocotyledones marines) sur l'estran des côtes de Loire-Atlantique et du Nord de la Vendée. *Bulletin Société des Sciences Naturelles de Ouest de la France* 74: 86-90.
13 Whelan PM, Cullinane JP [1985]. The algal flora of a subtidal *Zostera* bed in Ventry Bay, south-west Ireland. *Aquatic Botany* 23: 41-51.
14 Hily C. Personal observations.
15 van Goor ACJ [1919]. Het zeegrass (*Zostera marina*) en zijn beteekenis voor het leven der visschen. *Rapp Verh Rijksinst Visscherij* 1(4): 415-498.
16 Martinet JF [1782]. *Verhandeling over wier der Zuiderzee. Verhandelingen Hollandsche Maatschappij der Wetenschappen* 20: 54-129.
17 van Katwijk MM [2000]. Possibilities for Restoration of *Z. marina* Beds in the Dutch Wadden Sea. PhD thesis, University of Nijmegen, Netherlands. 160 pp.
18 de Jonge VN, de Jong DJ [1992]. Role of tide, light and fisheries in the decline of *Zostera marina* in the Dutch Wadden Sea. *Netherlands Institute for Sea Research Publications Series* 20: 161-176.
19 den Hartog C, Polderman PJG [1975]. Changes in the seagrass populations of the Dutch Waddenzee. *Aquatic Botany* 1: 141-147.
20 Philippart CJM [1994]. Eutrophication as a Possible Cause of Decline in the Seagrass *Zostera noltii* of the Dutch Wadden Sea. PhD thesis, University of Wageningen, Netherlands. 157 pp.
21 Dijkema KS, van Tienen G, van Beek JG [1989]. Habitats of the Netherlands, German and Danish Wadden Sea 1:100,000. Research Institute for Nature Management, Texel and Veth Foundation, Leiden.
22 Nienhuis PH, de Bree BHH, Herman PMJ, Holland AMB, Verschuure JM, Wessel EGJ [1996]. Twenty-five years of changes in the distribution and biomass of eelgrass, *Zostera marina*, in Grevelingen Lagoon, the Netherlands. *Netherlands Journal of Aquatic Ecology* 30: 107-117.
23 Herman PMJ, Hemminga MA, Nienhuis PH, Verschuure JM, Wessel EGJ [1996]. Wax and wane of eelgrass *Zostera marina* and water column silicon levels. *Marine Ecology Progress Series* 144: 303-307.
24 Covey R, Hocking S [1987]. Helford River Survey. A report to the Helford River Steering Group. 121 pp.
25 Holme NA, Turk SM [1986]. Studies on the marine life of the Helford River: Fauna records up to 1910. *Cornish Biological Records* No. 9. 26 pp.
26 Le Gall J, Larsonneur C [1972]. Séquences et environnements sédimentaires dans la Baie des Veys (Manche). *Revue de Géographie Physique et Géologie Dynamique* 14: 189-204.
27 Auby I [1991]. Contribution à l'étude des herbiers de *Zostera noltii* dans le bassin d'Arcachon: Dynamique, production et dégradation, macrofaune associée. Thèse de doctorat, Université de Bordeaux. 234 pp.
28 RSPB [1995] Annual Report of the Royal Society for the Protection of Birds. RSPB, Sandy, Bedfordshire.
29 Hawker D [1993]. Eelgrass in the Solway Firth. Report for Scottish Natural Heritage.
30 Percival SM, Sutherland WJ, Evans PR [1997]. Intertidal habitat loss and wildfowl numbers: Application of a spatial depletion model. *Journal of Applied Ecology* 35(1): 57-63.
31 Laborda AJ, Cimadevilla I, Capdevila L, Garcia JR [1997]. Distribución de las praderras de *Zostera noltii* Hornem., 1832 en el litoral del norde de Espana. *Publ Espec Inst Esp Oceanogr* 23: 273-282.
32 den Hartog C, Hily C [1997]. Les herbiers de *Zostères*. In: Dauvin JC (ed) *Les Biocénoses marines et littorales françaises des côtes Atlantiques Manche et Mer du Nord: Synthèse, menaces et perspectives*. MNHN, Paris. pp 140-144.
33 Chauvaud L, Jean F, Ragueneau O, Thouzeau G [2000]. Long-term variation of the Bay of Brest ecosystem: Benthic-pelagic coupling revisited. *Marine Ecology Progress Series* 200: 35-48.
34 den Hartog C [1994]. Suffocation of a littoral *Zostera* bed by *Enteromorpha radiata. Aquatic Botany* 47: 21-38.
35 den Hartog C [1997]. Is *Sargassum muticum* a threat to eelgrass beds? *Aquatic Botany* 58: 37-41.
36 Wynne DW, Avery M, Campbell L, Gubbay S, Hawkswell S, Juniper T, King M, Newbery P, Smart J, Steel C, Stones T, Taylor J, Tydeman C, Wynde R [1995]. Proposed targets for habitat conservation. In: *Biodiversity Challenge*. 2nd edn. RSPB, Sandy, Bedfordshire. 285 pp.
37 van Katwijk MM, Hermuus DCR, de Jong DJ, Asmus RM, de Jonge VN [2000]. Habitat suitability of the Wadden Sea for restoration of *Zostera marina* beds. *Helgoland Marine Research* 54: 117-128.
38 van Katwijk MM, Schmitz GHW, Gasseling AP, Van Avesaath PH [1999]. Effects of salinity and nutrient load and their interaction on *Zostera marina. Marine Ecology Progress Series* 190: 155-165.
39 Giesen WBJT, van Katwijk MM, den Hartog C [1990]. Eelgrass condition and turbidity in the Dutch Wadden Sea. *Aquatic Botany* 37: 71-85.
40 Glémarec M, Le Faou Y, Cuq F [1996]. Long-term changes of seagrass beds in the Glenan Archipelago (South Brittany). *Oceanologica Acta* 20(1): 217-227.
41 Castric A. Personal communication.

3 The seagrasses of
THE WESTERN MEDITERRANEAN

G. Procaccini
M.C. Buia
M.C. Gambi
M. Perez
G. Pergent
C. Pergent-Martini
J. Romero

Studies on seagrasses in the Mediterranean basin date back to the beginning of the 19th century, when the most widespread and well-known species, *Posidonia oceanica*, was described for the first time. Since then thousands of papers have detailed different aspects of seagrass distribution, ecology, physiology, faunal and algal assemblages and, recently, genetics. Two international workshops in the early 1980s were dedicated to the endemic *Posidonia oceanica* and led to joint research programs among European countries to study the structure and functioning of the *Posidonia oceanica* ecosystem. Less information exists on the other Mediterranean seagrass species, although some of them are quite common and widespread in the basin. A significant contribution to the synthesis of the work conducted on Mediterranean seagrasses was offered by the organization of the Fourth International Seagrass Biology Workshop held in Corsica in 2000[1].

SPECIES DISTRIBUTION

Six seagrass species are present in the Mediterranean Sea, forming an almost continuous belt all along the coasts: *Posidonia oceanica*; *Cymodocea nodosa*, also present along the North Atlantic African coasts and Portugal; *Zostera marina* and *Zostera noltii*, both with a wide temperate distribution; *Halophila stipulacea*, probably a recent introduction from the Red Sea; and *Ruppia* spp., with a wide temperate distribution (Table 3.1). Extremely limited information is available on *Ruppia* in the Mediterranean and it will not be considered further.

Posidonia oceanica forms continuous meadows from the surface to a maximum depth of some 45 m and is common on different types of substrate, from rocks to sand, with the exception of estuaries where the input of freshwater and fine sediment is high. *Posidonia oceanica* beds have classically been considered one of the climax communities of the Mediterranean coastal area[2]. Meadows are very dense with over 1 000 shoots/m^2 [3] although this varies from year to year[4]. The horizontal and vertical growth of rhizomes, and the slow decay of this material, causes *Posidonia oceanica* to form a biogenic structure called "matte", that arises from the bottom up to a few meters and can be thousands of years old[5]. *Posidonia oceanica* is a monoecious species, with male and female flowers in the same inflorescence. Sexual reproduction is sporadic, especially in some areas. *Posidonia oceanica* has low genetic variability and meadows represent genetically distinct populations, even at a scale of a few kilometers[6]. A clear genetic distinction exists between northwestern, southwestern and eastern populations. Meadows are composed of a mosaic of large and ancient clones[6,7].

Cymodocea nodosa most commonly occurs in shallow water but exceptionally can reach a depth of 30-40 m. Shallow and deep stands are generally discontinuous. *Cymodocea nodosa* is usually found on sandy substrate and sheltered sites[8]. In France, the most important beds are known in coastal lagoons. Shoot density reaches almost 2 000 shoots/m^2 [9]. It has classically been considered to be a pioneer species in the succession leading to a *Posidonia oceanica* climax system[2]. However it also grows in areas previously colonized by *Posidonia oceanica* and characterized by dead matte. *Cymodocea nodosa* is a dioecious species. Seeds remain for a long time in the sediment, attached to the mother plant. The only existing analysis of Mediterranean meadows showed high genetic diversity. In fact, plants 5 m apart within a meadow were genetically distinct individuals[10].

Zostera marina is considered to be a relict species in the Mediterranean, where it forms perennial meadows distributed from the intertidal to a few meters deep. It can grow on sandy and muddy substrate and is also present in lagoons[8], though it is

Map 3.1
The western Mediterranean

rare throughout the Mediterranean. Shoot density in *Zostera marina* beds is almost 1 000 shoots/m^2[11]. *Zostera marina* is a monoecious plant. Studies on the genetic diversity of this species have never been performed in the Mediterranean.

Zostera noltii grows from the intertidal to depths of a few meters on sandy and muddy substrate[8]. It is also present in enclosed and sheltered areas, where it can form mixed beds with *Cymodocea nodosa*, at densities up to almost 1 300 shoots/m^2. *Zostera noltii* is a monoecious species and no information is available about the genetic variability of its populations.

Halophila stipulacea was recently introduced to the western Mediterranean Sea and was reported for the first time in 1988. In the eastern Mediterranean basin this species has been observed from the beginning of the 19th century and is believed to have been transported from the Red Sea, through the Suez Canal, an example of Lessepsian migration. In the Mediterranean it is distributed from the intertidal zone to 25 m[12]. It can grow on sandy and muddy substrate, and is present in enclosed areas. Shoot density is extremely high, up to almost 19 000 shoots/m^2 in shallow water[9]. *Halophila stipulacea* is a dioecious

Table 3.1
Examples of general features of Mediterranean seagrass meadows

	Posidonia oceanica shallow	*Posidonia oceanica* deep	*Cymodocea nodosa*	*Halophila stipulacea* shallow	*Halophila stipulacea* deep	*Zostera noltii*	*Zostera marina*
Density (shoots/m^2)	700	161	925-1 925	19 728	13 000	269-1 246	216-1 093
Leaf area index (m^2/m^2)	6.16-29	1.1-2.6	0.2-3.5	5	5.9	0.2-0.4	1.7-6.7
Leaf biomass (g dw/m^2)	175-670	52-94	17-159	157.8	–	13-79	45-775
Below biomass (g dw/m^2)	6 526	324	300-750	–	–	31-62	21-161
Epiphyte biomass (g dw/m^2)	7 147	3 21.3	3.4 12.5	–	–	–	–
Animal biomass (g dw/m^2)	0.077-0.4	–	1.0-2.6	–	–	–	0.7-3.2
Number of algal species	36	50	35	30	–	8	25
Number of animal species	38-60	22-84	83	53	–	–	–
Animal density (individuals/m^2)	380-1 100	210-680	1 486	2 035	–	–	–
Leaf production (g dw/m^2/y)	162-722	71.3-232	23.6-1 623	–	–	70-949	109-2 299
Rhizome elongation (cm/y)	1.1-7.4	–	3.6-57.8	–	–	91-168	18-91
Leaf lifespan (months)	11	11	5	–	–	–	1-3

Source: Modified from Buia et al.[9]. Values derived from key studies listed in Buia et al.[9].

Case Study 3.1
ITALY

LIGURIAN COAST
The Ligurian area is one of the best among the Italian coasts for information on the distribution and general status of seagrasses, in particular for *Posidonia oceanica*. Almost 50 *Posidonia oceanica* main meadows have been recorded and mapped[20]. Their extension ranges from a few to several hundred hectares, covering in total about 48 km². In general all the prairies are in different states of degradation due to coastal modifications for harbor and town development. In addition some *Posidonia oceanica* beds were impacted in the early 1990s by a crude oil spill following the wrecking of the oil tanker *Haven*, considered to be one of the worst Mediterranean oil spills[21].

Posidonia oceanica banquette on the Cava dell'Isola beach, Ischia Island, Italy.

TYRRHENIAN COAST (ISLAND OF ISCHIA)
At a smaller spatial scale, the best-known *Posidonia oceanica* meadows are those surrounding the Island of Ischia, in the northern part of the Gulf of Naples. *Posidonia oceanica* covers about 17 km² of the seafloor, and its meadows, forming a continuous belt around the island, were mapped in detail in 1979[22]. The different exposure of the coasts of the island, coupled with different environmental conditions and bottom type, give rise to meadows extremely diversified in terms of physiognomy (continuous and patchy beds), depth range (from 0 down to 38 m in depth), shoot density (from a mean of 900 shoots/m² at 1 m to 80 shoots/m² at 30 m depth) and biodiversity of associated communities (more than 800 associated species), and with low intrinsic genetic variability, coupled with a degree of isolation between shallow and deep stands[9, 6].

A recent monitoring of beds around the island (in the year 2000) demonstrates a substantial stability of distribution and the presence of other settlements not previously reported[23]. However, long-term studies carried out since 1979 in beds off Lacco Ameno have detected a reduction in shoot density, as a result of anchoring, the impact of the local fishery and a nearby wastewater outfall.

NORTH ADRIATIC COAST (VENICE LAGOON)
Zostera marina is present on the Italian coasts of the north Adriatic Sea – it was first recorded here in the 14th century. *Posidonia oceanica* has experienced a strong decrease in this area, being now limited to a few patches in the Gulf of Trieste[24]. The worst decline of *Posidonia oceanica* has occurred in the Venice lagoon. In 1990 *Zostera marina* covered an area of 36.5 km², forming pure and continuous beds of 2.4 km² and beds mixed with *Zostera noltii* over the other 34 km²[25, 26]. *Zostera noltii* was the most widespread species (42.5 km²) and *Cymodocea nodosa* was also present (15.6 km²). Monitoring results four years later from the southern part of the lagoon[27] showed an increase of about 7.6 percent in the overall extent of seagrass beds, but with more *Zostera marina* (an increase of 13.5 percent), a decrease of 10.1 percent in *Cymodocea nodosa* and a large decline in *Zostera noltii* (24.7 percent). The monospecific and discontinuous beds have increased while mixed species beds have declined. A high survival rate for *Zostera marina*, *Cymodocea nodosa* and *Zostera noltii* has been achieved in transplanting experiments using sods and rhizomes at various sites in the lagoon[28, 29].

species. Male flowers are frequent in the Mediterranean Sea but female flowers were only observed for the first time in 1998, in Sicily[13]. Studies on the genetic variability of two populations located along the Sicilian coasts showed that each shoot represents a genetically distinct individual. Genetic relatedness was higher among individuals collected at the same depth[12].

ASSOCIATED SPECIES

Seagrass ecosystems of the western Mediterranean are extremely rich in a number of associated plant and animal species. However, complete lists of associated species have been compiled only in a few cases, such as the *Posidonia oceanica* and *Cymodocea nodosa* meadows of the island of Ischia, where more than 800 and 250 species have been listed, respectively[14], or in the Medes Islands[15]. *Posidonia oceanica* beds are the exclusive habitat for many algal and animal species, such as the coralline red algae *Pneophyllum fragile* and *Hydrolithon farinosum*, the brown algae *Castagnea cilindrica*, *Giraudia sphacelarioides* and *Myrionema orbiculare*, the bryozoan *Electra posidoniae*, the hydroids *Aglaophenia harpago*, *Sertularia perpusilla*, *Campanularia asymmetrica*, *Cordylophora pusilla* and *Laomedea angulata*[16-18].

Posidonia oceanica meadows are nursery grounds for the juveniles of many commercially important species of fishes and invertebrates, such as several species of the family Sparidae (e.g. *Diplodus sargus* and *Diplodus annularis*), Serranidae (e.g. *Serranus cabrilla*), Labridae (e.g. *Coris julis* and *Crenilabrus maculatus*) and Scorpaenidae (e.g. *Scorpaena scrofa* and *Scorpaena porcus*), and the sea urchin *Paracentrotus lividus*. Among the rare or endangered associated species are the endemic sea star *Asterina pancerii*, the sea horse *Hippocampus hippocampus* and the bivalve *Pinna nobilis*: these species are protected, in both Italy and France, or are included among species requiring a specific legislation for protection[19].

EXTENT OF COVERAGE

Information on the distribution of seagrasses is scattered and therefore an estimate of the total area covered by seagrasses is difficult to make. However the beds in some areas are well known.

The Italian coastline is 7 500 km long, without taking into account the numerous small islands scattered all around the peninsula. It is almost entirely surrounded by seagrass meadows that, considering the three most abundant species (*Posidonia oceanica*, *Cymodocea nodosa* and *Zostera noltii*), extend from 0.2 to 45 m. Clearly the potential area covered by seagrass is enormous. Some 2 350 km^2 of seagrass are known to occur in Liguria, Lazio, Sardinia, Veneto and Friuli (Table 3.2). France has approximately 1 150 km^2 of *Posidonia oceanica* beds, but estimates for other species are not available. On the Mediterranean coasts of Spain, some regions have mapped their seagrass meadows in great detail allowing an estimate of more than 1 000 km^2 to be made.

Case Study 3.2
FRANCE

MARSEILLE-CORTIOU REGION

A long-term monitoring study of *Posidonia oceanica* beds in the Marseille-Cortiou region has recorded fluctuations over the 1883-1987 period, and the impact of a sewage treatment plant. In the period 1890 to 1898, when a sewage outlet was first set up in Marseille, the seagrass bed had reached a depth of 30 m and occupied an area of about 6.32 km^2. A number of authors noted loss of *Posidonia oceanica* between 1900 and 1970 as the city of Marseille expanded[30]. At the end of the 1970s the bed covered a smaller area, with a loss of 5-6 percent per decade.

When the wastewater treatment plant was set up in 1987, the lower limit of the seagrass bed was just 10 m and it included vast stretches of dead matte. Since then there have been further, much greater, losses amounting to 40 percent of the 1970 area. This is most likely due to the high levels of suspended matter, ammonium and phosphate coming from the treatment plant. After 1994, a natural recolonization of *Posidonia oceanica* was observed in certain areas, due to increased water clarity.

CORSICA

Corsican coasts experience low human impact with many marine protected areas; almost 71 percent of the Corsican coastline is still in its natural state. *Posidonia oceanica* beds occupy a total surface area of 624 km^2 [31] mainly along the eastern side of the island, where the continental shelf is very wide. Their distribution is limited on the steep and indented west coast. Upper limits are generally between 1 and 10 m depth, while the lower limit at several sites on the east coast is situated below 40 m. The lower limit rises to a depth of 15-20 m near large cities such as Ajaccio.

The beds of *Posidonia oceanica* are among the most important Mediterranean ecosystems, and their conservation is a high national and international priority (e.g. EU Habitats Directive 92/43/CEE, 21 May 1992). *Posidonia oceanica* beds exert a multifunctional role within coastal systems, comparable to that of other seagrasses in temperate and tropical areas, offering substrate for settlement, food availability and shelter, as well as participating in key biogeochemical and geological processes.

PRODUCTION AND BIOMASS

Both below-ground and above-ground biomass values of *Posidonia oceanica* exceed those of other seagrasses, including the Australian *Posidonia* species[40]. A striking feature is the distinct partitioning of the biomass, mainly directed into the lignified rhizomes, which can account for up to 90 percent of total biomass[41, 42] and production where leaves account for more than 90 percent[43]. In an extensive study[44] net primary production was estimated to range from 130-1 284 g dry weight/m^2/year. However,

Table 3.2
Distribution of seagrasses throughout the western Mediterranean (Italy, France and Spain)

Sites	Po	Cn	Zm	Zn	Hs	Area (km^2)	Comments
ITALY							
Liguria*	✓	✓				48	On rocky and sandy bottom, from 0 to 35 m[20].
Tuscany*	✓	✓		✓		–	On rocky and sandy bottom, from 0 to 40 m[69**].
Lazio	✓	✓		✓		200	Large extensions of dead "matte". Meadows in regression at north of the Tevere River due to sedimentation from construction works. Illegal trawling within the depth of 40 m[70, 71].
Campania*	✓	✓		✓	✓	–	Beds with different typology, extension and morphological features, due to the highly variable environmental conditions and sea bottom topography[22, 72].
Calabria*	✓	✓		✓	✓	–	Beds with different typology, extension and morphological features, due to the highly variable environmental conditions and sea bottom topography.
Apulia	✓	✓		✓		–	*Posidonia oceanica* is frequent along the southern Adriatic and the Ionian coasts. Meadows grow on old "matte" remains, in the Gulf of Taranto, while they grow on sand or rocks along the Adriatic side of Apulia. *Posidonia oceanica* is also present at the Tremiti Islands[73, 74].
Central Adriatic coasts			✓			–	*Posidonia oceanica* is absent from the Po River delta to the northern Apulian coasts. No information on other seagrasses except for *Zostera marina* (Numana Harbor, south of Ancona).
Veneto and Friuli V.G.	✓	✓	✓	✓		96	Seagrasses are not abundant along the northern coasts of the Adriatic Sea, which is influenced by the freshwater inflow and fine sediment coming from the Po River. *Posidonia oceanica* is present only in a few patches in the Gulf of Trieste and in the Venice lagoon, where *Zostera marina* is present in one of the few spots of the Mediterranean Sea[24-26, 75].
Sicily*	✓	✓		✓	✓	–	*Posidonia oceanica* is present all along the Sicilian coast. Dense prairies are present along the southeast and northwest coasts of the island on calcareous sediments. Illegal trawling within the 40-m zone has caused significant loss of *Posidonia oceanica* meadows in recent years, together with the damage caused by anchoring and recreational activities[76].
Sardinia	✓	✓				2 000	*Posidonia oceanica* extends all along the Sardinian coast, from a few meters to 30 m, and occasionally 40 m, depth. Prairies on the southern and northern coasts of the island are more fragmented (author's unpublished data).
FRANCE							
Provence Alpes Côte d'Azur	✓	✓	✓	✓		3	*Posidonia oceanica* is the most abundant species. *Cymodocea nodosa*: dense monospecific meadow from 0 to 15 m depth and mixed beds with *Zostera noltii* and *Caulerpa prolifera*. *Zostera marina*: dense meadows present in the Gulf of Fos, while small beds occur in the Bay of Toulon. *Zostera noltii* is present in small patches in the Berre lagoon[77-82].

this production is only minimally used for direct consumption by herbivores[45]. The very high biodiversity found in *Posidonia oceanica* beds is mostly due to the primary role of this seagrass as a multidimensional habitat for organisms directly participating in the system's trophic dynamics[46].

The *Posidonia oceanica* matte not only represents a net sink of carbon and other elements[5, 47] but also, when growing near the surface, can attenuate the wave action. Under such conditions, it has been estimated that the removal of 1 m^3 of matte can cause 20 m of coastal regression[48]. Moreover, the deposition of dead leaves ashore gives rise to a typical structure called "banquette" which, mixed with sand, can in some areas develop up to 2-3 m high[49]. The banquette has an important role in attenuation of waves and in the protection of beaches from erosion[50]. In addition, the banquette is hosting a reduced, but highly specialized fauna (isopods, amphipods and interstitial flatworms) that contribute to the decomposition of the seagrass material.

Sites	Po	Cn	Zm	Zn	Hs	Area (km^2)	Comments
Languedoc-Roussillon*	✓		✓	✓		26	*Posidonia oceanica* is present only in small patches between 7 and 15 m depth, with dead and living beds 1-4 km from the coast (extent not available). The region is characterized by the presence of many coastal lagoons with monospecific *Zostera marina* beds (e.g. Salse lagoon) or mixed beds with *Zostera noltii* (e.g. Thau lagoons). In open sea *Zostera noltii* occurs in small patches (e.g. Harbor of Banyuls)[83-87].
Corsica	✓	✓		✓		624	*Posidonia oceanica* meadows on sandy bottom on the east coast and on rocky bottom on the west coast. Dense *Cymodocea nodosa* meadows on sand or muddy bottom in shallow bays and in lagoons. *Zostera noltii* is only present in lagoons, often in association with *Cymodocea nodosa*[88, 89].
SPAIN							
Catalonia	✓	✓	✓	✓		40	Mostly on sandy bottom, but also on rocky bottom. From near the surface to 25 m. Conspicuous regressions have been reported, but most meadows seem to be stabilized nowadays[32].
Valencia	✓	✓		✓		270	This region has extensive meadows of *Posidonia oceanica* from near the surface to 25 m, exceptionally 30 m, generally on sandy bottom. The deep limit has suffered a significant regression due to illegal trawl fishing (Sanchez-Lisaso, unpublished data).
Murcia	✓	✓		✓		95	The main meadows are dominated by *Posidonia oceanica*, extending from the surface to 25-30 m. Conspicuous regressions have been observed near the deep limit due to illegal trawl fishing. *Cymodocea nodosa* and *Zostera noltii* appear in shallow waters[90].
Andalucia	✓	✓	✓	✓		-	*Posidonia oceanica* is abundant in the eastern part of the area, with extensive meadows on sandy and rocky substrata. The western limit of *Posidonia oceanica* is near Malaga; from this point westwards (to the Gibraltar strait), *Zostera marina* dominates[91].
Balearic Is*	✓	✓		✓		750	Extensive and dense meadows occur all around these islands, reaching up to 40 m depth, with some locally degraded sites, mainly due to tourism (moorings, sewage, etc.). One locality has been invaded by *Caulerpa taxifolia*. In shallow bays, dense *Cymodocea nodosa* meadows are frequent. *Cymodocea nodosa* is also found below 30 m[92].

Notes:
Po *Posidonia oceanica*; Cn *Cymodocea nodosa*; Zm *Zostera marina*; Zn *Zostera noltii*; Hs *Halophila stipulacea*.
✓ species present. - insufficient data.
* Interactions with *Caulerpa taxifolia* and *Caulerpa racemosa*. ** http://gis.cnuce.cnr.it/posid/html/posid.html

Case Study 3.3
SPAIN

CATALAN COAST

The main seagrass species on the Catalan coast is *Posidonia oceanica*. In the sandy coasts of the southern part of the country this species forms a large and continuous green belt of meadows only interrupted by rivers. This seagrass belt used to extend from 10 to approximately 25 m depth, although significant regressions have been detected and in many areas the deep limit is now between 17 and 20 m. Along the northern rocky coast, the meadows occur from near the surface to 20-25 m. With the publication of an edict protecting seagrasses in 1991, the autonomous government (Generalitat de Catalunya) has taken several actions for a proper management of these plants and, more specifically, of the *Posidonia oceanica* meadows. This includes a monitoring network, launched in 1998. This network consists of a total of 28 permanently marked sites (nearly one every 15 km) from which basic data on the vitality of *Posidonia oceanica* (e.g. shoot density, cover, etc.) are collected every year[32]. Underwater work is performed by volunteers (more than 400 for the whole project), trained and supervised by expert scientists. This monitoring network, after the first four years, has allowed a general diagnosis of both the status and the recent trends of seagrasses on the Catalan coast. The results obtained so far indicate that 42 percent of the studied meadows are in a normal or healthy state, while the rest show light (36 percent) or strong (22 percent) evidence of degradation. During the four-year period of the survey there have been no net changes in the *Posidonia oceanica* beds. Only in 15 percent of the sites has a negative, although slight, trend been detected from a decrease in water transparency, illegal trawl fishing and oversedimentation. Overall the Catalan seagrass beds appear to have remained remarkably stable over the period 1998-2001.

MEDES ISLANDS

The Medes Islands are a small and deserted archipelago situated 1.6 km off Spain's main coast, in the northern part of Catalonia. A large *Posidonia oceanica* meadow, extending from 5 to 15-20 m depth, and covering about 9 ha, is found in the sedimentary bottom of the southwest face of the main islands[33]. This meadow has been extensively studied[34, 35] in the course of the monitoring program of the marine reserve established there in 1990. The dataset has one of the longest series for this species, and the results show significant interannual differences. From the first observations in 1984 and 1987, density and cover decreased sharply (e.g. at the 5 m depth station, density decreased from 628 ±19 shoots/m^2 in 1984 to 481 ±14 shoots/m^2 in 1994, while cover decreased from 76 percent in 1984 to 48 percent in 1994) probably due, at least in the shallow station, to very high mooring activity on the seagrass bed. However, after the establishment of the marine reserve in 1990 anchoring was no longer allowed, and a system of low-impact mooring was deployed between 1992 and 1993. The density and cover values subsequently recovered (e.g. at the 5 m depth station, density reached 708 ±24 shoots/m^2 and cover 73 percent in 2001). Moreover, it would also seem that meteorological conditions (e.g. incoming irradiance) in these later years have been optimal, probably contributing to the observed increase.

ALFACS BAY (EBRO DELTA)

Although *Posidonia oceanica* is the most abundant seagrass species on the Catalan coast, in some specific habitats other marine angiosperms can dominate. This is the case in the two bays at each side of the Ebro Delta, the southern one of which (Alfacs Bay) has been extensively studied and mapped. In this bay, 50 km^2 in extent, dense meadows extend from very near the surface to 2-3 m and, more rarely, 4 m in depth. This narrow bathymetric range is due to high water turbidity[36]. *Cymodocea nodosa*, with the green alga *Caulerpa prolifera* interspersed in some places, dominates these meadows. Some patches of *Zostera noltii*, as well as *Ruppia cirrhosa*, exist in shallow areas. The presence of *Zostera marina* was detected in the early 1980s, but it has never been seen again. A detailed map was produced in 1986 revealing a total surface of seagrass beds of 3.5 km^2, including 1 km^2 of patchy beds in the southern zone. In the last ten years, the bay has undergone some remarkable vegetation changes[37]. *Cymodocea nodosa* has greatly expanded in the southern part of the bay, covering now about 2.5 km^2 which represents, for this southern area, an increase of approximately 15 percent a year. This increase may be associated with work performed to stabilize the sandbar, since sand instability was one of the main processes controlling seagrass abundance in this area[38]. In the northern parts of Alfacs Bay, the most remarkable change is the replacement of a mixed *Zostera noltii* and *Ruppia cirrhosa* bed, described in 1982[39], by *Cymodocea nodosa* with abundant drifting macroalgae, such as *Ulva* spp. and *Chaetomorpha linum*, by 1997.

Mediterranean seagrass meadows host many commercially important fish species. As well as nurseries they provide essential feeding grounds for cephalopods, crustaceans, shellfish and finfish[51]. Although specific fisheries legislation does not allow destructive fishing (e.g. trawling) in seagrasses, such restrictions are often violated. The only fishery allowed in the *Posidonia oceanica* meadows are small fisheries based on the use of standing nets and cages.

Posidonia oceanica detritus is used as fertilizer in agriculture in Tunisia[52] and the leaves have also been used in small proportions in chicken feed[53], with an increase in egg production and weight. More recently, different attempts to exploit the banquette were focused on production of methane[54], conversion of detritus into fungal biomass[55] and formation of dried pellet for preparation of light bricks for buildings. Further anecdotal uses of air-dried leaf detritus to protect glass objects in transport, and to fill pillows and mattresses, have been reported. *Posidonia oceanica* detritus is used in Corsica as thermal insulation material on roofs[56] and as soundproof material[57]. The ability of *Posidonia oceanica* leaves to produce active substances, which accelerate the growth of bacteria such as *Staphylococcus aureus*, has been demonstrated. This seems to be related to the presence of chicoric acid, one of the most abundant metabolites found in this seagrass[58].

THREATS

Beds of *Posidonia oceanica* have suffered a progressive regression throughout the Mediterranean due to trawling, fisheries and sand extraction and development of coastal infrastructure[59,60], such as harbors and artificial beaches, and associated enhanced turbidity and sedimentation. The damming of rivers has caused changes in sedimentation in the littoral zone, either exposing or burying seagrass habitats. One dramatic example occurred in Port-Man Bay (southeast Spain), where a seagrass meadow was buried under a large amount of highly toxic mining debris. Eutrophication, which decreases water transparency and promotes epiphyte overgrowth, is a serious regional threat. Sometimes associated with fish cages, the most common causes are sewage and industrial waste discharge.

Caulerpa taxifolia is a tropical green seaweed accidentally introduced in the Monaco area in 1984. After its introduction, *Caulerpa taxifolia* spread through France, to Italy and Spain (the Balearic Islands) by 1992, and to Croatia in 1994[61]. The area colonized has now reached more than 60 km² along the French and Italian coasts. *Caulerpa taxifolia* grows throughout the entire depth range of the Mediterranean seagrass species and, in some places, is progressively overwhelming them. Another strong competitor with seagrass beds is the introduced congeneric species *Caulerpa racemosa*,

Posidonia oceanica growing on rocks and forming matte, Porto Conte, Sardinia.

which has become widespread in the last ten years. Experimental work on the interactions between introduced *Caulerpa* species and local seagrasses show that dense meadows of both *Posidonia oceanica* and *Cymodocea nodosa* are likely to be less affected by seaweed invasion. The competitive success of *Caulerpa racemosa* with *Posidonia oceanica* meadows is a function of seagrass density and edge-meadow orientation. Competition between *Caulerpa racemosa* and *Cymodocea nodosa* seems to favor the expansion of *Zostera noltii*[62-64]. The locations of interactions between seagrasses and *Caulerpa* spp. are listed in Table 3.2.

Although Mediterranean seagrasses are now being increasingly well monitored, reliable estimates, made by direct observation, of the area of seagrass lost or degraded by the various pressures are not available for most of the western Mediterranean coastline. In fact only in the last few years have maps of distribution been produced. In the future the application of aerial cartography techniques may supply important information on seagrass status throughout the Mediterranean[65,66].

In general, for *Posidonia oceanica*, the following statement by the European Union for Coastal Conservation is probably accurate: "The situation in the Western Mediterranean is serious. Shoot density is rapidly decreasing, up to 50 percent over a few decades. Besides, increased turbidity and pollution have resulted in a squeeze of the beds; in various places living beds have withdrawn between 10 and 20 m depth. Dead beds occur abundantly, even in waters which have already been protected for 35 years. For the French mainland coast habitat loss is estimated at 10-15 percent; but taking into account the decrease of shoot density the overall decline of the resource will be between 30 and 40 percent. This is probably a good estimate for most Western Mediterranean coastlines, although the situation around the islands and in the Eastern Mediterranean is better".

In France a disappearance of *Posidonia oceanica* beds between 0 and 20 m has been observed in the last 30 years for 13 percent of the seafloor in the Alpes Maritime department, 6.6 percent in Var and 18.4 percent in Bouches du Rhône[67, 68]. In Spain, a comparison of old marine charts with present distribution data in Catalonia indicates that meadow area is now about 75 percent of that at the beginning of the 20th century.

AUTHORS

G. Procaccini, M.C. Buia and M.C. Gambi, Stazione Zoologica "A. Dohrn", Laboratorio di Ecologia del Benthos, 80077 Ischia (Naples), Italy. **Tel:** +39 (0)81 5833508. **Fax:** +39 (0)81 984201. **E-mail:** gpro@alpha.szn.it

M. Perez and J. Romero, Departament d'Ecologia, Universitat de Barcelona, Av. Diagonal 645, Barcelona, Spain.

G. Pergent and C. Pergent-Martini, Equipe Ecosystèmes Littoraux, Faculty of Science, BP 52, 20250 Corte, France.

REFERENCES

1. Pergent G, Pergent-Martini C, Buia MC, Gambi MC [eds] [2000]. Proceedings Fourth International Seagrass Biology Workshop. *Biologia Marina Mediterranea* 7(2): 1-443.
2. Molinier R, Picard J [1952]. Recherches sur les herbiers de phanérogames marines du littoral méditerranéen français. *Annales Institut Océanographique* 27(3): 157-234.
3. Pergent-Martini C, Pergent G [1996]. Spatio-temporal dynamics of *Posidonia oceanica* beds near a sewage outfall (Mediterranean-France). In: Kuo J, Phillips RC, Walker DI, Kirkman H [eds] *Seagrass Biology*. Sciences UWA Press, Nedlands, Western Australia. pp 299-306.
4. Buia MC, Mazzella L [2000]. Diversity in seagrass ecosystems: Biological descriptors at different temporal scales. *Biologia Marina Mediterranea* 7(2): 203-206.
5. Mateo MA, Romero J, Pérez M, Littler M, Littler D [1997]. Dynamics of millenary organic deposit resulting from the growth of the Mediterranean seagrass *Posidonia oceanica*. *Estuarine Coastal and Shelf Science* 44: 103-110.
6. Procaccini G, Orsini L, Ruggiero MV, Scardi M [2001]. Spatial patterns of genetic diversity in *Posidonia oceanica*, an endemic Mediterranean seagrass. *Molecular Ecology* 10: 1413-1421.
7. Procaccini G, Ruggiero MV, Orsini L [in press]. Genetic structure and distribution of microsatellite diversity in *Posidonia oceanica* over the whole Mediterranean basin. *Bulletin of Marine Science*.
8. Buia MC, Marzocchi M [1995]. Dinamica dei sistemi a *Cymodocea nodosa, Zostera marina* e *Zostera noltii* nel Mediterraneo. *Giornale Botanico Italiano* 129(1): 319-336.
9. Buia MC, Gambi MC, Zupo V [2000]. Structure and functioning of Mediterranean seagrass ecosystems: An overview. *Biologia Marina Mediterranea* 7(2): 167-190.
10. Procaccini G, Mazzella L [1996]. Genetic variability and reproduction in two Mediterranean seagrasses. In: Kuo J, Phillips RC, Walker DI, Kirkman H [eds] *Seagrass Biology*. Sciences UWA Press, Nedlands, Western Australia. pp 85-92.
11. Rigollet V, Laugier T, Casabianca ML De, Sfriso A, Marcomini A [1998]. Seasonal biomass and nutrient dynamics of *Zostera marina* L. in two Mediterranean lagoons: Thau (France) and Venice (Italy). *Botanica Marina* 41(2): 167-179.
12. Procaccini G, Acunto S, Famà P, Maltagliati F [1999]. Structural, morphological and genetic variability in *Halophila stipulacea* (Hydrocharitaceae) populations in the Western Mediterranean. *Marine Biology* 135: 181-189.
13. Di Martino V, Marino G, Blundo MC [2000]. Qualitative minimal area of a macroalgal community associated with *Halophila stipulacea* from South Eastern Sicilian coasts (Ionian Sea). *Biologia Marina Mediterranea* 7(1): 677-679.
14. Gambi MC, Dappiano M, Iannotta A, Esposito A, Zupo V, Buia MC [in press]. Aspetti storici e attuali della biodiversità del benthos mediterraneo: un esempio in alcune aree del Golfo di Napoli. *Biologia Marina Mediterranea* 9(1).
15. Ballesteros E, García A, Lobo A, Romero J [1984]. L'alguer de *Posidonia oceanica* de les illes Medes. In: Ros J, Olivella I, Gili JM [eds] *Els Sistemes naturals de les Illes Medes*. Institut d'Estudis Catalans.
16. Van der Ben D [1971]. Les épiphytes des feuilles de *Posidonia oceanica* Delile sur les côtes françaises de la Méditerranée. *Mémoires, Institut Royal des Sciences Naturelles de Belgique* 168: 1-101.
17. Boero F [1981]. Bathymetric distribution of the epifauna of a *Posidonia* meadow of the Island of Ischia (Naples): Hydroids. *Rapports et Procès-Verbaux des Réunions Commission Internationale pour l'Expl Scientifique de la Mer Méditerranée Monaco* 27(2): 197-198.
18. Piraino S, Morri C [1990]. Zonation and ecology of epiphytic hydroids in a Mediterranean coastal lagoon: The "Stagnone" of Marsala (north-west Sicily). PSZNI *Marine Ecology* 11(1): 43-60.
19. Boudouresque CF, Avon M, Gravez V [1991]. *Les espèces à protéger en Méditerranée*. GIS Posidonie, Marseille. 449 pp.
20. Bianchi CN, Peirano A [1995]. Atlante delle fanerogame marine della Liguria: *Posidonia oceanica* e *Cymodocea nodosa*. ENEA, Centro Ricerche Ambiente Marino, La Spezia. pp 1-37.
21. Sandulli R, Bianchi CN, Cocito S, Morgigni M, Peirano A, Sgorbini S, Silvestri C, Morri C [1994]. Status of some *Posidonia oceanica* meadows on the Ligurian coasts influenced by the "Haven" oil spill. In: Albertelli G, Cattaneo-Vietti R, Piccazzo M [eds] *Proceedings of the Congress of the Italian Society of Limnology and Oceanology, Alassio 4-6 Nov. 1992*. pp 277-286.
22. Colantoni P, Gallignani P, Fresi E, Cinelli F [1982]. Patterns of *Posidonia oceanica* (L.) Delile beds around the island of Ischia (Gulf of Naples) and adjacent waters. PSZNI *Marine Ecology* 3(1): 53-74.
23. Gambi MC [2001]. Studio propedeutico per l'istituzione dell'area Marina Protetta "Regno di Nettuno" (Isole di Ischia, Procida e Vivara). Data Report, Ministero dell'Ambiente, Rome.
24. Caressa S, Ceschia C, Orel G, Treleani R [1995]. Popolamenti attuali e pregressi nel Golfo di Trieste da Punta Salvatore a Punta Tagliamento (Alto Adriatico). In: Cinelli F, Fresi E, Lorenzi C, Mucedola A [eds] *Posidonia oceanica. A contribution to the preservation of a major Mediterranean marine ecosystem. Rivista Marittima* 12(suppl): 160-187.
25. Rismondo A, Guidetti P, Curiel D [1997]. Presenza delle fanerogame marine nel Golfo di Venezia: un aggiornamento. *Bolletino del Museo Civico de Storia Naturale di Venezia* 47: 317-327.
26. Caniglia G, Borella S, Curiel D, Nascibeni P, Paloschi AF, Rismondo A, Scarton F, Tagliapietra D, Zanella L [1992]. Distribuzione delle fanerogame marine (*Zostera marina* L., *Zostera noltii* Hornem., *Cymodocea nodosa* (Ucria) Aschers. in Laguna di Venezia. *Lavori Società Veneziana di Scienze Naturali* 17: 137-150.
27. Scarton F, Curiel D, Rismondo A [1995]. Aspetti della dinamica temporale di praterie a fanerogame marine in

28. Rismondo A, Curiel D, Solazzi A, Marzocchi M, Chiozzotto E, Scattolin M [1995]. Sperimentazione di trapianto di fanerogame marine in laguna di Venezia: 1992-1994. In: Ravera, O, Anelli, A (eds) *Proceedings of the Società Italiana di Ecologia* (SITE) 16: 699-701.
29. Tagliapietra D [1995]. Uso di vasi di torba pressata nel trapianto di *Zostera noltii*. *Lavori Società Veneziana di Scienze Naturali* 20: 165-166.
30. Pergent-Martini C, Pasqualini V [2000]. Seagrass population dynamics before and after the setting up of a waste-water treatment plant. *Biologia Marina Mediterranea* 7(2): 405-409.
31. Meinesz A, Genot I, Hesse B [1990]. Données quantitatives sur les biocénoses littorales de la Corse et impact de l'aménagement du littoral. GIS Posidonie/DRAE Corse. pp 1-22.
32. Renom P, Romero J [2001]. Xarxa de vigilància dels herbassars de fanerògames marines. Informe Técnico. Generalitat de Catalonia.
33. Manzanera M, Romero J [1998]. Cartografia de la praderia de *Posidonia oceanica* de les Illes Medes. Informe Técnico. Generalitat de Catalonia.
34. Romero J [1985]. Estudio ecologico de las fanerogamas marinas de la costa Catalana: produccion primaria de *Posidonia oceanica* (L.) Delile en las islas Medes. Doctoral thesis, Departamento de Ecologia, Facultad de Biologia, Universidad de Barcelona. pp 1-261.
35. Alcoverro T, Duarte CM, Romero J [1995]. Annual growth dynamics of *Posidonia oceanica*: Contribution of large-scale versus local factors to seasonality. *Marine Ecology Progress Series* 120: 203-210.
36. Pérez M, Romero J [1992]. Photosynthetic response to light and temperature of the seagrass *Cymodocea nodosa* and the prediction of its seasonality. *Aquatic Botany* 43: 51-62.
37. Pérez M, de Pedro X, Renom P, Camp J, Romero J [in press]. Changes in Benthic Vegetation in a Mediterranean Estuarine Bay.
38. Marba N, Duarte CM [1995]. Coupling of seagrass (*Cymodocea nodosa*) patch dynamics to subaqueous dune migration. *Journal of Ecology* 83(3): 381-389.
39. Pérez M, Camp J [1986]. Distribución espacial y biomasa de las fanerógamas marinas de las bahías del delta del Ebro. *Investigaciones Pesquera* 50: 519-530.
40. Mazzella L, Zupo V [1995]. Reti trofiche e flussi di energia nei sistemi a fanerogame marine. *Giornale Botanico Italiano* 129(1): 337-349.
41. Pirc H [1983]. Below ground biomass of *Posidonia oceanica* (L.) Delile and its importance to the growth dynamics. *Proceed Int Symp Aquat Macrophytes, Nijmegen*. pp 177-181.
42. Mateo MA, Romero J [1997]. Detritus dynamics in the seagrass *Posidonia oceanica*: Elements for an ecosystem carbon budget. *Marine Ecology Progress Series* 151: 43-53.
43. Wittmann K [1984]. Temporal and morphological variations of growth in a natural stand of *Posidonia oceanica* (L.) Delile. PSZNI *Marine Ecology* 5: 301-316.
44. Pergent-Martini C, Rico-Raimondino V, Pergent G [1984]. Primary production of *Posidonia oceanica* in the Mediterranean basin. *Marine Biology* 120: 9-15.
45. Cebrián J, Duarte CM, Marbà N, Enríquez S, Gallegos M, Olesen B [1996]. Herbivory on *Posidonia oceanica*: Magnitude and variability in the Spanish Mediterranean. *Marine Ecology Progress Series* 130: 147-155.
46. Mazzella L, Buia MC, Gambi MC, Lorenti M, Russo GF, Scipione MB, Zupo V [1992]. Plant-animal trophic relationships in the *Posidonia oceanica* ecosystem of the Mediterranean Sea: A review. In: John DM (ed) *Plant-Animal Interactions in the Marine Benthos*. Special Volume No. 46. Systematics Association, Clarendon Press, Oxford. pp 165-187.
47. Gacia E, Duarte CM, Middelburg JJ [2002]. Carbon and nutrient deposition in a Mediterranean seagrass (*Posidonia oceanica*). *Limnology and Oceanography* 47(1): 23-32.
48. Jeudy de Grissac A [1984]. Effets des herbiers à *Posidonia oceanica* sur la dynamique marine et la sédimentologie littorale. In: Boudouresque CF, Jeudy de Grissac A, Olivier J (eds) *International Workshop* Posidonia oceanica *Beds*. GIS Posidonie Publication 1: 437-443.
49. Mateo MA, Sánchez-Lizaso JL, Romero J [in press]. *Posidonia oceanica* "banquettes": A preliminary assessment of the relevance for meadow carbon and nutrients budget. *Estuarine Coastal and Shelf Science*.
50. Boudouresque CF, Meinesz A [1982]. Découverte de l'herbier de Posidonie. *Cahier Parc National de Port-Cros* 4: 1-79.
51. Harmelin-Vivien M, Francour P [1992]. Trawling or visual censuses? Methodological bias in the assessment of fish population in segrass beds. PSZNI *Marine Ecology* 13(1): 41-52.
52. Saïdane A, De Waele N, Van De Velde [1979]. Contribution à l'étude du compostage de plantes marines en vue de la préparation d'un amendement organique et d'un substrat horticole. *Bulletin Institut international scientifique et Technique d'Océanographie et de Pêche de Salammbô* 6(1-4): 133-150.
53. Baldissera-Nordio C, Gallarati-Scotti GC, Rigoni M [1967]. Valore nutritivo e possibilità di utilizzazione zootecnica di *Posidonia oceanica*. *Atti Convegno nazionale sulla Attività subacquee ital* 1: 21-28.
54. Fresi, E. Personal communication.
55. Cuomo V, Vanzanella F, Fresi E, Cinelli F, Mazzella L [1985]. Fungal flora of *Posidonia oceanica* and its ecological significance. *Transactions of the British Mycological Society* 84(1): 35-40.
56. Boudouresque CF. Personal communication.
57. Molinier and Pellegrini. Personal communication.
58. Cariello L, Zanetti L [1979]. Distribution of chicoric acid during leaf development of *Posidonia oceanica*. *Botanica Marina* 22: 359-360.
59. Astier JM [1984]. Impact des aménagements littoraux de la rade de Toulon, liés aux techniques d'endigage, sur les herbiers à *Posidonia oceanica*. In: Boudouresque CF, Jeudy de Grissac A, Olivier J (eds) *International Workshop* Posidonia oceanica *Beds*. GIS Posidonie Publication 1: 255-259.
60. Martin MA, Sánchez-Lizaso JL, Ramos-Esplá AA [1997]. Cuantificación del impacto de las artes de arrastre sobre la pradera de *Posidonia oceanica* (L.) Delile. *Publicaciones Especiales Instituto Espanol de Oceanografia* 23: 243-253.
61. Gravez V, Ruitton S, Boudouresque CF, Le Direac'h L, Meinesz A, Scabbia G, Verlaque M (eds) [2001] *Fourth International Workshop on* Caulerpa taxifolia. GIS Posidonie Publication, Marseille. pp 1-406.
62. Ceccherelli G, Cinelli F [1999a]. Effect of *Posidonia oceanica* canopy on *Caulerpa taxifolia* size in a north-western Mediterranean bay. *Journal of Experimental Marine Biology and Ecology* 240: 19-36.
63. Ceccherelli G, Cinelli F [1999b]. A pilot study of nutrient enriched sediment in a *Cymodocea nodosa* bed invaded by the introduced alga *Caulerpa taxifolia*. *Botanica Marina* 42: 409-417.
64. Ceccherelli G, Piazzi L, Cinelli F [2000]. Response of the non-indigenous *Caulerpa racemosa* (Forsskal) J. Agardh to the native seagrass *Posidonia oceanica* (L.) Delile: Effect of density of shoots and orientation of edges of meadows. *Journal of Experimental Marine Biology and Ecology* 243: 227-240.
65. Cancemi G, Pasqualini V, Piergallini G, Baroli M, De Falco G, Pergent Martini C [1997]. Indagine cartografica sulla prateria a *Posidonia oceanica* (L.) Delile di Capo S. Marco (Golfo di Oristano), mediante elaborazione di immagini fotoaeree. *Biologia Marina Mediterranea* 4(1): 472-474.
66. Keegan BF (ed) [1992]. Space and time series data analysis in coastal benthic ecology. An analytical exercise organized within the framework of the COST 647 Project (EC) on Coastal Benthic Ecology. European Communities, Luxembourg. pp 1-581.

67. Meinesz A, Astier JM, Lefevre JR [1981]. Impact de l'amènagement du domaine maritime sur l'étage infralittoral du Var, France (Méditerranée occidentale). *Annales Institut Océanographique* NS 57(2): 65-77.
68. Meinesz A, Astier JM, Bodoy A, Lefevre JR [1982] Inventaire des restructurations des rivages et leurs impacts sur la vie sous-marine littorale du département des Bouches-du-Rhône. Contrat N° 81.01.075 Mission interministérielle et UDUN 13 Fr. pp 1-55.
69. Cinelli F, Fresi E, Lorenzi C, Mucedola A (eds) [1995]. *Posidonia oceanica*. A contribution to the preservation of a major Mediterranean ecosystem. *Rivista Marittima*, suppl. 12: 1-271.
70. AAVV [1996]. *Il Mare del Lazio*. University of Roma "La Sapienza" and Regione Lazio. pp 1-328.
71. Diviacco G, Spada E, Lamberti CV [2001]. Le fanerogame marine del Lazio. Descrizione e cartografia delle praterie di *Posidonia oceanica* e dei prati di *Cymodocea nodosa*. ICRAM, Rome. pp 1-113.
72. Buia MC, Mazzella L, Russo GF, Scipione MB [1985]. Observations on the distribution of *Cymodocea nodosa* (Ucria) Aschers. prairies around the island of Ischia (Gulf of Naples). *Rapports et Proces-Verbaux des Réunions Commission Internationale pour l'Expl. Scientifique de la Mer Méditerranée* 29(6): 205-208.
73. Viel M, Zurlini G (eds) [1986]. Indagine ambientale del sistema marino costiero della regione Puglia. ENEA, Santa Teresa, La Spezia. pp 1-277.
74. Guidetti P, Buia MC, Mazzella L [2000]. The use of lepidochronology as a tool of analysis of dynamic features in the seagrass *Posidonia oceanica* of the Adriatic sea. *Botanica Marina* 43: 1-9.
75. Turk R [2000]. Main phenological characteristics of *Posidonia oceanica* (L.) Delile in the Gulf of Koper (Gulf of Trieste), North Adriatic. *Biologia Marina Mediterranea* 7(2): 139-142.
76. Calvo S, Fradà Orestano C, Tomasello A [1995]. Distribution, structure and phenology of *Posidonia oceanica* meadows along Sicilian coasts. *Giornale Botanico Italiano* 129(1): 351-356.
77. Pergent G, Pergent C [1988]. Localisation et état de l'herbier de Posidonies sur le littoral PACA: Bouches-du-Rhône. DRAE-PACA et GIS Posidonie Publication, Marseille. pp 1-53.
78. Sinnassamy JM, Pergent-Martini C [1990]. Localisation et état de l'herbier de Posidonies sur le littoral PACA: Var. DRAE-PACA et GIS Posidonie Publication. pp 1-75.
79. Meinesz A, Laurent R [1980]. Cartes de la limite inférieure de l'herbier de *Posidonia oceanica* dans les Alpes-Maritimes (France). Campagne Poséidon 1976. *Annales Institut Océanographique* 56(1): 45-54.
80. Meinesz A, Boudouresque CF, Jeudy De Grissac A, Lamare JP, Lefevre JR, Manche A [1985]. Aménagement et préservation du milieu marin littoral en région Provence-Alpes-Côte d'Azur: bilan et perspectives. In: Ceccaldi HJ, Champalbert G (eds) Les amènagements côtiers et la gestion du littoral. *Colloque pluridisciplinaire Franco-Japonais Océanographie* 1: 133-142.
81. Boudouresque CF, Meinesz A, Ledoyer M, Vitiello P [1994]. Les herbiers à phanérogames marines. In: Bellan-Santini D, Lacaze JC, Poizat C (eds) *Les biocénoses marines et littorales de Méditerranée, synthèse, menaces et perspectives*. Museum national d'Histoire naturelle, Paris. pp 98-118.
82. Pergent-Martini C, Semroud R, Rico-Raimondino V, Pergent G (1997a). Localisation et évolution des peuplements de phanérogames aquatiques de l'étang de Berre (Bouches du Rhône – France). In: 39ème Congrès Nationaux, Association Française de Limnologie édit, Univ Corse & Office de l'Environnement publ, Corte. pp 169-179.
83. Bodoy A, Nicolas F, Vaulot D [1982]. Inventaire préliminaire des formations des posidonies dans le Golfe d'Aigues-Mortes (Méditerranée Nord-Occidentale). *Téthys* 10(4): 382-383.
84. Avril A, Dutrieux E, Nicolas F, Vaxelaire A [1984]. Etude des fonds marins des Aresquiers (Languedoc): état des herbiers de Posidonies. In: Boudouresque CF, Jeudy de Grissac A, Olivier J (eds) *International Workshop* Posidonia oceanica *Beds*. GIS Posidonie Publication 1: 173-177.
85. Pergent G, Boudouresque CF, Thelin I, Marchadour M, Pergent-Martini C [1991]. Cartography of benthic vegetation and sea-bottom types in the harbour of Banyuls-sur-Mer (Pyrénées-orientales, France). *Vie Milieu* 41(2/3): 165-168.
86. Ballesta L, Pergent G, Pergent-Martini C, Pasqualini V [2000]. Distribution and dynamics of *Posidonia oceanica* beds along the Albères coastline. *Comptes Rendus de l'Académie des Sciences Série III Sciences de la Vie* 323: 407-414.
87. Plus M [2001]. Etude et modélisation des populations de macrophytes dans la lagune de Thau (Hérault, France). Thèse Océanologie biologique & Environnement marin, Université de Paris VI. pp 1-369.
88. Pergent-Martini C, Fernandez C, Agostini S, Pergent G [1997b]. Les étangs de Corse, Bibliographie-Synthèse 1997. Contrat Eq. E.L. – Université de Corse/Office de l'Environnement de la Corse & IFREMER. pp 1-269.
89. Pergent-Martini C, Fernandez C, Pasqualini V, Pergent G, Segui C, Tomaszewski JE [2000]. Les étangs littoraux de Corse: Cartographie des peuplements et types de fonds. |Contrat Eq. E.L. – Université de Corse & IFREMER, N° 99 3 514004. pp 1-33.
90. Calvin JC (ed) [1999]. El litoral sumergido de la Región de Murcia: cartografía bionómica y valores ambientales. Consejería de Medio Ambiente, Agricultura y Agua. Región de Murcia.
91. Barrajon A, Cuesta S, Gonzalez MI, Larrad A, Lopez E, Moreno D, Templado J, Luque AA [1996]. Cartografía de las praderas de fanerógamas marinas del litoral de Almería (SE de España). Libro de Resumenes: IX Symposium Iberico de Estudios del Bentos Marino.
92. Mas J, Franco I, Barcala E [1993]. Primera aproximación a la cartografía de las praderas de *Posidonia oceanica* de las costas mediterráneas españolas. Factores de alteración y regresión. Legislación. *Publicaciones Especiales Instituto Espanol de Oceanografia* 11: 111-119.

4 The seagrasses of
THE BLACK, AZOV, CASPIAN AND ARAL SEAS

N.A. Milchakova

The principal characteristic of the temperate Euro-Asian seas is their total (Aral and Caspian) or near-total (Azov and Black) isolation from open ocean systems. These temperate seas have many common environmental features especially with regard to variable salinity and levels of pollutants. Geographical proximity and isolation determine a number of special features inherent in these seas: they have a distinctly continental climate, no tides (but considerable long-term fluctuations of sea level, and both coastal upwelling and downwelling, have occurred), minimal or zero water exchange with other seas and seawater of unusual chemical composition[1-3]. All four seas contain the estuaries of major rivers and are consequently dependent on the influx of freshwater. In nearly every case these rivers have been severely disrupted from their natural state by the construction of dams, and upriver pollution and water extraction, as well as changes in rainfall across their watersheds. Therefore long-term changes in seawater chemical composition and concentration have occurred near the river mouths. Distinguishing hydrological characteristics of the coastal shelves, the main potential habitat for seagrasses, are their shallow depth (about 20 m) and large area, marked seasonal and interannual fluctuations in productivity, winter ice cover, predominant wind-induced seawater circulation, and fast water exchange owing to the small capacity.

Highest productivity is found in the brackish areas of the north Caspian Sea, the northwestern Black Sea, and the Sea of Azov and Kerch Strait[2, 4, 5]. This high productivity is due to massive freshwater influx from the Rivers Volga, Danube and Don, and the correspondingly high nutrient input, fast turnover of these nutrients, intense summer warming, high dissolved oxygen content of the brackish water and the longer summer daily growth period at higher latitudes. However, at the same time anthropogenic pollution substantially reduces the biological diversity and productivity of the water bodies, and has been especially damaging to the traditional fisheries of these four seas. In the Caspian and Azov Seas, which are of the greatest significance for commercial fishing in the region, the usable fish stock has been reduced by more than half. The largest sturgeon stock in the Caspian Sea has dramatically decreased. In the last two years, the commercial stock (approximately 250 000 metric tons) of kilka, *Clupeonella cultriventris*, has been reduced by 40 percent. Twenty thousand Caspian seals and about 10 million birds have died. High concentrations of heavy metals and oil products were detected in the dead animals[6].

Seagrasses play a key role in the coastal ecosystems of the seas and occupy vast areas in the shallow bays and gulfs of the Black, Azov, Caspian and Aral Seas. The diversity of algae, invertebrates and fishes in seagrass communities is astonishing. The condition and distribution of seagrasses are strongly influenced by freshwater influx, industrial, municipal and agricultural sewage, shipping, sea-bottom dredging, dumping, and oil and gas extraction on the shelf. Fluctuations in the sea level of the Caspian and Aral Seas also influence the coastal ecosystems.

BLACK SEA
There are four seagrasses, two seagrass associates and about 300 macroalgae in the flora of the Black Sea[7]. Communities of *Zostera marina*, *Zostera noltii*, *Potamogeton pectinatus* and *Ruppia cirrhosa* occupy vast areas in shallow bays and gulfs, especially in the northwestern part of the sea[7-10]. The distribution and ecology of Black Sea seagrass was first reported at the beginning of the 20th century[11, 12], with further details of seagrass biology and community structure in the Black Sea, and environmental impacts on the seagrasses, being obtained in subsequent decades[8, 10, 13-18]. During 1934-37 communities of *Zostera marina* were seriously damaged due to a wasting disease

Early symptoms of wasting disease in *Zostera marina* – an epidemic in the 1930s seriously damaged communities of this seagrass in the Black Sea.

epidemic similar to that registered along the North Atlantic population of this species[19]. Fortunately, *Zostera noltii*, also widely found in the shallow bays and coves, was not affected[19, 20].

During the 1970s and 1980s, the stock of *Zostera* spp. growing in the four largest bays of the Black Sea, the Tendrovsky, Dzharilgatsky and Yagorlitsky Bays and the Karkinitsky Gulf, was estimated at 633 000 metric tons[10, 21]. After 1982 the coasts of these bays accumulated considerable cast-off of *Zostera* spp., estimated at 35 000 metric tons dry weight[21]. However, according to previously calculated data on the annual leaf production of *Zostera* spp., the actual annual estimate was about 4 000 metric tons dry weight.

Many researchers have noted that *Zostera* spp. usually grow in the coastal salt lakes and sometimes in the deltas of the rivers[22-24]. There is no information on the distribution of *Zostera* spp. along the shores of Georgia and Bulgaria. *Zostera noltii* is found in Sinop Bay, on the Anatolian coast of Turkey.

Seagrasses in the Black Sea grow in single species and mixed communities, located on silt and sandy sediments, often with a portion of shell grit. The depths at which they are found range from 0.5-17 m, across a salinity gradient of 0.3-19.5 psu. Some 115 algal species have been identified growing in *Zostera marina* communities, and 62 in communities of *Zostera noltii*[10]. The majority of the algae are epiphytes encrusting the leaves and occasionally the rhizomes and roots. *Cladophora*, *Enteromorpha*, *Ceramium*, *Polysiphonia* and *Kylinia* spp. predominate. There are more than 70 species of invertebrates, 34 fishes and 19 fish larvae in seagrass meadows, among which shrimps, scad and perch predominate[25].

The average biomass of *Zostera marina* in Karkinitsky Gulf is 1 109 g wet weight/m^2 with a density of 105 shoots/m^2 [26]. Lower biomass and higher density occur in the Donuzlav Salt Lake (836 g wet weight/m^2 and 218 shoots/m^2)[22]. Near the mouth of the Chernaya River, where the salinity is less than in other parts of the Black Sea (11-17 psu), seagrass biomass reaches 2 986 g wet weight/m^2 and density 1 136 shoots/m^2 [27], although the maximum biomass values recorded for the Black Sea occur in Kamysh Burun Bay in the southern part of the Kerch Strait (5 056 g wet weight/m^2)[10]. In the Kerch Strait, which links the Black and Azov Seas, *Zostera marina* biomass ranges from 2 008 g wet weight/m^2 at Cape Fonar[28] to 3 958 g wet weight/m^2 in Kerch Bay[29], and plant density from 916 to 600 shoots/m^2 respectively. The longest shoots of *Zostera marina*, at more than 2 m, have also been found in the Kerch Strait[29], though in other areas of the Black Sea their length more typically varies between 25 and 100 cm[15,16]. For *Zostera noltii*, biomass estimates vary from 0.5 to 2 kg wet weight/m^2; the highest values were registered at depths down to 1 m in summer[26, 28, 30, 31].

The total sea-bottom area occupied by *Zostera* spp. in the bays of the northwestern Black Sea is more than 950 km^2, or 40 percent of the total area of all the bays[10]. *Zostera* spp. communities cover a similar portion, about 50 percent, of the sea bottom in shallow bays of Sevastopol region, the Kerch Bay and Kerch Strait.

Though most investigators have acknowledged that the recent eutrophication of coastal ecosystems of the Black Sea has led to degradation of key benthic and plankton communities[5, 9], the dynamics of the long-term changes observed in *Zostera* spp. communities have revealed that there are many localities, including dumping grounds, where recovery of the seagrass beds is occurring. Estimates of *Zostera marina* biomass have increased two- to threefold in Laspi, Kazachaya, Kamyshovaya, Streletskaya, Severnaya, Holland and Kerch Bays, and in the Kerch Strait, over a period from 1981-83 to 1994-99[15, 16, 27, 29]. The greatest increase in biomass, from 1 185 to 3 958 g wet weight/m^2, has occurred in Kerch Bay, and the greatest increase in shoot density, from 252 to 936 shoots/m^2, in Streletskaya Bay. This increase in the yield of *Zostera* spp. biomass is probably due to several factors, the most significant of which is likely to be reduced industrial pollution and the natural resilience of *Zostera marina* and *Zostera noltii* to environmental changes. According to unpublished data obtained by the Southern Research Institute for Fishery and Oceanography in Kerch, the amount of *Zostera* spp. in meadows in the Karkinitsky Gulf has also increased, despite extensive annual excavation of sand.

My own observations indicate that self-restoration and enlargement of *Zostera* beds is occurring in Sevastopol, Kerch and Yalta Bays, all of

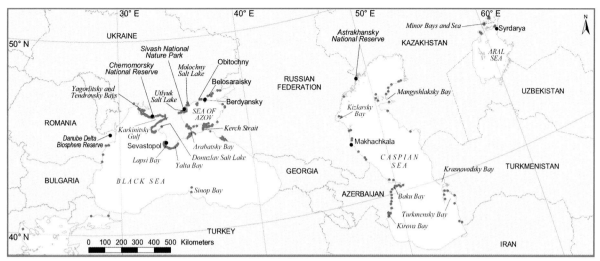

Map 4.1
The Black, Azov, Caspian and Aral Seas

which have been subject to considerable disturbance from recreational activities. Indeed not only seagrass but all Black Sea benthic macrophytes are stabilizing and recovering from recreational pressures over wide areas. In contrast seagrass and algae communities are most degraded in areas with heavy sedimentation loads. This decline has occurred particularly along the deepest boundary of macrophyte growth.

Black Sea *Zostera* spp. are traditionally used in local agriculture as a forage additive and for winter insulation for barns for livestock[32]. It has been proved experimentally that the daily yield of milk of cows whose fodder was mixed with *Zostera marina* increased by 15-20 percent. Weight increases of 20-30 percent in sheep fed *Zostera*, and of 10-15 percent in pigs[33], have also been observed. Seagrass additives appear to increase milk quality and fat in dairy cows and provide better quality and quantity of sheep wool. *Zostera* spp. are a valuable source of pectins, aquatic solutions which produce firm gels. Being rich in hemicellulose and pectin substances, seagrasses are also used as a gluing component in mixed fodder granulation and packaging.

In the Black Sea, seagrasses have been placed under protection in ten nature reserves under the national control of Ukraine and Romania[5, 34, 35]. The largest of them are the Danube Delta Biosphere Reserve and the Chernomorsky National Reserve.

SEA OF AZOV

There are four species of seagrass, three seagrass associates and 64 macroalgae in the flora of the sea[36]. The *Zostera* spp. have a Mediterranean origin and are believed to have appeared in the Sea of Azov in the Paleocene[37].

The meadows of *Zostera noltii* are the most extensive and dense compared with other seagrasses. This species grows on silt-sandy sediments with shell grit from 0.2 to 8 m[14, 36] and across a salinity gradient of 2-26 psu. *Zostera marina* inhabits the same depths but covers a considerably smaller area. *Zostera* spp. grow in single-species beds and also in mixed communities with other seagrasses, mostly *Potamogeton pectinatus* and *Ruppia cirrhosa*, and with algae such as *Ceramium*, *Polysiphonia*, *Cladophora* and *Enteromorpha* spp. *Zostera noltii* and *Zostera marina* are found almost everywhere along the shoreline of the sea[38], and also in the coastal salt lakes, river mouths and floodplains. This is due to their tolerance to salinity fluctuations. *Zostera noltii* in the Sea of Azov has shoots 15-70 cm long, while those of *Zostera marina* measure 20-90 cm.

The vast meadows of *Zostera noltii* and *Zostera marina* predominate in the northern part of the sea, close to sandy spits, and in the coastal salt lakes. Communities of *Zostera noltii* are also widely prevalent in the eastern Sea of Azov, while *Zostera marina* is found here only in patches[36, 39]. In the western part of the sea, *Zostera* spp. are rare, being usually found as solitary sparse seagrass beds. Along the southern coast *Zostera* spp. are dispersed. *Zostera* spp., washed ashore after the leaf fall, abundantly cover the coast. The annual commercial after-storm harvest amounts to about 1 200 metric tons dry weight.

Recent field measurements have recorded the biomass of both *Zostera* species around the Sea of Azov. *Zostera noltii* biomass in the bays (wet weights: Arabatsky 1 197 g/m², Kasantyp 284 g/m², Tamansky 374 g/m², Belosaraisky 860 g/m², Obitochny 1 180 g/m² and Berdyansky 400 g/m²) is generally comparable with the salt lakes (wet weights: Sivash 1 157-1 400 g/m², Molochny 378 g/m² and Utlyuk 667 g/m²) but higher than the seaward coasts of the large

sand spits which are such a feature of the Sea of Azov (wet weights: Belosaraiskaya 28 g/m², Fedotov 30 g/m², Obitochnaya 45 g/m² and Berdyanskaya 30 g/m²)[18, 28, 30, 31, 36, 40, 41]. The biomass of *Zostera marina* was also measured at three of these locations: in Sivash (2 000 g wet weight/m²) and Molochny (592 g wet weight/m²) salt lakes and Tamansky Bay (219 g wet weight/m² at 1 m depth but more than ten times this at 3.5 m).

Analysis of the long-term dynamics of the structure of *Zostera* communities indicates that, despite changes in the environment and increased eutrophication, the recent 60 years have not been marked with radical changes. For example, in the late 1930s, the biomass of *Zostera marina* in Utlyuk Salt Lake was estimated to range from 213 to 2 242 g wet weight/m²[20] and in the early 1970s from 333 to 1 024 g wet weight/m²[14]. Furthermore, over the past 30 years, the biomass of *Zostera noltii* in Utlyuk Salt Lake has increased from 260 to 667 g wet weight/m²[14].

The local population traditionally uses *Zostera* spp. cast-off to insulate housing, for livestock during winter and as an efficient means of deterring rodents in barns. The high silica content of this material reduces its flammability and therefore its risk as a fire hazard. It has been experimentally proved that dried *Zostera marina* mixed with urea is a valuable forage additive for livestock.

Seagrasses of the Sea of Azov have been placed under the protection of many international conventions and the state laws of Ukraine[35]. They are the object of protection in seven nature reserves, the largest of which are Sivash National Nature Park and the coastal Molochny Salt Lake.

CASPIAN SEA

Three species of seagrass, two seagrass associates and 65 macroalgae make up the submerged flora of the Caspian Sea[36]. The earliest work on the composition and distribution of Caspian Sea seagrasses was produced in 1784[42] but it was not until the 1930s that the most comprehensive reviews on the topic were published[42, 43]. This was the first time that the hypothesis about *Zostera noltii*'s penetration into the Caspian Sea from the Black and Azov Seas was advanced. Presumably *Zostera noltii* was introduced from the Mediterranean to the Caspian Sea in the Paleocene, 36-65 million years ago. At that time the Black Sea, the Sea of Azov and the Caspian Sea were connected by the Kumo-Manych Strait.

Zostera noltii communities were then widely distributed throughout the Caspian Sea[4, 36, 44, 45], typically at depths of 2.5-4.5 m along eastern shores, though occasionally as shallow as 0.5 m and as deep as 18 m, across a narrow range of salinity, 12-13 psu. Single species and mixed communities of *Zostera noltii* were found on sand sediments with shell grit but never on a silt bottom. Highest productivity takes place in mixed communities of *Zostera noltii* and Charophyceae, *Ruppia* and *Potamogeton* spp. Species of *Chara*, *Ceramium*, *Polysiphonia*, *Laurencia*, *Enteromorpha* and *Cladophora* are common algae in seagrass beds.

The distribution of seagrasses and macrophytes in the Caspian Sea changed markedly in the period 1934-61 to 1967-81. In the 1930s there were extensive *Zostera noltii* beds along western coasts, principally Baku and Kirova Bays in present-day Azerbaijan, with records indicating the presence of this species at Derbent, Izerbesh and Makhachkala, and in Astrakhansky and Kizlarsky Bays in modern Russia. Along the eastern coast Kaidak, Mangyshlaksky, Kazakhsy and Turkmensky Bays, and the Mangyshlak Peninsula, were the main locations of mixed *Zostera*, *Ruppia* and *Potamogeton* beds. Estimates of the biomass of *Zostera noltii* and *Ruppia cirrhosa* in the Caspian Sea at this time were much higher than in the present day. In Kaidak Bay, with the salinity of the seawater ranging from 25 to 51 psu, the highest salinity level ever documented for *Zostera noltii*, the biomass of *Zostera noltii* was estimated to be 7 000-8 000 g wet weight/m² and that of *Ruppia cirrhosa* 10 000-12 000 g wet weight/m²[42]. The shoots of *Zostera noltii* from Kaidak Bay were 75-100 cm long, while in the open sea the length was 25-30 cm. In comparison with Kaidak Bay, in the open sea the biomass of this species was substantially less, varying from 100 to 1 500 g wet weight/m². The total stock of *Zostera noltii* for the Caspian Sea was estimated at approximately 700 000 metric tons (wet weight), with about 500 000 metric tons for the eastern and 200 000 metric tons for the western coast. The area covered by the seagrass in just the northeastern Caspian Sea was 1 650 km².

During the 1950s coastal configurations changed and many shallow bays such as Kaidak Bay, in which *Zostera noltii* and other macrophytes formerly flourished, vanished and the area of others such as Krasnovodsky Bay decreased substantially[36, 44-49]. Ever since that time, *Zostera noltii* communities have been degrading, having almost completely vanished along the western coast and becoming seriously depleted in the east. In 1935-38 the biomass of *Zostera noltii* along the eastern coast ranged from 50 to 8 000 g wet weight/m²[44, 45]. By 1971-74 the range had decreased to 50-1 300 g wet weight/m²[46, 49], and in the early 1980s it was 127-1 340 g wet weight/m²[36]. Despite the decline in biomass the area of some beds in Krasnovodsky Bay enlarged considerably, so much so that in the early 1970s different experts evaluated the stock of *Zostera noltii* in Krasnovodsky Bay to be 200 000-440 000 metric tons wet weight. Apparently, such an expansion may be

due to environmental changes and the drop in sea level which brought about the extinction of competing algae such as the Charophyceae.

At present, available data indicate that *Zostera noltii* is only rarely found in the western Caspian Sea at Makhachkala and in Kizlarsky Bay. Seagrasses have completely disappeared from the southern Caspian Sea[4, 36]. Single and mixed communities of *Ruppia* spp. are found growing in Astrakhansky Bay in the west and in Komsomolez, Kazakhsy, Krasnovodsky and Turkmensky Bays in the east, on silt sediments at depths from 0.5 to 3 m.

Though the areas of sea bottom covered with seagrasses have substantially declined, they are still important in the ecology of the Caspian Sea. Seagrasses play an important role in the nutrition of invertebrates on which the state of commercial fish stocks depends[4, 6, 48]. In the northern Caspian Sea, *Zostera noltii* growth is of special significance, because this is where wild carp, Caspian roach, bream and other valuable fish spawn and feed[4]. Other seagrasses are the usual food item for waterfowl. *Ruppia* spp. constitute up to 25 percent of the intestinal content of swans and gray geese and 54-84 percent of that of ducks.

The seagrass communities have been placed under protection in two national nature reserves (Astrakhansky and Krasnovodsky National Reserves).

ARAL SEA

There are two seagrasses, *Potamogeton pectinatus* and 16 macroalgae in the flora of the Aral Sea[50]. Presumably *Zostera noltii* was introduced from the Mediterranean to the Aral Sea, also through the Kumo-Manych Strait. The most extensive knowledge about seagrass distribution had been acquired prior to the severe anthropogenic disruption of the Amudarya and Syrdarya river systems in the mid-1950s[51] that caused catastrophic changes to the ecosystem of the Aral Sea and adjacent water bodies.

Zostera noltii grew from 0.1 to 10 m deep, with most growth being concentrated at 0.1-2 m depth in the northern shallow bays[51, 52]. In the mid-1950s, the biomass of *Zostera noltii* was estimated to be 17 to 800 g wet weight/m^2, with the largest values registered near the mouth of the Syrdarya River. In recent years, the environmental crisis which is wiping out a large part of the Aral Sea has manifested itself in drastic increases in salinity which, in turn, have led to changes in the biological components of all ecosystems. However, the areas occupied by *Zostera noltii* and estimates of its biomass have increased in the northern bays, while in the more brackish area near the Syrdarya's mouth biomass has considerably decreased. Records from the early 1990s show that biomass is now apparently positively correlated with salinity. In the Syrdarya Estuary at salinity of 7 psu biomass was just 42 g wet weight/m^2, whereas in Tshe-Bas Bay *Zostera noltii* not only tolerates salinity as high as 45 psu but thrives on silt-sandy and sandy sediments supporting beds with biomass of 2258 g wet weight/m^2. Intermediate values were observed in the Berg Strait (417 g wet weight/m^2 at 23 psu), Butakov Bay (899 g wet weight/m^2 at 36 psu) and Shevchenko Bay (1076 g wet weight/m^2 at 30 psu)[53]. As the Aral Sea continues to disintegrate, *Zostera noltii* communities are expected to persist mostly in bays of the Minor Sea, where the sea level has remained constant for the past decade.

Total macrophytic stock in the sea is estimated at 1.34 million metric tons wet weight. The share contributed by *Zostera noltii* is about 8.1 percent (109000 metric tons wet weight), while algae such as Charophyceae and *Vaucheria dichotoma* contribute 77.6 and 13.4 percent, respectively[51].

Compared to phytoplankton, macrophytes such as *Zostera noltii* are of little importance in the food chains of the Aral Sea. However they are ecologically important. The meadows of *Zostera noltii* are the spawning location of diverse invertebrates and fish. Benthic invertebrates and fish predominantly feed on diatoms (*Navicula* spp. and *Merismopedia* spp.) and are found in abundance[51, 52]. However, during the past 50 years, the catches of commercial fish have collapsed to the point where the Aral Sea has almost lost its significance for fisheries.

There are no data regarding nature reserves along the coastal zones of the Aral Sea.

ACKNOWLEDGMENTS

I am grateful to Prof. R.C. Phillips (Marine Research Florida Institute), O.A. Akimova, G.F. Guseva, M.Yu. Safonov (IBSS), Dr I.I. Serobaba (YugNIRO), Dr I.I. Maslov (Nikita Botanical Garden) and Ms Olga Klimentova for their help in preparing this chapter.

AUTHOR

Nataliya A. Milchakova, Department of Biotechnologies and Phytoresources, Institute of Biology of the Southern Seas, National Academy of Sciences of Ukraine, 2 Nakhimov Ave., Sevastopol 99011, Crimea, Ukraine. **Tel:** +38 (0)692 544110. **Fax:** +38 (0)692 557813. **E-mail:** milcha@ibss.iuf.net

REFERENCES

1. Kosarev AN, Yablonskaya EA [1994]. *The Caspian Sea*. Backhuys Publishers, Hague.
2. Matishov GG, Denisov VV [1999]. *Ecosystems and bioresources of the European seas of Russia in the late XXth – early XXIst centuries*. Murmansk.
3. Zenkevich LA [1963]. *Biology of the Seas of the USSR*. Nauka Publishing, Moscow (in Russian).

4. Kasymov AG [1987]. *The Caspian Sea*. Hydrometizdat Publ, Leningrad.
5. Zaitsev YuP, Mamayev V [1997]. *Marine Biological Diversity in the Black Sea: A Study of Change and Decline*. United Nations Publications, New York.
6. Ivanov VP [2000]. *Biological Resources of the Caspian Sea*. CaspNIRO, Astrakhan.
7. Kalugina-Gutnik AA [1975]. *Phytobenthos of the Black Sea*. Naukova Dumka Publishing House, Kiev (in Russian).
8. Morozova-Vodyanitskaya NV [1959]. Aquatic plant associations in the Black Sea. *Proc Sevastopol Biol Station* 11: 3-28.
9. Zaitsev YuP, Alexandrov BG [1998]. *Black Sea Biological Diversity: Ukraine*. Black Sea Environmental Series 7. United Nations, New York.
10. Milchakova NA [1999]. On the status of seagrass communities in the Black Sea. *Aquatic Botany* 65: 21-32 (in Russian).
11. Savenkov MY [1910]. *Materials of the Study of Oikology and Morphology of Zostera spp. near Sevastopol*. Pechatnik Publishing House, Kharkov.
12. Zernov SA [1913]. On studies of the life of the Black Sea. *Reports of the Imperial Academy of Sciences* 32(1): 1-299 (in Russian).
13. Morozova-Vodyanitskaya NV [1973]. About the biology and distribution of Zostera in the Black Sea. In: *Hydrobiological Studies of the Northeastern Black Sea*, Rostov University Press, Rostov-on-Don. pp 5-19.
14. Kulikova NM [1981]. *Zostera* phytocenoses in the Black and Azov seas. In: *Commercial Algae and their Use*. VNIRO, Moscow. pp 74-80.
15. Milchakova NA [1988a]. The Spatiotemporal Description of the Structure of Phytocenoses and Populations of *Zostera marina* L. in the Black Sea. Synopsis of PhD thesis (Biology), Sevastopol. pp 1-22.
16. Milchakova NA [1988b]. The composition and distribution of *Zostera marina* L. phytocenoses in some bays of the Black Sea. *Plant Resources* 1: 41-47.
17. Sadogursky SE [1998]. Ecobiological features of *Zostera* spp. growing along the Crimean southern coast. *Bull Nikitsky Botanical Garden* 80: 27-36.
18. Sadogursky SE [1999]. Vegetation of soft bottom sediments of the Arabatsky Bay (Azov Sea). *Algologiya* 9(3): 49-55.
19. Morozova-Vodyanitskaya NV [1938]. Epidemic disease of eelgrass in the Black Sea. *Priroda* 1: 94-98.
20. Generalova VN [1951]. Aquatic vegetation of Utlyuk coastal salt lake and Arabatskaya spit (Azov Sea). *Proc Az Cher NIRO* 15: 331-337.
21. Kaminer KM [1981]. *Phyllophora* and *Zostera* in the bays of the northwestern Black Sea and the prospects of exploitation. In: *Commercial Algae and their Use*. VNIRO, Moscow. pp 81-87.
22. Milchakova NA, Aleksandrov VV [1999]. Bottom vegetation at some sites of coastal salt lake Donuslav (the Black Sea). *Ekologiya Morya* 49: 68-72.
23. Pogrebnyak II [1965]. Bottom Vegetation of the Lagoons in the Northwestern Black Sea and at the Adjacent Sea Areas. Synopsis of DSc thesis (Biology), Odessa. pp 1-46.
24. Shelyag-Sosonko YuR [ed] [1999]. *Biodiversity of Danube Biosphere Reserve, Conservation and Management*. Naukova Dumka Publishing House, Kiev.
25. Makkaveeva EB [1979]. *Invertebrates of the Macrophyte Beds of the Black Sea*. Naukova Dumka Publishing House, Kiev.
26. Southern Research Institute for Fishery and Oceanography in Kerch [1994]. Unpublished data.
27. Alexandrov VV [2000]. The evaluation of *Zostera marina* L. coenopopulations state in the Sevastopol region (the Black Sea). *Ekologiya Morya* 52: 26-30.
28. Milchakova N [2001]. Unpublished data.
29. Maslov II, Sadogursky SE [2000]. Ecological description of *Zostera marina* L. in Kerch Strait. *Bull Nikitsky Botanical Garden* 76: 26-27.
30. Sadogurskaya SA [2000]. *Zostera noltii* Hornem. in the coastal sea water of the Kerch Strait near the Crimea. *Bull Nikitsky Botanical Garden* 76: 34-35.
31. Maslov [1992]. Personal communication.
32. Morozova-Vodyanitskaya NV [1939]. *Zostera* as a commercial object in the Black Sea. *Priroda* 8: 49-52.
33. Lukina GD [1986]. Polysaccharide Seagrasses of the Black Sea: The Chemical Composition, Structure, Properties and Practical Use. Synopsis of PhD thesis (Chemistry), Odessa. pp 1-22.
34. Borodin AM, Syroechkovsky EE (eds) [1980]. *Reserves in the USSR: A Guidebook*. Lesprom, Moscow.
35. Leonenko VB et al. (eds) [1999]. *National Nature Reserves of Ukraine: A Guidebook*. Kiev.
36. Gromov VV [1998]. Bottom Vegetation of Upper Shelf Sections of the Southern Seas of Russia. Synopsis DSc (Biology), St Petersburg. pp 1-45 (in Russian).
37. Kuzmichev AI [1992]. *Hydrophylous Flora of the Southwestern Russian Plain and its Genesis*. Hydrometizdat, St Petersburg.
38. Dubyna DV [1989]. Comparative structural analysis of the Kuban river flood-plain and littoral flora. North Caucasian Res Centre News. *Natural Sciences* 2: 28-36.
39. Matishov GG (ed) [2000]. *Regularities found in Oceanographic and Biological Processes in the Azov Sea*. Kor Res Centre Rus Ac Sci Press, Apatites (in Russian).
40. Isikov VP, Kornilova NV, Pasin YG et al. [1999]. The Project of Territorial Management and Natural Formations Protection in the Kazantip Nature Reserve, Yalta.
41. Gubina GS, Shevchenko VN, Pilyuk VN [1991]. The material about vegetation of the northern coast of the Azov Sea. *Abstract 6th Congr Hydrobiol Soc, Murmansk* 1: 47-48.
42. Kireyeva MS, Shchapova TF [1939]. Bottom vegetation of the northeastern Caspian Sea. *Bull Soc Nat Moscow, Biol* 48(2-3): 3-14.
43. Shchapova TF [1938]. Bottom vegetation of the notheastern bays Komsomolez (Dead Kultuk) and Kaidak. *Botanical Journal* 23(2): 122-144.
44. Kireyeva MS, Shchapova TF [1957a]. The material about taxonomic composition and biomass of algae and high aquatic vegetation of the Caspian Sea. *Proceedings of the Institute of Oceanology* 23: 125-137 (in Russian).
45. Kireyeva MS, Shchapova TF [1957b]. Bottom vegetation of Krasnovodsk Bay. *Proceedings of the Institute of Oceanology* 23: 138-145.
46. Blinova EI [1974]. Phytobenthos of the East Caspian Sea. Abstract. In: *All-Union Conference Marine Algology and Macrophytobenthos*. VNIRO Publishing House, Moscow. pp 12-14.
47. Petrov KM [1967]. Vertical distribution of aquatic vegetation in the Black and Caspian seas. *Oceanology* 7(2): 314-320.
48. Salmanov MA [1987]. *The Role of Microflora and Phytoplankton in Production Processes in the Caspian Sea*. Nauka Publishing, Moscow.
49. Zaberzhinskaya EB, Shakhbazi ChT [1974]. Bottom Vegetation of Krasnovodsky Bay. Abstract. In: *All-Union Conference Marine Algology and Macrophytobenthos*. VNIRO Publishing House, Moscow. pp 51-53.
50. Dobrokhotova KV, Roldugin II, Dobrokhotova OV [1982]. *Aquatic Plants*. Kainar Publishing House, Alma-Ata.
51. Yablonskaya EA [1964]. About the role of phytoplankton and benthos in food chains of organisms inhabiting the Aral Sea. In: *The Stock of Marine Plants and their Use*. Nauka Publishing, Moscow. pp 71-91.
52. Yablonskaya EA [1960]. The present state of benthos of the Aral Sea. *Proc VNIRO* 43(1): 115-149.
53. Orlova MI [1993]. Material contributing to total assessment of production and destruction processes in coastal zone of the northern Aral Sea.1. Results of the field studies and experiments, 1992. *Proceedings of the Zoological Institute RAS* ("Ecological crisis in the Aral Sea") 250: 21-37 (in Russian).

5 The seagrasses of
THE EASTERN MEDITERRANEAN AND THE RED SEA

Y. Lipkin
S. Beer
D. Zakai

This chapter is divided into two sections and considers the seagrasses of the eastern Mediterranean and the Red Sea. While the eastern Mediterranean has a relatively restricted range of species, the Red Sea is home to 11 species, all of tropical origin.

EASTERN MEDITERRANEAN

Early contributions from the eastern Mediterranean reported on the presence of *Cymodocea nodosa* (also reported as *Cymodocea aequorea* or *Cymodocea major*), *Posidonia oceanica* (also as *Zostera oceanica*), *Zostera marina* and *Zostera noltii* (also as *Zostera nana*), in Greece, Syria and Egypt. *Halophila stipulacea*, a migrant from the Red Sea, was first reported in the Mediterranean from the island of Rodos late in the 19th century[1]. Lipkin[2, 3] summarized its distribution in the Mediterranean through the early 1970s; during the last three decades *Halophila stipulacea* has spread further, mostly in the eastern basin (e.g. Methoni and Paxoi Islands, Ionian Sea[4, 5]; Marmaris[6]; Korinthiakos Kolpos[7-9], but also at and near Sicily[10, 11].

The most common seagrass in the eastern Mediterranean is *Cymodocea nodosa*. It occurs on all coasts of this basin on sandy and, less frequently, muddy bottoms. The next most prevalent seems to be *Posidonia oceanica*, a climax seagrass. In many regions in the northern part of the basin the balance between the two is inverted, with *Posidonia oceanica* becoming the more common. The third in abundance appears to be *Zostera noltii* and the least common *Halophila stipulacea*. *Zostera marina*, a species common in the western Mediterranean, seems to be rare in the eastern part, if it still exists there at all. Collections of the latter species reported by den Hartog[12] were from the northern parts of the Aegean Sea, made in 1854, 1891 and 1910. Publications later than 1930, by Greek and Turkish authors[13-27], reported *Zostera marina* from the same area. Interestingly, more recent papers about seagrasses on the Turkish and Greek northern Aegean coasts[8, 28] do not include *Zostera marina* among the seagrasses of this area. The only report of this seagrass from Egypt[29, 30] – the latter reference being based on the former – was probably a case of misidentification; in later papers on Egyptian Mediterranean seagrasses, Aleem did not mention *Zostera marina*. However, Täckholm et al.[31] reported that the filling of an ancient Egyptian mummy was composed of *Zostera marina*, which indicates that the plant must have occurred, or was even common, in shallow Egyptian waters some 2000 years ago, and seems to have gradually disappeared, first from the warmer southeastern corner, then from wider and wider areas in the eastern and central parts of the eastern Mediterranean, and remained until rather recently on its coldest, northernmost coasts. A similar retreat from a former, wider range seems also to have occurred with *Zostera noltii*, which is concentrated mainly in the Aegean Sea with considerably less representation in other parts of the northeastern Mediterranean, and almost none in the south. It has disappeared, or become very rare, even in the south Aegean Sea[32].

Ecosystem description

Cymodocea nodosa and *Zostera noltii* usually grow in shallow water, from a few centimeters to a depth of 2.5-3 m (it has been reported that *Cymodocea nodosa* occupied a depth range of 5-10 m in the Bay of Limassol, Cyprus[33]). *Posidonia oceanica* is found from the shallows, where the tips of the leaves reach the surface, down to 35-40 m. *Halophila stipulacea*, many beds of which also occur in the shallows, e.g. at Rodos, penetrates much deeper water. Bianchi et al.[33] reported it as the deepest seagrass in the Bay of Limassol, growing at 25-35 m (for *Posidonia oceanica* they reported a range of 10-30 m). Fresh, seemingly *in situ*, material was dredged from around 145 m off

Boundary of *Posidonia oceanica* meadow with *Cymodocea nodosa* growing on sand.

Cyprus; however, below about 50 m it was rather scarce[3].

All four seagrasses grow in the eastern Mediterranean on soft bottoms, quartz sand in shallow waters and mud at greater depths. *Cymodocea nodosa* frequently occurs in small sandy pockets that accumulate in crevices or small depressions on rocky flats, and *Posidonia oceanica* is often found on rough substrates such as pebbles and gravel and even solid rock. It is noteworthy that *Halophila stipulacea*, growing in a wide range of environmental conditions in the northern Red Sea, including all kinds of coastal substrates[34], has a much narrower ecological range in the eastern Mediterranean, being restricted in this basin to soft substrates only. The form with bullate leaves, the so-called "bullata" ecophene, so common in extreme conditions in the northern Red Sea, has not been reported from the eastern Mediterranean. Several ecotypes of *Halophila stipulacea* occur in the northern Red Sea[35]. Probably only one of them has penetrated and spread into the Mediterranean.

Seagrasses occupy extensive areas in Greek waters[8]. Clusters of *Cymodocea nodosa* appear in very shallow water only a few centimeters deep, mostly in sheltered areas, and to a lesser extent on beaches exposed to winds and waves. In sheltered areas, *Cymodocea nodosa* tends to occupy deeper bottoms and form larger beds. In the southern Ionian Sea, such beds appear from a depth of 60 cm down to 1.5-2 m. In shallower water, the plant is found on sandy bottoms, and a little deeper on muddy ones. *Posidonia oceanica* occurs at greater depths, on sandy bottoms. *Zostera noltii* (reported as *Zostera nana*) was represented by scattered plants; no beds are recently reported. *Halophila stipulacea* was rare in Ionian Greece around 1990, occurring at two sites only, at about 2.5 m depth at Methoni, together with the siphonous green alga *Caulerpa prolifera*, and at Paxoi[4,5].

Most seagrass beds in the eastern Mediterranean are composed of one seagrass species only. Beds of *Posidonia oceanica* are usually very dense. Only when they start to deteriorate, for example when affected by pollution, do other marine plants, usually algae, invade. In *Cymodocea nodosa* beds, the seagrass is occasionally accompanied by *Caulerpa prolifera*, which may reach 20 percent of the plant cover[34]. Mixed populations of *Posidonia oceanica* and *Cymodocea nodosa*[36] or *Zostera noltii* and *Cymodocea nodosa*[37] also occur.

Egypt

Reports on seagrass habitats and community structure in the eastern Mediterranean are scanty, compared with the information available on these subjects in the western basin. The seagrass vegetation of the bay at Marsa Matrûh Harbor (western end of the Mediterranean coast of Egypt) and its close vicinity was described by Aleem in the early 1960s[38]. He reported healthy beds of *Cymodocea nodosa*, *Halophila stipulacea* and *Posidonia oceanica* and provided a distribution map. In 1957-60, *Cymodocea nodosa* beds covered a continuous belt 10-40 m wide and about 750 m long in the inner part of the bay at a depth of around 50 cm to 2 m, on the slope between the watermark and the horizontal bottom that starts at 2 m. At the innermost part of the bay, the belt broke into scattered patches. In extremely sheltered areas, the seagrass was absent. Right below this belt, on the lower parts of the slope at 2-8 m depth, a belt of *Halophila stipulacea* of similar size occurred. *Posidonia oceanica* formed a bed 70-120 m wide and about 300 m long, outside the inner bay, in an area more exposed to winds and waves.

At El Dab'a, about 160 km west of Al Iskandarîya (Alexandria), *Posidonia oceanica* covered a small area along with a few small patches of *Cymodocea nodosa*. *Halophila stipulacea* did not occur at this site. The macrofauna and algal macroflora were also scarce, both in numbers of species and in numbers of individuals. For example, only 12 epiphytic algae were found on the leaves and rhizomes of *Posidonia oceanica*[39, 40].

Aleem[41] described in detail the establishment, development and stabilization of the seagrass beds at Al Iskandarîya. His description was later incorporated by den Hartog[12] into his account of the ecology of *Posidonia oceanica*, and therefore will not be repeated

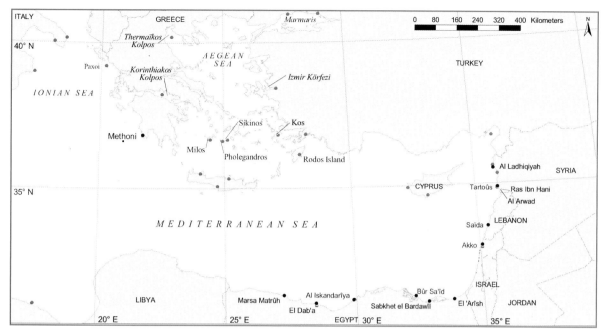

Map 5.1
The eastern Mediterranean

here. At that time (the 1950s), however, the *Posidonia oceanica* beds, once established, persisted for long periods of time. Later, as in some other Mediterranean sites, they became affected by domestic and industrial pollution and started to dwindle[42].

The coast of Sinai and the southern part of the Israeli coast are mostly covered with pure quartz sand, with only a few rocky outcrops here and there. This part of the eastern Mediterranean coast, lacking bays and coves, is highly exposed to wind and wave action. Wide *Cymodocea nodosa* beds occur at depths of 2 m and more, along the Sinai coast, below the littoral belt in which the bottom sediment is intensively worked by the breakers. A few small stands of *Posidonia oceanica* were reported by Aleem[41] from the several rocky habitats between Bûr Sa'îd and El 'Arîsh; the status of these sites has not been reported since 1955.

Sabkhet el Bardawîl, the large lagoon on the Mediterranean coast of Sinai, harbors a large bed of *Ruppia cirrhosa*, which covers up to a third of the lagoon[34]. The size of this bed fluctuates considerably seasonally, and during severe winters it may disappear completely.

Israel

Along the generally exposed Israeli coast, rich beds of *Cymodocea nodosa* are found on sandy bottoms at sheltered sites. The best developed is at Akko, at the northern end of Haifa Bay. Small patches of the seagrass also occur in sand-filled depressions on submerged horizontal platforms just below mean sea

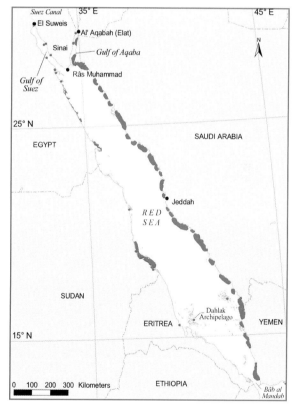

Map 5.2
The Red Sea

level. All *Cymodocea nodosa* populations are subject to large seasonal and year-to-year fluctuations in size, on occasion disappearing completely, eventually to renew from the seed stocks in the sediment[34]. Area estimates for Israel are very approximate since no exact mapping has been carried out. We estimate the total Israeli

Mediterranean coast populations of *Cymodocea nodosa* to be no more than a few hundred square meters.

Lebanon

From the Lebanese coast there is no information about seagrass beds except that gathered by J.H. Powell on the occurrence of a *Cymodocea nodosa* and *Halophila stipulacea* bed some 800 m off Saïda (Sidon), in which the former comprised 70 percent of the seagrass cover[3]. To judge from the very few records of seagrasses from the Lebanese coast, seagrass beds are uncommon.

Syria

On the Syrian coast, too, seagrass beds are uncommon[43]. *Cymodocea nodosa* and *Zostera noltii* beds are found near the river mouths between Tartoûs and Banias, in the vicinity of Jable, in small, relatively calm embayments north of Al Ladhiqiyah (Latakia) and near the harbors of Tartoûs, Al Arwad and Al Ladhiqiyah, where they grow intermingled with *Caulerpa scalpelliformis* and/or *Caulerpa prolifera*[43]. *Zostera noltii* appears also as an accompanying species in the plant community dominated by *Caulerpa scalpelliformis* at Tartoûs and Al Arwad.

These seagrasses grow on this coast on clayey sand rich in organic matter. They seem to tolerate considerable variations in salinity. *Posidonia oceanica* is rare on the Syrian coast; Mayhoub[43] found it in only two localities: northwest of Al Arwad islet, and in a bay near Ras Ibn Hani. In both cases, the beds were not well developed; in his opinion they were in the process of disappearing. He assumed that the rapid degradation of the *Posidonia oceanica* beds northwest of Al Arwad Island, a great part of which were already replaced by *Caulerpa*, was the result of large sewage installations that had been constructed a short while previously at nearby Tartoûs[43].

Cyprus

Seagrass beds are widespread around the island of Cyprus. Rich stands of *Posidonia oceanica* and of *Cymodocea nodosa* are common at different depths, *Posidonia oceanica* beds descending much deeper than those of *Cymodocea nodosa*. Mixed populations are found, but less often. *Halophila stipulacea* beds are not as plentiful as those of the other two; they also descend to considerable depths[3]. The quickly expanding green alga *Caulerpa racemosa* is considered a threat to the *Posidonia oceanica* beds. Since first noticed in the island in 1991, it has spread, unchecked, into a wide range of habitats from the shallows to depths of at least 60 m, on sandy as well as muddy bottoms, competing directly with *Posidonia oceanica*[44, 45].

Turkey

Along the Turkish coasts, at the eastern part of the Mediterranean coast, meadows of *Posidonia oceanica* dominate the lower levels of the infralittoral zone, but no further information about them is available, except that *Cymodocea nodosa* and *Zostera noltii* have also been found in the area[46]. On the Aegean coast, monospecific beds of *Posidonia oceanica* and of *Cymodocea nodosa* were reported from Izmir Bay (Izmir Körfezi), as were mixed beds of the two. *Cymodocea nodosa* beds were 20-50 m in diameter, whereas those of *Posidonia oceanica* were much larger – 150 m and more in diameter[36]. Similar meadows are probably common on the Aegean Turkish coast.

Greece

From Greek waters, Bianchi and Morri[47] reported dense monospecific stands of *Cymodocea nodosa* and of *Posidonia oceanica* at the island of Kos, in the eastern Aegean, the latter seagrass appearing to be more common. In the western part, large seagrass beds were reported from the islands of Sikinos, Milos and Pholegandros. Vast beds of *Cymodocea nodosa* were found on mud in shallow bays at the latter two localities, at 0.3-4 m depth. *Posidonia oceanica* was common on sandy deposits around the entire coast of all three islands studied, not just in bays. In shallower water they formed isolated tufts and in deeper water, 2-8 m or more, they formed quite large beds[48].

RED SEA

Historical and present distribution

The Red Sea harbors 11 seagrass species, all of tropical origin, which penetrated through its relatively narrow mouth at Bâb al Mandab. These are: *Halodule uninervis*, *Cymodocea rotundata*, *Cymodocea serrulata*, *Syringodium isoetifolium*, *Thalassodendron ciliatum*, *Enhalus acoroides*, *Thalassia hemprichii*, *Halophila ovalis*, *Halophila ovata*, *Halophila stipulacea* and *Halophila decipiens*[12, 49, 50]. Only a single plant of *Halophila decipiens* has hitherto been reported from the Red Sea, grabbed from 30 m[50]. For early records and distribution see Lipkin[51].

Enhalus acoroides seems not to reach much beyond the Tropic of Cancer, whereas the other ten species continue to the northwestern part of the Red Sea proper, but only seven (the above listed species excluding *Enhalus acoroides*, *Cymodocea serrulata*, *Halophila ovata* and *Halophila decipiens*) penetrate into most of the Gulf of Elat (Gulf of Aqaba) and only five (*Halodule uninervis*, *Halophila stipulacea*, *Halophila ovalis*, *Halophila decipiens* and *Thalassodendron ciliatum*) into much of the Gulf of Suez. Hulings and Kirkman[52] reported *Cymodocea serrulata* from "a shallow lagoon on the west coast of the Gulf of Aqaba

40 km south of Eilat", but this record should be confirmed. *Halophila stipulacea*, *Halodule uninervis* and *Halophila ovalis* appear at present to be the only seagrasses that reach the tips of these gulfs[34,50,53], although old records also listed *Thalassodendron ciliatum* and *Syringodium isoetifolium* from El Suweis (Suez), at the tip of the Gulf of Suez, and from Al' Aqabah at the tip of the Gulf of Elat, and in addition listed *Cymodocea rotundata* and *Cymodocea serrulata* from El Suweis[51]. Notably, Aleem[54] did not find any seagrass at Bûr Taufiq, near El Suweis.

Halophila stipulacea, very common in the northern part of the Red Sea, is rather scarce at its central and southern parts[49,55], as well as at the tropical east African coast south of the Horn of Africa. It becomes common again on the east African coast near the Tropic of Capricorn[56]. Thus, Lipkin[57] concluded that this species is of subtropical affinity rather than tropical.

Some of the Red Sea seagrasses occur in the intertidal zone and most species usually grow at the shallow subtidal, not deeper than 5 m, but may be found as deep as 10 m[34,55]. However, *Halophila stipulacea* is widely found in the Gulf of Elat at depths down to 50 and even 70 m and *Thalassodendron ciliatum* down to 30 m[34,53]. In the Gulf of Suez, *Halophila decipiens* was found at 30 m, *Halophila ovata* down to 20 m and one of the populations of *Halophila ovalis* at 23 m[50]. On the Jordanian coast of the Gulf of Elat, two *Halophila ovalis* stands were found at 15 and 28 m, respectively[52].

Most seagrasses in the Red Sea grow on mud, silt or fine coralligenous sand, or mixtures of them. The eurybiontic *Halophila stipulacea* and, to a lesser extent, *Halodule uninervis* thrive on a wide variety of substrates. *Thalassodendron ciliatum* and *Thalassia hemprichii*, however, seem to prefer coarser substrata, that is coarse sand admixed with coral and shell debris or even rather large pieces of coral from the surrounding fringing reefs or coral knolls at sites exposed to considerable water movement[34,55].

Almost all beds of *Thalassodendron ciliatum* and *Enhalus acoroides* are monospecific, whereas *Syringodium isoetifolium*, *Thalassia hemprichii* and *Halodule uninervis* often occur in multispecific seagrass communities. This tendency also changes geographically, e.g. *Syringodium isoetifolium* forms monospecific stands as well as occurring in multispecific communities on the central Saudi Arabian coast, whereas in the Gulf of Elat it was found only in mixed populations.

Although seagrass beds are common in the Red Sea, information about the seagrass habitats and plant communities in this basin is very limited. A general account of Red Sea seagrass beds was given by Lipkin[55], including information about the typical accompanying fauna. Below is a summary of the few available descriptions of the seagrass vegetation in some Red Sea localities.

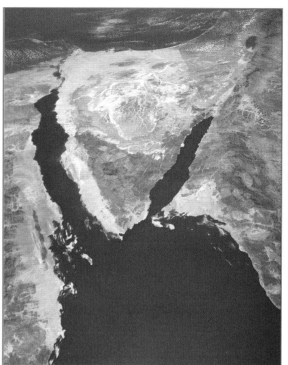

The northern Red Sea taken from the Space Shuttle. The Red Sea harbors 11 seagrass species – all of tropical origin.

Eritrea
In the south, within the Dahlak Archipelago, on the Eritrean coast, seagrasses are not common. A sparsely vegetated *Caulerpa racemosa-Thalassia hemprichii* community was reported from sandy patches at the lowermost intertidal zone[58]. Small patches of *Halophila stipulacea* and *Halophila ovalis* were also found in the archipelago[59].

Saudi Arabia
For the central part of the Saudi Arabian coast, in the Jeddah area, Aleem[49] reported in the late 1970s that *Thalassodendron ciliatum*, *Syringodium isoetifolium*, *Enhalus acoroides*, *Halophila ovalis* and *Halophila stipulacea* grew predominantly as pure stands, but were sometimes mixed with other seagrasses. He remarked that *Thalassia hemprichii*, *Cymodocea rotundata* and *Halodule uninervis* tended to form mixed communities. *Thalassodendron ciliatum* beds, to 20-30 m² in size, grew on coarse coralligenous sand with shell debris and sometimes on dead corals that were covered by a thin layer of sand. These stands of the seagrass appeared at about 2 m or a little deeper[49]. Beds of *Thalassia hemprichii*, to 100 m² in size, were

plentiful on the central coast of Saudi Arabia; they appeared at 1-2 m depth as mixed vegetation in which *Thalassia hemprichii* constituted 60-70 percent of the plant cover, *Cymodocea rotundata* 20-30 percent and *Halodule uninervis* 10-20 percent. Pure stands of *Halodule uninervis* were common on this coast in shallow water. In very shallow lagoons, a thin-leaved form appears, whereas on open coasts, a little deeper, the beds are composed of the wide-leaved form. *Cymodocea serrulata* dominates in seagrass beds between 0.5-2 m deep, making up 70 percent of the plant cover. In the shallower beds (0.5-1 m), it is accompanied by *Halodule uninervis* and *Halophila ovalis* and in the deeper beds (1-2 m) by *Cymodocea rotundata* and *Halodule uninervis*. Small, 0.5-4 m² in size, almost pure patches of *Syringodium isoetifolium* occurred at one site along this coast at depths of 0.5-1 m. The green alga *Caulerpa serrulata* accompanied the dominant seagrass in these patches. In another site, *Syringodium isoetifolium* was mixed with *Thalassia hemprichii*, *Cymodocea rotundata* and *Halodule uninervis*. Beds of *Enhalus acoroides* were unusual on the central Saudi Arabian Red Sea coast. Pure patches, about 30 m² in size, grew at 1-2 m on coarse sand with shell debris on top and black mud below, in one site on this coast. *Halophila ovalis* formed small patches, 0.5-2 m in diameter, of sparse growth in shallow water in most localities visited[49].

Gulfs of Suez and Elat

At the Gulf of Suez and the Gulf of Elat, in the north, thin-leaved *Halodule uninervis* formed sparse monospecific prairies in the lower intertidal zone of muddy coasts. In the subtidal zone, pure stands of this seagrass were much denser, and the plants were larger. Mixed stands of *Halodule uninervis* with *Halophila stipulacea*, and sometimes also *Halophila ovalis*, were common in the two gulfs as well[34, 50]. Four other communities dominated by *Halodule uninervis* were reported from the Sinai coast of the Gulf of Elat: the *Halodule uninervis-Syringodium isoetifolium* community, the *Halodule uninervis-Syringodium isoetifolium-Halophila stipulacea* community, the *Halodule uninervis-Cymodocea rotundata* community and the *Halodule uninervis-Halophila ovalis* community. Vegetation types dominated by *Halophila stipulacea* occupy a wide range of habitats. Mostly *Halophila stipulacea* is represented by rather dense monospecific beds that extend between the lower intertidal zone and depths of 50-70 m at the Gulf of Elat.

Density in these beds decreases below 10 m[35]. Here and there mixed stands occur, in which *Halophila stipulacea* is accompanied by *Halodule uninervis* or *Halophila ovalis*, and in one small patch near Zeit Bay (Ghubbel ez-Zeit) at the mouth of the Gulf of Suez, also with *Thalassodendron ciliatum*[34, 50]. The *Thalassodendron ciliatum* community is the most complex of Red Sea seagrass communities, and probably the most important for other life forms. The roomy space under the seagrass canopy and between its woody vertical stems harbors larvae of many pelagic animals, as well as its own assemblage of sciaphilic plants and animals. The height of these vertical stems, varying with depth

---

Case Study 5.1
ISRAELI COAST OF THE GULF OF ELAT

Along the Israeli coast of the Gulf of Elat, at the northwestern end of the gulf, *Halophila stipulacea* is the only seagrass found at all sites but one (the middle of the site south of the Marine Laboratory, where a small bed of *Halodule uninervis* is also present). In 2001 the distribution of *Halophila stipulacea* was follows:

o Along the northern shore of the Gulf of Elat. An extensive bed of *Halophila stipulacea* occurs along the northern shore of the Gulf of Elat, probably extending towards and beyond the nearby Jordanian town of Al' Aqabah. The plants grow at depths from 5 m to more than 45 m, with the highest densities (and the largest-leaved shoots) occurring from 18 to 25 m. The extent of the bed, as well as biomass within the bed, has been observed to fluctuate during the last few years, with a general decline during the last six years.

o Several sites are located near the navy base and the commercial harbor. Plants grow at depths from 8 to more than 25 m.

o A further site is near the harbor where oil and petrol are unloaded. Plants grow at 20-30 m depth.

o A substantial site extends from just south of the Steinitz (Interuniversity) Marine Laboratory to the Egyptian border. Plants grow at depths from 7 m to over 30 m. Between this and the site near the harbor, there are sporadically occurring smaller (<100 m²) beds.

from around 15-20 cm at the shallows to more than 1 m at 30 m depth, determines the volume of this under-canopy space.

Thalassodendron ciliatum is unique among Red Sea seagrass communities in extending right up to coral reefs, without the usual "halo" zone that typically separates reefs from seagrass beds in their proximity. This halo is formed by reef fishes grazing on the other seagrasses. Standing stock of the *Thalassodendron ciliatum* community is by far the highest among Red Sea seagrass communities; its productivity, however, is among the lowest. This seeming contradiction stems from the extremely low consumption of most of the organic matter produced by the seagrass and by epiphytic algae in the under-canopy space. The only highly productive and quickly consumed element in this community is that of photophilic epiphytic algae of the upper, well-illuminated surface of the canopy, on which many herbivorous fishes and invertebrates, mainly snails, graze[34, 55, 60].

The *Syringodium isoetifolium* community is rare in the Gulf of Elat, where it forms small patches. However, the plant accompanies other seagrasses in communities they dominate. Monospecific stands of sparse vegetation of *Halophila ovalis* usually appear in the Gulf of Elat as a narrow belt at the lee margins of larger stands of *Halophila stipulacea*, or in clearings within wide beds of the latter. Mixed stands of *Halophila ovalis*, *Halophila stipulacea* and *Halodule uninervis* appear in wider areas. *Cymodocea rotundata* beds are the second least common seagrass community in the Gulf of Elat, forming monospecific dense stands down to 2 m.

Thalassia hemprichii, although the least common seagrass on the Sinai coast of the Gulf of Elat, dominates in four communities at the southern part of this coast. The first is represented by dense mono-specific beds growing on a layer, about 30 cm thick, of very coarse-grained substrate made of gravel-sized coral debris covering the underlying rock. Beds of *Thalassia hemprichii* and *Halophila stipulacea* in equal proportions occurred on the same type of substrata, but the unconsolidated layer was somewhat thicker. Wide areas of *Thalassia hemprichii* with *Thalasso-dendron ciliatum* appeared at Râs Muhammad, on the tip of the Sinai Peninsula, to the seaward of mono-specific *Thalassia hemprichii* stands on a thin 20-cm layer of even coarser unconsolidated material. Finally, large areas of dense vegetation of *Thalassia hemprichii*, with 20-40 percent *Halodule uninervis*, covered large stretches of wide reef flats between Marsa abu Zabad and Shorat el Manqata', growing on coarse coralligenous sand at 0-30 cm below the low water of spring tides[34].

The total populations of *Halophila stipulacea* on

Butterfly goby (*Amblygobius albimaculatus*) in seagrass in the Red Sea, Jordan.

the Israeli Red Sea coast (only about 5 km long) probably occupy some 0.5-1.0 km².

EFFECTS OF POLLUTION

Most of the few reports on pollution effects on seagrass beds in the eastern Mediterranean and the Red Sea refer to chemical pollution. Haritonidis *et al.*[37] in 1990 reported considerable declines in the sizes of beds of *Posidonia oceanica* and *Cymodocea nodosa* in the Thermaïkos Kolpos (northern Aegean Sea) during the preceding two decades, with the former suffering greatest losses. They also remarked that the density of the shoots had decreased, and that marked changes in the seagrass epiphytic communities had taken place. The authors attributed these phenomena to the increased amounts of domestic and industrial pollutants discharged into the gulf during that period. In contrast, *Zostera noltii*, the least common of the three seagrasses that occur in the gulf, seemed to have benefited from the increased discharge of sewage, as the area covered by its beds had increased.

A similar decline in the area occupied by *Posidonia oceanica* beds, and their thinning, was reported for Cyprus[45], but here the authors attributed these phenomena to competition with the invading green alga *Caulerpa racemosa*. Between 1992 and 1997, dense stands of the latter replaced *Posidonia oceanica* in part of the area it had covered at the beginning of this period (total plant cover in the *Posidonia oceanica* beds decreased from 70-90 percent to 40-60 percent), and a number of algae, not previously found in the thinned beds, penetrated into them, not replacing *Caulerpa prolifera*, an accompanying species in some of the *Posidonia oceanica* beds during the earlier period. Similarly, Fishelson *et al.*[61] reported that *Halophila stipulacea* meadows,

formerly widespread, dramatically retreated at the northern end of the Gulf of Elat, in the northern Red Sea. Here the source of pollution was fish culture in cages in the gulf.

Dando et al.[62] dealt with the effects of thermal pollution. They reported that *Cymodocea nodosa* replaced *Posidonia oceanica* near hydrothermal discharge vents at the bottom of the Aegean Sea.

REFERENCES

1. Fritisch C [1895]. Über die Auffindung einer marinen Hydrocharidee im Mittelmeer. *Verhandlungen der k.k. Zoologisch-Botanischen Gesellschaft in Wien* 45: 104-106.
2. Lipkin Y [1975a]. *Halophila stipulacea*, a review of a successful immigration. *Aquatic Botany* 1: 203-215.
3. Lipkin Y [1975b]. *Halophila stipulacea* in Cyprus and Rhodes, 1967-1970. *Aquatic Botany* 1: 309-320.
4. Haritonidis S, Tsekos I [1976]. Marine algae of the Greek west coast. *Botanica Marina* 19: 273-286.
5. Tsekos I, Haritonidis S [1977]. A survey of the marine algae of the Ionian Islands, Greece. *Botanica Marina* 20: 47-65.
6. Çirik S [1989]. Espèces rares ou nouvelles pour la flore marine égéenne. *Pelagos* 7: 60-102.
7. Malea P, Haritonidis S [1989]. Concentration of aluminium in *Halophila stipulacea* (Forsk.) Aschers. and the substrate of the Antikyra Gulf, Greece. *Toxicological and Environmental Chemistry* 20/21: 241-248.
8. Haritonidis S, Diapoulis A [1990]. Evolution of Greek marine phanerogam meadows over the last 20 years. *Posidonia Newsletter* 3: 5-10.
9. Zibrowius H [1993]. Records of *Halophila stipulacea* from "Calypso" cruises in Greek and Turkish waters, 1955-1977. *Posidonia Newsletter* 4: 7-10.
10. Biliotti M, Abdelahad N [1990]. *Halophila stipulacea* (Forsk.) Aschers. (hydrocharitaceae): espèce nouvelle pour l'Italie. *Posidonia Newsletter* 3: 23-26.
11. Rindi F, Maltagliati F, Rossi F, Acunto S, Cinelli F [1999]. Algal flora associated with a *Halophila stipulacea* (Forsskål) Ascherson (Hydrocharitaceae, Helobiae) stand in the western Mediterranean. *Oceanologica Acta* 22: 421-429.
12. den Hartog C [1970]. The sea-grasses of the world. *Verhandingen der Koninmlijke Nederlandse Akademie van Wetenschappen, Afd. Natuurkunde, Tweede Reeks* 59(1): 1-275.
13. Politis J [1930]. Plantes marines de la Grèce. *Rapp Comm Int Mer Méditer* NS 5: 195-205.
14. Politis J [1932]. On the marine flora of the island of Crete. *Proceedings of the Academy of Athens* (3) [2(3)]: 1-30 (in Greek).
15. Diannelidis T [1935]. Algues marines du Golfe de Pagassai. *Praktika of Akademy Athenon* 10: 249-254.
16. Diannelidis T [1940]. Algues marines du Golfe de Pagassai. *Actes de l'Institut Botanique de l'Université d'Athenes* 1: 205-209. [translation of the 1935 Greek paper (reference 15)].
17. Diannelidis T [1950]. Greek marine flora and its utilisation. *Praktika of the Hellenic Hydrobiological Institute* 3(2): 71-84 (in Greek, with English summary).
18. Diannelidis T [1953]. Contribution à la connaissance des algues marines des Spordes du Nord (Cyanophyceae, Chlorophyceae, Phaeophyceae, Rhodophyceae) *Praktika of the Hellenic Hydrobiological Institute* 6(2): 41-84.
19. Rechinger KH [1943]. *Flora Aegaea*. Flora der Inseln und Halbinseln des ägäischen Meeres. Denkschriften der Oesterreichischen Akademie der Wissenschaften. Mathematisch-naturwissenschaftliche Klasse, Wien 105. 924 pp.
20. Diannelidis T, Tsekos I [1968]. CH-Schwellen der Uraninfarbbarkeit des Protoplasmas pflanzlicher Zellen. *Protoplasma* 66: 231-240.
21. Tsekos I, Haritonidis S, Diannelidis T [1972]. Protoplasmaresistenz von Meeresalgen und Meeresanthophyten gengen Schwermetallsalze. *Protoplasma* 75: 45-65.
22. Tsekos I, Haritonidis S, Diannelidis T [1973]. La résistance du protoplasme d'algues marines et de phanérogames marines aux sels de métaux lourds. *Rapp Comm Int Mer Méditer* 22(4): 57-58.
23. Haritonidis S, Tsekos I [1974]. A survey of the marine algae of Thassos and Mytilene islands, Greece. *Botanica Marina* 17: 30-39.
24. Yayinta A [1977]. Studies on the distribution, growth and development of *Acetabularia mediterranea* Lamour. From the Aegean coast of Turkey. Ege Üniversitesi, Fen Fakültesi, Yüksek Lisans Tezi, Izmir. 32 pp (in Turkish).
25. Aysel V, Güner H [1978]. In bucht Izmir einige befindliche Punctaria Arten und ihre verbreitungsgebiete. Ege Üniversitesi, Fen Fakültesi Dergisi [Ege University, Faculty of Sciences Journal, ser. B] 1: 375-384 (in Turkish with German abstract).
26. Güner H, Aysel V [1978]. Taxonomische untersuchungen uber die einigen Ulva Arten (Chlorophyta) im Golf von Izmir. Ege Üniversitesi, Fen Fakültesi Dergisi, Seri B [Ege University, Faculty of Sciences Journal, ser. B] 1: 241-251 (in Turkish, with German abstract).
27. Aysel V [1979]. Studies on some of the Polysiphonia Grev. Species (Rhodomelaceae, Rhodophyta) from the Bay of Izmir. Ege Üniversitesi, Fen Fakültesi Dergisi [Ege University, Faculty of Sciences Journal, ser. B] 3: 19-42 (in Turkish, with English summary).
28. Çirik S, Zeybek N, Aysel V, Çirik S [1990]. Note préliminaire sur la végétation marine de l'ile de Gökçeada (Mer Egée nord, Turquie). *Thalassographica* 13: 33-37.
29. Aleem AA [1945]. Contribution to the study of the marine algae of Alexandria and its vicinities. MSc thesis, Farouk University, Alexandria. 182 pp.
30. Nasr AH, Aleem AA [1949]. Ecological studies of some marine algae from Alexandria. *Hydrobiologia* 1: 251-281.
31. Täckholm V, Täckholm G, Drar M [1941]. Flora of Egypt, Vol 1. *Bull Fac Sci Egypt Univ* 17: 1-574.
32. Giaccone G [1968]. Raccolte di fitobenthos nel Mediterraneo orientale. *Giornale Botanico Italiano* 102: 217-228.
33. Bianchi TS, Argyrou M, Chippett HF [1999]. Contribution of vascular-plant carbon to surface sediments across the coastal margin of Cyprus (eastern Mediterranean). *Organic Geochemistry* 30: 287-297.
34. Lipkin Y [1977]. Seagrass vegetation of Sinai and Israel. In: McRoy CP, Helfferich C (eds) *Seagrass Ecosystems: A Scientific Perspective*. Marcel Dekker, New York. pp 263-293.
35. Lipkin Y [1979]. Quantitative aspects of seagrass communities, particularly of those dominated by *Halophila stipulacea*, in Sinai, northern Red Sea. *Aquatic Botany* 7: 119-128.
36. Pergent G [1985]. Florasion des herbiers à *Posidonia oceanica* dans la region d'Izmir (Turquie). *Posidonia Newsletter* 1: 15-21.

AUTHORS

Yaacov Lipkin and Sven Beer, Department of Plant Sciences, Tel Aviv University, Tel Aviv 69978, Israel. **Tel:** +972 (0)3 640 9848. **Fax:** +972 (0)3 640 9380. **E-mail:** lipkin@post.tau.ac.il

David Zakai, Israel Nature and Parks Protection Authority, P.O. Box 667, Elat 88105, Israel. The Interuniversity Institute for Marine Sciences, P.O. Box 469, Elat 88103, Israel.

37. Haritonidis S, Diapoulis A, Nikolaidis G [1990]. First results on the localization of the herbiers of marine phanerogams in the gulf of Thermaikos. *Posidonia Newsletter* 3: 11-18.
38. Aleem AA [1962]. The occurrence of the sea-grass: *Halophila stipulacea* (Forsk.) Asch. on the west coast of Egypt. *Bulletin of the Faculty of Science, University of Alexandria* 4: 79-84.
39. Vitiello P, Boudouresque CP, Carries JC, Hassan EMA, Maubert H, Sourenian B, Thelin I [1985]. Le benthos littoral d'El Dabaa (Méditerranée, Egypte). I Données générales sur le domaine benthique. *Rapp Comm Int Mer Méditer* 29: 245-246.
40. Thelin I, Mosse RA, Boudouresque CP, Lion R [1985]. Le benthos littoral d'El Dabaa (Méditerranée, Egypte). II. L'herbier a *Posidonia oceanica*. *Rapp Comm Int Mer Méditer* 29: 247-248.
41. Aleem AA [1955]. Structure and evolution of the sea grass communities *Posidonia* and *Cymodocea* in the southeastern Mediterranean. In: Allan Hancock Foundation, Miscellaneous Publications. Essays in the Natural Sciences in honor of Captain Allan Hancock on the occasion of his birthday, July 26, 1955, University of Southern California, Los Angeles. pp 279-298.
42. Mostafa HM [1991]. Ecological study of the marine phanerogam *Posidonia oceanica* and some of the associated communities in the Mediterranean Sea of Alexandria. PhD thesis, Faculty of Science, Alexandria University. 288 pp.
43. Mayhoub H [1976]. Recherches sur la végétation marine de la côte syrienne. Etude expérimentale sur la morphogénèse et la développement de quelques espèces peu connues. Thèse de Doctorat d'Etat (Docteur des Sciences naturelles), Université de Caen. 288 pp.
44. UNEP [1998]. Report of the workshop on invasive *Caulerpa* species in the Mediterranean. UNEP, Athens. 16 pp.
45. Argyrou M, Demetropoulos A, Hadjichristophorou M [1999]. Expansion of the macroalga *Caulerpa racemosa* and changes in softbottom macrofaunal assemblages in Moni Bay, Cyprus. *Oceanologia Acta* 22: 517-528.
46. Çirik S [1999]. A propos de la végétation de la baie d'Akkuyu (Mersin, Turquie). *Flora Mediterranea* 1: 205-212.
47. Bianchi CN, Morri C [1983]. Note sul benthos marino costiero dell'isola di Kos (Egeo sud-orientale). *Natura* 74: 96-114.
48. Coppejans E [1974]. A preliminary study of the marine algal communities of the islands of Milos and Sikinos (Cyclades, Greece). *Bulletin de la Societé Royale de Botanique de Belgique* 107: 387-406.
49. Aleem AA [1979]. A contribution to the study of seagrasses along the Red Sea coast of Saudi Arabia. *Aquatic Botany* 7: 71-78.
50. Jacobs RPWM, Dicks B [1985]. Seagrasses in the Zeit Bay area and at Ras Gharib (Egyption Red Sea Coast). *Aquatic Botany* 23: 137-147.
51. Lipkin Y [1975c]. A history, catalogue and bibliography of Red Sea seagrasses. *Israel Journal of Botany* 24: 89-105.
52. Hulings NC, Kirkman H [1982]. Further observations and data on seagrasses along the Jordanian and Saudi Arabian coasts of the Gulf of Aqaba. *Tethys* 10: 218-220.
53. Hulings NC [1979]. The ecology, biometry and biomass of the seagrass *Halophila stipulacea* along the Jordanian coast of the Gulf of Aqaba. *Botanica Marina* 22: 425-430.
54. Aleem AA [1980]. Contribution to the study of the marine algae of the Red Sea. IV – The algae and seagrasses inhabiting the Suez Canal (systematic part.) *Bulletin of the Faculty of Science, King Abdul Aziz University, Jeddah* 4: 31-89.
55. Lipkin Y [1991]. Life in the littoral of the Red Sea (with remarks on the Gulf of Aden). In: Mathieson AC, Nienhuis PH (eds) *Intertidal and Littoral Ecosystems of the World*. Ecosystems of the World. Vol 24. Elsevier, Amsterdam. pp. 391-427.
56. Pichon M [1974]. Dynamics of benthic communities in the coral reefs of Tuléar (Madagascar): Succession and transformation of the biotopes through reef tract evolution. *Proceedings of 2nd International Coral Reef Symposium* 2: 55-68.
57. Lipkin Y [1987a]. Seagrasses of the Sinai coast. In: Gvirtzman G, Shmueli A, Gardus Y, Beit-Arieh I, Har-El M (eds) *Sinai, Pt 1: Sinai – physical geography*. Eretz, Ministry of Defence Publishing House, Israel. pp 495-504 (in Hebrew).
58. Lipkin Y [1987b]. Marine vegetation of the Museri and Entedebir Islands (Dahlak Archipelago, Red Sea). *Israel Journal of Botany* 36: 87-99.
59. Lipkin Y, Silva PC [in press]. Marine algae and seagrasses of the Dahlak Archipelago, southern Red Sea. *Nova Hedwigia*.
60. Lipkin Y [1988]. *Thalassodendron ciliate* in Sinai (northern Red Sea) with special reference to quantitative aspects. *Aquatic Botany* 31: 125-139.
61. Fishelson L, Bresler V, Abelson A, Stone I, Gefen E, Rosenfeld M, Mohady O [2002]. The two sides of man-induced changes in littoral marine communities; eastern Mediterranean and the Red Sea as an example. *The Science of the Total Environment* (in press) online uncorrected proof.
62. Dando PR, Stuben D, Varnavas SP [1999]. Hydrothermalism in the Mediterranean Sea. *Progress in Oceanology* 44: 333-367.

6 The seagrasses of THE ARABIAN GULF AND ARABIAN REGION

R.C. Phillips

The seagrass ecosystem of the Arabian Gulf (hereafter called "the Gulf") is a unique biotope. The Gulf is a shallow semi-enclosed sea measuring ca 1000 km by 200-300 km[1, 2]. The average depth is only 35 m. The maximum depth of 100 m occurs near the entrance to the Strait of Hormuz. There are vast areas in some of the Gulf States, such as the United Arab Emirates (UAE), Saudi Arabia and Bahrain, with shallow areas less than 15 m deep suitable for seagrass growth.

Seagrass habitats have been designated a critical marine resource in the Gulf[1-9]. They have also been listed as a key renewable resource[2, 6].

There are only three species of seagrass in the Gulf. It is considered to be a very stressful habitat for seagrasses[1], characterized by large seasonal air and water temperature variations, fluctuating nutrient levels and high salinities. The three species found are considered to be tolerant of such conditions (Table 6.1). Outside the Gulf, as many as 11 seagrass species have been described for the Red Sea area[10]. Seven species are known in the Arabian Sea[2, 10-14], seven species in the Gulf of Aqaba[2, 13, 15] and eight in the Gulf of Suez[2].

Jones[4] observed that seagrasses occur at only six locations in Iran. He stated that the Iranian coastline was mainly rocky. No seagrasses have been reported for Iraq. Seagrass occurrence in Kuwait is quite sparse[16]. Jones[4] stated that *Halodule uninervis* was the principal species in Kuwait and reported that large beds of seagrasses extended along the coasts of Saudi Arabia. However, IUCN–The World Conservation Union[6] diagrammed seagrasses along the entire Gulf coastline occurring in scattered locations. As a whole, the resulting report stated that seagrasses were of only limited occurrence along the Saudi Arabian coast. Price[17] sampled at 53 sites along the entire coastline and found seagrasses at only 15 sites. The largest beds of seagrass occurred in the north between Safaniyah and Manifah, in Al-Musallamiyah, south of Abu'Ali, in Tarut Bay[3], in the Dawhat Zalum (Halfmoon Bay), parts of Al Uqayr and in the Gulf of Salwah[6]. Seagrass occurrence around Bahrain is extensive[2, 5]. Sheppard et al.[2] stated that seagrasses were extensive along the coasts of Qatar, but failed to provide documentation or maps. The seagrass occurrence in the UAE is also extensive[18]. An estimated seagrass occurrence of 5500 km^2 occurs in Abu Dhabi Emirate alone.

Jones[4] stated that while the coastline of Iran was mainly rocky, the western and southern coastlines of the Gulf were soft sediments. It also appears that the Gulf has its most extensive shallow flats on the western and southern coastlines. From an analysis of the largest seagrass beds within the Gulf, the beds increase in size as one proceeds eastward along the southern shoreline.

BIOGEOGRAPHY

The Arabian Gulf is characterized by large seasonal temperature variations. The area is arid and very hot for many months of the year. There are few rivers that drain into the Gulf. There is little rainfall and very little freshwater runoff. In addition, the evaporation from Gulf waters leads to salinities averaging 40 psu, but which exceed 70 psu in the Gulf of Salwah[4]. Price and Coles[19] reported that inshore waters of the Gulf vary seasonally in temperature from 10°C to 39°C and offshore from 19°C to 33°C, with salinities varying from 38 psu to 70 psu. The three species which are found in the Gulf can tolerate these extreme conditions: *Halodule uninervis*, *Halophila ovalis* and *Halophila stipulacea*.

Very few studies on seagrasses in the Gulf have been produced reporting density, biomass and primary production values. Basson et al.[3] calculated the average dry weight of seagrass leaves in Tarut Bay (Saudi Arabia) to be 128 g/m^2. They doubled this value for an

annual average. They calculated the energy content of the 175-km² seagrass bed in the bay to be 1.4×10^{11} kcal, an energy equivalent of about 95 000 barrels of oil.

Price and Coles[19] took samples from a series of sites along the entire Gulf coast of Saudi Arabia. They took triplicate samples at eight stations during four seasons in 1985 and three seasons in 1986. Seagrass biomass values ranged from 6.0 to 435 g dry weight/m² (means for each station ranged from 53.3 to 234.8 g/m²). They reported significant correlations between seagrass biomass and depth, sediment hydrocarbons and sediment grain size, but no significant correlations between biomass and season, salinity, or nutrient concentrations and heavy metals.

Kenworthy et al.[20] reported total biomass of Halodule uninervis from two heavily oiled sites at Ad Dafi and Al-Musallamiyah (northern Saudi Arabia), ranging from 50 to 116 g dry weight/m². At one non-

Map 6.1
The Arabian Gulf and Arabian region

oiled outer bay site nearby, the biomass was 188 g dry weight/m². For Halophila ovalis, the lowest values were observed in the oiled inner bay stations (12 and 17 g dry weight/m²), while the largest biomass was found in the non-oiled outer bay site (39 g dry weight/m²). The biomass of Halophila ovalis was nearly three times greater at another heavily oiled site (Jinnah Island) as compared to an unoiled site (Tanequib) (34 compared with 12 g dry weight/m²). Densities of Halodule uninervis varied from a high of 5 879 shoots/m² at oiled inner bay and mid-bay stations at Dawhat Al-Musallamiyah to the lowest densities recorded for oiled inner bay and mid-bay sites at Dawhat Ad Dafi (1 960 to 3 250 shoots/m²). Densities for Halophila ovalis at heavily oiled inner and mid-bay sites at Dawhat Al-Musallamiyah and Dawhat Ad Dafi ranged between 1 530 and 2 533 leaf pairs/m². A similar range of values existed for Halophila stipulacea. At oiled sites at Tanequib and Jinnah Island, densities ranged from 1 721 to 3 776 leaf pairs/m² for Halophila ovalis, with a single value of 2 772 leaf pairs/m² for Halophila stipulacea at Tanequib. This study was conducted in 1991, one year after the Gulf War oil spill.

In Tarut Bay, Basson et al.[3] derived tentative productivity values by converting from biomass values of seagrass leaves. They estimated the production

Table 6.1
Seagrass species in the Arabian region

Arabian Gulf	Number of species	Species
Iran	1	Halodule uninervis
Iraq	0	No seagrass
Kuwait	2	Halodule uninervis
		Halophila ovalis
Saudi Arabia	3	Halodule uninervis
		Halophila ovalis
		Halophila stipulacea
Bahrain	3	Halodule uninervis
		Halophila ovalis
		Halophila stipulacea
Qatar	3	Halodule uninervis
		Halophila ovalis
		Halophila stipulacea
United Arab Emirates	3	Halodule uninervis
		Halophila ovalis
		Halophila stipulacea
Arabian Sea	**Number of species**	**Species**
Oman	4	Halodule uninervis
		Halophila ovalis
		Syringodium isoetifolium
		Thalassodendron ciliatum
Yemen	7	Cymodocea serrulata
		Enhalus acoroides
		Halodule uninervis
		Halophila ovalis
		Syringodium isoetifolium
		Thalassia hemprichii
		Thalassodendron ciliatum

value of the leaves to be 100 g carbon/m^2/year. These calculations did not include the productivity of roots and rhizomes. These values included many assumptions and were largely hypothetical[2].

Kenworthy et al.[20] determined the net leaf productivity of *Halodule uninervis* at a single oiled station at Dawhat Ad Dafi. Values ranged from 0.094 to 0.250 g dry weight/m^2/year.

Durako et al.[21] exposed plants of all three species from a well-flushed area seaward of the south of

---

Case Study 6.1
THE BAHRAIN CONSERVANCY

In the summer of 1982, civil engineers won a contract to build a roadway from Saudi Arabia across 25 km of sea to the island state of Bahrain[27]. The causeway, consisting of five bridges linked by seven solid embankments, carries two parallel roads from Jasrah on the northeast coast of Bahrain, across Umm Na'san Island, over the Gulf of Bahrain to join the Saudi coastline at Al Aziziyah. The largest of the bridges weighs 1200 metric tons, and passes 28.5 m above the water. It can carry 3000 vehicles per hour. Halfway between Umm Na'san and Al Aziziyah, an artificial island was built to house coastguards, customs and immigration offices. Madany et al.[31] stated that the total cost of the project was US$564 million, and was one of the largest projects undertaken in the Middle East during the 1980s.

At least half of the causeway's span consists of embankments of dredged rocks and fine mud spoil. In consideration of the massive negative impacts which this project could have on the extensive seagrass beds of Bahrain, the Regional Organization for the Protection of the Marine Environment from Pollution (ROPME) arranged to cooperate with Bahrain's Directorate of Environmental Affairs to carry out an ecological study on the possible effects of building the causeway. A team from the Tropical Marine Research Unit at the University of York carried out the study through IUCN and the United Nations Environment Programme (UNEP).

SAMPLING THE SITES
Sampling was done in three visits in 1983, each of about one month. Flora and fauna, and temperature, salinity, turbidity and chlorophyll concentrations, and zooplankton were sampled at six coastal and six offshore sites.

Many of the sites sampled were deemed to be critical habitats. In the immediate vicinity of the causeway between Umm Na'san and the main island of Bahrain, the water was found to be 9 psu more saline in the now partly enclosed bay on the south side of the road than just to the north, less than 100 m away. Of greater concern were the more obvious physical impacts of dredging and reclamation. The water became more turbid the closer the team moved to the construction work. Long plumes of sediment stretched downstream from the construction areas. Just to the north of the causeway, it found a layer of very fine silt mud 20 cm thick, with a complete absence of surface flora and fauna, but a high abundance of infaunal polychaete worms dominated by *Ceratonereis*. Below the silt was a once healthy bed of seagrass, now almost completely dead. This was probably once a part of a large nursery of shrimp within the bay area, as large numbers of juveniles and fish thrived in the nearby intertidal flats. These flats also supported a rich fauna, including crabs and mollusks. At two sites south of the causeway on the west coast of Bahrain there was no clear evidence of damage.

SYMPTOMS OF STRESS
At the offshore sites on the east coast, where relatively unspoiled conditions were expected, the team instead found turbid water. On calm days, visibility was only 60 cm into the water, and a layer of fine silt covered the seabed out to the coral reefs. These reefs were already displaying the classic symptoms of sedimentary stress, with horizontal faces showing bleached skeletons devoid of polyps. The round heads of *Platygyra* looked like a monk's head with a tonsured haircut, with the top bleached white. Dead and dying branches of *Acropora* were covered with sea urchins or were turned green by colonies of epiphytic algae.

Madany et al.[31] did not state whether this project was regulated by the Environmental Protection Committee of Bahrain. This committee has established specific rules and regulations to control dredging and reclamation projects before implementation. However, the authors found that some projects were carried out without the permission of the committee, due to the lack of legislation to support the regulations.

Dawhat Al-Musallamiyah, Saudi Arabia, to unweathered Kuwait crude oil. The treatment duration was 12 to 18 hours. There were no significant photosynthesis as against irradiance response effects, nor were there any effects noted on respiration rates. Their conclusion was that the Gulf War oil spill primarily impacted intertidal communities, rather than the submerged plant communities of the northern Gulf region.

Phillips et al.[18] found density values of *Halodule uninervis* from five sites in the UAE to vary from 1 745 to 21 590 shoots/m^2, while leaf pair densities of *Halophila ovalis* from two sites varied from 166 to 1 108/m^2.

Outside the Gulf, Wahbeh[22] and Jones et al.[23], using oxygen release methods, estimated values of 1 326 g carbon/m^2/year for *Halodule uninervis* in the Gulf of Aqaba, 617 g carbon/m^2/year for *Halophila stipulacea* and 11 g carbon/m^2/year for *Halophila ovalis*[2].

The studies of Basson et al.[3] and Price et al.[24] suggested that primary production from seagrass and shallow water benthic algae may be of greater importance in the Gulf than that from phytoplankton.

Coles and McCain[25] identified a total of 834 species associated with seagrass and sand/silt substrates at seagrass stations north of Al Aziziyah (Saudi Arabia). Mean numbers of benthic organisms in the seagrass beds averaged nearly 52 000/m^2, an average of 36 000/m^2 in the Manifah-Safaniyah area[26] and up to around 67 000/m^2 in Tarut Bay.

Basson et al.[3] reported a total of 530 floral and faunal species associated with the seagrass beds in Tarut Bay, an area of 410 km^2. McCain[26] found 369 species of benthic organisms in the seagrass beds on the Saudi Arabian coast between Manifah and Bandar Mishab.

The major associated animal species of the seagrass beds of the Gulf are dugongs, green sea turtles, pearl oysters and shrimp[1, 3, 5, 6, 9, 27]. Jones[4] stated that the collapse of the shrimp fishery in the northern Gulf has been largely attributed to the loss of the critical seagrass habitat.

The preliminary studies done by Basson et al.[3] suggested that the annual production of the Tarut Bay seagrass beds may be about 230 million kg wet weight per year. This, in turn, might be expected to yield 2.3 million kg of fish, at a value of US$8 million annually, or the same quantity of shrimps at a value of US$12 million per year (conversion rate of 1 percent efficiency for use of seagrass for fish and shrimps).

IUCN[6] estimated that the industrial shrimp fishery in the Saudi Arabian Gulf and the Red Sea totaled some 6 800 metric tons, with a value of US$35.28 million. The net profit was US$13.94 million, an increase of US$4.78 million from 1982. The IUCN report concluded that this economic value of the Saudi Arabian fisheries would be maintained and increased, but only if managed on a sustainable basis. It also noted that shrimp production had dropped over the five years previous to the date of the report. This was the result of either the resource being overexploited, or a destruction of the critical seagrass habitat, or both.

Case Study 6.2
RAPID ASSESSMENT TECHNIQUE

Price[17] devised a simple, rapid assessment technique for coastal zone management requirements. The method was based on semi-quantitative (ranked) data on coastal resources, uses and environmental impacts. He recorded data at 53 geographically discrete sites, at intervals of usually less than 10 km along virtually the entire 450-km Saudi Arabian Gulf shoreline. Each sampled site comprised a quadrat 500 m x 500 m, bisecting the beach. Within each quadrat, the abundance or magnitude of the mangroves, seagrasses, halophytes, algae and freshwater vegetation were estimated and recorded semi-quantitatively. The attributes were scored using a ranked scale of 0-6 (0 was no impact; 6 was the greatest impact). For the resources, abundance scores were based on estimates of areal extent (m^2) for flora, or of estimated number of individuals for fauna, both within each sample area of the quadrat. Cluster analysis was applied after the scores were recorded. The method chosen for analyzing the biological resource data and the resource uses/impacts was the Bray-Curtis similarity index, followed by a hierarchical clustering of sites, using the arithmetically determined centroid. The results of the cluster analyses were depicted as dendrograms.

Correlations were determined between, and within, the following groups of variables: biological resources and latitude/salinity; and biological resources and uses/environmental impacts. Price[17] concluded that the method can be of value to managers and scientists alike, to determine associations between different environmental variables, and is especially useful for management.

Dugong feeding on seagrasses.

Vousden[5] linked the extensive seagrass beds surrounding Bahrain to juvenile stages of commercially important penaeid shrimp and to a number of adult fish species, e.g. *Siganus* spp., a popular local food resource. Seagrasses also provided a habitat for the settlement of high densities of pearl oyster spat (*Pinctada* sp.), an important commercial species in Bahrain. Vousden[5] reported a herd of 700 dugongs at one location over seagrass beds in Bahrain.

HISTORICAL AND PRESENT DISTRIBUTION

Jones[4] stated that seagrasses were sometimes present in the upper subtidal zone (2-3 m deep) along the Saudi Arabian coast as a band some 1-20 m wide. In these situations, seagrasses were recorded at 57 percent of the shore sites inspected, but seldom in luxuriant stands. The report estimated an areal extent of seagrasses along Saudi Arabia of 370 km^2. De Clerck and Coppejans[28] studied seagrass distribution in the Gulf sanctuary between Ras az-Zaur and the northeast point of Abu Ali. They found that *Halodule uninervis* formed extensive meadows from the low-water mark to 3 m deep. Locally, it was replaced by *Halophila ovalis* and *Halophila stipulacea*. In some places near Dawhat Ad Dafi, seagrass cover declined rapidly below 3 m deep.

At the Jubail Marine Wildlife Sanctuary (Saudi Arabia), Richmond[29] found that *Halodule uninervis* was again the dominant species, with the best developed beds at 3-4 m deep. Both species of *Halophila* were also found. Seagrasses were not found below 5 m.

Vousden[5] mapped the seagrasses of Bahrain using satellite imagery. He reported that, as far as percentage cover was concerned, seagrass beds were the major soft-bottom habitat type within the 2-12 m subtidal zone. He found areas where the seagrasses went to 14 m deep. Seagrass distribution was widespread around the islands, covering most of the east coast, from south of Fasht Adhm to the Hawar Islands. Seagrasses also covered significant areas around Fasht Jarim and along the west coast, south and north of the Saudi-Bahrain causeway and along the southwestern coast. He also reported that the seagrass beds died back to low cover in winter, but found the beds to be healthy in March 1986. He concluded that the majority of well-developed beds occurred to the southeast of Bahrain. *Halodule uninervis* was the most common species. In summer, *Halodule uninervis* cover was as high as 90 percent. More than 50 percent of the sites at which seagrasses occurred supported 40 percent or greater cover.

In the UAE, Phillips et al.[18] performed an extensive study of seagrass distribution and extent of growth in 1999 and 2000. *Halodule uninervis* was the most abundant species in the Gulf waters of the UAE. Seagrasses occurred from 1.5 m to 15 m deep. Even though *Halodule uninervis* was occasionally found at 15 m deep, *Halophila ovalis* tended to become the dominant species in depths greater than 11 m. Extensive continuous meadows were found wherever water depths were suitable (for *Halodule uninervis* from 1.5 m to 11 m deep). Digitized estimates show that there were 5 500 km^2 of seagrasses in the Gulf waters of Abu Dhabi Emirate.

PRESENT THREATS AND LOSSES

Sheppard et al.[2] listed a variety of coastal and marine uses and their major environmental impacts which affect or could affect seagrasses in the Gulf. They ranked them as short-term to medium-term impacts, medium-term to long-term impacts, and possible longer-term impacts.

Except for my own observation on the effects of oil globules and oily black films over the bottom near an oil processing plant west of Jabel Dannah (seagrasses absent under the films), none of the literature records any negative impacts from oil-related pollution in the Gulf.

Vousden[5] stated that the agricultural industry was one of the major sources of organic non-petrochemical pollution to the marine environment. He found that the agricultural sector contributed 50 percent of the total biological oxygen demand (BOD) loading to the waters around Bahrain. An oil refinery discharged 19 percent of the loading, with domestic discharges amounting to some 25 percent. The remaining discharges came from other industries.

Sheppard et al.[2] stated that coastal reclamation and dredging represented one of the most significant impacts on the coastal and marine environments of the Arabian region. They reported that coastal development and infilling have been far greater along

Case Study 6.3
MARINE TURTLES AND DUGONGS IN THE ARABIAN SEAGRASS PASTURES

The seagrass beds in the Gulf are home to the world's second largest assemblage of endangered dugongs (*Dugong dugon*) – upwards of 7 000 individuals[9] (the largest population is off the coast of Australia), distributed mostly in the southern and southwestern regions of the Gulf. The dugongs belong to the monotypic order Sirenia and are the only herbivorous marine mammals, feeding directly on seagrasses. They can live to be 70 years of age and grow to over 3 m in length and 400 kg in weight. Their nearest living non-sirenian relative is believed to be the elephant. Dugongs have extremely low reproductive capacities as they do not become sexually mature until about ten years of age, with subsequent calving only occurring at intervals of seven or more years.

The most important foraging habitats for dugongs in the Gulf are on either side of Bahrain, off Saudi Arabia between Qatar and the UAE, and off Abu Dhabi[9]. Outside the Gulf, the nearest population is in the Gulf of Kutch, northern India, suggesting the Gulf population is genetically and physically isolated. Until some 30 years ago, dugongs formed the staple diet of many Gulf-bordering villages, and had been used for their leathery skin and fats rendered into oils[9]. This suggests that populations were significantly larger than at present, and further reduction in population size might adversely impact their chances of survival.

FEEDING GROUNDS

Significant populations of herbivorous green turtles (*Chelonia mydas*) also depend on the seagrasses of the Gulf. They nest on Karan and Jana Islands off the Saudi Arabian coast (ca 1 000 females/year)[34], outside the Gulf at Ras Al-Hadd, Oman (ca 4 000 females/year)[35], and to a smaller extent off the southern coast of Iran, and are believed to feed among the seagrass pastures bordering the southern Gulf. Evidence of this is supported by recent tag returns from Saudi Arabia and Oman[36, 37]. The green turtles in the Gulf also have low reproductive capacities, with estimates of sexual maturation periods of 15-40 years, and a survival rate of hatchlings of roughly only one in a thousand. These turtles have several key physiological features that set them apart from other Testudines, such as non-retractile limbs, extensively roofed skulls, limbs converted to paddle-like flippers, and salt glands to excrete excess salt. As with other reptiles, the sex of hatchlings is dependent on temperature during incubation[38]. Adults can reach over 1 m in length and weigh over 150 kilograms, and feed nearly exclusively on seagrasses. The Gulf green turtles exhibit strong nesting site fidelity, returning to the same beaches to nest within and over several seasons[5]. This fidelity coupled with a relatively low emigration rate from the Gulf, other than to the Omani nesting site, suggests that populations which nest and feed within the Gulf are, much as the dugongs, genetically and physically isolated.

Threats to the turtle populations in the Gulf include moderate egg and adult harvesting, mortality in commercial and artisanal fishing gears, loss of nesting habitats, and significant loss or alteration of foraging grounds. While most Gulf-bordering nationals do not generally eat turtles or their eggs, many fishing boat crews are being replaced with a number of other nationalities who do, and unless the nesting beaches are patrolled the fishermen frequently dig up clutches of eggs. Fishermen are also known to take adults on an opportunistic basis[39]. An important modern impact is the extensive dredging and landfilling projects of several Gulf-bordering nations, which are altering or completely destroying foraging (seagrass) pastures. As in the case of the dugongs, the seagrasses upon which the green turtles in the Gulf depend are of supreme importance to the survival of these isolated, regionally important populations.

CALL FOR PROTECTION

Based on the genetic isolation and population sizes of these two species, a recent meeting of experts in Hanoi, Viet Nam, concluded that the Gulf seagrass habitats are of outstanding universal value at a global level, and recommended they should be protected through international instruments such as the World Heritage Convention. Although there are a number of national conservation programmes and regional initiatives, they tend to be species-specific and not, as yet, directed at preserving marine habitats other than coral reefs. There is a need for focused attention on the remaining habitats, particularly seagrass pastures, if the populations of dugongs and green turtles are to survive.

Nicolas J. Pilcher
Community Conservation Network, P.O. Box 1017, Koror, Republic of Palau

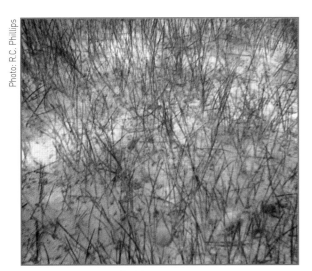

Halodule uninervis, Abu Dhabi area.

the Gulf coast than in the Red Sea or other parts of the Arabian region[6].

IUCN[6] and Sheppard and Price[30] reported that approximately 40 percent of the Saudi Arabian coast had been developed, involving extensive infilling and reclamation. They found that conditions were similar in other Gulf States, such as Bahrain and Kuwait. More than 30 km² (3 306 ha) of Bahrain was either reclaimed or artificial land[31]. In the late 1980s, there were plans for further infilling on an area of almost 200 km² in Bahrain[32]. I have observed extensive dredging activities around the UAE. These activities involved maintenance channel dredging, dredging for new channels and land reclamation. They were being carried out inshore in the most extensive continuous seagrass beds in Abu Dhabi Emirate.

Price[6] noted that dredging and coastal infilling projects were occurring throughout Saudi Arabia, e.g. Tarut Bay and the Jubail area, and also in Bahrain and Kuwait. He conjectured that such activity was likely to affect not only the shrimp and fish stocks, but also the ecology of coastal habitats generally.

Vousden[5] stated that the effects of coastal development represented a significant problem to the marine environment of Bahrain. He noted that the shallow intertidal flats next to a reclamation site became smothered in a thick glutinous silt often many centimeters deep and of little biological value due to its anoxic nature. Offshore, the benthic communities became choked by the anoxic sediments. Primary productivity was reduced drastically by the high sediment loads and consequent increase in water turbidities. Price *et al.*[24] noted that seagrass had become smothered as a result of the sedimentation caused by dredging.

Thus, many studies have recorded the large-scale and continuing dredging and land reclamation projects throughout the Gulf States. However, no one has documented the amount of historical loss of seagrasses as a result of this activity. Such studies are needed. The study of Phillips *et al.*[18] in the UAE appears to be the only study that has precisely documented the extent of seagrass distribution in any of the Gulf States. Such studies are also needed.

POLICY AND MANAGEMENT

Each Gulf State has a varying number of authorities designed to study and/or protect seagrasses. However, one can still see massive and continuing dredging and land reclamation in all countries. Since there is so little effective cooperation between the states as concerns marine conservation of seagrasses, the feeling within the Gulf is that this conservation and protection effort would be best accomplished at the regional level. There is a plan, the Kuwait Action Plan (KAP), based on the Kuwait Regional Convention for Cooperation on the Protection of the Marine Environment from Pollution. All countries within the KAP region are signatories of the convention. IUCN/UNEP[1] reported that the priority concern was the current extensive loss or severe degradation of seagrass habitats, and the probable reduction in natural resources associated with this habitat.

The reports of Price[6], the Coral Reef and Tropical Marine Research Unit[33] and Price *et al.*[24] contained detailed recommendations for conserving seagrass beds in the Gulf area. These focused largely on preventing further uncontrolled habitat destruction and widespread pollution. IUCN/UNEP[1] concluded that any legislation aimed at preventing impacts must be followed by enforcement. Little has been done to implement these recommendations. Except for the UAE, none of the countries has taken any steps to implement the beginning of an effective management program that would start with baseline mapping, followed by periodic monitoring and mapping efforts. The distribution and rate of seagrass loss needs to be determined in the various KAP countries. As of 1985, the conservation status of seagrass habitats had been considered in Bahrain[24] and Saudi Arabia[3, 33], but not in detail in any of the other KAP countries.

Sheppard *et al.*[2] stated that in addition to the UNEP Regional Seas Programme, there were other regional agreements, including those of the GCC (Gulf Cooperative Council), the GAOCMAO (Gulf Area Oil Companies Mutual Aid Organisation) and others. These agreements relate to environmental management and pollution control.

AUTHOR

Ronald C. Phillips, Florida Marine Research Institute, 100 Eighth Avenue, S.E., St Petersburg, Florida 33701, USA. **Tel (home):** +38 (0) 692 413086. **E-mail:** ronphillips67@hotmail.com

REFERENCES

1. IUCN/UNEP [1985]. The Management and Conservation of Renewable Marine Resources in the Indian Ocean Region in the Kuwait Action Plan Region. UNEP Regional Seas Reports and Studies No. 63. 63 pp.
2. Sheppard CRC, Price ARG, Roberts C [1992]. *Marine Ecology of the Arabian Region*. Academic Press, London. 359 pp.
3. Basson PW, Burchard JE, Hardy JT, Price ARG [1977]. *Biotopes of the Western Arabian Gulf*. Aramco, Dhahran. 284 pp.
4. Jones DA [1985]. The biological characteristics of the marine habitats found within the ROPME Sea Area. *Proceedings of ROPME Symposium on Regional Marine Pollution Monitoring and Research Programmes* (ROPME/GC-4/2). pp 71-89.
5. Vousden DHP [1988]. The Bahrain Marine Habitat Survey. Vol. 1. The Technical Report. ROPME. 103 pp.
6. Price ARG [1982]. Conservation and Sustainable Use of Natural Resources. Part II. Marine. Report for IUCN/MEPA for the Expert Meeting of the Gulf Coordinating Council to review environmental issues.
7. Price ARG, Chiffings TW, Atkinson MJ, Wrathall TJ [1987]. Appraisal of resources in the Saudi Arabian Gulf. In: Magoon OT, Converse H, Miner D, Tobin LT, Clark D, Domurat G (eds) *5th Symposium on Coastal and Ocean Management*. Vol. 1. American Society of Coastal Engineers, New York. pp 1031-1045.
8. Vine PJ [1986]. *Pearls in Arabian Waters*. Immel Publishing, London. 59 pp.
9. Preen A [1989]. The Status and Conservation of Dugongs in the Arabian Region. Vol. 1. MEPA Coastal and Marine Management Series Report No. 10. Meteorological and Environmental Protection Administration. Jeddah. 200 pp.
10. Jupp BP, Durako MJ, Kenworthy WJ, Thayer GW, Schillak L [1996]. Distribution, abundance and species composition of seagrasses at several sites in Oman. *Aquatic Botany* 53: 199-213.
11. Lipkin Y [1977]. Seagrass vegetation of Sinai and Israel. In: McRoy CP, Helfferich C (eds) *Seagrass Ecosystems: A Scientific Perspective*. Marcel Dekker, New York. pp 263-293.
12. Aleem AA [1979]. A contribution to the study of seagrasses along the Red Sea coast of Saudi Arabia. *Aquatic Botany* 7: 71-78.
13. Hulings NC [1979]. The ecology, biometry, and biomass of the seagrass *Halophila stipulacea* along the Jordanian coast of the Gulf of Aqaba. *Botanica Marina* 22: 425-430.
14. Jacobs RPWM, Dicks B [1985]. Seagrasses in the Zeit Bay and at Ras Gharib (Egyptian Red Sea coast). *Aquatic Botany* 23: 137-147.
15. Hulings NC, Kirkman H [1982]. Further observations and data on seagrasses along the Jordanian and Saudi Arabian coasts of the Gulf of Aqaba. *Tethys* 10: 218-220.
16. Jones DA [2002]. Personal communication.
17. Price ARG [1990]. Rapid assessment of coastal zone management requirements: Case study in the Arabian Gulf. *Ocean and Shoreline Management* 13: 1-19.
18. Phillips RC, Loughland RA, Youssef A [Submitted manuscript]. Seagrasses of Abu Dhabi Emirate, United Arab Emirates, Arabian Gulf. *Tribulus*.
19. Price ARG, Coles SL [1992]. Aspects of seagrass ecology along the western Arabian Gulf coast. *Hydrobiologia* 234: 129-141.
20. Kenworthy WJ, Durako MJ, Fatemy SMR, Valavi H, Thayer GW [1993]. Ecology of seagrasses in northeastern Saudi Arabia one year after the Gulf War oil spill. *Marine Pollution Bulletin* 27: 213-222.
21. Durako MJ, Kenworthy WJ, Fatemy SMR, Valavi H, Thayer GW [1993]. Assessment of the toxicity of Kuwait crude oil on the photosynthesis and respiration of seagrasses of the northern Gulf. *Marine Pollution Bulletin* 27: 223-227.
22. Wahbeh MI [1980]. Studies on the Ecology and Productivity of the Seagrass *Halophila stipulacea*, and Some Associated Organisms in the Gulf of Aqaba (Jordan). D.Phil. thesis, University of York.
23. Jones DA, Ghamrawy M, Wahbeh MU [1987]. Littoral and shallow subtidal environments. In: Edwards A, Head SM (eds) *Red Sea*. Pergamon Press, Oxford. pp 169-193.
24. Price ARG, Vousden DHP, Ormond RFG [1983]. Ecological Study of Sites on the Coast of Bahrain, with Special Reference to the Shrimp Fishery and Possible Impact from the Saudi-Bahrain Causeway under Construction. IUCN Report to the UNEP Regional Seas Programme. Geneva.
25. Coles SL, McCain JC [1990]. Environmental factors affecting benthic communities of the western Arabian Gulf. *Marine Environmental Research* 29: 289-315.
26. McCain JC [1984]. Marine ecology of Saudi Arabia. The nearshore, soft bottom benthic communities of the northern area, Arabian Gulf, Saudi Arabia. *Fauna of Saudi Arabia* 6: 102-126.
27. Vousden DHP, Price ARG [1985]. Bridge over fragile waters. *New Scientist* No. 1451: 33-35.
28. De Clerck O, Coppejans E [1994]. The marine algae of the Gulf Sanctuary. In: Establishment of a Marine Habitat and Wildlife Sanctuary for the Gulf Region. Final Report for Phase III. Jubail and Frankfurt. CEC/NCWCD. pp 254-280.
29. Richmond MD [1996]. Status of subtidal biotopes of the Jubail Marine Wildlife Sanctuary with special reference to soft-substrata communities. In: Krupp F, Abuzinada AH, Mader IA (eds) *A Marine Wildlife Sanctuary for the Arabian Gulf. Environmental Research and Conservation Following the 1991 Gulf War Oil Spill*. NCWCD, Riyadh and Seneckenberg Research Institute, Frankfurt.
30. Sheppard CRC, Price ARG [1991]. Will marine life survive in the Gulf? *New Scientist* 1759: 36-40.
31. Madany IM, Ali SM, Akter MS [1987]. The impact of dredging and reclamation in Bahrain. *Journal of Shoreline Management* 3: 255-268.
32. Linden et al. [1990]. State of the Marine Environment in the ROPME Sea Area. UNEP Regional Seas Reports and Studies. No. 112. Rev. 1. UNEP, Nairobi.
33. TMRU [1982]. Management Requirements for Natural Habitats and Biological Resources on the Arabian Gulf Coast of Saudi Arabia. IUCN Report to MEPA prepared by Coral Reef and Tropical Marine Research Unit. University of York.
34. Pilcher NJ [2000]. Reproductive biology of the green turtle *Chelonia mydas* in the Arabian Gulf. *Chelonian Conservation & Biology* 3: 730-734.
35. Ross JP and Barwani MA [1982]. Review of sea turtles in the Arabian Area. In: Bjorndal KA (ed) *Biology and Conservation of Sea Turtles*. Smithsonian Institution Press, Washington, DC. pp 373-382.
36. Al-Ghais. Personal communication.
37. As-Saady. Personal communication.
38. Miller JD [1985]. Embryology of marine turtles. In: Gans C, Billett F and Maderson PEA (eds) *Biology of the Reptilia*, Vol. 14. John Wiley & Sons. pp 269-328.
39. Miller JD [1989]. Marine Turtles, Volume 1: An Assessment of the Conservation Status of Marine Turtles in the Kingdom of Saudi Arabia. Coastal and Marine Management Series Report No. 9. MEPA, Jeddah. 289 pp.

7 The seagrasses of KENYA AND TANZANIA

C.A. Ochieng
P.L.A. Erftemeijer

Seagrasses are a major component of the rich and productive coastal and marine ecosystems in the East African region. The Kenyan (600 km) and Tanzanian (800 km) coastlines have a shallow and relatively narrow continental shelf bordering the Indian Ocean and are characterized by extensive fringing coral reefs, several sheltered bays and creeks, limestone cliffs, mangrove forests, sand dunes and beaches[1, 2]. The tidal amplitude is rather large – up to 4 m near Mombasa[3] – and therefore there is a fairly extensive intertidal zone between the fringing reefs and the coast in many places. The substrate in this zone consists mainly of carbonate sands derived from eroding reefs. The productivity of these intertidal areas is determined predominantly by the presence of seagrasses and macroalgae, which grow wherever shallow depressions retain a covering of water during low tide.

The most extensive seagrass meadows occur in back-reef lagoons, which are found between the beaches or cliffs and the adjacent fringing reefs. Narrow channels connect the lagoons with the sea during low tide, but high-tide waters pass over the reef crest into the lagoon. Apart from many fish species that reside permanently inside such lagoons, many other species feed there during high tide, leaving for deeper offshore waters during the ebbing tides.

At several places along the East African coast, these lagoons grade into sheltered semi-enclosed bays (e.g. at Mida, Kilifi, Mtwapa, Tudor, Gazi and Funzi in Kenya, and at Tanga, Bagamoyo, Mohoro, Kilwa and Mtwara in Tanzania) where mangroves, seagrass meadows and coral reefs occur as adjacent and interrelated ecosystems. Where the supply of terrigenous sediments is limited, seagrass vegetation is also common in the creeks and channels that run through the mangroves, possibly functioning as traps and reducing the extent of the fluxes of particulate matter and nutrients between the mangroves and the ocean. In Gazi Bay (Kenya), for example, it is possible to snorkel in creeks and small rivers inside the mangroves, where the water is very clear and the bottom is covered in a luxuriant growth of seagrasses. In the delta areas of major rivers, such as the Tana River in Kenya and the Rufiji River in Tanzania, seagrass growth is minimal.

BIOGEOGRAPHY

The following 12 seagrass species have been encountered during several studies in Kenya and Tanzania[4-14]: *Halodule uninervis*, *Halodule wrightii*, *Syringodium isoetifolium*, *Cymodocea rotundata*, *Cymodocea serrulata*, *Thalassodendron ciliatum*, *Zostera capensis*, *Enhalus acoroides*, *Halophila minor*, *Halophila ovalis*, *Halophila stipulacea* and *Thalassia hemprichii*. All species appear to be widely distributed along the entire coastline of both countries, even those with only a limited number of observations, such as *Zostera capensis*, *Halophila minor* and *Halophila stipulacea*. Seagrasses often occur in mixed communities consisting of two to several of the 12 species. *Thalassodendron ciliatum* is often the most dominant species, forming pure stands with high biomass. Three additional seagrass species (*Halodule pinifolia*, *Halophila ovata* and *Halophila beccarii*) have been reported for the region[4, 15], but these observations may constitute misidentifications and need further confirmation.

There is some controversy over the occurrence of the species *Halodule wrightii* in East Africa. Most authors have included *Halodule wrightii* in their species descriptions for the region based on leaf width and tip morphology[4, 8]. However, field observations in Florida[16] indicated that leaf tips in *Halodule* spp. vary widely from bicuspidate to tridentate on shoots of the same rhizome. Experimental culture[17] revealed that leaf tips of *Halodule* are environmentally variable, related to nutrient variability or tidal zone. Furthermore, isozyme analyses of diverse collections throughout the tropical

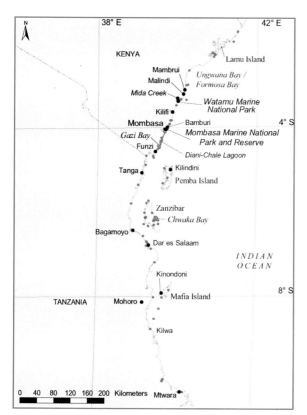

Map 7.1
Kenya and Tanzania

western Atlantic as well as the Indo-Pacific revealed a clear genetic difference between the two ocean systems, but genetic uniformity within each of the two ocean systems[18]. Based on these results it was concluded that all plants (with this morphology) from the Indo-Pacific are *Halodule uninervis* while those in the tropical western Atlantic are *Halodule wrightii*[19]. Nevertheless, *Halodule wrightii* continues to be reported in literature despite these field, culture and isozyme findings[16, 17, 19]. It appears therefore that there is a need for further analyses of chromosomal differences and physiological studies to determine the relationship between nutrients and leaf morphology of *Halodule* species.

The seagrass beds in East Africa, as indeed elsewhere, harbor a diverse array of associated plant and animal species. Detailed studies on seagrass associates in this region have identified over 50 species of macroalgae and 18 species of algal epiphytes[5, 9, 14, 20], at least 75 species of benthic invertebrates[9, 21, 22] – especially gastropods and bivalves – several species of sea cucumbers[22] and at least seven sea urchin species[23-25], various shrimp, lobster and crab species[26-28] and over 100 fish species[26-29] in association with seagrass beds. This clearly underscores the importance of seagrass meadows for biodiversity conservation.

Seagrass beds in the region also support sizeable populations of two endangered species, i.e. the green turtle *Chelonia mydas*[30-33] and the dugong *Dugong dugon*[31, 33-37], both of which feed on seagrasses. In 1994, a total of 443 sea turtles was recorded along the Kenyan coast, among which the green turtle was by far the most common species[33]. In Tanzania, there are no recent population studies[31]. Similar surveys along the Kenyan coast revealed ten dugongs during November 1994 and six dugongs during February-March 1996, representing a significant decline in comparison to earlier counts of over 50 animals in the 1960s and 1970s[33, 37]. The most important dugong habitat in Kenya can be found in the Lamu Archipelago. In Tanzania, the main centers of dugong population have been reported along the Pemba-Zanzibar channels and in the Rufiji-Mafia area[31]. The need for protection and management of sea turtle and dugong habitats (seagrass beds) has been stressed[37].

The importance of East African seagrass ecosystems for fisheries is gradually emerging from an increasing research effort on the role of the seagrass meadows in this region as nursery, breeding and feeding grounds for marine fish and crustacean species of economic importance such as shrimps (*Penaeus*) and spiny lobster (*Panulirus*)[27, 28, 38]. Several fish species graze on seagrasses, notably rabbitfishes (Siganidae) and surgeonfishes (Acanthuridae), while parrotfishes (*Leptoscarus* spp.) preferentially graze the epiphytes on the seagrass. Adult fishes, such as snappers, groupers, grunts and barracuda, feed on the infauna of seagrass beds while the diet of their juvenile stages is mainly seagrass-derived detritus. *Portunus pelagicus*, an important contributor to the crab fishery in Bagamoyo and Dar es Salaam, is said to inhabit shallow coastal habitats such as estuaries, sheltered bays and open sublittoral waters (all of which may include seagrass), where all stages of its life cycle are found[26].

Significantly higher fish abundance and catch rates were found in seagrass beds in comparison to bare sand areas in a study of dema trap fishery in the coastal waters of Zanzibar, Tanzania[28]. Similarly, 11 of the 99 fish species of Tudor Mangrove Creek (Kenya) are typically associated with seagrass (6 percent of the total catch)[29], while 74 species of fish (in a total of 39 families) and 15 species of macro-crustaceans were reported for the seagrass beds of Chwaka Bay and Paje on Zanzibar[27]. At both of these latter sites, *Gerres oyena* was the dominant fish species in the seagrass beds (>60 percent of the total catch).

Harvesting of bivalves (notably *Anadara antiquata*, *Anadara natalensis* and *Anadara uropigilemana*, *Gardium assimile*, *Gardium pseudolina*, *Gardium flavum* and *Scapharca erythraeonensis*) and gastropods (including *Murex ramosus*, *Pleuroploca*

Case Study 7.1
GAZI BAY, KENYA: LINKS BETWEEN SEAGRASSES AND ADJACENT ECOSYSTEMS

Gazi Bay, a semi-enclosed bay (15 km^2) ca 50 km south of Mombasa, is characteristic of the creeks and bays along the East African coastline. Mangroves, seagrass meadows and coral reefs occur here as adjacent ecosystems. Mangroves are found along small seasonal rivers on the landward side of the bay and are drained by two tidal creeks. Extensive seagrass vegetation is common among and between the mangroves, where it functions as a trap reducing the flux of sediment, organic material and nutrients from the mangroves to the ocean. Snorkeling in creeks and small rivers among the mangroves can be a sensational experience: the water can be very clear and the bottom is covered in a luxuriant growth of seagrasses, traversed by mangrove roots where schools of juvenile fish hide from predators. Adjacent to the mangroves on the seaward side are intertidal flats, intersected by some channels, and shallow subtidal areas which stretch to the fringing reef. Most of this area is covered by various species of seagrasses and macroalgae, with the exception of a few sandy patches[5, 13]. Seagrasses in Gazi Bay cover an estimated total area of approximately 8 km^2. The maximum tidal range in Gazi Bay is 250 cm.

All the 12 seagrass species of eastern Africa are found in Gazi Bay. Macroalgae are among their most conspicuous floral associates. Sixteen species of Chlorophyta, 4 species of Phaeophyta and 31 species of Rhodophyta associated with seagrass beds in Gazi Bay have been identified[5]. Among these were *Euchema*, *Gracilaria*, *Ulva* and *Sargassum*, all of which include species of potential economic value. Average leaf production of *Thalassodendron ciliatum*, the most dominant seagrass species, ranges from 4.9 to 9.5 g/m^2/day[7, 11]. A separate study of the growth and population dynamics of *Thalassodendron ciliatum* has shown that its vertical growth is the fastest reported for any seagrass to date (42 internodes, i.e. 42 leaves/year), whereas the horizontal growth rate (16 cm/year) is among the slowest[41]. As a result of the slow horizontal rhizome growth, shoot recruitment through branching of vertical shoots is an important part of the clonal growth of this population and so an essential component of the production of *Thalassodendron ciliatum*.

Seagrass meadows are open systems subject to nutrient impoverishment due to export processes mediated by tidal inundation. The intriguing feature of the occurrence, in very close proximity to one another, of mangroves, seagrasses and corals in Gazi Bay attracted scientists to study the interlinkages between these systems in terms of dissolved nutrients and seston fluxes as well as shuttle movements of fish[71]. Analysis of the stable isotope signature of the sediment carbon in the seagrass zone revealed significant carbon outwelling from the mangroves, but deposition of particulate organic matter rapidly decreased with distance from the forest, with most litter trapped within 2 km of the mangroves[7]. However, marked decreases in the carbon signature of seston flowing over the seagrass zone during flood tides pointed to a reverse flux of organic particles from the seagrass zone to the mangroves, with the nearby coral reefs existing in apparent isolation. Direct flux measurements of both mangrove and seagrass litter showed that trapping of mangrove litter by adjacent seagrasses is reciprocated by a retention of seagrass litter in the mangrove, and this give-and-take relationship is mediated by tides. Further research has indicated that the detrital cycling in the inner parts of the mangrove forest forms part of a rather closed system based on local inputs, whereas cycling in the outer parts of the forest is tightly connected with the adjacent seagrass ecosystem[13]. Despite the presence of tide-mediated chemical fluxes which allow one system to influence another, the input of mangrove carbon did not coincide with enhanced leaf production of the dominant subtidal seagrass *Thalassodendron ciliatum*[48]. Presumably, carbon outwelling from the mangrove coincides with only limited export of nitrogen and phosphorous, and the restricted effects of these nutrients on the seagrass (if any) are masked by other local factors.

Gazi Bay is typical of the major fishing grounds in Kenya, most of which are located in shallow near-coastal waters due to a lack of sophisticated gear and motorized boats which would allow exploitation of deeper waters. Carbon isotope and delta ^{15}N studies into trophic relationships in Gazi Bay allowed the identification of three trophic levels, i.e. herbivores, zoobenthiplanktivores and piscivores/benthivores. Seagrass beds were found to be the main feeding grounds providing food for all fish species studied in Gazi Bay, Kenya[49]. Seagrass plants were the major source of carbon for four fish species studied in the bay. They also contribute (together with mangroves) to the particulate organic carbon for prawn larvae, zooplankton, shrimps and oysters, hence their support for food webs[72].

trapezium and *Oliva bulbosa*) for food is common on many of the intertidal areas (with or without seagrass) in Tanzania. No data currently exist on the quantities collected from seagrass areas. *Strombus gibberulus*, *Strombus trapezium* and *Cypraea tigris*, all of which are popular curio goods, are common in seagrass areas around Dar es Salaam[9]. Twenty species of sea cucumbers, the most common of which are *Holothuria scabra*, *Holothuria nobilis*, *Bohadschia vitiensis*, *Bohadschia argus*, *Thelenota anax*, *Stichopus chloronotus*, *Stichopus variegatus* and *Stichopus hemanni*, are harvested from intertidal areas (including seagrass beds) along the Tanzania coast for export[22, 39].

PRODUCTIVITY AND VALUE

Studies on the ecological processes and functioning of seagrass ecosystems in Kenya and Tanzania have provided a better understanding of the natural factors limiting the growth and geographical distribution of seagrasses, environmental stresses and indirect values of seagrass ecosystems in this region.

Leaf productivity of *Thalassodendron ciliatum* ranges from 4.9 to 9.5 g/m^2/day[7, 11, 12, 40]. Vertical growth rates of *Thalassodendron ciliatum* (42 internodes, i.e. 42 leaves/year) measured in Kenya are among the fastest reported for any seagrass species to date, whereas its horizontal growth rates (16 cm/year) rank among the slowest[41]. Shoot recruitment rates measured in seagrass meadows along the coasts of Kenya and Zanzibar were either the same as or larger than shoot mortality rates, suggesting that the environmental quality in this region is still suitable for sustaining vigorous seagrass vegetation[42].

Most factors that govern primary production, including light and temperature, are relatively constant throughout the year in this region. However, the composition of the oceanic water and the amount of freshwater which enters the coastal areas are variable. At several sites along the coast substantial seepage of freshwater occurs, as a result of which brackish water is often found in areas of seagrass beds[42]. Using nitrogen stable isotope signatures, groundwater was found to influence seagrass species diversity and abundance where *Thalassodendron ciliatum* dominated high groundwater outflow areas as opposed to *Thalassia hemprichii*.

Photosynthetic studies carried out in Zanzibar, Tanzania, indicate that seagrasses may respond favorably to any future increases in marine carbon dioxide levels due to global climate change[43, 44]. The enhanced photosynthetic rates by *Halophila ovalis* and *Cymodocea rotundata* in the high, frequently air-exposed, intertidal zone may have been related to a capacity to take up the elevated HCO_3^- levels directly[43]. Furthermore, these tropical intertidal seagrasses were found to be more sensitive to desiccation than subtidal seagrasses with the exception of the species *Syringodium isoetifolium*[45]. Desiccation tolerance, however, may not be a trait that determines the vertical zonation of tropical seagrasses. The ability to tolerate high irradiances, as well as the high nutrient inputs from the shore, apparently allows the shallow species to occupy the uppermost intertidal zone.

Seagrass beach cast material may contribute significantly to beach stability, as implied by a study along the Kenyan coast[12, 46] (see Case Study 7.2).

Detailed studies in Gazi Bay, Kenya, revealed significant carbon outwelling from the mangroves into the adjacent seagrass meadows and a reverse flux of organic particles from the seagrass zone to the mangroves, with nearby coral reefs existing in apparent isolation[7] as far as particulate organic matter is concerned. Export of organic matter from mangroves in Chwaka Bay (Tanzania) was also limited to a narrow fringe of seagrasses immediately adjacent to the mangroves[47]. Despite the presence of tide-mediated chemical fluxes, which allow one system to influence another, the input of mangrove carbon did not coincide with enhanced leaf production of the dominant subtidal seagrass *Thalassodendron ciliatum*[48].

Carbon isotope and delta ^{15}N studies on trophic relationships showed that seagrass beds were the main feeding grounds for all fish species studied in Gazi Bay, Kenya[49]. An experiment on feeding preference showed that *Calotomus carolinus* (Scaridae), the second most abundant fish in Watamu Marine National

The catch from a trap fishing trip in the seagrass beds – mainly the seagrass parrotfish *Leptoscarus vaigiensis*, some pink ear emperor *Lethrinus lentjan*, and a grouper *Epinephelus flavocaeruleus*.

Park, preferred pioneering short-lived seagrass species to climax species. The study also highlighted the role of grazing fish in influencing seagrass abundance[50].

Sea urchins mediate the competitive success of different seagrass and fish species, in terms of distribution and abundance. Sea urchins can reduce grazing rates of some species of parrotfish[25], while the relative dominance of some of the sea urchin species indicates a high fishing pressure on herbivorous fish species[23]. *Tripneustes gratilla*, for instance, can graze at a rate of 1.8 seagrass shoots/m^2/day at fronts that support a sea urchin abundance of 10.4 individuals/m^2 [24]. The species composition of seagrass communities in reef environments appears to be partially affected by prey choices of the dominant grazers. Parrotfishes and the sea urchin *Echinothrix diadema* appear to favor seagrass beds dominated by *Thalassodendron ciliatum*, while other sea urchin species such as *Diadema setosum*, *Diadema savignyi* and *Echinometra mathaei* favor areas high in *Thalassia hemprichii*[25].

There have been few studies on western Indian Ocean seagrasses to date. A recent bibliographic survey of marine botanical research outputs from East Africa between 1950 and 2000 yielded only 44 papers and reports that dealt with seagrasses[51]. Even baseline data on distribution are largely lacking[52]. In recent years, however, the number of seagrass publications from studies in the region has increased and efforts are under way for integrated coastal zone management and participatory management of marine protected areas including seagrass beds, indicating a growing recognition of the important value of seagrass ecosystems.

Massive beaching of seagrass litter was reported as early as 1969 by an expedition to Watamu on the Kenyan coast[53], rendering it unlikely that these accumulations have increased over past or recent years[46].

No direct utilization of seagrasses in East Africa has been reported[51] except for anecdotal reference to the small-scale use of the leaves of *Enhalus acoroides* for weaving mats and thatching huts, and the harvesting of their rhizomes by people of the Lamu Archipelago in Kenya, who dry and then grind them into flour for cooking what is locally known as *mtimbi*[54]. Quantitative data on such direct uses as well as catch statistics of the seagrass-associated fisheries in this region are lacking, making it impossible to draw any conclusions regarding trends. There are no published data on estimates of area loss or degradation from the East African region[52, 55]. At present, there are insufficient data for even a crude estimate.

ESTIMATED COVERAGE

There are very few area estimates for seagrasses in this region. Distribution maps of seagrasses are only available for Mida Creek, Gazi Bay, Diani-Chale Lagoon and Chwaka Bay. The recent *UNEP Atlas of Coastal Resources* shows that seagrass beds occur throughout the 600-km-long Kenyan coastline in sheltered tidal flats, lagoons and creeks, with the exception of the coastal stretch adjoining the Tana Delta[1]. The testing of a remote-sensing methodology for seagrass mapping in southern Kenya estimated the net area of vegetation cover to be approximately 33.63 km^2 within a stretch of around 50 km of coastline[6]. Ground-truthing revealed that most of these areas were dominated by pure stands of *Thalassodendron ciliatum*.

Chwaka Bay on the eastern side of Unguja Island (Zanzibar), which covers more than 100 km^2, has extensive mixed seaweed-seagrass areas, with seagrasses representing between 50 and 80 percent of the macroflora biomass[5]. In Gazi Bay, which covers approximately 15 km^2, seagrass beds cover an area of approximately 8 km^2 from the lower margin of the mangrove forest through the intertidal and subtidal flats up to the fringing reef, with the exception of a few sandy patches[5, 13]. The Diani-Chale Lagoon along the Kenyan coast measures roughly 6 km^2 with seagrass beds covering up to 75 percent[14]. The Nyali-Shanzu-Bamburi Lagoon, with a total area of approximately 20 km^2, is 60 percent covered by seagrass beds[56].

At present, there are insufficient data for even a "best guess" of total seagrass coverage in Kenya and Tanzania, but new mapping data are expected to become available from a recently started regional seagrass research project under the Marine Science for Management Programme (MASMA).

THREATS

The lack of a true continental shelf, stretching out no more than a few kilometers from the Kenyan and Tanzanian shores, makes the coastal resources all the more vulnerable to overexploitation and influences from activities on land[39]. In general, seagrasses appear to have experienced fewer direct negative impacts than mangroves or coral reefs in the region, but this may merely reflect the lack of any reliable (quantitative) data. Deepening of channels for ships at harbors results in uprooting and burial of seagrass plants by dredge-spoil[54].

Several beaches and adjacent coastal areas in Kenya and Zanzibar are under increasing pressure from expanding tourism development[57]. High hotel density in close proximity to the beach is common[46]. Seagrass beds are locally damaged by motor boat propellers and anchoring in the waters near these highly intensive tourist areas[54]. While mooring buoys

have been deployed within the marine park to protect the coral reef, the seagrass beds remain unguarded. In some areas very popular with tourists stretches of seagrass meadow (deemed a nuisance to swimmers) are cleared by cutting and/or uprooting[54]. In addition, the cumulative effects of raking, burying and removing seagrass beach cast material may have negative impacts on the functioning of the adjacent seagrass meadows[46].

Direct destruction of seagrass vegetation occurs by trawling activities. Commercial trawlers operating in the Rufiji Delta, Mtwara and coastal areas between Bagamoyo and Tanga (Tanzania), as well as Ungwana Bay in Kenya (where they reportedly have a fishing effort well beyond the potential sustainable yield[58]), are non-selective and are destructive to the seabed. Illegal trawling – even during the closed season – occurs in Bagamoyo, Tanzania, where up to 80 percent of prawn bycatch is seagrass[22]. Trawling has also been reported as a major cause of mortality of the green turtle along the Kenyan coast[33]. Artisanal fishermen often connect separate navigable channels by digging through intertidal flats in order to make way for their canoes, causing damage to seagrass, albeit at a small scale[54]. Overfishing could pose a likely threat to seagrass communities, as has been reported for the coral reefs in this region[59], although there are no direct reports to confirm this.

Recent agricultural activities in the Sabaki catchment have resulted in accelerated soil erosion and a tremendous increase in river sediments from some 58 000 tons/year in 1960 up to as much as 7-14 million metric tons/year at present[60]. Considerable amounts of sediment brought down by the river to the coral reefs and seagrass beds have been implicated in the low seagrass species composition at Mambrui[54]. The apparent absence of seagrass beds in Ungwana Bay and northern Rufiji Delta[1, 2] might also be related to siltation by the Tana and Rufiji Rivers, but no studies have been conducted here.

Oil pollution is one of the potential threats to seagrasses in East Africa owing to spillage of crude oil in harbors and the risk posed by a large fleet (over 200 oil tankers per day) from the Middle East across the coastal waters. There have been no major oil spills to date, except in 1988 when 5 000 metric tons from a pierced fuel tank in Mombasa destroyed a nearby area of mangroves and associated biotopes. The seagrass species *Halophila stipulacea* and *Halodule wrightii* have not reappeared at the site since the spill[54]. In Tanzania, oil pollution along the coast – though not severe – is heaviest during the southwest monsoon[61]. The extent and specific effects of oil pollution on seagrass ecosystems in East Africa's largest harbors, Kilindini and Dar es Salaam, especially in creeks and

Impacts of tourism industry on seagrasses. Seagrass cover has declined in front of a Mombasa north coast beach hotel.

sheltered lagoons, may be high but remain uninvestigated to date.

Increasing populations in coastal towns and cities, such as Mombasa, Malindi and Dar es Salaam, present a potential (but localized) threat to the coastal seagrass resources from domestic solid waste, sewage disposal and dredge spoil dumping, all of which are responsible for the declining water quality[62]. Seasonal blooming of *Enteromorpha* and *Ulva* species occurs locally, especially in areas close to sewage discharge points from hotel establishments and municipal sewage[9, 14, 63]. Although low organic loading is a feature of the well-flushed lagoon system, eutrophic conditions and high bacterial contamination in the sheltered and semi-enclosed creeks have been reported[63]. Significant heavy metal pollution from urban and industrial effluents has been reported in coastal waters around Dar es Salaam[64], affecting edible shellfish populations[65]. Reclamation of tidal flats, such as proposed by the Selander Bridge coastal waterfront reclamation project in Dar es Salaam, constitutes another potential threat to seagrass ecosystems.

The expanding open-water mariculture farms of the seaweed *Eucheuma spinosa* currently cover around 1 000 ha of intertidal area on Zanzibar (Tanzania). The various adverse effects that seaweed farming has on intertidal areas could well mar the positive picture of its socioeconomic benefits to coastal people. A marked decline in seagrass cover from physical clearing of seagrass vegetation by seaweed farmers has been reported[66]. Seaweed farming areas on Zanzibar appear

Case Study 7.2
SEAGRASS BEACH CAST AT MOMBASA MARINE PARK, KENYA: A NUISANCE OR A VITAL LINK?

The Mombasa Marine National Park and Reserve (Kenya) encompasses a major part of the Nyali-Shanzu-Bamburi Lagoon (20 km^2) which has a maximum depth of 6 m. It is bordered by white sandy beaches on the landward and a fringing reef on the seaward sides. Mixed seagrass communities (dominated by *Thalassodendron ciliatum*) and associated seaweeds cover 60 percent of the lagoon, which is typical for most of such lagoons along the Kenyan coast. Turbulent water motion (exposure) in these areas is relatively high compared with the sheltered creeks and bays and, due to hydrodynamic forcing, the spatial and temporal concentrations of nutrients and chlorophyll *a* do not reach eutrophic levels because the lagoon is well flushed[63].

Large banks of macrophytes, 88 percent of which is seagrass, are deposited on the beaches (beach cast) as the plants become detached from the sea bottom by the surge effect from waves. This phenomenon is seasonal and controlled by tides and monsoon winds[12, 46]. The most intense accumulations – as much as 1.2 million kg dry weight along a 9.5-km stretch of beach – are washed ashore during the southeast monsoons when wind and current speeds, water column mixing and wave height are usually greatest[3].

The Mombasa Marine Park is "fenced" by a stretch of about 30 hotels, whose guests enjoy the white sandy beaches and other water sports within a stone's throw of their rooms. Although the beach cast phenomenon is seasonal and only peaks during the low season, the burgeoning tourism industry in this area considers it a nuisance and would prefer its removal. Some of the hotels employ staff to rake the seagrass material from the beach in the immediate vicinity of the hotel and bury it under the sand. A detailed study on the beach cast phenomenon showed that burying the material does not significantly affect decomposition rates[12].

The same study, however, also pointed to the role of seagrass beach cast in contributing to beach stability. By filtering out wave action, the beach cast material can reduce erosion of beaches caused by swash/backwash processes. The beach cast material may also reduce beach erosion due to wind. Furthermore, the cumulative effects of removing seagrass beach cast may intensify beach erosion either through the export of sand in the process or the loosening up of compact sand, or through removal of the protecting material that slows wave action. The potential rate of beach erosion in this study was estimated at 492 450 kg of beach sand (per removal) if beach cast material at any given

to have a lower abundance of meio- and macrobenthos than unvegetated sandy areas, and may cause declining seagrass productivity due to shading[66, 67].

POLICY RESPONSES
Kenya has been one of the most active countries in marine conservation in Africa. The first marine protected area was gazetted as early as 1968. Kenya's guidelines for establishing parks and reserves, safeguarding marine ecosystems and preserving rare species have been adopted from the United Nations Environment Programme's (UNEP's) Action Plan for the East African Regional Seas Programme[68].

There are no existing management practices to protect existing seagrass beds from overexploitation or pollution *per se*. However, concern for the marine environment is demonstrated by the establishment of six marine protected areas covering a total area of 850 km^2 while an additional marine reserve has been proposed. All protected areas are under the custodianship of the Kenya Wildlife Service, a well-equipped parastatal organization that has received much donor support. Most, if not all, of the marine protected areas in Kenya contain seagrass beds, but detailed distribution maps of seagrasses in these protected areas are not available.

In addition, several legal and administrative instruments address aspects related to the protection and management of marine protected areas and thus (indirectly) of seagrass ecosystems. These include the protection of wildlife species, regulation of fisheries, land planning and coastal developments, research and tourism. Dugongs and turtles are both listed as "protected animals" under the Wildlife Conservation and Management Act and various initiatives for their conservation have been implemented. By working closely with respective local authorities, the Kenya Wildlife Service may avoid approval of activities that could impact negatively on marine parks, as provided for under the Land Planning (1968) and the Physical

moment was removed from the entire beach (9.5-km stretch).

The total annual deposition of seagrass beach cast was estimated to be in the order of 6.8 million kg dry weight, indicating that about 19 percent of the annual production of seagrass meadows (14.7 million kg carbon/year) in the lagoon passes through the beach, where decomposition is accelerated through exposure to oxygen availability, drying and vigorous fragmentation by wave action[12]. These processes speed up the release of dissolved nutrients and particles back into the adjacent ecosystems and thus contribute to the detrital or energy pathways. The material was further found to contain over 23 000 amphipods/m^2, 3 100 isopods/m^2 and various other faunal groups, providing an important food source for fishes during high tide. The role of seagrass beach cast accumulations in nutrient regeneration processes and beach stability, and as nursery sites and a source of food for fish, crabs and shorebirds in the nearshore zone, is thus highly significant. Removing seagrass beach cast, though desirable for tourism, would have negative impacts on the health and functioning of adjacent seagrass beds, on which artisanal fisheries and tourism itself rely.

Seagrass beach cast in Mombasa Marine National Park, 1995 – significant amounts of seagrass litter are washed up on the Kenyan beaches with each tide.

Planning (1996) Acts. Efforts are being made to encourage environmentally sensitive tourism as one of the measures to achieve protection goals. A draft national strategy for sea turtle conservation is currently under review while seagrasses have been considered in the most recent management plan of the Mombasa Marine National Park and Reserve.

Outside marine protected areas, however, management and control over the exploitation of coastal and marine resources are virtually non-existent. Conservation of coastal and marine systems has concentrated its attention on either tourism-related or directly exploitable marine resources such as shells and coral reefs. Therefore it would seem that the important functions of seagrass beds related to fisheries nursery grounds, or to marine primary production, their contribution to energy pathways (involving a diversity of organisms), or linkages with land-based activities are not at the top of the conservation agenda.

Since the recommendations to establish marine protected areas in Tanzania[69], the first two marine parks, of which seagrass ecosystems are part, were only recently gazetted in 1995. Despite considerable effort, the management of protected areas in Tanzania, as in many developing countries, suffers from insufficient capacity and law enforcement. One of the stated objectives of the National Fisheries Policy and Strategy, provided for by the Fisheries Act (1970), is to protect the productivity and biological diversity of coastal and aquatic ecosystems by preventing habitat destruction, pollution and overexploitation.

Tanzania's Coastal Management Partnership, whose goal is to establish a foundation for effective coastal zone governance, has produced the first national programme for Integrated Coastal Management. Among the first outputs of this program are a State of the Coast Report, a National Mariculture Issue Profile, guidelines and a conflict resolution forum dealing with such issues as trawling and dynamite fishing. A draft national Integrated Coastal Manage-

ment Strategy is awaiting government approval. These initiatives and, more so, the process have raised the profile and level of understanding of the importance of coastal and marine resources, including seagrass beds. Implementation of integrated coastal zone management initiatives in Tanzania is currently under way in Tanga (by IUCN–The World Conservation Union), Zanzibar (Menai Bay Conservation Project), Mafia Marine Park (by WWF – the World Wildlife Fund), and in Kinondoni (Kinondoni Coastal Area Management Programme)[70].

AUTHORS

C.A. Ochieng and P.L.A. Erftemeijer, WL | delft hydraulics, P.O. Box 177, 2600 MH Delft, the Netherlands.. **Tel:** +31 (0)15 285 8924. **Fax:** +31 (0)15 2858718. **E-mail:** paul.erftemeijer@wldelft.nl

REFERENCES

1. UNEP [1998]. *Eastern Africa Atlas of Coastal Resources. Kenya.* United Nations Environment Programme, Nairobi. 119 pp.
2. UNEP [2001]. *Eastern Africa Atlas of Coastal Resources. Tanzania.* United Nations Environment Programme, Nairobi. 111 pp.
3. McClanahan TR [1988]. Seasonality in East Africa's coastal waters. *Marine Ecology Progress Series* 44(2): 191-199.
4. Aleem AA [1984]. Distribution and ecology of seagrass communities in the Western Indian Ocean. In: Angel MV (ed) Marine Science of the North-West Indian Ocean and Adjacent Waters. *Deep Sea Research* 31(6-8): 919-933.
5. Coppejans E, Beeckman H, de Wit M [1992]. The seagrass and macroalgal vegetation of Gazi Bay (Kenya). *Hydrobiologia* 247: 59-75.
6. Dahdouh-Guebas F, Coppejans E, van Speybroeck D [1999]. Remote sensing and zonation of seagrasses and algae along the Kenyan coast. *Hydrobiologia* 400: 63-73.
7. Hemminga MA, Slim E, Kazungu J, Ganssen GM, Nieuwenhuize J, Kruyt NM [1994]. Carbon outwelling from a mangrove forest with adjacent seagrass beds and coral reefs (Gazi Bay, Kenya). *Marine Ecology Progress Series* 106: 291-301.
8. Isaac FM [1968]. Marine botany of the Kenya coast. 4. Angiosperms. *Journal of East African National History Society* 27: 29-47.
9. Lugendo BR, Mgaya YD, Semesi AK [2001]. The seagrass and associated macroalgae at selected beaches along Dar es Salaam coast. In: Richmond M, Francis J (eds) *Marine Science Development in Tanzania and Eastern Africa. Proceedings of the 20th Anniversary Conference on Advances in Marine Science in Tanzania. 28th June-1st July 1999, Zanzibar, Tanzania.* IMS/WIOMSA. pp 359-374.
10. McMillan C [1980]. Flowering under controlled conditions by *Cymodocea serrulata, Halophila stipulacea, Syringodium isoetifolium, Zostera capensis* and *Thalassia hemprichii* from Kenya. *Aquatic Botany* 8(4): 323-336.
11. Ochieng CA [1995]. Productivity of seagrasses with respect to intersystem fluxes in Gazi Bay (Kenya). In: Interlinkages between Eastern-African Coastal Ecosystems. Contract No. TS3*-CT92-0114, Final Report. pp 82-86.
12. Ochieng CA, Erftemeijer PLA [1999]. Accumulation of seagrass beach cast along the Kenyan coast: A quantitative assessment. *Aquatic Botany* 65(1-4): 221-238.
13. Slim FJ, Hemminga MA, Cocheret de la Morinière E, van der Velde G [1996]. Tidal exchange of macrolitter between a mangrove forest and adjacent seagrass beds (Gazi Bay, Kenya). *Netherlands Journal of Aquatic Ecology* 30(2-3): 119-128.
14. Uku JN, Martens EE, Mavuti KM [1996]. An ecological assessment of littoral seagrass communities in Diani and Galu coastal beaches, Kenya. In: Björk M, Semesi AK, Pedersén M, Bergman B (eds) *Current Trends in Marine Botanical Research in the East African Region. Proceedings of the 3-10 December 1995 Symposium on the Biology of Microalgae, Macroalgae and Seagrasses in the Western Indian Ocean.* SIDA Marine Science Programme, Department for Research Cooperation, SAREC. pp 280-302.
15. Akil JM, Jiddawi NS [2001]. A preliminary observation of the flora and fauna of Jozani-Pete mangrove creek, Zanzibar, Tanzania. In: Richmond M, Francis J (eds) *Marine Science Development in Tanzania and Eastern Africa. Proceedings of the 20th Anniversary Conference on Advances in Marine Science in Tanzania. 28th June-1st July 1999, Zanzibar, Tanzania.* IMS/WIOMSA. pp 343-357.
16. Phillips RC [1967]. On species of the seagrass *Halodule* in Florida. *Bulletin of Marine Science* 17: 672-676.
17. McMillan C, Williams SC, Escobar L, Zapata O [1981]. Isozymes, secondary compounds and experimental cultures of Australian seagrasses in *Halophila, Halodule, Zostera, Amphibolis* and *Posidonia. Australian Journal of Botany* 29: 247-260.
18. McMillan C [1983]. Morphological diversity under controlled conditions for the *Halophila ovalis – Halophila minor* complex and *Halodule uninervis* complex from Shark Bay, Western Australia. *Aquatic Botany* 17: 29-42.
19. Phillips RC, Menez EG [1988]. *Seagrasses.* Smithsonian Contributions to the Marine Sciences No. 34. 104 pp.
20. Semesi AK [1988]. Seasonal changes of macro-epiphytes on the seagrass *Thalassodendron ciliatum* (Forssk) den Hartog at Oyster Bay, Dar es Salaam, Tanzania. In: Mainoya JR (ed) *Proceedings of a Workshop on Ecology and Bioproductivity of Marine Coastal Waters of Eastern Africa, Dar es Salaam, Tanzania, 18-20 January 1988.* Faculty of Science, University of Dar es Salaam. pp 51-58.
21. Kayombo NA [1988]. Ecology and fishery of gastropods and molluscan species along the Dar es Salaam coast. In: Mainoya JR (ed) *Proceedings of a Workshop on Ecology and Bioproductivity of Marine Coastal Waters of Eastern Africa, Dar es Salaam, Tanzania, 18-20 January 1988.* Faculty of Science, University of Dar es Salaam. pp 59-65.
22. Semesi AK, Mgaya YD, Muruke MH, Francis J, Mtolera M, Msumi G [1998]. Coastal resources utilization issues in Bagamoyo, Tanzania. *Ambio* 27: 635-644.
23. McClanahan TR, Shafir SH [1990]. Causes and consequences of sea urchin abundance and diversity in Kenyan coral reef lagoons. *Oecologia* 83: 362-370.
24. Alcoverro T, Mariani S [2000]. Effects of sea urchin grazing over a Kenyan mixed seagrass bed. *Biologia Marina Mediterranea* 7(2): 195-198.
25. McClanahan TR, Nugues M, Mwachireya S [1994]. Fish and sea urchin herbivory and competition in Kenyan coral reef lagoons: The role of reef management. *Journal of Experimental Marine Biology and Ecology* 184: 237-254.
26. Chande AI, Nikundiwe AM, Kyomo J [1997]. The Status of the Crab Fishery at Mtoni Creek, Dar es Salaam. A paper presented at IMS/CIDA National Workshop on the Artisanal Fisheries Sector, Zanzibar, 22-24 September 1997. 10 pp.
27. Muhando CA [1995]. Species composition and relative abundance of fish and macro-crustaceans in the mangrove and seagrass habitats. In: Interlinkages between Eastern-African Coastal Ecosystems. Contract No. TS3*-CT92-0114, Final Report. pp 158-166.
28. Mgimwa F, Mgaya YD, Ngoile M [1997]. Dynamics of the dema trap fishery in the coastal waters of Zanzibar, Tanzania. In: Jiddawi N, Stanley R (eds) *Fisheries Stock Assessment in the Traditional Fishery Sector: Information Needs. Proceedings of the National Workshop on the Artisanal Fisheries Sector, Zanzibar, September 22-24, 1997.* Zanzibar. pp 125-134.

29. Little MC, Reay PJ, Grove SJ [1988]. The fish community of an East African mangrove creek. *Journal of Fish Biology* 32: 729-747.
30. Frazier J [1975]. *The Status of Knowledge on Marine Turtles in the Western Indian Ocean.* East African Wildlife Society. 16 pp.
31. Howell KM [1988]. The conservation of marine mammals and turtles in Tanzania. In: Mainoya JR (ed) *Proceedings of the Workshop on Ecology and Bioproductivity of Marine Coastal Waters of Eastern Africa, Dar es Salaam 18-20 January 1988.* Faculty of Science, University of Dar es Salaam. pp 154-161.
32. Clark F, Khatib A [1994]. Preliminary Report on the Status of Sea Turtles in Zanzibar and Pemba Islands. 10 pp.
33. Wamukoya GM, Mirangi JM, Ottichillo WK [1996]. Report on the Marine Aerial Survey of Marine Mammals, Sea Turtles, Sharks and Rays. Kenya Wildlife Service Technical Series Report No. 1. pp 1-22.
34. Jarman PJ [1966]. The status of the dugong, *Dugong dugon* (Muller) in Kenya. *East African Wildlife Journal* 4: 82-88.
35. Ligon SH [1982]. Aerial survey of the dugong, *Dugong dugon* in Kenya. In: Clark JG (ed) *Mammals in the Seas, Volume IV: Small Cetaceans, Seals, Sirenians and Otters.* Food and Agriculture Organization of the United Nations, Fisheries Series No. 5. pp 511-513.
36. Kendall B [1986]. The Status and Conservation Needs of the Dugong in the East African Region. A partially completed report of a study for UNEP/IUCN under the Global Plan of Action for the Conservation, Management and Utilization of Marine Mammals. Nairobi. 67 pp.
37. Wamukoya GM, Ottichilo WK, Salm RV [1997]. Aerial Survey of Dugongs (*Dugong dugon*) in Ungwana Bay and the Lamu Archipelago, Kenya. Kenya Wildlife Service Technical Series Report No. 2. pp 1-13.
38. Brakel WH [1981]. Alteration and destruction of coastal habitats: Implications for marine fisheries. In: *Proceedings of the Workshop of the Kenya Marine and Fisheries Research Institute on Aquatic Resources of Kenya, July 13-19, 1981.* Kenya Marine and Fisheries Research Institute and Kenya National Academy for Advancement of Arts and Science. pp 247-255.
39. Richmond MD (ed) [1997]. *A Guide to the Seashores of Eastern Africa and the Western Indian Ocean Islands.* SIDA, Department for Research Cooperation, SAREC. 448 pp.
40. Kamermans P, Hemminga MA, Marba N, Mateo MA, Mtolera M, Stapel J [2001]. Leaf production, shoot demography, and flowering of the seagrass *Thalassodendron ciliatum* (Cymodoceaceae) along the East African coast. *Aquatic Botany* 70(3): 243-258.
41. Duarte CM, Marba N, Hemminga MA [1996]. Growth and population dynamics of *Thalassodendron ciliatum* in a Kenyan back-reef lagoon. *Aquatic Botany* 55: 1-11.
42. Kamermans P, Hemminga MA, Tack JF, Mateo MA, Marba N, Mtolera M, Stapel J, Verheyden A [in press]. Groundwater Effects on Diversity and Abundance of Lagoonal Seagrasses in Kenya and on Zanzibar Island (East Africa). *Marine Ecology Progress Series.*
43. Björk M, Weil A, Semesi A, Beer S [1997]. Photosynthetic utilisation of inorganic carbon by seagrasses from Zanzibar, East Africa. *Marine Biology* 129: 363-366.
44. Schwarz AM, Björk M, Buluda T, Mtolera M, Beer S [2000]. Photosynthetic utilisation of carbon and light by two tropical seagrass species as measured *in situ. Marine Biology* 137: 755-761.
45. Björk M, Uku J, Weil A, Beer S [1999]. Photosynthetic tolerances to desiccation of tropical intertidal seagrasses. *Marine Ecology Progress Series* 191: 121-126.
46. Ochieng CA [1996]. Environmental Impact Statement on the Removal of Seagrass Beach Cast from the Beaches along the Mombasa Marine Park and Reserve. Technical Report to the Kenya Wildlife Service. Kenya Marine and Fisheries Research Institute, Mombasa. 58 pp.
47. Mohammed SM, Johnstone RW, Widen B, Jordelius E [2001]. The role of mangroves in the nutrient cycling and productivity of adjacent seagrass communities, Chwaka Bay, Zanzibar. In: Richmond M, Francis J (eds) *Marine Science Development in Tanzania and Eastern Africa. Proceedings of the 20th Anniversary Conference on Advances in Marine Science in Tanzania. 28th June-1st July, 1999, Zanzibar, Tanzania.* IMS/WIOMSA. pp 205-226.
48. Hemminga MA, Gwada P, Slim FJ, de Koeyer P, Kazungu J [1995]. Leaf production and nutrient content of the seagrass *Thalassodendron ciliatum* in the proximity of a mangrove forest (Gazi Bay, Kenya). *Aquatic Botany* 50: 159-170.
49. Marguillier S, van der Velde G, Dehairs F, Hemminga MA, Rajagopal S [1997]. Trophic relationships in an interlinked mangrove-seagrass ecosystem as traced by $\delta^{13}C$ and $\delta^{15}N$. *Marine Ecology Progress Series* 151: 115-121.
50. Mariani S, Alcoverro T [1999]. A multiple-choice feeding-preference experiment utilising seagrasses with a natural population of herbivorous fishes. *Marine Ecology Progress Series* 189: 295-299.
51. Erftemeijer PLA, Semesi AK, Ochieng CA [2001]. Challenges for marine botanical research in East Africa: Results of a bibliometric survey. *South African Journal of Botany* 67: 411-419.
52. Bandeira S, Bjork M [2001]. Seagrass research in the eastern Africa region: Emphasis on diversity, ecology and ecophysiology. *South African Journal of Botany* 67: 420-425.
53. Anonymous [1969]. Report of the Watamu Expedition by the Bangor University, Wales, UK.
54. Wakibya JG [1995]. The potential human-induced impacts on the Kenyan seagrasses. *UNESCO Report in Marine Sciences* 66: 176-187.
55. Johnstone R [1995]. Community production and nutrient fluxes in seagrass beds (Unguja island, Tanzania). In: Interlinkages between Eastern-African Coastal Ecosystems. Contract No. TS3*-CT92-0114, Final Report. pp 88-93.
56. Muthiga NA [1996]. *A Survey of the Coral Reef Habitats of the Mombasa Marine Park and Reserve with a Review of the Existing Park Boundaries and Reserve Areas Restricted from Fishing.* Kenya Wildlife Service – Marine Research Unit, Reports and Publications Series No. 2, Mombasa. 13 pp.
57. Coast Development Authority [1995]. Towards Integrated Management and Sustainable Development of Kenya's Coasts: Initial Findings and Recommendations for an Action Strategy in the Nyali-Bamburi-Shanzu Area. Draft report, August 1995. Coast Development Authority, Mombasa. 74 pp.
58. Sanders MJ, Gichere SG, Nzioka RM [1990]. Report of Kenya Marine Fisheries Subsector Study. FAO RAF/87/008/DR/65/E. 44 pp.
59. Samoilys M [1988]. Abundance and species richness of coral reef fish on the Kenyan coast: The effects of protective management and fishing. *Proceedings of the 6th International Coral Reef Symposium.* pp 261-266.
60. Katwijk van MM, Meier NF, van Loon R, van Hove EM, Giesen WBJT, van der Velde G, den Hartog C [1993]. Sabaki River sediment load and coral stress: Correlation between sediments and condition of the Malindi-Watamu reefs in Kenya (Indian Ocean). *Marine Biology* 117: 675-683.
61. Ngoile MA [1988]. Marine pollution in Tanzania: Sources, dispersion and effects. In: Mainoya JR (ed) *Proceedings of a Workshop on Ecology and Bioproductivity of Marine Coastal Waters of Eastern Africa, Dar es Salaam, Tanzania, 18-20 January 1988.* Faculty of Science, University of Dar es Salaam. pp 133-143.
62. Kenya-ICAM [1996]. *Towards Integrated Management and Sustainable Development of Kenya's Coast. Findings and Recommendations for an Action Strategy in the Nyali-Bamburi-Shanzu area.* 74 pp.
63. Mwangi S, Kirugara D, Osore M, Njoya J, Yobe A, Dzeha T [2001]. Status of Marine Pollution in Mombasa Marine National Park, Reserve and Mtwapa Creek, Kenya. Kenya Wildlife Service Report, Coastal Regional Headquarters, Mombasa.

64 Daffa JM [1996]. Land-based pollutants to the coastal and marine waters of Dar es Salaam and the effects to the marine plants. In: Björk M, Semesi AK, Pedersen M, Bergman B (eds) *Current Trends in Marine Botanical Research in the East African Region. Proceedings of the 3-10 December 1995 Symposium on the Biology of Microalgae, Macroalgae and Seagrasses in the Western Indian Ocean.* SIDA Marine Science Programme, Department for Research Cooperation, SAREC. pp 315-331.

65 Pratap HB [1988]. Impact of heavy metal pollution on the bioproductivity of marine coastal waters. In: Mainoya JR (ed) *Proceedings of a Workshop on Ecology and Bioproductivity of Marine Coastal Waters of Eastern Africa, Dar es Salaam, Tanzania, 18-20 January 1988.* Faculty of Science, University of Dar es Salaam. pp 121-132.

66 Msuya FE [1995]. Environmental and Socio-economic Impact of Seaweed Farming on the East Coast of Unguja Island, Zanzibar, Tanzania. Paper presented at the National Workshop on Integrated Coastal Zone Management (MNTRE, SAREC, World Bank), Zanzibar, 8-12 May 1995. 24 pp.

67 Olafsson E, Johnstone RW, Ndaro SGM [1995]. Effects of intensive seaweed farming on the meiobenthos in a tropical lagoon. *Journal of Experimental Marine Biology and Ecology* 191(1): 101-117.

68 Pertet F [1982]. Kenya's experience in establishing coastal and marine protected areas. In: McNeely JA, Miller KR (eds) *National Parks, Conservation and Development: The Role of Protected Areas in Sustaining Society.* Smithsonian Institution Press, Washington. pp 101-108.

69 Ray C [1968]. *Marine Parks for Tanzania. Results of a survey of the coast of Tanzania by invitation of the trustees of Tanzania National Parks.* The Conservation Foundation & New York Zoological Society, Baltimore, Maryland, October 1968. 47 pp.

70 Coughanowr CA, Ngoile MN, Linden O [1995]. Coastal zone management in Eastern Africa including the island states: A review of issues and initiatives. *Ambio* 24(7-8): 448-457.

71 SOZ [1990]. *Netherlands Indian Ocean Programme, Part 3: Kenya Shelf and Coastal Ecosystems.* Netherlands Marine Research Foundation. pp 1-28.

72 Woitchik AF (ed) [1993]. Dynamics and Assessment of Kenyan Mangrove Ecosystems. (Project No. TS2-0240-C, GDF). Final Report to the EC. Free University of Brussels. 239 pp.

8 The seagrasses of MOZAMBIQUE AND SOUTHEASTERN AFRICA

S.O. Bandeira

F. Gell

Thirteen seagrass species occur in Mozambique and 12 in the remaining southeastern African region. Madagascar has nine common species. Five species occur in South Africa, seven in Mauritius and up to ten in Comoros and Seychelles[1, 2, 3]. *Ruppia maritima*, recently defined as a seagrass[4], is also a dominant species in the southeastern Africa region.

BIOGEOGRAPHY
Mozambique
The Mozambican coast can be divided into three regions: a sandy coastline from the southern end of the country to the Save River; an estuarine coastline from the Save River up to around 500 km north of the Zambezi River; and a rocky limestone coastline, typically surrounded by coral reefs, which runs from the Zambezia province up to the northern end of the country, and also covers the Tanzanian and Kenyan coasts[5, 6]. Seagrasses abound in the sandy and limestone areas.

Seagrasses in general occur in mixed seagrass stands, especially in intertidal areas. The three dominant mixed-seagrass communities on the sandy substrates of southern Mozambique consist of combinations of *Thalassia hemprichii*, *Halodule wrightii*, *Zostera capensis*, *Thalassodendron ciliatum* and *Cymodocea serrulata*[7].

In contrast, the seagrass communities of the more northerly limestone areas are quite different, with seagrasses tending to occur intermingled with seaweed species[8]. Here, the dominant botanical communities also include *Thalassia hemprichii* and *Halodule wrightii*, but species such as *Gracilaria salicornia*, *Halimeda* spp. and *Laurencia papillosa* occur mixed with *Thalassia hemprichii*, and *Sargassum* spp. with *Thalassodendron ciliatum*. Elsewhere *Zostera capensis* and *Halodule wrightii* also form mixed beds.

In general *Thalassodendron ciliatum* and *Thalassia hemprichii* are the dominant subtidal seagrass species in Mozambique. A detailed comparison has been made of the former growing along the rocky and sandy coasts of southern Mozambique[9]. Leaves appear to grow faster on plants in the rocky areas (20-26 g/m^2/day and up to 57 mm^2/day) than in sandy (8-10 g/m^2/day and up to 22 mm^2/day). Leaf biomass in rocky areas is more than twice that of sandy (258 g/m^2 and 124 g/m^2 respectively) and beds are characterized by a much higher shoot density (4 561 shoots/m^2 and 888 shoots/m^2 respectively).

The underground biomass of *Thalassodendron ciliatum* is presumably relatively high in sandy environments because total biomass (862 g/m^2), while significantly lower than in the rocky seagrass beds (1 070 g/m^2), is comparable. Although possibly slower growing, the *Thalassodendron ciliatum* plants in the sandy habitat have wider (1.4 cm ±0.1) and longer (12.51 cm ±0.6) leaves than in the rocky habitat (0.7 cm ±0.1 and 8.2 ±0.5 respectively). The biomass of epiphytes on *Thalassodendron ciliatum* plants is an order of magnitude higher in the rock (512 g/m^2) than in the sand (40 g/m^2), and consequently these organisms account for nearly half (48 percent) of the combined seagrass and epiphyte biomass, compared with just 5 percent in the southern sandy-bottom beds.

Enhalus acoroides, *Halophila stipulacea* and *Halophila minor* are found only in northern Mozambique while pure stands of *Zostera capensis* are found only in the south[10]. Pioneer species observed in Mozambique include *Halodule wrightii*, *Halophila ovalis* and *Cymodocea serrulata*. The first two species act as pioneers in exposed sandy areas close to the coastline, whereas *Cymodocea serrulata* is a pioneer in silted channels.

South Africa
Zostera capensis is most widespread and one of the dominant seagrass species in South Africa. It occurs

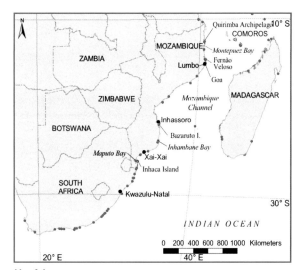

Map 8.1
Mozambique and southeastern Africa

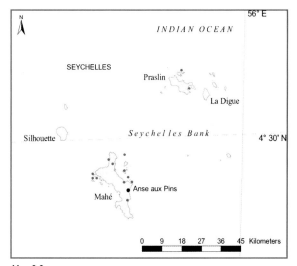

Map 8.2
The Seychelles

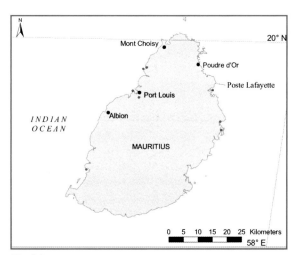

Map 8.3
Mauritius

Table 8.1
Area cover and location for the seagrass *Zostera capensis* in South Africa

Estuary name	Area (km²)	Climate	Estuary classification
St Lucia	1.81	Subtropical	Estuarine lake
Mbashe	0.01	Subtropical	Permanently open
Mlalazi	0.04	Subtropical	Permanently open
Mngazana	0.02	Subtropical	Permanently open
Mtakatye	0.04	Subtropical	Permanently open
Xora	0.01	Subtropical	Permanently open
Knysna	3.48	Warm temperate	Estuarine bay
Klein	0.37	Warm temperate	Estuarine lake
Swartvlei	0.23	Warm temperate	Estuarine lake
Keiskamma	0.12	Warm temperate	Permanently open
Keurbooms	0.64	Warm temperate	Permanently open
Krom Oos	0.02	Warm temperate	Permanently open
Qora	0.08	Warm temperate	Permanently open
Swartkops	0.16	Warm temperate	Permanently open
Hartenbos	0.01	Warm temperate	Temporarily closed
Kabeljous	0.02	Warm temperate	Temporarily closed
Ngqusi\Inxaxo	0.01	Warm temperate	Temporarily closed
Total	**7.07**		

Source: Colloty[12].

mostly in estuarine waters along a number of estuaries from Kwazulu-Natal to the western Cape region.

Another important location with seagrass species is found off Kwazulu-Natal. Here, a number of rocky protuberances into the sea are mostly dominated by *Thalassodendron ciliatum* adapted to live in rocky habitat together with seaweeds[11]. These rocky areas generally experience strong water dynamics and winds similar to those of southern Mozambique[9].

The distribution of *Zostera capensis* in southeast South Africa is well recorded. It grows in 17 estuaries (Table 8.1). Individual beds are small, generally only a few hectares, and the total area covered by seagrass is about 7 km².

Madagascar

Little is known about the relative dominance of seagrass species in Madagascar although it is likely that in the southwest of the country they are similar to the species from the limestone areas of Mozambique, with most of the meadows being dominated by *Thalassodendron ciliatum* and *Thalassia hemprichii*. Seaweeds are also a common feature in the intertidal and subtidal seagrass areas of Madagascar[13].

Mauritius

Thalassodendron ciliatum, *Halodule uninervis* and *Syringodium isoetifolium* appear to be the most common seagrass species in Mauritius[14].

Comoros

Little is known about the seagrass meadows of Comoros. Being located less than 400 km east of the coastline of Mozambique and sharing a similar climate, Comoros may have similar meadows to northern Mozambique with mixed seagrass species in intertidal areas and subtidal seagrass species dominated by broad-leaved species such as *Thalassodendron ciliatum*.

Seychelles

Seychelles is composed of 115 granite and coral islands. Seagrass meadows are dominated by *Cymodocea serrulata*. *Syringodium isoetifolium* and *Thalassia hemprichii* occur at Anse aux Pins[3] on the main island of Mahé. Shoot density varies from 1 093 to 1 107 shoots/m^2 in *Cymodocea serrulata* plants, 1 123 to 1 761 in *Syringodium isoetifolium* and 540 to 627 in *Thalassia hemprichii*[3]. *Thalassodendron ciliatum* is also common in subtidal areas down to depths of 33 m throughout the Seychelles[15].

SEAGRASS FISHERIES IN MOZAMBIQUE

The Quirimba Archipelago is a chain of 32 islands off the coast of northern Mozambique, running from north of the town of Pemba up to the Tanzanian border. One of the largest and most populated islands in the chain is Quirimba. Quirimba is 6 km long by 2 km wide and has a population of 3 000. This island is separated from the mainland by the Montepuez Bay. The island's main fishery is located in the shallow seagrass beds of the bay, and this seagrass fishery is the main source of income and protein for people on the island. In 1996 and 1997 part of the Darwin/Frontier Quirimba Archipelago Marine Research Programme studied the Quirimba fishery which is dependent on a diverse seagrass ecosystem[16, 17].

Montepuez Bay is between 1 and 10 m deep and has extensive intertidal flats and banks, large areas of which are covered in seagrass. The bay takes its name from the Montepuez River, which enters the southwest of the bay from the Mozambique mainland. Ten species of seagrass are present in the bay: *Enhalus acoroides*, *Thalassodendron ciliatum*, *Cymodocea rotundata*, *Cymodocea serrulata*, *Syringodium isoetifolium*, *Halodule uninervis*, *Halodule wrightii*, *Halophila ovalis*, *Halophila stipulacea* and *Thalassia hemprichii*.

The intertidal seagrass beds are dominated by *Thalassia hemprichii*. Subtidally, the most abundant species are *Enhalus acoroides* and *Thalassodendron ciliatum*, both of which can grow to over 1 m in height, and the smaller species *Cymodocea rotundata* and *Cymodocea serrulata*. Small seagrass species (e.g. *Syringodium isoetifolium*, *Halophila* spp. and *Halodule* spp.) are present in small quantities, often forming an understorey in stands of larger species. In quadrat surveys of the Montepuez Bay seagrass beds, the most common seagrass types were stands dominated by *Enhalus acoroides*. The most common combination of seagrasses found was *Enhalus acoroides* with *Halophila ovalis*, which were mainly found together in areas that were exposed to the air at very low tides.

Such a predominance of *Enhalus acoroides* is unusual in the region and, even within the Quirimba Archipelago which has extensive seagrass beds, Montepuez Bay was the only area dominated by *Enhalus acoroides*. Other subtidal seagrass beds in the Quirimba Archipelago were dominated by *Thalassodendron ciliatum*. Dense meadows of tall (often between 50 and 100 cm) *Enhalus acoroides* were home to a diverse range of invertebrates and fish, and the seagrass itself was covered in epiphytes, altogether constituting a complex habitat. Over 30 species of algae were identified living on or in association with the seagrass[18]. Fishers in Montepuez Bay target shallow areas of *Enhalus acoroides* for their main fishing activities, whilst women collect invertebrates in the intertidal seagrass beds dominated by *Thalassia hemprichii*. Although the direct use of seagrass plants has been reported from other places, this was not observed in the Quirimba Archipelago.

Fishing methods

Seine nets were set from small sail-powered boats by teams of between five and 12 men. Fishermen in the water kept the net in place and drove the fish into the net as it was hauled into the boat. Seine net fishing was carried out in shallow water, from 1 to 8 meters deep, in areas of *Enhalus acoroides*. The nets used were approximately 100 m in length with a main mesh size of 4 cm stretch and a cod-end of 2 cm stretch, or less. The mean duration of a fishing trip was about five hours and the mean catch per trip was 75 kg. Catch per unit effort was 3.6 kg of fish per man-hour spent fishing or 2 kg of fish per man-hour spent at sea. The net catches were highly diverse, with a total of 249 fish species in 62 families identified from more than 46 600 fish sampled[19]. The fishers also caught invertebrates in the seine nets, particularly squid. Approximately 30 fish species were common in the catch. The most important species in the net fishery in terms of weight were the African whitespotted rabbitfish *Siganus sutor* (24 percent); the pink ear emperor *Lethrinus lentjan* (12.2 percent); the seagrass parrotfish *Leptoscarus vaigiensis* (11 percent);

A *marema* fish trap in an *Enhalus acoroides* bed, Mozambique.

the variegated emperor *Lethrinus variegatus* (7.4 percent); the blacktip mojarra *Gerres oyena* (6.3 percent) and the spinytooth parrotfish *Calotomus spinidens* (3.2 percent). The majority of fish caught in seine nets were less than 15 cm long and many were juveniles. Virtually all the fish caught in the seine net fishery were eaten. Amongst the more unusual food species that were common in the fishery were the tailspot goby *Amblygobius albimaculatus* and the three-ribbon wrasse *Stethojulis strigiventer*.

The traps used in Montepuez Bay are known locally as *marema*. They are of an arrowhead design and constructed from woven bamboo panels secured together with palm fibers. *Marema* were set by fishermen from outrigger canoes in shallow areas of *Enhalus acoroides* at low tide, and were hauled the following day at low tide. The traps were sometimes baited with crushed *Terebralia* snails collected from mangrove areas, or with squid, but often the traps were not baited at all. The traps were weighed down with stones and placed amongst long, densely growing *Enhalus acoroides* (the fishermen said that this seagrass was very important for keeping the traps in place in the strong tidal currents). The mean daily catch for a fisherman setting 40 traps was nearly 7 kg of fish, although catches could be as high as 27 kg per trip. The catch per unit effort for the trap fishery was 2.2 kg of fish per hour spent fishing and the mean catch for a trap set for 24 hours was 0.2 kg of fish. Trap catches

Case Study 8.1
INHACA ISLAND AND MAPUTO BAY AREA, SOUTHERN MOZAMBIQUE

Seagrasses at Inhaca Island and Maputo Bay area cover more than 80 km². At Inhaca Island the seagrasses alone cover around 50 percent of the entire intertidal area[7]. The diversity of seagrasses is very high, especially at Inhaca Island where eight seagrasses species can be found in just one hectare[7]. Nine seagrass species have been identified in the area namely: *Cymodocea rotundata*, *Cymodocea serrulata*, *Halodule uninervis*, *Halodule wrightii*, *Halophila ovalis*, *Syringodium isoetifolium*, *Thalassia hemprichii*, *Thalassodendron ciliatum* and *Zostera capensis*[7]. These seagrass species are grouped in three main dominant seagrass communities: *Thalassodendron ciliatum/Cymodocea serrulata*, *Thalassia hemprichii/Halodule wrightii* and *Zostera capensis*[2,22]. *Zostera capensis* shoot density is higher at Inhaca (2 880 shoots/m²) than at Maputo Bay (1 285 shoots/m²) as are leaf, rhizome and root biomass[23].

The species *Thalassia hemprichii* and *Thalassodendron ciliatum* tend to occupy deeper areas far from the coastline whereas *Halodule wrightii* and *Cymodocea serrulata* tend to occupy shallow areas closer in. *Thalassodendron ciliatum*, in the subtidal fringe, occurs in homogeneous stands except in some areas where it is also accompanied by a band of *Syringodium isoetifolium* parallel to the coastline. When *Thalassia hemprichii* and *Halodule wrightii* co-occur in mixed species communities, *Thalassia hemprichii* is found in small depressions whereas *Halodule wrightii* occupies elevated areas which are exposed at low spring tides.

Seagrass meadows in Maputo Bay region are widely used by the people who collect, by hand, seafood from them, the most common being mussels (*Anadara natalensis*, *Cardium flavum*, *Modiolus phillipinarum*), oysters (*Pinctada capensis*), gastropods (*Conus betulinus*, *Strombus gibberulus*) and sea urchins (e.g. *Salmacis bicolores*, *Tripneustes gratilla*). They also use the meadows for fishing using traditional techniques, for species such as *Crenidens crenidens*, *Gerres acinaces*, *Leiognathus equulus*, *Lithognathus aureti*, *Liza macrolepis*, *Lutjanus fulviflamma*, *Platycephalus indicus*, *Pseudorhombus arsius*, *Rhabdosargus sarba*, *Scarus ghobban*, *Siganus sutor* and *Terapon jarbua*, and crustaceans such as *Matuta lunaris* and *Portunus pelagicus*[2,24,25]. The sea cucumber were dominated by the parrotfish *Leptoscarus vaigiensis*, which accounted for over 74 percent of the fish caught by weight. Other important species included the parrotfish *Calotomus spinidens* (5 percent by weight), the rabbitfish *Siganus sutor* (4 percent), the dash-dot goatfish *Parupeneus barberinus* (4 percent), the blackspot snapper *Lutjanus fulviflamma* (3 percent) and the flagfin wrasse *Pteragogus flagellifera* (2.5 percent). A total of 61 species of fish were identified from 3500 fish sampled from the trap fishery, with about 16 of these species appearing commonly in the fishery. A wide variety of invertebrates entered the traps, including swimming crabs (*Portunidae*) which were kept for use as food.

Spatially referenced catch data from the seine net fishery was used to identify the fishing sites in Montepuez Bay with the highest mean fish catch per unit effort. These were also the sites with the highest mean percentage cover of seagrass and highest seagrass biomass. This suggests that seagrass cover and biomass may influence fish biomass and fishery productivity. In experimental trap fishing, the preference of trap fishermen for areas of *Enhalus acoroides* to other species of seagrass was shown to be well founded. In these experiments the mean catch per trap for *Enhalus acoroides* greatly exceeded that for other common seagrass species, *Thalassodendron ciliatum* or *Cymodocea* spp. Catch compositions from the three different seagrass species were also different. Catches from *Enhalus acoroides* beds were dominated by the parrotfish *Leptoscarus vaigiensis* as the fishermen's catches were, but catches from *Thalassodendron ciliatum* beds were dominated by the file fish *Paramonacanthus barnardi*, and catches from *Cymodocea* spp. by the snapper *Lutjanus fulviflamma*.

Invertebrate fishery

Women did most of the collecting of seagrass invertebrates that could be achieved without a boat. They walked out over the seagrass beds at low tides and collected bivalves by hand. On spring tides, some women traveled in groups by boat to some of the larger banks in the bay that become exposed at low tide. The main species they collected were the ark shell *Barbatia fusca* and the pinna shell *Pinna muricata*, found in sand in seagrass beds, and the oyster *Pinctada nigra* which grows on the seagrass plant itself. These shellfish were dried and most of them were sold on the mainland for higher prices than they would fetch on the island, particularly the pinna shell which is a local delicacy[20].

Holothuria scabra, presently endangered in many parts of the country, was earlier heavily collected by the local people at Inhaca and north of Maputo city and sold for export to Asia. The same is true of *Holothuria atra*, but to a lesser extent.

More than 20 nets are set daily around Inhaca Island from boats and by people walking on the beach, and around 100 people may be seen collecting edible organisms during the spring low tides[7]. To the north of Maputo city, at the fishing village of Bairro dos Pescadores, close to 50 people dug up seagrass meadows at spring low tide for collection of invertebrates, mainly bivalves, in the mid-1990s[7]. Recent counting estimated around 200 people involved in this activity, which includes digging the intertidal areas for the same purpose. Seagrasses have also been reported as being used for alluring and bewitching at Inhaca Island[7] and the dried detached leaves of *Thalassodendron ciliatum* as being used to fill pillows.

The seagrasses of the area are under considerable stress from a variety of sources. Sewage disposal along the Maputo coastline threatens seagrasses there, with polluted areas tending to be covered by seaweeds *Ulva* spp. and *Enteromorpha* spp. instead of seagrass. Additional pollution, especially oil spills, comes from the city harbor and industrial area. Sedimentation due to erosion and floods further diminishes local seagrass coverage around Maputo. Trampling and the heavy concentration of fishing and tourist activities directly disturb seagrass meadows at Inhaca Island's main village. Fishing in very shallow water is another disturbance. The combination of all these factors places heavy pressures on the extensive seagrass meadows, and has already caused a disappearance of *Zostera capensis* from in front of Inhaca's main village[2].

Some priority areas for intervention to reduce these disturbances include increased monitoring and reduction of sewage disposal, industrial pollutants and port activities. At Inhaca Island, only the seagrass beds located close to coral reefs are under protection. This protection should be reviewed to target the conservation of seagrass areas with a high concentration of threatened and depleted invertebrate species such as holothurians and seastars, e.g. *Holothuria scabra* and *Holothuria atra*. The *Thalassodendron ciliatum* communities occurring in rocky protuberances in the sea habitats have only recently been described. This form of seagrass only occurs in sandstone rocks facing the strong waves of the Indian Ocean[9]. Few similar areas exist in Mozambique and therefore some kind of protection should be put in place for their conservation.

Some fishermen went out on the seagrass beds in small canoes to dive with a mask to collect invertebrates, mainly sea cucumbers, and mollusks such as the tulip shell *Pleuroploca trapezium* and the murex *Chicoreus ramosus*. Sea cucumbers were one of the more valuable seagrass residents and were dried and sold across the border in Tanzania, to be exported to markets in the Far East. During the study period fishermen reported the virtual disappearance of sea cucumbers from the seagrass beds of Montepuez Bay, and attributed this to overexploitation by local and itinerant fishers. Fishermen involved in the seine net and trap fishery would also collect murex, tulip shells and sea cucumbers when they got the opportunity. Tulip shells were collected for their opercula which were sold to traders in Tanzania. Murex were eaten and the shells of these and other mollusks were collected and burnt for lime that was used locally in building.

Subtidal surveys identified 34 species of large invertebrates which were associated with the seagrass beds. Commonly observed invertebrates that were not collected locally included the sea urchins *Diadema setosum* and *Tripneustes gratilla*, the sea cucumber *Synapta maculata* and the starfish *Pentaceraster tuberculatus* and *Protoreaster lincki*.

Local value of seagrass resources

The seagrass fisheries of Montepuez Bay supported over 400 fishermen on Quirimba Island alone and many more in the mainland villages and from other islands in the vicinity. More than a hundred women from Quirimba also collected invertebrates in the seagrass beds. In total over 500 people were involved in the seagrass fisheries of Quirimba, out of a total population of 3000. The total fish catch from the 35 km^2 seagrass beds of the whole bay was estimated at around 500 metric tons per year, or 14.3 metric tons per km^2 per year. This figure does not include invertebrates, but is still high compared with many tropical reef and estuarine fisheries. A minimum estimate for annual invertebrate collection from seagrass beds around Quirimba was 40 metric tons per year.

In the study period, the fish caught in Montepuez Bay had an estimated annual saleable value of ca US$120000, based on prices paid for fish locally. Roughly half the fish caught was consumed by the fishers and their families or exchanged for other goods or services. The other half was dried and traded on the mainland by the owners of the net fishing boats, or other traders who buy the surplus from trap fishermen.

Management issues

During the study period the local fishery seemed to have a relatively low impact on the seagrass beds and was apparently sustainable. The seine net fishery did appear to have some negative effects on the seagrass beds. The nets were often dragged along the bottom and substantial amounts of seagrass, sponges and small corals were sometimes brought up with the nets. Trampling of intertidal seagrass was kept to a minimum by the use of small paths across the seagrass that restricted the trampling damage to a small area. The main threats to the sustainability of the seagrass fishery came from external sources, mainly unregulated itinerant fishers and commercial sea cucumber fishing for international trade. On a larger scale, potential threats came from upstream activities in the catchment of the Montepuez River – particularly deforestation leading to changes in sedimentation rates.

With so many people relying on the seagrass beds of Montepuez Bay for their livelihoods, and with the paucity of alternative employment or sources of protein, their conservation and sustainable use is vital. One of the reasons that the resources of the Montepuez Bay seagrass beds are so widely used is that the habitat is so accessible, even to those with the most limited resources. Much of the seagrass can be reached on foot at low tide, and even the deeper areas are close to shore and are sheltered from the heavy seas that the eastern coast of the island is subject to. At the time of this study Quirimba Island had rich and diverse marine resources including mangrove forests and extensive coral reefs on the east coast. However, few fishers utilized the reef resources because of the difficulties of accessing the exposed reefs in their traditional fishing vessels. This issue of the accessibility of seagrass beds is seen in other places where fishers with small boats, or on foot, are able to fish in seagrass beds in shallow sheltered bays or lagoons. A priority for seagrass research should be to look at how best to manage open-access, multi-user seagrass systems such as Montepuez Bay, to ensure their sustainable use, and to conserve biodiversity. The Quirimba Island seagrass fishery is a clear, and rare, example of the direct value of seagrasses to local communities.

HISTORICAL PERSPECTIVES AND LOSS

The digging of *Zostera capensis* beds to collect bivalves has dramatically depleted the seagrass cover at Bairro dos Pescadores (near Maputo, Mozambique) from a cover of around 60 percent or more to 10 percent or less in the last ten years (Figure 8.1). This activity lasts for the entire spring tide period spanning about 15 days each month. The bivalves are collected mainly for food. It is expected that this activity will eventually completely destroy the *Zostera capensis* beds at Bairro dos Pescadores, and that the food security of the local population will suffer as a consequence.

Sedimentation due to floods has buried seagrasses in Maputo Bay and Inhassoro. In the heavy floods in southern Mozambique in 2000 around 24 km² of seagrasses may have been buried here. Harbor development, sewage and coastal development in areas of southern Mozambique have further diminished seagrass coverage. Heavy concentrations of artisanal fishing boats in combination with intense trampling in low tides have also caused reduction of seagrass species at Inhaca Island.

In Mauritius seagrasses are threatened by the high use of fertilizers in the sugar cane industry, and specifically by the eutrophication of coastal lagoons that is caused when they leach into these shallow contained areas. Seagrass beds are being dredged and destroyed to provide bathing and skiing areas for tourists. Sedimentation, sewage disposal and sand mining are among other threats to Mauritius seagrasses.

In Anse aux Pins, Seychelles, sedimentation,

Figure 8.1
Digging of *Zostera capensis* meadows at Vila dos Pescadores, near Maputo city

Photos: S.O. Bandeira

a. Photo taken in 1994 – hunks of plants being lifted, washed and then placed upside down – plant cover is still high.

b. Photo taken in 2002 – plant cover is very low and in most areas the seagrass has already disappeared.

Table 8.2
Seagrass cover and area lost in Mozambique

Site name	Main seagrass species	Area (km²)	Area lost (km²)
Quirimba Archipelago	Cr, Cs, Ea, Hm, Ho, Hs, Hw, Tc, Th	45	0
Mecúfi-Pemba	Hm, Ho, Hs, Hu, Hw, Si, Tc, Th, Zc	30	0.02
Fernão Veloso	Cr, Cs, Ea, Hm, Ho, Hu, Hw, Si, Tc, Th	75	0
Quissimajulo	Th	2	0
Relanzapo	Tc, Th, macroalgae	8	0
Matibane-Quitagonha Island	Cr, Hw, Tc, Th	34	
Chocas Mar-Cabaceira Grande-Sete Paus Island	Tc, Th	19	0
Mozambique Island-Lumbo-Cabaceira Pequena	Cr, Cs, Hm, Ho, Hu, Hw, Si, Tc, Th, Zc	15	3
Goa Island	Tc	1	0
Inhassoro-Bazaruto Island	Cs, Tc, Th	25	10
Inhambane Bay	Hw	30	0.5
Xai-Xai	Tc, macroalgae	0.04	0
Bilene	Rm, Hu	3	0
Maputo Bay	Ho, Hw, Tc, Th, Zc	37	14
Inhaca Island	Cr, Cs, Ho, Hu, Hw, Si, Tc, Th, Zc	46	0.03
Inhaca-Ponta do Ouro	Tc, macroalgae	69	0
Total		**439.04**	**27.55**

Notes: Cr *Cymodocea rotundata*; Cs *Cymodocea serrulata*; Ea *Enhalus acoroides*; Hm *Halophila minor*; Ho *Halophila ovalis*; Hs *Halophila stipulacea*; Hu *Halodule uninervis*; Hw *Halodule wrightii*; Rm *Ruppia maritima*; Si *Syringodium isoetifolium*; Tc *Thalassodendron ciliatum*; Th *Thalassia hemprichii*; Zc *Zostera capensis*.

salinity and decreased water quality associated with a river effluent discharge have adversely affected seagrasses[3]. Flooding in estuaries is the main threat to the survival of *Zostera capensis* on the South African east coast.

Other areas where seagrass cover has been lost include Pemba, Mozambique Island, Inhambane Bay and Inhaca Island (Table 8.2). The total known historical loss of seagrasses in Mozambique is 27.55 km^2, although some of the areas affected by the 2000 floods have already regained seagrass cover.

In Mauritius, seagrasses have diminished from areas such as Albion (*Halodule uninervis*), Poudre d'Or, Mont Choisy and Poste Lafayette (*Syringodium isoetifolium*) though the actual area lost is unknown. Similarly areas covered by *Zostera capensis* in estuaries in Kwazulu-Natal, South Africa, are believed to have been seriously depleted by periodic heavy floods[21] without measurements being available. Nothing is known about the loss of seagrass from Madagascar, Comoros or Seychelles.

PRESENT COVERAGE

Mozambique has a total of 439 km^2 of seagrasses (Table 8.2). There are 25 km^2 around Inhassoro and Bazaruto Island, 30 km^2 at Mecúfi-Pemba and 45 km^2 in the southern Quirimba Archipelago. The largest seagrass beds occur at Fernão Veloso, Quirimba and Inhaca-Ponta do Ouro. Additional inventories are needed, particularly in remote coastal areas. In South Africa *Zostera capensis* covers a total area of just over 7 km^2: other seagrasses species cover smaller areas. While extensive seagrass meadows do occur in Madagascar, Mauritius, Comoros and Seychelles, the exact area is unknown.

AUTHORS

Salomão O. Bandeira, Department of Biological Sciences, Universidade Eduardo Mondlane, P.O. Box 257, Maputo, Mozambique. **Tel:** +258 (0)1 491223. **Fax:** +258 (0)1 492176. **E-mail:** sband@zebra.uem.mz

Fiona Gell, Environment Department, University of York, Heslington, York YO10 5DD, UK.

REFERENCES

1. Titlyanov E, Cherbadgy I, Kolmakov P [1995]. Daily variations of primary production and dependence of photosynthesis on irradiance in seaweeds and seagrass *Thalassodendron ciliatum* of the Seychelles Islands. *Photosynthetica* 31: 101-115.
2. Bandeira SO [2000]. Diversity and Ecology of Seagrasses in Mozambique: Emphasis on *Thalassodendron Ciliatum* Structure, Dynamics, Nutrients and Genetic Variability. PhD thesis, Göteborg University.
3. Ingram JC, Dawson TP [2001]. The impacts of a river effluent on a coastal seagrass habitat of Mahé, Seychelles. *South African Journal of Botany* 67: 483-487.
4. Short FT, Coles RG, Pergent-Martini C [2001]. Global seagrass distribution. In: Short FT, Coles RG (eds) *Seagrass Research Methods*. Elsevier Publishing, Amsterdam. pp 5-30.
5. Hartnoll RG [1976]. The ecology of some rocky shores in tropical East Africa. *Estuaries Coastal Marine Science* 4: 1-21.
6. Gove DZ [1995]. The coastal zone of Mozambique. In: Lindén O (ed) *Workshop and Policy Conference on Integrated Coastal Zone Management in Eastern Africa including the Island States. Conference Proceedings*. Coastal Management Center (CMC), Metro Manila. pp 251-273.
7. Bandeira SO [1995]. Marine botanical communities in southern Mozambique: Sea grass and seaweed diversity and conservation. *Ambio* 24: 506-509.
8. Bandeira SO, António CM [1996]. The intertidal distribution of seagrasses and seaweeds at Mecúfi Bay, northern Mozambique. In: Kuo J, Phillips RC, Walker DI, Kirkman H (eds) *Seagrass Biology: Proceedings of an International Workshop*. University of Western Australia, Nedlands. pp 15-20.
9. Bandeira SO [2002]. Leaf production rates *Thalassodendron ciliatum* from rocky and sandy habitats. *Aquatic Botany* 72: 13-24.
10. Bandeira SO, Björk M [2001]. Seagrass research in the eastern Africa region: Emphasis to diversity, ecology and ecophysiology. *South African Journal of Botany* 67: 420-425.
11. Barnabas AD [1991]. *Thalassodendron ciliatum* (Forsk.) den Hartog: Root structure and histochemistry in relation to apoplastic transport. *Aquatic Botany* 40: 129-143.
12. Colloty BM [2000]. Botanical Importance of Estuaries of the Former Ciskei/Transkei Region. PhD thesis, University of Port Elizabeth. 202 pp.
13. Rabesandratana RN [1996]. Ecological distribution of seaweeds in two fringing coral reefs at Toliara (SW of Madagascar). In: Björk M, Semesi AK, Pedersén M, Bergman B (eds) *Current Trends in Marine Botanical Research in East African Region*. SIDA/SAREC, Uppsala. pp 141-161.
14. Dulymamode. Personal communication.
15. Pärnik T, Bil K, Kolmakov P, Titlyanov E [1992]. Photosynthesis of the seagrass *Thalassodendron ciliatum:* Leaf anatomy and carbon metabolism. *Photosynthetica* 26: 213-223.
16. Gell FR [1999]. Fish and Fisheries in the Seagrass Beds of the Quirimba Archipelago, Northern Mozambique. PhD thesis, University of York.
17. Whittington MW, Antonio CM, Corrie A, Gell FR [1997]. Technical Report 3: Central Islands Group – Ibo.
18. Antonio MC. Unpublished data.
19. Gell FR, Whittington MW [2000]. Diversity of fishes in seagrass beds in the Quirimba Archipelago, northern Mozambique. *Marine and Freshwater Research* 53: 115-121.
20. Barnes DKA, Corrie A, Whittington M, Carvalho MA, Gell FR [1998]. Coastal shellfish resources use in the Quirimba Archipelago, Mozambique. *Journal of Shellfish Research* 17(1): 51-58.
21. Adams JB, Bate GC, O'Callaghan M [1999]. Estuarine primary producers. In: Allanson BR, Baird D (eds) *Estuaries of South Africa*. Cambridge University Press, Cambridge. pp 91-118.
22. Martins ARO, Bandeira, SO [2001]. Biomass and leaf nutrients of *Thalassia hemprichii* at Inhaca island, Mozambique. *South African Journal of Botany* 67: 439-442.
23. Martins AR [1997]. Distribuição, estrutura, dinâmica da erva marinha *Zostera capensis* e estudo de alguns parametros fisicos em duas áreas da Baía de Maputo. Licenciatura thesis, Eduardo Mondlane University, Maputo. 49 pp.
24. Kalk M [1995]. *A Natural History of Inhaca Island, Mozambique*. Witwatersrand University Press, Johannesburg. 395 pp.
25. de Boer WF, Longomane FA [1996]. The exploitation of intertidal food resources at Inhaca bay, Mozambique by shorebirds and humans. *Biological Conservation* 78: 295-303.

9 The seagrasses of INDIA

T.G. Jagtap
D.S. Komarpant
R. Rodrigues

India has coastal wetlands of ca 63 630 km^2, mostly consisting of estuaries, bays, lagoons, brackish waters, lakes and salt pans[1]. The intertidal and supralittoral shallow sheltered regions of these wetlands harbor various marine macrophytic ecosystems such as seaweed, seagrass, mangrove and other obligate halophytes. Coastal wetland habitats are of a productive nature, and are of immense ecological and socioeconomic importance. Marine macrophytes support various kinds of biota, and produce a considerable amount of organic matter, a major energy source in the coastal marine food web; they play a significant role in nutrient regeneration and shore stabilization processes.

The major seagrass meadows in India exist along the southeast coast (Gulf of Mannar and Palk Bay) and in the lagoons of islands from Lakshadweep in the Arabian Sea to Andaman and Nicobar in the Bay of Bengal (Table 9.1). The flora comprises 14 species and is dominated by *Cymodocea rotundata*, *Cymodocea serrulata*, *Thalassia hemprichii*, *Halodule uninervis*, *Halodule pinifolia*, *Halophila beccarii*, *Halophila ovata* and *Halophila ovalis* (Table 9.2). Distribution occurs from the intertidal zone to a maximum depth of ca 15 m. Maximum growth and biomass occur in the lower littoral zone to a depth of 2-2.5 m. Greatest species richness and biomass of seagrass occur mainly in open marine sandy habitats.

Seagrasses, though one of a predominant and specialized group of marine flora, are poorly known in India, compared to other similar ecosystems such as mangroves. Earlier studies dealt mainly with the distributional and taxonomic aspects of Indian seagrasses[2-5]. Over the last 20 years, efforts have been made to understand the community structure and function of seagrass ecosystems in India[6-14]. However, the structure and function of Indian seagrass ecosystems remain poorly understood[15-17]. Inadequate information and almost total lack of awareness might be the reasons for this lack of knowledge in India.

Surprisingly, seagrasses have not been introduced even at the level of plant science education programs. Hence, large number of students, researchers and coastal zone managers in India may be unaware of the existence of seagrass ecosystems. Here, we present an overall account of seagrass habitats from India.

Epiphytes form an important constituent of seagrass ecosystems in India, though very limited information is available[9, 10, 14]. The floral epiphytes comprise a few species of marine algae belonging to Cyanophyceae, Chlorophyceae, Rhodophyceae and Bacillariophyceae. The Rhodophyceae, particularly *Melobesia* sp., occur frequently and are a dominant part of epiphytic biomass[9, 10]. Cyanophycean members such as species of *Microcoleus*, *Mastogocoleus* and *Oscillatoria* were observed to be dominant epiphytes. Ten species of diatoms have been reported on seagrass blades and roots. The oldest leaves and roots were found to be more infested and *Navicula*, *Nitzschia* and *Pleurosigma* form the characteristic diatoms associated with seagrasses[9, 10]. Large numbers of fungi have also been reported in association with seagrass[18]. Nine species of fungi have been recorded in association with *Thalassia hemprichii* in India[14]. Microbial flora actively mineralize seagrass litter and constitute about 1-3 percent of detrital biomass[14]. The epiphytes contribute 7.5-52 percent of total seagrass ecosystem biomass in shallower (1-3 m) depths. The higher epiphytic biomass (Figure 9.1b) results mostly from algal genera such as *Melobesia*, *Hypnea*, *Ceramium* and *Centroceros*. The intensity of epiphytization increases with shoot age and decreases at depths of more than 3 m.

Epifauna mostly consist of protozoans, nematodes, polychaetes, rotifers, tardigrades, copepods, amphipods and chironomid larvae. Very few attempts have been made to explore the faunal diversity of the seagrass beds of India. Harpacticoids, nauplii and nematodes are rarely found on seagrasses from India[9, 10].

Most of the algal groups in marine seagrass beds grow on coral or shell debris, and on seagrass stems and roots, in their earlier stages. Later stages of these algae become detached and float in the waters overlying the meadows. Some 100 species of algae have been reported from seagrass in various regions of India (Table 9.3). The algal flora in general is dominated by *Ulva lactuca*, *Ulva fasciata*, *Boodlea composita*, *Chaetomorpha linum*, *Halimeda* spp., *Chnoospora implexa*, *Chnoospora minima*, *Dictyota bartayresiana*, *Dictyota dichotoma*, *Dictyota divaricata*, *Hydoclathrus clathratus*, *Gracilaria edulis*, *Hypnea musciformes*, *Amphiroa fragillissima*, *Amphiroa rigida*, *Centroceros clavilatum* and *Centroceros* spp. Coralline algae, particularly *Halimeda* spp., contribute substantially to the formation of sediments suitable for the growth of the seagrasses[19]. Most of the associated algal biomass contributes organic matter to the seagrass environment.

Phytoplankton in the water column over the seagrass beds largely belong to Bacillariophyceae and Dinoflagellata; their occurrence is mostly patchy and the population density remains very low. The phytoplankton from the seagrass beds of Lakshadweep was reported to comprise 13 species (Table 9.3), commonly represented by *Achnanthes longipes*, *Asterionella japonica*, *Diploneis weisfloggi*, *Navicula hennedyii*, *Pinnularia* sp., and *Trichodesmium* sp. Absolutely no information exists on nanoplankton and picoplankton from the seagrass environment of India.

The regions of India that are colonized by seagrasses support rich and diverse fauna[12, 20, 21]. Hard corals, sea anemones, mollusks, sea cucumbers, starfishes and sea urchins are common invertebrates. Vertebrates such as fish and turtles commonly occur in seagrass beds; however, *Dugong dugon*, the marine mammal (dugong), has been very rarely reported in recent years[20]. The fish fauna is reported to consist of 192 species, dominated by sardine, mullet, eel, cat- and parrotfishes and grouper[22]. The mollusks (143 species), crustaceans (150 species) and echinoderms (77 species) are also found in large numbers (Table 9.3). Mollusks are mostly represented by *Acanthopleura spiniger*, *Acniaea stellaris*, *Conus generalis*, *Cypraea figris* and *Nerita costata*. There are four species of sea turtle, with *Chelonia mydas* and *Lepidochelys olevacea* being common.

The biomass and species richness of meiofauna and macrofauna in general is relatively very high in the seagrass beds compared to unvegetated areas in the vicinity[12]. Sediment organic content from seagrass beds varies from 4 to 13 percent, ten times higher than the sediments from unvegetated areas.

The textural characteristics of the sediment may be of great significance in determining density of seagrass growth. Well-established seagrass meadows influence mean size, sorting, skewness and shape of the accumulated sedimentary particles[23]. The sediments from seagrass beds of the Lakshadweep Islands show a significant correlation coefficient ($r = 0.85$, $p<0.05$) between kurtosis and total biomass, indicating prevalence of a relict environment[24]. This means that the depositional environment, which developed from coral reef biota over geological time, is most suitable for seagrass growth, a concept supported by the occurrence of major seagrass beds in association with coral reef regions[8-10, 25, 26]. *Halophila beccarii*, an estuarine seagrass, acts as pioneer species in the succession process leading to mangrove formation in India[27, 28]. Thus, seagrasses play a very important role as basic land builders and shore stabilizers, in a similar way to sand dunes and mangroves.

BIOGEOGRAPHY

Seagrass habitats are mainly limited to mud flats and sandy regions from the lower intertidal zone to a depth of ca 10-15 m along the open shores and in the lagoons around islands[5, 8]. The major seagrass meadows in India occur along the southeast coast (Gulf of Mannar and Palk Bay), and a number of islands of Lakshadweep in the Arabian Sea and of Andaman and Nicobar in the Bay of Bengal. The largest area (30 km²) of seagrass occurs along the Gulf of Mannar and Palk Bay, while it is estimated that ca 1.12 km² occur in the lagoons of major islands of Lakshadweep[21] (Table 9.1). A total 8.3 km² of seagrass cover has been reported from the Andaman and Nicobar Islands, a large portion of which is confined to islands like Teressa, Nancowry, Katchall and Great Nicobar[20, 29]. Seagrasses have been reported to occur in long or broken stretches, or small

Table 9.1
Quantitative data for major seagrass beds in Indian waters

Region	No. of species	Biomass (g dry weight/m²)	Area (km²)
Southeast coast (Gulf of Mannar and Palk Bay)	14	2.5-21.8	30
Lakshadweep group of islands	8	15-72	1.12
Nicobar group of islands	9	–	8.3
West coast	4	–	Patchy

Notes: – no data available.

Source: Various sources[5, 9, 20, 21].

Regional map: Africa, West and South Asia

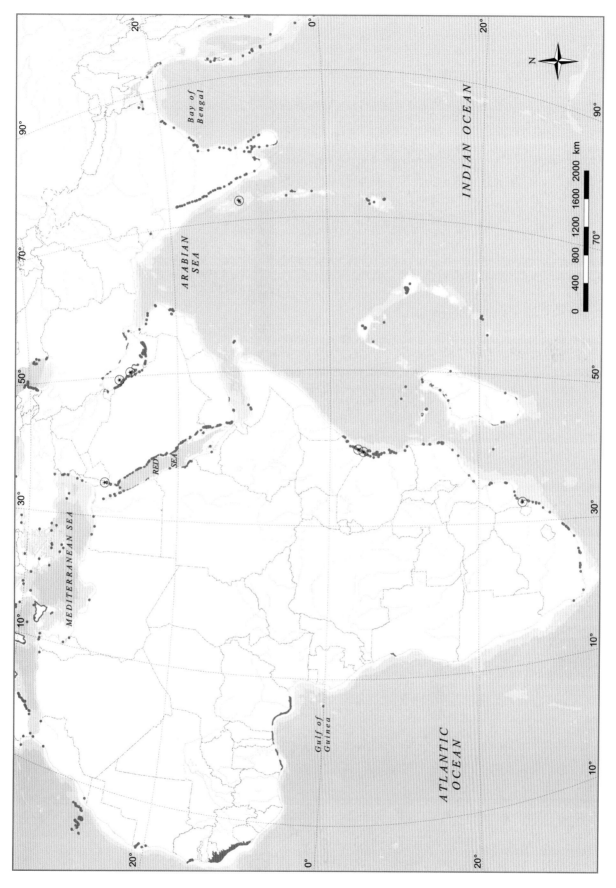

IMPACTS TO SEAGRASS ECOSYSTEMS

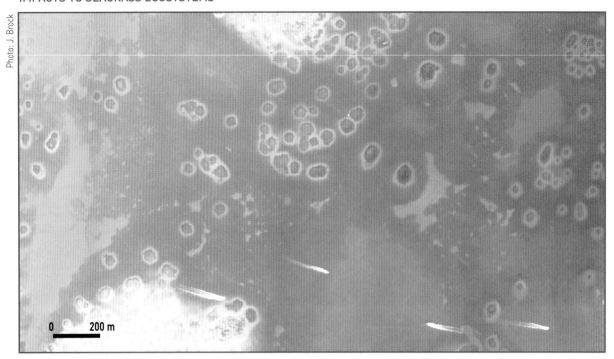

Patch reef in Florida (USA). Dark areas are seagrass meadows. Light areas around the coral heads (25-50 m diameter) are haloes created by herbivorous fish which live in the corals and graze on the seagrass.

Commercial ship aground on a seagrass flat in Australia's Great Barrier Reef.

Epiphytic algae growing on *Zostera marina*, Ninigret Pond, Rhode Island, USA.

Seagrass beds on the flats adjacent to an Indonesian community are being destroyed by boat traffic, fishing activities and waste discharge, in contrast to the healthy seagrasses across the channel (lower left).

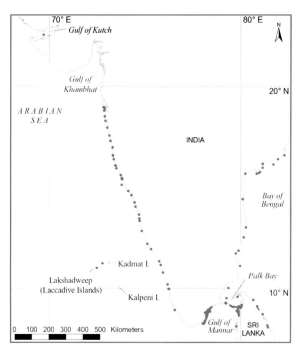

Map 9.1
India

to large patches[9, 20, 21]. The maximum seagrass cover, abundance and species richness are generally found in the sandy regions along the seashores, and in the lagoons of islands, where salinity of overlying waters remains above 33 psu throughout the year (Table 9.2). The estuaries, bays, lakes and gulf regions harbor a limited number of seagrass species in the lower intertidal mud flats in regions of moderate to high (10-40 psu) salinity during pre-monsoon (March-June) and post-monsoon (November-February) periods[27]. During the monsoon itself (July-October) the seagrass beds, particularly estuarine seagrasses, are subject to freshwater flooding and become silted and decay[27]. The new growth of estuarine seagrasses starts during August-September with a gradual increase in salinity, and attains maximum growth during November-December, and May-June[6].

PRESENT DISTRIBUTION

The seagrasses of India consist of 14 species belonging to seven genera (Table 9.2). The Tamil Nadu (southeast) coast harbors all 14 species, while eight and nine species have been reported from the Lakshadweep and Andaman-Nicobar groups of islands, respectively. The mainland east coast supports more species than the west coast of India. The main seagrasses are *Thalassia hemprichii*, *Cymodocea rotundata*, *Cymodocea serrulata*, *Halodule uninervis* and *Halophila ovata*. Species such as *Syringodium isoetifolium* and *Halophila* spp. occur in patches as mixed species. Meadows are mostly heterospecific. However,

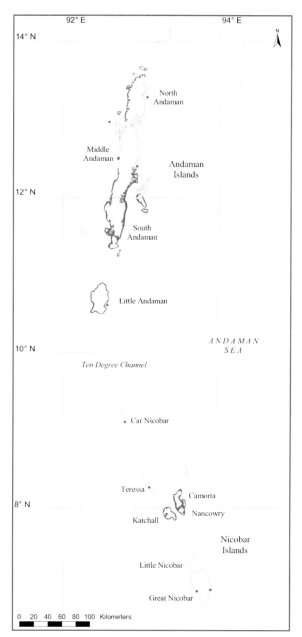

Map 9.2
Andaman and Nicobar Islands

from Kalpeni and Kadmat Islands of Lakshadweep, plant composition is bispecific and monospecific, respectively[10]. Gulf and bay estuaries mostly harbor low numbers of species, dominated by *Halophila beccarii* in the lower intertidal regions, and by *Halophila ovalis* in the lowest littoral zones. *Enhalus acoroides* has restricted distribution in the mid-intertidal swampy regions and shallow brackish waters[5, 11, 29].

Seagrasses grow from the regularly inundated intertidal zone to ca 15 m depth in the sandy subtidal zones[9, 10]. Unlike other species, *Halophila beccarii* is found in the upper intertidal. The maximum number of species and highest biomass usually occur at the depth of 1-2.5 m (Figure 9.1). The biomass of major seagrass beds has been reported to be significantly ($r = -0.63$ and -0.71, $p<0.05$) correlated with depth[9, 10]. *Thalassia hemprichii*, *Cymodocea rotundata*, *Cymodocea serrulata* and *Halophila ovata* are well adapted to the poor ambient light at greater depths (>3 m).

Biomass of Indian seagrasses varies from 180 to 720 g wet weight/m^2 (see also Table 9.1). *Halodule uninervis* and *Cymodocea rotundata* in the shallower depths (0.5-2.5 m), and *Thalassia hemprichii* and *Cymodocea serrulata* from the deeper (>3 m) waters, are the main contributors to biomass along the southeast coast (Figure 9.1). A similar trend of distribution and abundance was observed from major seagrass beds of Lakshadweep Islands in the Arabian Sea[9]. The lower biomass and reduced number of taxa in seagrasses deeper than 2 m is mainly attributed to insufficient ambient light. The older plants provide substratum for colonization by epiphytes, which make a considerable contribution to total seagrass system biomass[9, 10]. Biomass of *Halophila beccarii* is reported to vary from 4 to 24 g wet weight/m^2 with a minimum in the month of August and a maximum in October[6].

PRESENT THREATS

The natural causes of seagrass destruction in India are cyclones, waves, intensive grazing and infestation of fungi and epiphytes, as well as "die-back" disease. Exposure at ebb tide may result in the desiccation of the bed. Strong waves and rapid currents generally destabilize the meadows causing fragmentation and loss of seagrass rhizome. The decrease in salinity due to excessive freshwater runoff also causes disappearance, particularly of estuarine seagrass beds in the confluence regions.

Anthropogenic activities such as deforestation in the hinterland or mangrove destruction, construction of harbors or jetties, and loading and unloading of construction material as well as anchoring and moving

Table 9.2
Occurrence of seagrasses in coastal states of India

Seagrass sp.	West						East				
	GJ	MH	G	KA	KL	LD	WB	OR	AP	TN	A&N
Cymodocea rotundata	-	-	-	-	-	++++	-	-	-	++++	+
Cymodocea serrulata	-	-	-	-	-	+	-	-	-	++++	+
Enhalus acoroides	-	-	-	-	+	+	-	-	-	+	+++
Halodule pinifolia	-	-	-	-	-	-	+	-	+	+++	+++
Halodule uninervis	+	-	-	-	-	+++	+	-	+	+++	+
Halodule wrightii	-	-	-	-	-	-	+	-	+	+	-
Halophila beccarii	+++	+++	+++	+++	-	-	+++	+++	+++	+++	-
Halophila decipiens	-	-	-	-	-	-	-	-	-	+	-
Halophila ovalis	+	-	+++	-	-	+++	++	++	++	+++	+++
Halophila ovalis var. *ramamurtiana*	-	-	-	-	-	-	-	-	-	+	-
Halophila ovata	+	-	-	-	-	+++	+	+	+	+++	+++
Halophila stipulaceae	-	-	-	-	-	-	-	-	-	+	-
Syringodium isoetifolium	-	-	-	-	-	+++	-	-	-	+++	+
Thalassia hemprichii	-	-	-	-	-	++++	-	-	-	+++	+++
Ruppia maritima	-	-	-	++	-	-	-	-	-	+++	-
Total no. of species	4	1	2	2	1	8	6	3	6	14	9
Status of seagrass ecosystem	D	G	G	D	MD	C	G	G	G	D	VG
Salinity (psu)	35-40	0-33	0-33	0-33	0-33	34-36	0-31	0-33	0-33	33-35	33-35

Notes:
States: GJ Gujarat; MH Maharashtra; G Goa; KA Karnataka; KL Kerala; LD Lakshadweep Islands; WB West Bengal; OR Orissa; AP Andhra Pradesh; TN Tamil Nadu; A&N Andaman and Nicobar Islands.
Frequency of occurrence: − absent; + very rare; ++ rare; +++ common; ++++ dominant.
Status of seagrass ecosystem: VG very good; G good; D degraded; MD most degraded; C in the process of formation.

Source: Various sources[3, 5, 9, 10, 20, 29].

Table 9.3
Associated biota of seagrass beds of India

Group	Number of species
Fauna	
Bait fishes	21
Ornamental fishes	138
Fin fishes	33
Crustaceans	150
Mollusks	143
Echinoderms	77
Turtles	4
Mammals	1
Flora	
Marine algae	100
Phytoplankton	13
Fungi	9

Source: Various sources[7, 8, 9, 20, 22].

of boats and ships, dredging and discharge of sediments, land filling and untreated sewage disposal, are some of the major causes of seagrass destruction in India. As a result of the above natural and anthropogenic activities, the sediment load in the overlying waters of seagrass beds increases, reducing the amount of ambient light, resulting in lower productivity because of a decline in photosynthetic processes and increased respiration. The excess sediment input in the region results in the siltation and decline of seagrass beds. The siltation of seagrass beds has been commonly observed in the Gulf of Kutch, Gujarat, Andaman and Nicobar Islands, and in most of the estuaries.

Seagrass beds in the lower intertidal region in the Gulf of Kutch and a number of islands have experienced decline. *Halophila decipiens*, reported earlier[30] along the west coast, has totally disappeared, which might be due to its elimination during natural succession. Overexploitation of fisheries, particularly sea cucumbers and sea urchins, has impacted the resources associated with seagrass beds. *Dugong dugon*, which was abundant five decades ago[31], has totally disappeared along the Indian coast. The last report of dugong sightings dates back to 1994-95 in Andaman waters[20]. The loss of this mammal from the Indian coast could be attributed to overexploitation for fat and meat, as well as the obvious declines in seagrass beds.

POLICY RESPONSES
In India, seagrass regions, along with mangroves and corals, have been categorized as ecologically sensitive ecosystems under the Coastal Regulation Zone Notification to the Environment (Protection) Act[32].

However, seagrasses in India have been largely left out of education, research and management compared to other ecologically sensitive habitats such as mangroves, sand dunes and corals. Considering the lack of awareness, limited distribution and rising anthropogenic pressures, it is imperative to develop a national educational and conservation management plan for the seagrass ecosystem with the following objectives:
o quantification, mapping and regular monitoring to evaluate changes over time;
o education, research and awareness programs;
o environmental impact assessments;
o mitigation of adverse impacts;
o identification and conservation of areas as germplasm centers;
o rehabilitation.

Figure 9.1
Abundance of seagrass species at various depths in the Gulf of Mannar (southeast coast)

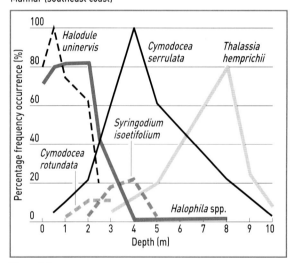

Figure 9.1a

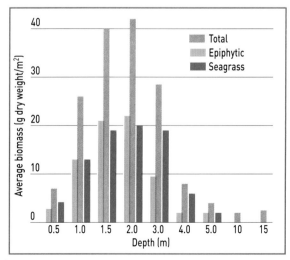

Figure 9.1b

Case Study 9.1
KADMAT ISLAND

Kadmat Island is located at 11°10'52"–11°15'20"N and 72°45'41"–72°47'29"E. It stretches ca 8 km from north to south, ranging in width from ca 50 to ca 400 m, with an area of 3.12 km². The lagoon is on the leeward (western) side, with a depth of 2-3 m. The storm beach along the eastern side has an average width of ca 100 m. A coralline algal ridge occurs along the breaking zone of the storm beach. The island is a submarine platform with a coral reef in the form of an atoll. It is crescent-shaped, having a north-south orientation. The western margin of the lagoon is a submarine bank marked by a narrow reef below.

Sampling and observations occurred along five fixed transects laid down from –10 m on the reef slope up to ca 150-200 m above high-tide line on the island. The length of the transect varied from ca 1 to 3.5 km depending upon the topography or the contour. The samplings were done during the post-monsoon (November 1998) and pre-monsoon (May 1999) seasons. The collections and observations were made from depths of –10 m and –5 m on the reef slope and from –1.5 and –2.5 m in the lagoon, and from exposed flats of reef and storm beach.

The seagrass bed in Kadmat Lagoon occurs in patches as well as longer stretches along the shore. A dense meadow occurs towards the northwest region of the lagoon covering some 0.14 km² and

Characterization of a seagrass meadow at Kadmat Island, Lakshadweep

Period	November 1998			March 1999		
Zone	Lagoon towards fore reef	Mid-lagoon region	Lagoon towards land	Lagoon towards fore reef	Mid-lagoon region	Lagoon towards land
Depth (m)	1-1.5	1.5-2.5	0-0.5	1-1.5	1.5-2.5	0-0.5
Substratum	S+CD	S	S	S+CD	S	S
Thickness of substratum (cm)	>2.5	5-10	>10	>2.5	5-10	>10
Sand % (range)	97.1-97.95	97.8-98	94.8-97.6	–	–	–
Silt % (range)	0.23-2.8	1.67-1.82	2.02-2.48	–	–	–
Clay % (range)	0.1-2.03	0.32-0.42	0.41-2.74	–	–	–
Organic carbon (%)	0.11-0.27	0.21-0.23	0.36-0.42	1.08-1.4	1.52-1.96	0.92-1
Nature of seagrass beds	SP	LP	BS	SP	LP	BS
Quantitative aspect of seagrasses						
Number of seagrass species	1	2	1	2	2	1
Thalassia hemprichii						
% frequency of occurrence	10-20	10-20	A	10-20	50-70	A
Biomass (g dry weight/m²)	N	5	NA	N	7.5	NA
Cymodocea rotundata						
% frequency of occurrence	A	10-20	50-70	–	50-70	>70
Biomass (g dry weight/m²)	NA	15	17	–	23	26
Total biomass (g dry weight/m²)	NA	20	17	N	30.5	26
Average total drifted biomass (g dry weight/m²)	NA	NA	N	NA	NA	195

Notes:
– data not collected.
S sandy; CD coral debris; SP small patches; LP large patches; BS broken stretches.
A absent; N negligible; NA not applicable.

Source: Desai et al.[36]

exhibiting marked zonation. Mostly sparse and small patches of *Thalassia hemprichii* occur in the shallow sandy regions towards the fore reef, while the mid-lagoon deeper region (1.5-2.5 m) harbors mixed dense beds of *Thalassia hemprichii* and *Cymodocea rotundata*. The shallow region (0.5-1.5 m) towards land supports intensive growth of *Cymodocea rotundata*. A similar kind of distribution trend has been reported from the other islands on the Laccadive Archipelago[10]. The seagrass flora of Kadmat comprises two species with higher biomass (20-35 g dry weight/m^2) occurring from the mixed zone in the mid-lagoon (see table, left). A biomass of drifting seagrasses (195 g dry weight/m^2) was recorded during March when the biomass of the seagrass standing crop was higher (26 g dry weight/m^2). The frequency of occurrence of drifting seagrass increased from 20 percent to 70 percent during March, reflecting seagrass maximum biomass; it is during this pre-monsoon period that high wind speeds cause disturbances in the state of the sea, including lagoon waters. Previously, five species of seagrasses were recorded from the lagoons of Kadmat[21]. It has been observed that the small-sized seagrasses, such as *Halophila* spp., commonly grow as pioneer species and form a suitable substratum for other larger-sized seagrasses to follow during the succession process[10,33]. The absence of such species from Kadmat Lagoon during this study might be due to competition by the existing species during succession.

A considerable amount of seagrass biomass contributes to the detrital food chain[34]. The benthic faunal population from the seagrass beds has been reported to be higher due to high organic carbon in the sediments[13]. The organic carbon in the sediments, particularly from the seagrass beds, varied from 0.11 to 1.96 percent (see table, left). Macrofauna from the seagrass bed of Kadmat Island consisted of eight groups (see table, right). Macrofauna were largely Oligochaeta (40.17 percent), but the maximum number of species (22) were from Polychaeta group[35]. It was reported earlier that Polychaeta (44.6 percent) and Crustacea (42 percent) constitute the major macro-invertebrates in the seagrass beds of India[12].

The composition of meiofauna in seagrasses varies seasonally[12]. The meiofauna from the seagrass bed of Kadmat[32] is represented by 19 groups dominated by Turbellaria (34.2 percent), Nematoda (37.3 percent) and herpacticoid copepods (10.1 percent).

Thalassia hemprichii.

Benthic macrofauna in the seagrass bed at Kadmat Island, Lakshadweep

Macrofauna group	No. of genera	No. of species	% composition	Dominant taxa
Polychaeta	20	22	18.96	Lumbriconeries, Syllis, Onuphis, Polydora
Nematoda	1	1	18.71	–
Oligochaeta	1	1	40.17	–
Pelecypoda	3	3	2.96	Mesodesma, Donax
Gastropoda	8	8	2.32	Cerithium, Cerithidea
Crustacea	6	6	11.36	Amphipoda, Isopoda
Ophiuroidea	1	1	0.61	Echiurida
Ascheliminthes	–	–	4.47	–
Unidentified	–	–	0.44	–

Note: – not identified to genus/species level.

Source: Branganza et al.[35]

The Ministry of Environment and Forests, Government of India, coordinates environment and biodiversity-related coastal zone management programs in the country. This department has a vital role in adapting and implementing educational and management plans for the seagrass environments of India, similar to those for mangrove and coral reef habitats. The necessary inputs based on research would be of great importance in the formation of a national seagrass management plan. Hence, the ministry must encourage universities and national laboratories to undertake investigations on the various aspects of seagrass ecosystems.

ACKNOWLEDGMENTS

The authors are grateful to the Director of the National Institute of Oceanography (CSIR), Donapaula, Goa, for his encouragement.

AUTHORS

T.G. Jagtap, D.S. Komarpant and R. Rodrigues, National Institute of Oceanography, Dona Paula, Goa – 403004, India. **Tel:** +91 (0)832 456700 4390. **Fax:** +91 (0)832 456702/456703. **E-mail:** tanaji@csnio.ren.nic.in

REFERENCES

1. Anon [1992]. 'Coastal Environment' A Remote Sensing Application Mission. A Scientific Note by Space Application Center (ISRO), Ahmedabad, funded by Ministry of Environment and Forests, Government of India, SAM/SAC/COM/SN/11/92. 100 pp.
2. Santapu H, Henry AN [1973]. *A Dictionary of the Flowering Plants in India*. CSIR Publication, New Delhi.
3. Untawale AG, Jagtap TG [1978]. A new record of *Halophila beccarii* (Aschers) from Mandovi estuary, Goa, India. *Mahasagar, Bulletin of the National Institute of Oceanography* 10: 91-94.
4. Lakshmanan KK [1985]. Ecological importances of seagrass in marine plants, their biology, chemistry and utilization. In: Krishnamurthy V (ed) *All India Symposium Marine Plants, Proceedings*. Donapaula, Goa. pp 277-294.
5. Ramamurthy K, Balakrishnan NP, Ravikumar K, Ganesan R [1992]. *Seagrasses of Coromandel Coast, India*. Flora of India, ser. 4. BSI, Coimbatore. 79 pp.
6. Jagtap TG, Untawale AG [1981]. Ecology of seagrass bed *Halophila beccarii* (Aschers) in Mandovi estuary, Goa. *Indian Journal of Marine Sciences* 4: 215-217.
7. Jagtap TG [1987]. Distribution of algae, seagrass and coral communities from Lakshadweep Islands, Eastern Arabian Sea. *Indian Journal of Marine Sciences* 16: 56-260.
8. Jagtap TG [1991]. Distribution of seagrasses along the Indian Coast. *Aquatic Botany* 40: 379-386.
9. Jagtap TG [1996]. Some quantitative aspects of structural components of seagrass meadows from the southeast coast of India. *Botanica Marina* 39: 39-45.
10. Jagtap TG [1998]. Structure of major seagrass beds from three coral reef atolls of Lakshadweep, Arabian Sea, India. *Aquatic Botany* 60: 397-408.
11. Untawale AG, Jagtap TG [1989]. Marine macrophytes of Minicoy (Lakshadweep) coral atoll of the Arabian Sea. *Aquatic Botany* 19: 97-103.
12. Ansari ZA [1984]. Benthic macrofauna of seagrass (*Thalassia hemprichii*) bed at Minicoy, Lakshadweep. *Indian Journal of Marine Sciences* 13: 126-127.
13. Ansari ZA, Rivonker CV, Ramani P, Parulekar AH [1991]. Seagrass habitat complexity and microinvertebrate abundance in Lakshadweep coral reef lagoons, Arabian Sea. *Coral Reefs* 10: 127-131.
14. Sathe V, Raghukumar S [1991]. Fungi and their biomass in detritus of the seagrass *Thalassia hemprichii* (Ehrenberg) Ascherson. *Botanica Marina* 34: 271-277.
15. Jacobs RPWM [1982]. *A Report: Component Studies in Seagrass Ecosystems along West European Coasts*. DRUK: Drukkerij verweij BV, Mijdercht. pp 11-215.
16. Bortone SA [1999]. *Seagrasses: Monitoring, Ecology, Physiology and Management*. CRC Press, New York. 309 pp.
17. Larkum AWD, McComb AJ, Shepherd SA [1989]. *Biology of Seagrasses*. Elsevier, New York. 841 pp.
18. Cuomo V, Jones EB, Grasso S [1988]. Occurrence and distribution of marine fungi along the coast of the Mediterranean Sea. *Progress in Oceanography* 21:189-200.
19. Siddique HN [1980]. The ages of the storm beaches of the Lakshadweep (Laccadive). *Marine Geology* 38: M11-M20.
20. Das HS [1996]. Status of Seagrass Habitats of Andaman and Nicobar Coast. SACON Technical Report No. 4, Coimbatore. 32 pp.
21. Jagtap TG, Inamdar SN [1991]. Mapping of seagrass meadows from the Lakshadweep Islands (India), using aerial photographs. *J Ind Soc Remote Sensing* 19: 77-81.
22. James PSBR [1989]. Marine living resources of the union territory of Lakshadweep – an indicative survey with suggestions for development, Cochin, India. *CMFRI Bulletin* 43: 256 pp.
23. Swinchatt JP [1965]. Significance of constituent composition, texture and skeletal breakdown in some recent carbonate sediments. *Journal of Sedimentary Petrology* 35: 71-90.
24. Rajamanickam GV, Gujar AR [1984]. Sediment depositional environment in some bays in the central west coast of India. *Indian Journal of Marine Sciences* 53-59.
25. Hackett HE [1977]. Marine algae known from Maldive Islands. *Atoll Research Bulletin* 210: 2-37.
26. Fortes MD [1989]. *Seagrass: A Resource Unknown in the Asian Region*. ICLARM, Manila. 46 pp.
27. Jagtap TG [1985]. Ecological Studies in Relation to the Mangrove Environment along the Goa Coast, India. PhD thesis, Shivaji University, Kolhapur. 212 pp.
28. Untawale AG, Jagtap TG [1991]. Floristic composition of the deltaic regions of India. In: Vaidyanadhan R (ed) *Quaternary Deltas of India*, Memoir 22. Publication GSI, Bangalore. pp 243-265.
29. Jagtap TG [1992]. Marine flora of Nicobar group of Islands, Andaman Sea. *Indian Journal of Marine Sciences* 22: 56-58.
30. Parthasarthy N, Ravikumar K, Ramamurthy K [1988]. *Halophila decipens* Ostenf. Southern India. *Aquatic Botany* 32: 179-185.
31. Nair RV, Lal Mohan RS, Roa KS [1975]. The Dugong (*Dugong dugon*). *CMFRI Bulletin*. 42 pp.
32. Anon [1990]. Coastal area classification and development regulations. Gazette Notification, Part II, Section 3 (ii), Govt of India, No. SC 595 (F) – Desk – 1/97.
33. Birch WR, Birch M [1984]. Succession and pattern of tropical intertidal seagrasses in Cockle Bay, Queensland, Australia: A decade of observation. *Aquatic Botany* 19: 343-368.
34. Mann KH [1988]. Production and use of detritus in various fresh water, estuarine and marine ecosystems. *Limnology and Oceanography* 33: 910-930.
35. Branganza C, Ingole BS, Jagtap TG. Unpublished data.
36. Desai VV, Komarpant DS, Jagtap TG [Accepted manuscript]. Distribution and diversity of marine flora in coral reef ecosystems of Kadmat Island in Lakshadweep Archipelago, Arabian Sea, India. *Atoll Research Bulletin*.

10 The seagrasses of WESTERN AUSTRALIA

D.I. Walker

The coastline of Western Australia extends 12500 km, from the temperate waters of the Southern Ocean at 35°S to the tropical waters of the Timor Sea at 12°S, with the contiguous coastline of the Northern Territory extending across to Queensland.

ECOSYSTEM DESCRIPTION
The long coastline has a diversity of environments that support seagrass, ranging from those tropical species associated with coral reefs and mangroves in the north to large temperate seagrasses, in the shelter of limestone reefs and in large embayments, on the west and south coasts. These are exposed to different tidal conditions (amplitudes 9 m in the north to less than 1 m on the west and south coasts[1]), substratum types and exposure to wave energy. Although some areas of the Western Australian coast, such as Cockburn Sound, have been the subject of much research, a great deal of the rest of the marine environment is poorly described or understood.

This chapter will provide a brief description of the coastal geomorphology, seagrass species and habitats, and their biogeography. Current uses will be described and current and potential threats to these habitats/uses considered. Extensive use of *Environment Western Australia 1998: State of the Environment Report*[2] and of *The State of the Marine Environment Report*[3,4] has been made in compiling the latter section of this review. Issues of seagrass management will also be discussed.

Geomorphology of the coast
The underlying geology of the coast consists of granitic rocks in the south and southwest, with extensive mantling of tertiary limestone, and sandstones in the northwest and north. In the southeast of the state, the vertical limestone cliffs of the southern edge of the Nullarbor Plain delimit a narrow coastal plain. For almost 300 km, offshore reefs protect sandy beaches and high foreshore sand dunes from oceanic swell, producing a calmer habitat between the reefs and the shore, suitable for seagrass growth. At Twilight Cove the cliffs again approach the sea and follow the coastline to just east of Israelite Bay. From there to Esperance, beaches and seagrass beds are sheltered by the granitic islands of the Recherche Archipelago, 5-50 km offshore.

From Esperance to Albany, sheltered beaches are broken by granite outcrops although occasionally limestone reefs and eroded cliffs occur. Small rivers flow into a number of bays along this 500-km coastline, but they have relatively low discharge rates, particularly during the summer dry season. Offshore of these estuaries, seagrasses of the *Posidonia ostenfeldii* group occur as they can withstand swell and sediment movement.

From Albany to Cape Naturaliste, limestone overlies granitic rocks for much of the coast. Seagrasses occur in this region in sheltered inshore lagoons protected by offshore reefs.

Geographe Bay, east of Cape Naturaliste, is north facing and the prevailing southwesterly swell is refracted into the relatively sheltered embayment. The embayment has a thin sediment veneer (mean thickness: 1 m) overlying Pleistocene limestone[5]. It provides an ideal habitat for seagrasses[6], and extensive meadows are found to depths of 25 m. A number of estuaries, larger than those further east, also afford habitat for seagrasses and other submerged aquatic plants such as *Ruppia*[7,8] and their associated invertebrates.

The western coastline, from Geographe Bay to Kalbarri, is relatively straight and continuous, as it has been eroded by the action of winds and currents which have built up sand dunes and bars parallel to the coast. There is also a fringe of limestone reefs running

parallel to the coast which are relict Pleistocene dune systems composed of aeolianite; these break the Indian Ocean swells, forming relatively calm, shallow (4-10 m deep) lagoons up to 10 km wide, in which the tidal range is small (<1 m), and the waters generally clear. These lagoons are dominated by seagrasses.

From Kalbarri to Steep Point (the most westerly point of the mainland), along Dirk Hartog Island, Bernier and Dorre Islands and up to Quobba Point, there are high cliffs composed of sandstone to the south and limestone to the north. These cliffs shelter Shark Bay, a large (13 000 km^2), shallow, semi-enclosed embayment (see Case Study 10.1). This is an area of intense carbonate sedimentation, which is affected by wind and tidal-driven water movement, leading to high turbidity. It also has relatively low water temperatures in winter (down to 13°C)[9].

North of Quobba Point, the Pilbara coastline has a low relief with gently sloping beaches, numerous headlands and many small offshore islands. Headlands are composed of isolated patches of very hard hematite-bearing quartzite, which is more resistant to erosion than the surrounding rocks. Normal erosion processes, combined with submergence, have led to a broken, rough coastline. Mangroves become conspicuous. Coral reefs and atolls occur north of Quobba Point (near the Tropic of Capricorn), where tropical seagrasses are found in lagoons, as well as in mangrove swamps and around islands[10]. There is a progressive increase in tidal amplitude with decreasing latitude. Large tides affect seagrass distributions by resuspending sediments; the high turbidity limits seagrass growth to shallow water. On broad intertidal flats, seagrasses are restricted to those species which can tolerate high temperatures and desiccation, as well as periodic freshwater inundation from rainfall.

The Kimberley coast is a typical ria (drowned river valley) system, characterized by resistant basement rock, with faults oriented at angles to the shore, creating a rugged coastline. The area is subject to large tidal amplitudes and is remote and sparsely populated, with little information available about the marine habitats. Embayments and sounds grade shorewards into mangrove-covered tidal flats, and there are many offshore islands. Extensive terracing of these expanses of the intertidal zone often results in seagrass, particularly *Enhalus acoroides*[10], high in the intertidal just below the mangroves.

Much of the Kimberley landscape is of extraordinary natural beauty, extending to its coastal regions. With a vast land area and a small population, the Kimberley has been, until recently, largely unexplored by biologists. Its isolated coastline is devoid of settlement along the 2 000-km stretch between Derby and Wyndham. The area is receiving growing attention from tourists, with increasing activity by small private boats and charter operators. As part of the development of a marine park and reserve system in Western Australia, several areas are being considered as potential marine parks. In addition, some of the areas have been designated as potential Aboriginal reserves. These designations have been based on severely limited data available from the few scientists and other people who have traveled in the area. The only substantial data on marine organisms in the Kimberley relate to salt water crocodile populations and turtles. Marine plants, fish and invertebrates are largely unknown. Recent surveys by the West Australian Museum, the University of Western Australia and the Northern Territory Museum, and by CSIRO (Australia's Commonwealth Scientific and Industrial Research Organisation), have yet to be published, but will help provide a basis for future research.

BIOGEOGRAPHY

Seagrasses recorded from Western Australia fall into two general distribution patterns. Twelve species are endemic to Western Australia or to the southern Australian coast, and are confined to temperate, clear waters (Table 10.1). Twelve species are tropical and are found throughout the Indian Ocean and tropical Pacific Ocean.

Australia's seagrasses can be divided into temperate and tropical distributions, with Shark Bay on the west coast and Moreton Bay on the east coast being located at the center of the overlap zones. Temperate

Table 10.1
Western Australian endemic seagrass species

Species	Distribution
Cymodoceaceae	
Amphibolis antarctica	Southern Australian endemic
Amphibolis griffithii	Southern Australian endemic
Cymodocea angustata	Tropical Western Australian endemic
Thalassodendron pachyrhizum	Southern Australian endemic
Hydrocharitaceae	
Halophila australis	Australian endemic
Posidoniaceae	
Posidonia angustifolia	Southern Australian endemic
Posidonia australis	Southern Australian endemic
Posidonia coriacea	Western Australian endemic
Posidonia denhartogii	Southern Australian endemic
Posidonia kirkmanii	Western Australian endemic
Posidonia ostenfeldii	Western Australian endemic
Posidonia sinuosa	Southern Australian endemic

species have been studied most extensively, particularly the large genera *Amphibolis*, *Posidonia* and *Zostera*, but there are other species which have been little studied. Temperate species are distributed across the southern half of the continent, extending northwards on both the east and west coasts. The highest biomasses, and highest regional species diversity, occur in southwestern Australia, where seagrasses are found in the coastal back-reef environments within the fringing limestone reef, or in semi-enclosed embayments.

In areas of northern Australia with a high tidal range, visibility is often poor, and conventional remote-sensing techniques are of limited value for mapping. The Northern Territory coastline is largely unexplored for seagrass distribution, and their associated animal communities, especially the Northern Territory prawn fisheries, remain largely unstudied. Recent research in the Kimberley region of Western Australia has provided some distribution information. Seagrasses in that region either occur sparsely in coral reef environments or can attain high biomasses within high intertidal lagoons, where seawater is ponded during the falling tide[11]. The environments are otherwise too extreme (tidal movements/turbidity/freshwater runoff in the wet season) for seagrass survival[12]. Again, the significance of these seagrass communities for any associated fisheries species is unknown.

In general, our knowledge of shallow water (<10 m) temperate seagrass distributions is reasonably good, but our understanding of deep water (>20 m) seagrasses throughout Australia is rudimentary. Areas subject to more extreme water movement, either tidal or wave induced, are also poorly studied compared with seagrasses in more protected areas.

The main habitats for seagrasses are very extensive shallow sedimentary environments that are sheltered from oceanic swell, such as embayments (e.g. Shark Bay, Cockburn Sound), protected bays (e.g. Geographe Bay, Frenchman's Bay) and lagoons enclosed by fringing reefs (e.g. Bunbury to Kalbarri). Seagrasses occupy approximately 20 000 km^2 on the Western Australian coast[13], ranging in depth from the intertidal to 45 m[14], making up a major component of nearshore ecosystems. The diversity of seagrass genera (10) and species (25) along this coastline is unequaled elsewhere in the world[10], mainly due to the overlap between tropical and temperate biogeographic zones, and the extent of suitable habitats.

Large, mainly monospecific meadows of southern Australian endemic species form about one third of the habitat in the coastal regions of Western Australia. These meadows have high biomasses (500-1 000 g/m^2) and high productivities (>1 000 g/m^2/year)[15]. Southern Australian seagrasses occur in

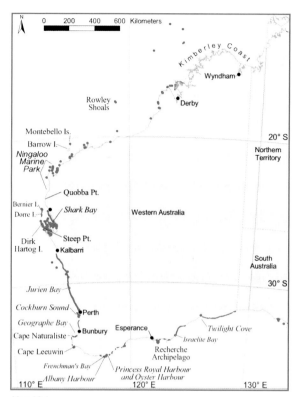

Map 10.1
Western Australia

water bodies exposed to relatively high rates of water movement. Nevertheless, Australian species also occur where there is some protection from extreme water movement and most are found in habitats with extensive shallow sedimentary environments, sheltered from the swell of the open ocean, such as embayments (e.g. Shark Bay and Cockburn Sound), protected bays (e.g. Geographe Bay and Frenchman's Bay) and lagoons sheltered by fringing reefs (e.g. the western coast from 33° to 25°S).

MECHANISMS OF SEAGRASS DECLINE IN WESTERN AUSTRALIA

Seagrass declines have been well documented from around Australia. There are a variety of mechanisms of seagrass loss, but the most ubiquitous and pervasive cause of decline is the reduction of light availability. Seagrasses are rather unique plants in that they have high minimum light requirements for survival compared with other plants[16]. These high minimum light requirements (10-30 percent incident light) are hypothesized to be related to the significant portions of seagrass biomass that can be in anoxic sediments. Reduction in light availability can occur as a result of three major factors: chronic increases in dissolved nutrient availability leading to proliferation of light-absorbing algae, either phytoplankton, macroalgae or algal epiphytes on seagrass leaves and stems; chronic

Intertidal *Enhalus acoroides*, Leonie Island, Kimberley, Western Australia.

Habitat removal

Coastal development in Western Australia is localized to centers of population, and takes the form of construction of ports, marinas and groynes. Housing developments impact on coastal water quality, whereas canal estates, such as in Carnarvon, have greater direct impact on the marine environment. All these developments have potential consequences for seagrass habitats and associated fauna.

Some developments have resulted in direct destruction of seagrass communities, by smothering or deterioration in water quality, e.g. construction of the causeway at the southern end of Cockburn Sound[19], where construction destroyed existing reef environments, and resulted in loss of seagrass habitat due to reduced flushing. The construction of ports and marinas in the Perth Metropolitan area has degraded existing seagrass and reef habitats, as well as fragmenting the remaining distributions. Subsequent dredging and sediment infill has often reduced the water quality and resulted in further losses.

Impacts of pollution

Pollution of coastal environments can result in major changes to water quality, either from point or diffuse sources which can influence marine community structure, especially in relation to seagrass. Marine disposal of sewage from the Perth Metropolitan region's three outfalls contributes excess nutrients to coastal areas[20]. The Kwinana Industrial Strip along the shores of Cockburn Sound still relies on marine disposal of the industries' effluents, although now under license conditions to regulate the amounts of toxicants.

increases in suspended sediments leading to increased turbidity; and pulsed increases in suspended sediments and/or phytoplankton that cause a dramatic reduction of light penetration for a limited time period.

Loss of habitat

Seagrasses are limited to the photic zone, extending up to 54 m[17]. Reductions in water quality can lead to a reduction in the depth of the photic zone[18], and hence to a direct loss of habitat. Seagrasses in Cockburn Sound, for example, are limited to a depth of less than 9 m, whereas in unpolluted areas the depth limit would be 11-15 m. Increasing population pressure in Western Australia leads to increasing pressure on the coast. Development of the coastal zone, all along the Western Australian coastline, in the form of construction of marinas, port facilities and canal estates, results in degradation of coastline causing direct destruction of seagrass communities as well as indirect changes in hydrodynamics and sedimentation.

Water quality, especially nutrients

The Western Australian coastal environment is particularly sensitive to nutrient enrichment from human activities. The effects of this anthropogenic eutrophication include an increase in frequency, duration and extent of phytoplankton and macroalgal blooms[21], low oxygen concentrations in the water column, shifts in species composition[22,23], loss of seagrass and benthic vegetation[6,24], decrease in diversity of organisms present[18] and an increase in diseases in fish and waterfowl. Western Australian marine waters are generally low in nutrients and biological productivity. Serious seagrass losses resulting from increased nutrient loading have occurred in the Albany Harbours, Cockburn Sound and parts of Geographe Bay. Cockburn Sound is the most degraded marine environment in Western Australia, having experienced the second largest loss of seagrass in Australia (more than two thirds)[3].

The major human-induced declines of seagrass

in Western Australia are summarized in Table 10.2, with suggested principal causes – in most cases, other factors interact to make the process of loss more complex. The general hypothesis for all these instances of seagrass decline is that a decrease in the light reaching seagrass chloroplasts reduces effective seagrass photosynthesis. The decrease may result from increased turbidity from particulates in the water, or from the deposition of silt or the growth of epiphytes on leaf surfaces or stems[18]. Seagrass meadows occur between an upper limit imposed by exposure to desiccation or wave energy, and a lower limit imposed by penetration of light at an intensity sufficient for net photosynthesis. A small reduction in light penetration through the water will therefore reduce the depth range of seagrass meadows, while particulates on leaves could eliminate meadows over extensive areas of shallower water (e.g. Princess Royal Harbour, Western Australia)[25-27].

Increasing turbidity of water above seagrasses may occur directly, by discharge or resuspension of fine material in the water column from, for example, sludge dumping. Indirect effects on attenuation coefficients occur through increased nutrient concentrations resulting from the discharge of sewage and industrial wastes, or from agricultural activity in catchments, which in turn increase phytoplankton biomass reducing light penetration significantly[22,28]. The extent of phytoplankton blooms associated with nutrient enrichment will be determined by water movement, and mixing will dilute nutrient concentrations. Deeper seagrass beds further from the sources of contamination may show no influence of turbidity.

Epiphytes

In Cockburn Sound, nutrient enrichment has led not only to enhanced phytoplankton growth but also to enhanced growth of macroscopic and microscopic algae on leaf surfaces[29]. Macroalgae dominate over seagrasses under conditions of marked eutrophication, both as epiphytes and as loose-lying species (e.g. the genera *Ulva, Enteromorpha, Ectocarpus*) which may originate as attached epiphytes[25]. Increased epiphytic growth results in shading of seagrass leaves by up to 65 percent[29], reduced photosynthesis and hence leaf densities[18]. In addition, the epiphytes reduce diffusion of gases and nutrients to seagrass leaves.

Light penetration

As photosynthetically active radiation passes through water, it is attenuated by both absorption and scattering. Attenuation is increased by the presence of suspended organic matter (e.g. phytoplankton) and inorganic matter, particularly in eutrophic systems when phytoplankton concentrations are high[17], thus reducing light penetrating to benthic primary producers. In Cockburn Sound, where this continued for extended periods of time, reduction in density and loss of benthic macrophytes resulted[19,29].

The requirement of light by benthic macrophytes makes the presence of submerged aquatic vegetation an indicator of water quality (adequate light penetration) and hence, nutrient status (i.e. low nutrient concentrations)[18]. Light reduction for extended periods, which is common in eutrophic systems, causes loss of benthic macrophyte biomass[29].

Siltation

Changes in landuse practices often result in increased sediments in runoff from land, e.g. in Oyster Harbour. Larger sediment loads reduce light penetration, as detailed above. Increased sedimentation can result in changes in the abundance and percentage cover of seagrass due to increased sediment deposition or scour[30].

Toxic chemicals

In general the Western Australian coastal environment is not subjected to large-scale inflows of toxic chemicals. The 1998 Western Australian *State of the Environment Report* does not consider them a threat[2]. Awareness of toxic, human-produced chemicals and

Table 10.2
Summary of major human-induced declines of seagrass in Western Australia

Place	Seagrass community	Extent of loss	Cause
Cockburn Sound, Western Australia	*Posidonia sinuosa* *Posidonia australis*	7.2 km² lost (more than two thirds)	Increased epiphytism blocking light
Princess Royal and Oyster Harbours, Western Australia	*Posidonia australis* *Amphibolis antarctica*	8.1 km² lost (46%)	Decreased light, increased epiphyte and drift algal loads

Source: Cockburn Sound: Cambridge et al.[19], Silberstein et al.[29]; Princess Royal and Oyster Harbours: Walker et al.[26], Wells et al.[27].

Underwater meadow of *Posidonia australis* abutting a limestone reef, Rottnest Island, Western Australia.

their impacts on marine organisms has increased, and such industrial inflows are controlled by Licence Conditions from the Western Australia Department of Environmental Protection. Urban runoff may include such chemicals, but in Western Australia the runoff is separated from the sewage system. Some direct runoff may still influence groundwater or the coastal environment, and increasing population pressure will result in increased risk of contamination.

Fortunately, the aquaculture industry in Western Australia has avoided the use of antibiotics in fish foodstuffs. The potential effects of antibiotics[31,32] may result in widespread changes in microbial activities, with consequences up the food web, as well as for nutrient recycling in coastal sediments.

The effects of antifouling compounds are also a concern. Tributyltin (TBT) has been recorded from Western Australian locations[33], highest near marinas and ports. TBT contamination is present at various levels in all major ports in Western Australia. TBT contamination is widespread throughout Perth Metropolitan marine environment[2]. The use of TBT has been banned in Western Australia on vessels longer than 25 m.

Introduction of exotic (alien) species

Exotic marine organisms have been introduced to Western Australia via ballast water and hull fouling from shipping, and threaten natural distributions of organisms, including seagrass. It is estimated that 100 million metric tons of ballast water are discharged into this region's marine waters each year. Currently, controls are only voluntary. Introduced marine species may threaten native marine flora and fauna and human uses of marine resources such as fishing and aquaculture. Knowledge of species introduced and their distribution has recently been updated. The risk of damage to marine biodiversity is largely unknown but international experience suggests that the potential for significant environmental impact is high[2]. Displacement of existing flora and fauna by introduced species, intentional or accidental, has been widely reported elsewhere[34].

The 1998 Western Australian *State of the Environment Report* estimated that over 27 exotic species have been introduced to Western Australia[2]. Twenty-one of these are known to have been introduced into Perth Metropolitan waters, the most highly visible being a large polychaete worm *Sabella spallanzani* (Sabellidae family). This worm occupied up to 20 ha of the seafloor and most of the structures in Cockburn Sound, outcompeting the native *Posidonia* species, but its incidence has been declining[35].

ESTIMATE OF PRESENT COVERAGE

Large, mainly monospecific meadows of southern Australian endemic species form about one third of the habitat in the coastal regions of Western Australia and amount to some 20 000 km^2. The tropical species are less abundant but add a further 5 000 km^2.

THREATS

Human utilization of seagrass in Western Australia is relatively restricted. Few commercial and recreational species are taken from seagrass habitats. According to the 1998 Western Australian *State of the Environment Report*[2], human activities most affecting coastal seagrass habitats in Western Australia are:

o direct physical damage caused by port and industrial development, pipelines, communication cables, mining and dredging, mostly in the Perth Metropolitan and Pilbara marine regions;

o excessive loads of nutrients, causing seagrass overgrowth and smothering by epiphytes, from industrial, domestic and agricultural sources, mostly in the Lower West Coast, Perth Metropolitan and South West Coast marine regions;

o land-based activity associated with ports, industry, aquaculture and farming, mostly in the Pilbara, Central West Coast, Lower West Coast, Perth Metropolitan and South West Coast marine regions;

o direct physical damage caused by recreational and commercial boating activities including anchor and trawling damage, mostly in the

Kimberley, Pilbara, Shark Bay, Perth Metropolitan and Geographe Bay areas. Trawling nets remove sponges and other attached organisms from the seafloor.

The marine environment receives most of the surface water from land. The quality of this water is affected by activities and the environment of the catchments through which it flows. Soil and nutrients can be carried by river discharges to coastal waters, causing water quality deterioration. Groundwater can also carry terrestrial pollution into the marine environment. Direct discharges such as sewage and/or treated wastewater and industrial outfalls, and accidental discharges such as spills and shipping accidents, also influence coastal water quality[2].

These land-based activities, their impacts on ground and surface water and the ultimate movement of these waters into nearshore marine environments are the major human influence on the Western Australian coast. They result in most pollution of the marine environment and the resulting chronic degradation of marine habitat. Degradation of the marine environment leads to reductions in the area of seagrass, as well as corals and mangroves.

Growing land- and marine-based tourism development in Western Australia and the centralization of population growth will cause these impacts to increase unless adequate protection and management of the coast occurs.

Fisheries impacts

Most fishing methods in Western Australia are suggested to have a limited effect on the shallow coastal environments where seagrasses occur[2]. Methods that may significantly affect the environment, for example dredging and pelagic drift gill-netting, are banned. Other methods, such as trawling, that alter the benthic environment are restricted to prescribed areas. Currently, many of these impacts cannot be quantified[2], but current assessments of the sustainability of fisheries practices suggest that damage to seagrass beds is minimal.

At present there are fewer than 100 trawlers operating in a series of discrete managed fisheries within the total Western Australian fishing fleet of around 2 000. The number of these trawl licenses will be reduced over time. Areas available to trawling within each trawl fishery management area are also restricted. There are significant demersal gill-netting closures in areas of high abundance of vulnerable species such as dugong (for example, Shark Bay and Ningaloo Reef).

Pollution, loss of habitat, sedimentation from dredge spoil and agricultural runoff can impact heavily on fish stocks, primarily in nearshore waters and estuaries. Nutrient enrichment of some Western Australian estuaries continues to be a problem. The introduction of exotic marine organisms from ballast water and via the aquarium industry remains an area of concern.

SEAGRASS MANAGEMENT
Protected areas

All the marine parks in Western Australia contain significant seagrass habitats. In particular the Shark Bay World Heritage Property (see Case Study 10.1) contains more than 4 000 km^2 of seagrass beds of high diversity[9], as well as a population of more than 10 000 dugong, and turtles.

Two marine parks in the Perth area, Marmion and Shoalwater Islands, contain about 20 percent seagrass. The Swan River has small sections of marine park, mainly declared for their migratory bird populations but also including areas of the paddleweed, *Halophila ovalis*. Two coral reef areas to the north of Western Australia, Ningaloo and Rowley Shoals Marine Parks, contain small but relatively diverse seagrass populations[10]. Three areas to be declared as marine

Divers airlifting sediment samples from a *Posidonia sinuosa* meadow, Princess Royal Harbour, Western Australia.

parks, Jurien Bay, Cape Leeuwin-Cape Naturaliste and Montebellos-Barrow Island, also have diverse seagrass ecosystems represented.

The establishment of the West Australian Marine Parks and Reserves Authority, in which marine conservation reserves are vested, should help facilitate the development of a comprehensive series of reserves. This process is, however, slow, and current

Case Study 10.1
SHARK BAY, WESTERN AUSTRALIA: HOW SEAGRASS SHAPED AN ECOSYSTEM

Shark Bay is a large (13 000 km^2), shallow (<15 m), hypersaline environment, dominated by seagrasses. Situated on the West Australian coastline, at about 26°S, it contains the largest reported seagrass meadows as well as the most species-rich seagrass assemblages. Shark Bay is also a World Heritage Property, one of only 11 World Heritage sites in the world to have been listed under all four categories for nomination:

o   outstanding examples representing the major stages of the Earth's evolutionary history;
o   outstanding examples representing significant ongoing geological processes, biological evolution and humans' interaction with their natural environment;
o   superlative natural phenomena, formations or features, for instance outstanding examples of the most important ecosystems, areas of exceptional natural beauty or exceptional combinations of natural and cultural elements;
o   the most important and significant natural habitats where threatened species of animals or plants of outstanding universal value still survive.

Although the area also has terrestrial significance, and is home to dolphins, the world's largest stable population of dugongs and living stromatolites, the seagrasses are responsible for some of the most impressive illustrations in the world of the interaction between seagrasses and their environment. Shark Bay provides an outstanding example of the role that seagrasses can play in influencing the physical, chemical and biological evolution of a marine environment.

DESCRIPTION OF SHARK BAY
Shark Bay is a semi-enclosed basin, with restricted exchange with the Indian Ocean, situated in an arid landscape where evaporation exceeds precipitation by a factor of ten. There are two gulfs, the eastern and western, formed by pleistocene dunes, creating a series of inlets and basins. Astronomical tides are less than 1 m, thus atmospheric conditions influence water levels. In summer, strong southerly winds transport about 1-1.5 m of water northwards out of the bay, exposing sand flats up to 2 km wide. There is a well-developed salinity gradient developed as the marine waters cross the shallow carbonate banks of the Faure Sill. Salinities in Hamelin Pool may reach

issues such as extensive plans for aquaculture developments being implemented by another section of government (Fisheries) may compromise the effectiveness of the Parks Authority. The development of marine conservation reserves within Western Australia must form part of the framework being developed federally for Australia, and it must be assessed to see if it provides the necessary comprehensiveness, adequacy and representativeness for marine conservation to be effective.

Policy

On an urgent basis, more detailed studies of the Western Australian marine environment are required if a sound basis for management is to be developed, both within the marine park and reserve system and outside it. There have been few coherent, broad-based studies (both in time and space) that have researched the cumulative impact of pollution, siltation, habitat fragmentation and introductions of invasive species on the community structure of marine communities[2]. Further effort is needed on the influence of these human activities on the whole community, although it will take a long-term commitment to fund these multidisciplinary studies.

A more coherent approach to managing the marine environment is required by government agencies. Some 15 different government agencies have some responsibility for management of the Western Australian marine environment. The 1998 *State of the Environment Report*[2] recommends that the state government should establish a formal framework to coordinate environmental management within Perth's Metropolitan marine region and between these waters and their land catchments. This should be used as a pilot program for expansion to other areas under pressure from domestic and rural discharges.

A recent change in state government in Western Australia has seen major changes to the structure of government departments that may alleviate some of the previous problems.

AUTHOR
Diana Walker, School of Plant Biology, University of Western Australia, WA 6907, Australia. **Tel:** +61 (0)8 9380 2089/2214. **Fax:** +61 (0)8 9380 1001. **E-mail:** diwalker@cyllene.uwa.edu.au

70 psu. Strong tidal currents, up to 8 knots, flow through channels in the Faure Sill.

Seagrasses, particularly the southern Australian endemic species *Amphibolis antarctica* and *Posidonia australis*, dominate the subtidal environment, to depths of about 12 m. The intertidal flats are composed of mixed *Halophila ovalis* and *Halodule uninervis*. The 12 species of seagrass in Shark Bay make it one of the most diverse seagrass assemblages in the world. Seagrass covers more than 4 000 km^2 of the bay, about 25 percent, with the 1 030-km^2 Wooramel Seagrass Bank being the largest structure of its type in the world.

A STABILIZING ROLE

The presence of extensive, monospecific beds of these large, lengthy (2 m) seagrasses, baffle the currents and modify the sediments underlying the seagrass. The plants trap and bind the sediments accreting from calcareous epiphytes and associated epifauna. The plants can significantly slow the rate of water movement over the bottom, and stabilize the otherwise unstable sediments. Rates of sediment accretion associated with *Amphibolis antarctica* are higher than those associated with coral reefs. This is related to high rates of leaf turnover, depositing more calcareous sediments. Over geological time, this had led to the build-up of banks under the seagrass, forming the Faure Sill, as well as the extensive sand flats.

The build-up of the banks underlying the seagrass, in turn, has restricted the circulation of oceanic seawater, which with high evaporation and low rainfall results in the hypersalinity gradient in the inner reaches of the bay. This makes the southern areas of Hamelin Pool unsuitable for seagrasses, but has allowed the development of stromatolites.

HIGH RATES OF PRODUCTION

The waters flowing over the seagrasses are depleted in phosphorus by the seagrasses themselves. For Shark Bay as a whole, the seagrass meadows represent an enormous pool, with some 86 million kg of nitrogen and 6 million kg of phosphorus being required to support the seagrass growth. Only about 10 percent of this can be supplied from the oceanic inflow, so the high rates of production must be supported by tight recycling, both from decomposition *in situ* and from internal retranslocation.

Seagrasses in Shark Bay thus represent "an outstanding example representing significant on-going geological processes, and biological evolution", demonstrating how important seagrasses are throughout the world.

REFERENCES

1. Anon [1994]. *Australian National Tide Tables 1995*. Australian Government Publishing Service, Canberra. 256 pp.
2. Anon [1998]. *Environment Western Australia 1998: State of the Environment Report*. Department of Environmental Protection, Western Australia. 135 pp.
3. Zann LP, Kailola P (eds) [1995]. *The State of the Marine Environment Report*. Technical Annex I. The Marine Environment.
4. Zann LP [1996]. *The State of the Marine Environment Report for Australia*. Technical Summary. Department of Environment, Sport and Territories, Commonwealth of Australia. 515 pp.
5. Searle JD, Logan BW [1978]. A Report on Sedimentation in Geographe Bay. Research Project R T 2595. Department of Geology, University of Western Australia. Report to Public Works Department, Western Australia. 72 pp.
6. McMahon K, Young E, Montgomery S, Cosgrove J, Wilshaw J, Walker DI [1997]. Status of a shallow seagrass system, Geographe Bay, south-western Australia. *Journal of the Royal Society of Western Australia* 80: 255-262.
7. Congdon RA, McComb AJ [1979]. Productivity of *Ruppia*: Seasonal changes and dependence on light in an Australian estuary. *Aquatic Botany* 6: 121-132.
8. Carruthers TJB, Walker DI, Kendrick GA [1999]. Abundance of *Ruppia megacarpa* Mason in a seasonally variable estuary. *Estuarine Coastal and Shelf Science* 48: 497-509.
9. Walker DI [1989]. Seagrass in Shark Bay – the foundations of an ecosystem. In: Larkum AWD, McComb AJ, Shepherd SA (eds) *Seagrasses: A Treatise on the Biology of Seagrasses with Special Reference to the Australian Region*. Elsevier/North Holland, Amsterdam. pp 182-210.
10. Walker DI, Prince RIT [1987]. Distribution and biogeography of seagrass species on the north-west coast of Australia. *Aquatic Botany* 29: 19-32.
11. Walker DI [1997]. Marine Biological Survey of the Central Kimberley, Western Australia. Report to the National Estates Committee. 159 pp.
12. Dennison WC, Kirkman H [1996]. Seagrass survival model. In: Kuo JJS, Phillips R, Walker DI, Kirkman H (eds) *Seagrass Biology: Proceedings of an International Workshop, Rottnest Island, Western Australia, 25-29th January 1996*. Faculty of Science, University of Western Australia, Perth. pp 341-344.
13. Kirkman H, Walker DI [1989]. Western Australian seagrass. In: Larkum AWD, McComb AJ, Shepherd SA (eds) *Biology of Seagrasses: A Treatise on the Biology of Seagrasses with Special Reference to the Australian Region*. Elsevier/North Holland, Amsterdam. pp 157-181.
14. Cambridge ML [1980]. Ecological Studies on Seagrass of South Western Australia with particular reference to Cockburn Sound. PhD thesis, University of Western Australia, Perth. 326 pp.
15. Hillman K, Walker DI, McComb AJ, Larkum AWD [1989]. Productivity and nutrient limitation. In: Larkum AWD, McComb AJ, Shepherd SA (eds) *Seagrasses: A Treatise on the Biology of Seagrasses with Special Reference to the Australian Region*. Elsevier/North Holland, Amsterdam. pp 635-685.

16 Dennison WC, Orth RJ, Moore KA, Stevenson JC, Carter V, Dollar S, Bergstrom PW, Batiuk RA [1993]. Assessing water quality with submersed aquatic vegetation. *Bioscience* 43: 86-94.
17 Kirk JTO [1994]. *Light and Photosynthesis in Aquatic Ecosystems.* Cambridge University Press, Cambridge.
18 Walker DI, McComb AJ [1992]. Seagrass degradation in Australian coastal waters. *Marine Pollution Bulletin* 25: 191-195.
19 Cambridge ML, Chiffings AW, Brittan C, Moore L, McComb AJ [1986]. The loss of seagrass in Cockburn Sound, Western Australia. II. Possible causes of seagrass decline. *Aquatic Botany* 24: 269-285.
20 Lord DA [1994]. Coastal eutrophication: Prevention is better than cure. *The Perth Coastal Water Study* 45: 22-27.
21 Lukatelich RJ, McComb AJ [1989]. *Seasonal Changes in Macrophyte Abundance and Composition in a Shallow Southwestern Australian Estuarine System.* Waterways Commission, Perth, Western Australia.
22 Lukatelich RJ, McComb AJ [1986]. Distribution and abundance of benthic microalgae in a shallow southwestern Australian estuarine system. *Marine Ecology Progress Series* 27: 287-297.
23 Lavery PS, Lukatelich RJ, McComb AJ [1991]. Changes in the biomass and species composition of macroalgae in a eutrophic estuary. *Estuarine Coastal and Shelf Science* 33: 1-22.
24 McMahon K, Walker DI [1998]. Fate of seasonal, terrestrial nutrient inputs to a shallow seagrass dominated embayment. *Estuarine Coastal and Shelf Science* 46: 15-25.
25 Bastyan G [1986]. *Distribution of Seagrasses in Princess Royal Harbour and Oyster Harbour on the Southern Coast of Western Australia.* Technical Series 1. Department of Conservation and Environment, Perth, Western Australia. 50 pp.
26 Walker DI, Hutchings PA, Wells FE [1991]. Seagrass, sediment and infauna – a comparison of *Posidonia australis, Posidonia sinuosa* and *Amphibolis antarctica,* Princess Royal Harbour, South-Western Australia I. Seagrass biomass, productivity and contribution to sediments. In: Wells FE, Walker DI, Kirkman H, Lethbridge R (eds) *Proceedings of the 3rd International Marine Biological Workshop: The Flora and Fauna of Albany, Western Australia.* Vol 2. Western Australia Museum. pp 597-610.
27 Wells FE, Walker DI, Hutchings PA [1991]. Seagrass, sediment and infauna – a comparison of *Posidonia australis, Posidonia sinuosa* and *Amphibolis antarctica* in Princess Royal Harbour, South-Western Australia III. Consequences of seagrass loss. In: Wells FE, Walker DI, Kirkman H, Lethbridge R (eds) *Proceedings of the 3rd International Marine Biological Workshop: The Flora and Fauna of Albany, Western Australia.* Vol 2. Western Australian Museum. pp 635-639.
28 Chiffings AW, McComb AJ [1981]. Boundaries in phytoplankton populations. *Proceedings of the Ecological Society of Australia* 11: 27-38.
29 Silberstein K, Chiffings AW, McComb AJ [1986]. The loss of seagrass in Cockburn Sound, Western Australia. III. The effect of epiphytes on productivity of *Posidonia australis* Hook. f. *Aquatic Botany* 24: 355-371.
30 Kendrick GA [1991]. Recruitment of coralline crusts and filamentous turf algae in the Galapagos archipelago: Effect of stimulated scour, erosion and accretion. *Journal of Experimental Marine Biology and Ecology* 147: 47-63.
31 Coyne R, Hiney M, O'Connor B, Kerry J, Cazabon D, Smith P [1994]. Concentration and persistence of oxytetracycline in sediments under a marine salmon farm. *Aquaculture* 123(1-2): 31-42.
32 Kerry J, Hiney M, Coyne R, Cazabon D, Nicgabhainn S, Smith P [1994]. Frequency and distribution of resistance to oxytetracycline in micro-organisms isolated from marine fish farm sediments following therapeutic use of oxytetracycline. *Aquaculture* 123(1-2): 43-54.
33 Kohn AJ, Almasi KN [1993]. Imposex in Australian *Conus. Journal of Marine Biology Association UK* 73: 241-244.
34 Sindermann CJ [1991]. Case histories of effects of transfers and introductions on marine resources – Introduction. *Journal du Conseil* 47: 377-378.
35 Chaplin G, Evans DR [1995]. The Status of the Introduced Marine Fanworm *Sabella spallanzanii* in Western Australia: A Preliminary Investigation. Technical Report 2. Centre for Research on Introduced Marine Pests, Division of Fisheries, Hobart, Tasmania. 34 pp.

Regional map: Australasia

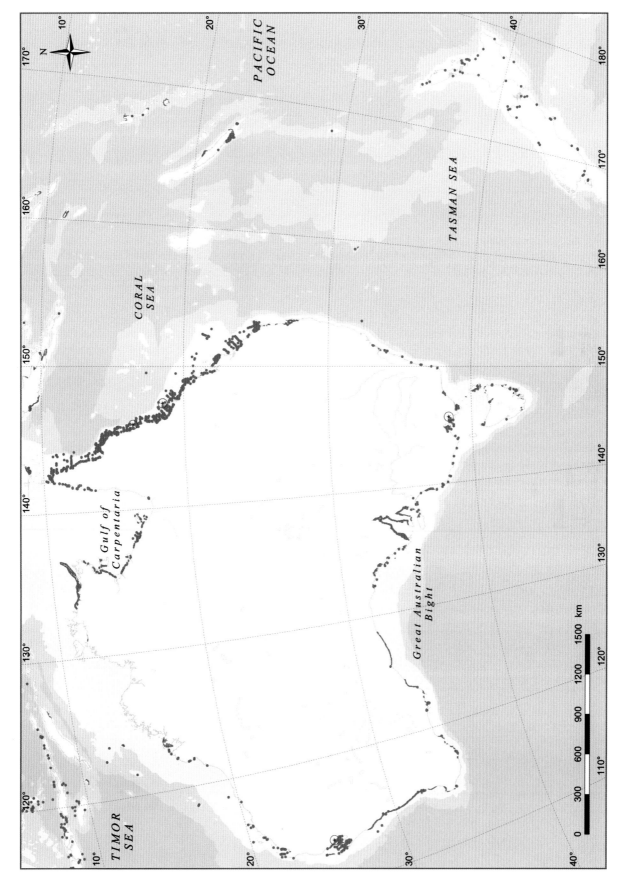

SEAGRASS ECOSYSTEMS

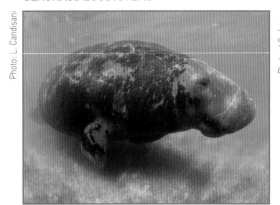

A manatee (*Trichechus manatus*), *feixe-boi* in Portugese, over a *Halodule wrightii* bed in Recife, Brazil.

A sea horse, *Hippocampus whitei*, amongst *Zostera capricorni* in Sydney Harbour, Australia.

Mediterranean *Posidonia oceanica* seagrass beds with saupe (*Sarpa sarpa*) and bream (*Diplodus* spp.)

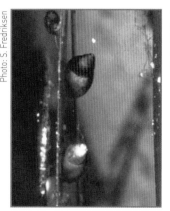

Snails grazing epiphytes on *Zostera marina* blades in southern Norway.

Sea star in *Enhalus acoroides* and *Thalassia hemprichii*, Micronesia.

Enhalus acoroides and soft coral in Komodo, Indonesia.

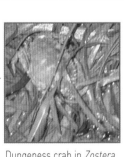

Dungeness crab in *Zostera marina*, Puget Sound, USA.

King helmet in *Thalassia testudinum*, Turks & Caicos.

Lizard fish in *Amphibolis antarctica*, Western Australia.

11 The seagrasses of EASTERN AUSTRALIA

R. Coles
L. McKenzie
S. Campbell

Seagrass meadows are a prominent feature of the eastern Australian coastline which extends from the tropics (10°S) to the cool temperate zone (44°S) and includes the Great Barrier Reef World Heritage Area. The area includes the Gulf of Carpentaria to the north and around the coastline of Australia to Tasmania and to Spencer Gulf. There are extensive seagrass habitats in this region including tropical and temperate seagrass assemblages. An overlap between these two zones occurs in Moreton Bay, southern Queensland[1]. Both tropical and temperate species in Australia are mostly found growing in water less than 10 m below mean sea level[2]. Some species of tropical *Halophila* can be found to depths of 60 m[3].

The eastern Australian coast includes areas of diverse physical characteristics. The tropical north coast and Gulf of Carpentaria are monsoon influenced, mostly with muddy sediments, low human population and low levels of disturbance. The tropical and most of the temperate subtropical Queensland east coast is sheltered by the Great Barrier Reef and is effectively a long lagoon. The temperate east and south coasts are sandier and more exposed and include the large (by Australian standards) population centers of Brisbane, Sydney, Melbourne and Adelaide with a standard suite of associated anthropogenic coastal disturbances.

The highest species diversity of seagrass is found near the tip of Cape York in the very north, with a gradual decline in diversity moving south down the east coast[4]. This is thought to be a result of geographic distance from a center of diversity in the Malaysian/Indonesian region driven by the east Australian current which runs roughly north to south[5], combined with changes in temperature, topography (available substrate), past changes in sea level and exposure to wave action[6].

The temperate species in the southern half of the region include members of the genera *Amphibolis*, *Posidonia* and *Zostera* which are found predominately in sheltered estuaries and bays. *Amphibolis* is an Australian endemic. They possibly had a much wider distribution in the early Paleocene (64 million years ago) with rapid climatic and tectonic changes since that time restricting their distribution to southern Australia[7]. *Posidonia* has a fractured distribution at the present time (southern Australia and the Mediterranean), also likely to be the result of localized extinctions in the past[7]. The genus *Zostera* has both temperate and tropical species in Australia.

The tropical meadows are highly diverse, but generally have lower biomass than those in temperate parts. While bays such as Hervey Bay and Moreton Bay have large areas of seagrass, most tropical seagrasses are found in the intertidal or shallow subtidal environments of the Gulf of Carpentaria and the central and southern Great Barrier Reef World Heritage Area lagoon with extension into deeper waters in the central and northern sections.

The importance of seagrass meadows as structural components of coastal ecosystems has resulted in research interest being focused on the biology and ecology of seagrasses and on the methods for mapping, monitoring and protection of critical seagrass habitats. Seagrasses of eastern Australia are important for stabilizing coastal sediments, providing food and shelter for diverse organisms, as a nursery ground for shrimp and fish of commercial importance, and for nutrient trapping and recycling[8]. In eastern Australia the marine mammal, *Dugong dugon*, and the green sea turtle, *Chelonia mydas*, feed directly on seagrasses. Both animals are used by traditional Australian communities for food and ceremonial use. Both species have declined in number, and protection of their habitat and food source is vital.

The extent of seagrass areas and the ecosystem values of seagrasses are the basic information required for coastal zone managers to aid planning and development decisions that will minimize impacts on

seagrass habitat. In general, our knowledge of intertidal and shallow subtidal (down to 10 m) distributions is good; however, we have only a basic understanding of deepwater (>10 m) seagrasses throughout the region. It is important to document seagrass species diversity and distribution and identify areas requiring conservation measures before significant areas and species are lost.

BIOGEOGRAPHY
Gulf of Carpentaria and Torres Strait

The Gulf of Carpentaria is a large, shallow, muddy marine bay. Extensive open coastline seagrass communities, mainly of the genera *Halodule* and *Halophila* intertidally, and *Syringodium* and *Cymodocea* subtidally, are found along the southern and western sides of the gulf[9]. Along the exposed eastern coast of the gulf, seagrasses are generally sparse and restricted to the leeside of islands, protected reef flats, and estuaries and protected bays. The coastline of the eastern gulf is extremely shallow and regularly disturbed by prevailing winds. Sediments throughout the gulf are predominately fine muds, and these are easily resuspended due to the shallow bathymetry resulting in increased turbidity, which restricts seagrass distribution and growth. Reef flat communities are dominated by *Thalassia*. Meadows in estuaries and sheltered bays are mostly of the genera *Halodule*, with *Cymodocea* and *Enhalus*.

The Torres Strait is a shallow (mostly 10-20 m depth) body of water 100 km long and 250-260 km wide (east-west), formed by a drowned land ridge extending from Cape York to Papua New Guinea. The area has a large number of islands, shoals and reefs. Reefs are generally aligned east-west, streamlined by the high-velocity tidal currents that pour through the inter-reef channels. Seagrass communities occur across the open seafloor, on reef flats and subtidally adjacent to continental islands. A well-defined line of large reefs runs northwards from Cape York, including the Warrior Reefs with extensive seagrass-covered reef flats. Mixed species occur on these flats, most commonly of the genera *Halodule*, *Thalassia*, *Thalassodendron* and *Cymodocea*. The large expanses of open water bottom are covered with either sparsely distributed *Halophila* or mixed species (*Halodule*, *Thalassia* and *Syringodium*) communities. Lush *Halophila ovalis* and *Halophila spinulosa* communities are also found in the deep waters (>30 m) of the southwestern Torres Strait.

Northeast coast

Tropical seagrass habitats in northeastern Australia are extensive, diverse and important for primary and secondary production[10]. A high diversity of seagrass habitats is provided by extensive bays, estuaries, rivers and the 2600-km Great Barrier Reef with its reef platforms and inshore lagoon.

Carruthers et al.[11] classified the northeast coast seagrass systems into river estuaries, coastal, deepwater and reef habitats. All but some of the reef habitats are significantly influenced by seasonal and episodic pulses of sediment-laden, nutrient-rich river flows, resulting from high-volume summer rainfall. Cyclones, severe storms and wind waves, as well as macrograzers (dugongs and turtles) influence all habitats in this region to varying degrees. The result is a series of dynamic, spatially and temporally variable seagrass meadows.

River estuary habitats include a wide range of subtidal or intertidal species and can be highly productive. The species mixture, growth and distribution of these seagrass meadows are influenced by terrigenous runoff as well as temperature and salinity fluctuations. Increased river flows in summer cause higher sediment loads and reduced light, creating potential light limitation for seagrass[12]. Associated erosion and unstable sediments make river and inlet habitats a seasonally stressful environment for seagrass growth. These meadows often have high shoot densities but low species diversity[2]. Differences in life history strategies, resilience to habitat variability, and the physical characteristics of the inlet act to control species assemblages in different river and inlet systems.

Coastal habitats also have extensive intertidal and subtidal seagrasses. Intertidal environments are impacted by sediment deposition, erosion, tidal fluctuations, desiccation, fluctuating and sometimes very high temperature, and variable salinity[13]. Tidal range can be as large as 6 m. These communities are affected rapidly by increased runoff with heavy rain or cyclone events[14], but a large and variable seed bank can facilitate recovery following disturbance[15]. Inshore seagrass communities are found in varying quantity along the eastern Queensland coastline, mostly where they are protected from the prevalent southeast winds by the Great Barrier Reef. Along the southern Queensland coast, the Great Barrier Reef offers little protection and coastal seagrass meadows are restricted to sheltered bays, behind headlands and in the lee of islands. Extensive coastal seagrass meadows occur in north-facing bays such as Moreton Bay, Hervey Bay and Shoalwater Bay.

Increasing distance from the coast decreases the impacts from pulsed terrigenous runoff, and in these regions clear inter-reef water at depth (>15 m) allows for deepwater seagrass growth. Throughout the Great Barrier Reef region, approximately 40 000 km² of lagoon and inter-reef area has at least some seagrass, most of low density (<5 percent cover)[3].

Deepwater seagrass areas are dominated by species of *Halophila*[2, 3, 16]. Large monospecific meadows of seagrass occur in this habitat composed mainly of *Halophila decipiens* or *Halophila spinulosa*. *Halophila* spp. display morphological, physiological and life history adaptations to survival in low-light environments. *Halophila* spp. can be annuals in the Great Barrier Reef region, have rapid growth rates and are considered to be pioneering species[17]. An important characteristic of this strategy is high seed production. Rates of 70 000 seeds/m^2/year have been estimated from field observations of *Halophila tricostata*[18].

The distribution of deepwater seagrasses appears to be mainly influenced by water clarity and a combination of propagule dispersal, nutrient supply and current stress. High-density deepwater seagrasses occur mostly on the inner shelf in the central narrow-shelf section of the east coast which experiences a moderate tidal range and is adjacent to high-rainfall rainforest catchments. Where there are large tidal ranges, just to the south of Mackay, no major deepwater seagrass areas exist, but some meadows occur further south in Hervey Bay where tide ranges moderate again[3]. Deepwater seagrasses are uncommon north of Princess Charlotte Bay, a remote area of low human population and little disturbance. This may be the result of the east Australian current diverging at Princess Charlotte Bay and the far northern section may not receive propagules for colonization from southern meadows. Much of this coast is also silica sand and low in rainfall and stream runoff, and it is possible that limited availability of nutrients restricts seagrass growth[3].

Reef seagrass communities support a high biodiversity and can be extensive and highly productive. Shallow unstable sediment and fluctuating temperature characterize these habitats. Low nutrient availability is a feature of reef habitats, and seagrasses are likely to be nitrogen limited[19]. Seagrasses are more likely to be present on reefs with vegetated cays than on younger reefs with highly mobile sand. Intermittent sources of nutrients arrive when seasonal runoff reaches the reef. In some localized areas, particularly coral cays, seabirds can add high amounts of phosphorus to reef environments. The more successful seagrass species in reef habitats of the Great Barrier Reef include *Thalassia hemprichii*, *Cymodocea rotundata*, *Thalassodendron ciliatum*, the colonizing species *Halophila ovalis*, and species of the genus *Halodule*.

New South Wales, Victoria and Tasmania

Ten species of seagrass (excluding *Lepilaena cylindrocarpa*) are recorded in this region[20]. Species of the genus *Zostera* (including the former *Heterozostera*) are

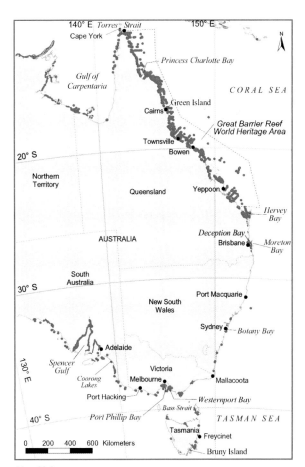

Map 11.1
Eastern Australia

the most common as they dominate in estuaries and coastal lagoons[21]. In Victoria and Tasmania, *Posidonia* and *Amphibolis* are also found, mainly near estuary entrances, or in sheltered bays adjacent to Bass Strait Islands.

The distribution and occurrence of seagrasses depends on the estuary type, i.e. drowned river valley, barrier estuary or coastal lagoon[21]. Seagrass species composition and distribution is associated mostly with sediment type and with differing exposure to wave energy from the open ocean. Seagrasses are generally more abundant several kilometers upstream from the estuary entrance due to lesser tidal and wave disturbance. Seagrasses in coastal lagoons may also be affected by the frequency with which the lagoon entrance is open to the ocean or closed by shifting sand banks, changing conditions from brackish to saline. Agricultural development and poor catchment practices in some regions have resulted in high sediment and nutrient loads reducing light availability and favoring species which can tolerate lower light levels. In other localities, reduced freshwater flows (due to industrial and agricultural extraction) have increased salinities.

In protected sites, mixed stands of *Zostera*

tasmanica, *Zostera capricorni* (formerly *Zostera muelleri*) and *Halophila ovalis* dominate. *Ruppia* meadows are common in areas of high freshwater input. A feature of estuarine habitats in this region is heavy winter-spring rains with associated high turbidity, followed by high salinity and low rainfall in summer.

In less-protected areas dominated by sandy sediments (e.g. north coast of Tasmania, Bass Strait islands) mixed seagrass communities consist of larger, slower growing species such as *Posidonia australis*, with small, faster growing species such as *Zostera tasmanica* and *Halophila ovalis* occupying the gaps between meadows and areas close to freshwater inputs. At the mouth of some bays and in areas dominated by sandy siliceous sediments and exposed to ocean swells in Victoria, the slow-growing seagrass *Amphibolis antarctica* (*Amphibolis griffithii* fills the same role in South Australia) forms patches of varying sizes rather than extensive monospecific meadows. In these areas, nutrient inputs are low and sediments are nutrient poor.

Large oceanic bays in southeast Tasmania have meadows of *Halophila* and *Zostera* species. Seagrass distribution is influenced by biogeography and geomorphology as well as wave energy. Deep, oceanic seagrass beds of *Posidonia australis* and *Amphibolis antarctica* are also present to depths of 22 m in clear non-polluted water. Their distributions are influenced by depth, bottom type, wave energy and geomorphology. Most seagrasses in southeastern Australia are restricted to depths of less than 20 m by light availability.

South gulf coast of South Australia

Seagrass distribution in South Australia is dependent on coastal topography, bathymetry and environment[21]. The most extensive meadows are found in the large expanses of sheltered shallow water in Spencer Gulf and Gulf St Vincent. These are predominantly *Posidonia* and *Amphibolis* meadows, with *Halophila* and *Zostera* species. Large seagrass meadows which are dominated by species of *Zostera* also occur along the southeastern coast of South Australia in coastal lagoons (e.g. Lake Alexandrina and Lake Albert).

HISTORICAL PERSPECTIVES

Australia has had a relatively stable climate with the northward movement of the continent compensating during past episodes of global cooling. The biomass and diversity of seagrass seen today is most likely to have remained relatively unchanged on a continental scale for tens of millions of years.

Agriculture and coastal development started in Australia with the arrival of European migrants only 200 years ago and coastal influences on seagrass before that date would have been almost entirely natural. Sediment and nutrient loads to estuaries and enclosed waters such as Moreton Bay and Westernport Bay have undoubtedly influenced the modern distribution of seagrasses, particularly in temperate waters. Less easy to determine is the likely effect in tropical waters where turbidities are already naturally high. Dramatic declines in grazer populations (turtle and dugong) from increased hunting would be expected to allow an increase in seagrass, particularly of biomass where climax communities can now develop.

Traditionally, the fruit of *Enhalus acoroides* was eaten in the northern islands and the leaf fibers were possibly used to make nets and cord. This use was likely to have been infrequent and of low importance to seagrass distribution as the human population was very small before European migration. Seagrasses were used to make matting and for bed mattresses during the Second World War. They were also used for fertilizer, such as in Lacepede Bay in southeastern South Australia, where *Posidonia angustifolia* leaf drift and wrack is still harvested from the beach for soil conditioner and compost mixes[22]. Such activities are now illegal in many parts of Australia where both live and dead seagrasses are protected.

There are reports from the southern and eastern Australian coastline that seagrass communities have declined in recent decades[23]. Anecdotal reports of 30 years ago from residents in the Hervey Bay and Great Sandy Strait region describe large long-leaved (>30 cm) *Zostera capricorni* meadows abundant over the intertidal banks. Long-time residents report abundant fishing and bird life (especially black swans) and say that the seagrass wrack was so plentiful that it was harvested from the beaches for garden mulch. Today, much of the seagrass on the intertidal banks in the region is sparse or low-cover *Zostera capricorni* with short (<10 cm) and narrow leaves. Fishing is reported to have declined and black swans no longer frequent the region. Unfortunately, accurate mapping programs were not instigated until the late 1980s so these types of report are impossible to verify and may well be overstated.

In Victoria there are unquantified reports of *Zostera* loss in Westernport Bay during the 1950s. The decline coincided with a reduction in fish catches. Anecdotal reports and photographs from local residents in the north and eastern regions of Westernport Bay, prior to the loss of seagrasses, describe "lush seagrass meadows". Similarly in Corner Inlet, Victoria, a decline in *Posidonia australis* in the 1960s was followed by a reduction in fishermen operating in the region[21].

AN ESTIMATE OF HISTORICAL LOSSES

More than 450 km² of seagrass have been lost from Australian coastal waters in recent years, largely

attributed to eutrophication, natural storm events, and reductions in available light due to coastal development. It is worth noting that there is a high probability of bias towards reporting decline, and that increases in biomass and area are often not reported.

There have been several well-documented cases of seagrass loss in eastern Australia over the past 50 years[22]. In Port Macquarie (New South Wales) 11.3 km² of seagrass was lost between 1953 and 1985 due to increased turbidity from human activity, resulting in declining fish stocks[24]. Similarly in Botany Bay, a loss of 2.5 km² of *Posidona sinuosa*, representing 58 percent of the bay's seagrass, was lost between 1942 and 1986, a consequence of dredging activities and eutrophication[25].

In South Australia there has been a significant decline in seagrasses on the eastern side of Gulf St Vincent due to sewage effluent. Approximately 60 km² of *Posidona sinuosa* and *Amphibolis antarctica* meadows were lost between 1935 and 1987[22]. In Spencer Gulf, fishermen and local residents have reported widespread loss of *Amphibolis* close to the intertidal zone. Recent loss (1992-93) of mixed meadows of *Posidonia australis*, *Zostera tasmanica* and *Zostera capricorni* were due to sediment accretion and desiccation caused by exposure to high air temperatures and low humidity[26].

In Victoria, the recorded loss of seagrass has been from the large marine bays of Westernport Bay, Port Phillip Bay and Gippsland Lakes. In Westernport Bay, persistent high turbidity and poor water quality due to agricultural runoff, sediment inputs and resuspension of sediments caused seagrass to decline from 196 km² in the early 1970s to 67 km² in 1984[27]. Seagrass recovery has occurred (154 km² by 2001)[28], but seagrass meadows have failed to recolonize the intertidal mud flats in north and western regions and in some areas seagrass meadows have at least 45 percent lower biomass compared to 25 years ago[29]. A near complete loss of seagrass (ca 31 km²) in the Gippsland Lakes from the 1920s to 1950s coincided with reduced commercial fish catches[30]. More recent estimates of seagrass abundance suggest that there has been little decline over the past 30 years[31] except for some localized replacement by algae. Similarly, in Port Phillip Bay, little change in seagrass area was recorded from 1957 (76 km²) to 1981 (96 km²)[32,33]. From 1981 to 2000, the area of seagrass in Port Phillip Bay declined from 96 km² to 68 km² [33], possibly due to increased turbidity and eutrophication in Corio Bay and Swan Bay (early 1990s) in the west and southwest of Port Phillip, respectively. Drifting algal communities have replaced some areas of seagrass vegetation.

In Queensland, declines of seagrass area resulted from flooding and sedimentation. In Moreton Bay, thousands of hectares of seagrass, which were present prior to the 1980s, have been destroyed by the effects of canal estate development. Deception Bay seagrasses have declined since 1996 in what may be a cyclic pattern. Both cases were due to low light and poor

Rob Coles visually estimating seagrass abundance and mapping distribution (using differential GPS), Shoalwater Bay, Queensland.

water quality associated with urban development and possibly agriculture[34]. Loss from climatic events (storms, flooding and cyclones) has occurred in a number of regions including Hervey Bay (1000 km² and 27.75 km² in separate events in the 1990s) and Townsville. Anecdotal evidence and evidence collected during lobster fishery surveys suggests that thousands of hectares have been lost in the northwest Torres Strait due to flooding and sedimentation from Papua New Guinea, but these are remote locations and difficult to track effectively.

In northeastern Australia, most seagrass losses have been followed by significant recovery. For example, approximately 1000 km² of seagrasses in Hervey Bay were lost in 1992 after two major floods and a cyclone within a three-week period added to pressures on the system from agricultural development and land development associated with increases in human populations[14]. The deepwater seagrasses died, apparently from light deprivation caused by a persistent plume of turbid water from the floods and the resuspension of sediments caused by the cyclonic seas. The heavy seas uprooted shallow-water and intertidal seagrass. Recovery of subtidal seagrass (at depth >5 m) began within two years of the initial loss[14], but recovery of intertidal seagrasses was much slower. These seagrasses only started to recover after four to five years and did not fully recover until December 1998[35].

The capacity of tropical seagrasses to recover appears to be a consequence of morphological,

physiological and life history adaptations; the plants can be fairly resilient in unstable environments. *Halodule uninervis* and *Halophila ovalis* are considered pioneer species, growing rapidly and surviving well in unstable or depositional environments[13, 17]. *Halophila tricostata* is an annual, only appearing in late September through to February and being sustained by a sizeable seed bank[18]. *Cymodocea serrulata* occurs in deeper sediments and has been linked to increased rates of sediment accretion[17]. *Zostera capricorni* meadows were found to recolonize through vegetative growth and can therefore survive small-scale disturbances[36].

Queensland's east and gulf coasts have areas where seagrass meadows have expanded. Little is known about long-term cycles in seagrass meadow size and biomass. The losses and gains being measured may fall within a natural range. In heavily grazed coastal waters with high dugong and green

Case Study 11.1
MAPPING DEEPWATER (15-60 M) SEAGRASSES AND EPIBENTHOS IN THE GREAT BARRIER REEF LAGOON

Seagrasses in waters deeper than 15 m in the Great Barrier Reef World Heritage Area were surveyed between 1994 and 1999. A real-time video camera and dredge were towed for four to six minutes on 1 426 sites to record bottom-habitat characteristics and seagrasses. In conjunction with the camera tow, a sled-net sample of benthos and a grab sample of the sediment were collected.

Sampling included the Great Barrier Reef province from the tip of Cape York Peninsula at 10°S to approximately 25°S, or 1 000 nautical miles of coastline. Sites were located from inshore out to the reef edge up to 120 km from the coast. Seagrass presence, species and biomass were recorded with depth, sediment, Sechii disk depth, associations with algae and epibenthos, and proximity to reefs.

Five seagrasses were present, all from the genus *Halophila*, in depths down to 60 m. Seagrasses were present at 33 percent of sites sampled. The species *Halophila ovalis, Halophila spinulosa, Halophila decipiens, Halophila tricostata* and *Halophila capricorni* were found. *Halophila tricostata* is a species endemic to northern Australia and *Halophila capricorni* is found only in the southern Indo-Pacific. All other species are broadly distributed throughout the Indo-Pacific region[71]. Most seagrass seen in video tows was of low density (<5 percent cover) and biomass ranged from less than 1 g to 45 g dry weight/m^2 (the highest was recorded from a *Halophila spinulosa*-dominant meadow in 21 m). Mean biomass was 3.26 ±0.36 g dry weight/m^2.

The map of seagrass was generated using generalized additive models incorporating Loess smoothers[72]. The degree of smoothness was minimized but sufficient to account for both spatial effects and spatial correlation. The location of the data points was recoded based on the proportion of the distance the point was located between the coast and the outer edge and the proportion from 10°S to the southern edge. The model estimated that as much as 40 000 km^2 of lagoon and inter-reef area may have at least some seagrass[3]. This type of map or statement of probability is necessary when factors such as depth make it impossible to plot around the edge of the meadow even if that could be defined. With areas of very low biomass and very large areas with patchy seagrass, the concept of a defined meadow is not always appropriate. Using probability to estimate the likelihood of seagrass presence must be explained with care as the

Probability of occurrence of deepwater seagrasses in the Great Barrier Reef Lagoon (contours obtained by spatial smoothing).

turtle populations, increases in meadow size and biomass may reflect simply changes (decreases) in herbivore populations and be an indicator of disturbance rather than a positive measure.

AN ESTIMATE OF PRESENT COVERAGE
Gulf of Carpentaria and Torres Strait

Approximately 779 km² of seagrass in the western Gulf of Carpentaria were mapped in 1984[9]. In 1986, Queensland Department of Primary Industries (QDPI) mapped 184 km² in the eastern gulf[37] and 225 km² around the Wellesley Island Group (southern gulf) in 1984[35].

Using probability models and ground-truthing, the Torres Strait is estimated to contain 13 425 km² of seagrass habitat on reef platforms and non-reef soft bottoms[35, 38], much of which is valuable habitat for juvenile commercial shrimp.

outcome may be scale dependent and the outcome is definitely not a "map" in the sense it is normally used. If you define the sampling unit as the entire Great Barrier Reef Region, the probability of finding seagrass in that sampling unit will be 100 percent. A smaller unit will have a lower probability. Typically the ability of a map drawn this way can be improved if physical factors such as light and bottom-type location can be incorporated in the model.

DEEPWATER SEAGRASSES

Deepwater seagrasses were most common in the central narrow shelf regions which experience a moderate tidal range and are adjacent to high-rainfall rainforest catchments. Highest densities occurred between Princess Charlotte Bay and Cairns, and south of 23°S. *Halophila tricostata* was found only between Princess Charlotte Bay and Mackay. Other species were spread throughout. Seagrasses (*Halophila ovalis*, *Halophila spinulosa*, *Halophila decipiens* and *Halophila capricorni*) occurred to 60 m depth. The frequency of occurrence of seagrasses declined below 35 m. *Halophila decipiens* was the most commonly found species at all depths. Dense algae beds (mainly *Caulerpa* and *Halimeda*) were found on the outer shelf north of Cooktown. Where there are large tidal velocities and ranges (4-6 m tidal range), just to the south of Mackay, no major deepwater seagrass areas occur. Some seagrass habitats were apparent further south in Hervey Bay where tidal ranges moderate.

The ecological role of inter-reef seagrasses and algae is not well understood. Some deepwater meadows (<25 m) of *Halophila ovalis* and *Halophila spinulosa* are important dugong feeding habitat. Commercial fish and crustacean species were uncommon in deep water compared to catches in coastal intertidal and shallow subtidal meadows[73].

This seagrass and benthic community information is one of the major databases supporting development of a multi-use marine park plan for maintaining the biodiversity of the Great Barrier Reef World Heritage Area based on the principles of comprehensiveness, adequacy, and representativeness. This "representative areas" program has used two processes: a data-based statistical approach and a delphic expert experience-based questionnaire approach. Thirty-eight relatively homogeneous inter-reef bioregions have been identified based on the presence and distribution of seagrasses, algae, other benthos, sediment and habitat descriptions. This information will be used to select areas to protect in "no-take" zones and to minimize the loss of economic use of reef areas by the tourist and fishing industries and by recreational users. The deepwater seagrass and epibenthos mapping is an excellent example of seagrass maps being used directly to support good management decisions.

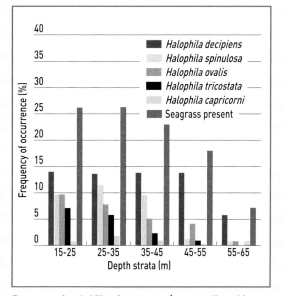

Frequency of probability of occurrence (percent adjusted for sampling frequency) of seagrasses within each depth stratum.

Note: Seagrass present = all species combined (including unidentified).

Source: Coles *et al.*[3].

Northeast coast

The northeastern Australia coastline is either within the Great Barrier Reef World Heritage Area with high conservation values or includes coastline with seagrass meadows supporting valuable shrimp fisheries, green turtle or dugong populations. The perceived importance of seagrasses in these regions, as well as concern about the downstream effects of agriculture, effects of fishing and the possibility of shipping accidents[10], have led to an extensive mapping program. Broad-scale surveys conducted between 1984 and 1989 mapped seagrass habitats down to 15 m depth in estuaries, shallow coastal bays and inlets, on some fringing reefs, barrier reef platforms, inner reef and Great Barrier Reef Lagoon[10]. Since 1989 there have been repeated surveys at finer scales of resolution in certain localities as a result of specific issues (e.g. port developments, dugong protection areas). Some studies have repeated surveys at a locality once or twice yearly for up to four or more years[39].

It is difficult to estimate the exact seagrass area as published information is from overlapping zones and information is being constantly updated as mapping improves. The most accurate estimates of seagrass meadows along the northeast coast are 5668 km² intertidal and shallow subtidal (down to 15 m water depth)[35, 39-50].

From Cape York to Cairns, seagrass communities are predominantly subtidal *Halophila* species with approximately equal area of sparse and dense cover. Species of *Cymodocea* and *Syringodium* are found in shallow subtidal areas where there is shelter from the southeast winds. Between Cairns and Bowen, around 70 percent of the area of seagrass is less than 10 percent cover and mostly a mixture of *Halodule* and *Halophila* species, both intertidal and subtidal. Between Bowen and Yeppoon approximately 50 percent of the area of the mainly intertidal *Halodule* communities is between 10 and 50 percent cover. South of Yeppoon, the seagrass communities are mostly

---

Case Study 11.2
WESTERNPORT BAY

Westernport Bay is a large estuarine tidal bay in southern Victoria. It encloses two large islands and has an area of 680 km² of which 270 km² is intertidal mud flat. Intersecting the mud flats is a series of complex channels where sediment movement is influenced by the water movement patterns in a net clockwise direction.

Westernport Bay is an area of high biological diversity because of its wide range of habitats, including seagrass meadows, mangroves, salt marsh and deepwater channels. It is an internationally significant coastal wetland acknowledged by nomination to the Ramsar Convention on Wetlands. The bay consists of extensive intertidal seagrass meadows, subtidal meadows and macroalgal communities. The dominant seagrasses are *Zostera tasmanica* and *Zostera capricorni*. The dominant macroalga associated with seagrass is *Caulerpa cactoides*, which, with other algae, comprises about 16 percent of the total marine vegetation.

The catchment to the north of Westernport Bay was cleared of vegetation in the late 1800s for agriculture, and the bay is now subject to inputs of nutrients and suspended particulates[74]. Change in seagrass distribution from 1956 to 2000 was examined using aerial photography at four sites in Westernport Bay[33]. The four sites were Rhyll (southern region), Corinella (eastern region), Stony Point (eastern region) and Point Leo (southwest region). From 1956 to 1974, there was a decrease in seagrass distribution at three (Rhyll, Corinella and Stony Point) of the four sites. From 1973-74 to 1983-84 an 85 percent reduction of seagrass and macroalgal biomass, from 251 km² to 72 km², was reported in the bay[27, 75], much of it on intertidal banks. A number of studies examined the causes of this dramatic loss of seagrass habitat[27, 76, 77] focusing on the effects of light reduction on seagrass communities as a result of increased sediment loads in the water column. These studies also examined the increased elevation of intertidal banks, the loss of pooling and the increased exposure of seagrasses to desiccation, a consequence of increased sediment inputs from catchment sources and resuspension of sediments in the water column. Annual sediment inputs from the northeastern catchment (>86 200 m³/year) were found to be six to seven times the loads of sediments into other regions of the bay (13 000 m³/year)[78] leading to decline in light availability in this region. Other causes such as the effects of industrial effluents on invertebrate fauna and subsequent reduced grazing of epiphyte loads were examined. No conclusive evidence was found that identified a single major factor as the cause of seagrass loss. The effects of seagrass loss on fish populations were also studied and findings suggested that Westernport Bay seagrass meadows play an important role in enhancing fish production and marine invertebrate numbers[77].

denser, with approximately 60 percent of the area of seagrass greater than 50 percent cover. These seagrass areas are dominated intertidally by *Zostera/Halodule* communities and subtidally by *Halophila* communities.

Waters of the Great Barrier Reef World Heritage Area deeper than 15 m have been surveyed and it is likely that as much as 40 000 km^2 of habitat that may support seagrass populations is present in the reef lagoon[3]. The map in this case was based on spatial probability and cannot be compared with a map drawn from global position system points taken on the edge of a meadow.

New South Wales, Victoria and Tasmania

Estimates of seagrass area in New South Wales from mapping exercises prior to 1985 were 155 km^2 in 111 estuaries. In New South Wales the Conservation Division of New South Wales Fisheries is presently mapping seagrasses in large estuaries of the Hawkesbury region and Port Hacking, but this information is not yet available.

The Victorian Department of Natural Resources and Environment has recently produced maps for bays and inlets of Victoria that include 470 km^2 of seagrass. Fine-scale maps (1:10 000) detailing seagrass species composition and estimates of abundance have been produced for large bays including Gippsland Lakes, Corner and Nooramunga Inlets, Westernport Bay and Port Phillip Bay. Smaller inlets that have been mapped include Anderson, Mallacoota, Shallow, Sydenham, Tamboon and Wingham Inlets. The dominant seagrass communities include sparse to dense meadows of *Zostera tasmanica*, *Zostera capricorni* and *Posidonia australis*[51].

The Tasmanian Aquaculture and Fisheries Institute mapped areas of seagrass in six bioregions of Tasmania. The total area of mapped seagrass was 845 km^2. The Boags, Flinders and Freycinet bioregions have been mapped primarily from aerial photographs and

---

Subsequent mapping of seagrass and macroalgal habitats in Westernport Bay in 1995 showed that seagrass and macroalgal cover in the bay had partly recovered, from an area of 72 km^2 in 1983-84 to 113 km^2 in 1995. A further increase to 154 km^2 was recorded in 1999[28]. Despite these increases, the seagrasses in the north and northeast regions of Westernport Bay either remain in poor condition or have not recovered[28,29]. This is likely to be a result of poor water quality as chlorophyll *a* and suspended sediment concentrations increased in the northeastern waters of Westernport Bay between 1975 and 2000[29]. This trend has reduced light availability and reduced the biomass and productivity of seagrasses.

In April 2000, the effects of poor catchment practices and water quality again resulted in hundreds of hectares of seagrasses being lost during a flood event. Although some recovery of seagrasses has occurred over the last 15 years, seagrass meadows in Westernport Bay would still appear to be threatened by flooding and high turbidity even over time periods as short as days to weeks.

The range of information from published papers and technical reports on Westernport Bay fails conclusively to attribute a single cause to the dramatic loss of seagrass from 1974 to 1984. By 1999, seagrasses in the region had shown some recovery, more than doubling the area present in 1984. Nevertheless, there are vast regions in Westernport Bay that have failed to recover, or they are at their threshold of survival during high turbidity[29].

These regions are closest to inputs of nutrients and sediments from a rapidly expanding urbanized catchment with extensive agricultural activities. Management strategies are being implemented to improve water quality in the northeast region by reducing flows of freshwater and loads of nutrients and sediments. These strategies are useful, but existing sediment resuspension issues and changes to intertidal bank topography will limit the possibility of full recovery of seagrass in this region.

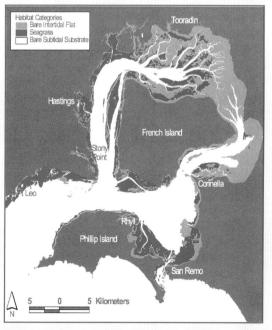

Distribution of estuarine habitats in Westernport Bay, Australia.

LANDSAT (1:100 000 or greater). The Bruny bioregion has been recently mapped in detail from aerial photography with extensive ground-truthing[52, 53]. No mapping has been conducted in the Davey, Franklin and Otway bioregions, but it is unlikely there is much seagrass in these because of exposure to ocean swells and because of high tannin loadings in estuaries (i.e. Port Davey and Macquarie Harbour)[54].

South gulf coast of South Australia

Seagrasses in South Australia cover an area of approximately 9 620 km^2. Shepherd and Robertson[55] recognize three seagrass zones: exposed coasts, gulfs and bays, and coastal lagoons, each with different species composition. The exposed coasts are mainly patchy *Posidonia* typically where islands or reefs give local protection. The two gulfs which are a main feature of the coast have a species gradient from entrance to head with *Posidonia* species being replaced with *Amphibolis* along the gradient. Three genera, *Zostera*, *Ruppia* and *Lepilaena*, are also found where intertidal mud flats occur. Coastal lagoons with a marine environment, such as the Coorong which is 100 km long and less than 2 m deep, are unique in this region. They feature an association of marine and brackish water genera such as *Ruppia*, *Zostera* and *Lepilaena* together with some marine algae[56].

USES

Seagrass habitats in this region are noted for their importance as nursery areas for juvenile fish and for the commercial penaeid shrimp fishery in northeastern Australia. Coles *et al.*[57] recorded 134 taxa of fish and 20 shrimp species in the seagrasses

---

Case Study 11.3
EXPANSION OF GREEN ISLAND SEAGRASS MEADOWS

Seagrasses are an integral and important part of coral reef systems. The Green Island seagrass meadows are one of many seagrass meadows found on reef platforms in the Great Barrier Reef waters[79]. At a time when declines in seagrass biomass and distribution have been widely reported, Green Island is one of the few localities in the eastern Australian region where expansion of seagrasses has been recorded.

Green Island is a vegetated coral cay located approximately 27 km northeast of Cairns. Ground-truthing and mapping of seagrass distribution was conducted in 1992, 1993 and 1994. Systematic mapping by transects was adopted on each occasion and vertical aerial photography (1:12 000) was used if captured within the same season that ground surveys were conducted. Transects were located along compass bearings from permanent markers. A theodolite was used to accurately determine geographic location of survey sites (±1.5 m). Estimates of above-ground seagrass biomass (three replicates of a 0.25 m^2 quadrat), species composition and sediment depth were collected every 20 m. Underwater video and still photography were used to provide permanent records. All data were entered onto a geographic information system. Boundaries of seagrass meadows were determined based on the geographic position of a ground-truthed site and aerial photograph interpretation. Digitally scanned and rectified vertical aerial photographs were used to map the past (1936, 1959 and 1972) seagrass distribution to the northwest of Green Island Cay.

From the interpretation of aerial photographs, a high-density seagrass meadow of 0.39 ±0.3 ha was first visible in 1936 as an isolated patch near the northwest tip of the cay. It appears to have expanded into the back-reef area northwest of Green Island in the 1950s to a small patch covering approximately 1.1 ±0.3 ha in 1959. It increased from the 1950s to 6.5 ±1.3 ha in 1972, 15.31 ±2.29 ha in 1992, 22.71 ±3.3 ha in 1993 and 22.9 ±2.4 ha in 1994. A survey in 1997 found little change[35].

In 1994 *Halodule uninervis* (average above-ground biomass, all sites pooled, 16.61 ±1.4 g dry weight/m^2) was the dominant species in the meadow. *Cymodocea rotundata* was the next most common species (3.95 ±1.6 g dry weight/m^2), with *Cymodocea serrulata* and *Syringodium isoetifolium* occurring in small patches of the meadow (4.12 ±0.7 g dry weight/m^2). *Halophila ovalis* (0.91 ±0.3 g dry weight/m^2) occurred intermixed with *Halodule uninervis* beyond the intertidal and subtidal edges of the main meadow. *Thalassia hemprichii* was uncommon in the meadow (0.03 ±0.02 g dry weight/m^2).

It has long been believed that the expansion of Green Island seagrass meadows was the result of biological and anthropogenic disturbances on the reef. It was first thought that the increases in area of the dense seagrass meadows to the northwest of Green Island Cay were linked to increases in tourist visitation and increased nutrients from the adjacent sewage outfall. This is because low nutrient availability dominates reef habitats such as Green Island and seagrasses are nitrogen limited[19].

of Cairns Harbour. Seagrasses also provide food for dugong and green sea turtle which are the subject of conservation measures.

Apart from licensed worm and bait collecting there is little or no gleaning activity on seagrasses in eastern Australia.

Larkum et al.[8] sum up the values of seagrass in six basic axioms:

o stability of structure;
o provision of food and shelter for many organisms;
o high productivity;
o recycling of nutrients;
o stabilizing effect on shorelines;
o provision of a nursery ground for fish.

In our view this remains an excellent summary of the uses and values of seagrass.

THREATS

Most Australian recorded losses of seagrass are probably the result of light reduction due to sediment loads in the water[14]. Quantifying loss of seagrass has been difficult in many locations as maps are often imprecise or unreliable and local change may be indistinguishable from map error[58]. Long-term data sets are not common so the extent to which loss of seagrass can be attributed to natural long-term cycles is impossible to estimate. Improved mapping of seagrass meadows will enable losses to be more accurately measured and tracked.

Coastal development, dredging and marina developments are generic threats to seagrass in the tourist regions of Australia's east coast. While these issues raise considerable public interest and concern they are usually closely managed through legislative

---

In 1972, a sewage system for hotel buildings and public toilets on Green Island was established[80]. Sewerage effluent from this was discharged onto the Green Island reef for 20 years, until December 1992 when a tertiary treatment facility was completed. It is estimated that approximately 70-100 m^3 of sewage was discharged per day[81]. With no treatment to the effluent, it was essentially raw sewage (nutrient loads unknown) being dumped onto the western edge of the reef platform.

It was, however, unlikely that sewage provided the major nitrogen source, as in September 1994, Udy et al.[19] measured leaf tissue ^{15}N and recorded values from 1.3 to 1.7 parts per thousand suggesting that the primary nitrogen source comes from either fertilizers or N_2 fixation[82]. If the primary nitrogen source was from sewage, the seagrass would have had a leaf tissue ^{15}N value closer to 10 parts per thousand[83]. It could be assumed that ^{15}N values would have been higher prior to the cessation of raw sewage discharge in 1992, but ^{15}N values tend to be highly conservative due to internal recycling of nitrogen in the seagrass.

Also, the expansion of the seagrass meadow before the sewage pipe was installed indicates that increased nutrient availability associated with the sewage outfall in 1972 was not a primary cause of the meadow expansion. This suggests other factors including water seepage and nutrient translocation from the cay, as well as regional changes (agriculture and urban development) in nutrient availability in Great Barrier Reef water may have caused the observed expansion prior to 1972. The continued expansion of the seagrass meadow after 1972 may have been influenced by the sewage discharge in addition to regional changes in nutrient availability.

Seagrass composition at Green Island continues to change, with a rapid increase in the area of *Syringodium isoetifolium* which was first recorded at the island in the mid-1980s. With detailed maps and geographic information system (GIS) formats, changes in the future can be readily quantified and the dynamics of reef island seagrass meadows better understood.

Source: Udy et al.[19] and Queensland Department of Primary Industries[35].

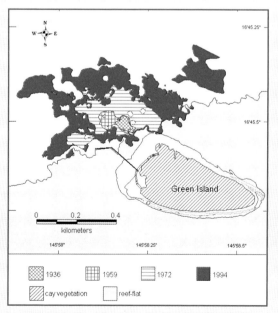

Seagrass distribution at Green Island in 1994, 1972, 1959, 1936.

Thalassia hemprichii meadow on flat adjacent to *Rhizophora* forest, Piper Reef, Queensland.

processes and the actual areas of seagrass destroyed are generally small.

Coastal agriculture may add to sediment loads in catchments and the presence of herbicides in seagrass sediments[59, 60] is a worrying trend, as unlike small-scale coastal developments, this has the potential to destroy large areas. Often the risk factors for the seagrass environment are many kilometers away in upper watersheds. The Coorong Lakes seagrasses are affected by changes in nutrients and freshwater flows in the Murray River catchment which extends from South Australia up to central Queensland thousands of kilometers north.

Port development and the management of risk can influence seagrass survival and many sheltered seagrass sites are also important port locations. The configuration of shipping lanes in northeastern Australia directs large ships transiting south of Papua New Guinea into Great Barrier Reef Lagoon waters. Shipping accidents remain a major concern for coastal habitats and, while infrequent, can be potentially devastating. Major programs exist in the western Pacific to provide advice on shipping-related incidents[61].

Estuarine seagrass communities are increasingly the most threatened of the seagrass habitats in eastern Australia[4, 62]. As provincial centers develop along the Queensland coast, rivers and inlets are often highly affected and need careful management to maintain these seagrass habitats and the fisheries they support[57].

Coastal habitats are threatened by coastal development as well as the impacts of runoff from poorly managed catchments, particularly when associated with large bays such as Botany Bay, Moreton Bay and Hervey Bay.

Reef seagrass habitats are the least threatened seagrass community with minor damage from boating and shipping activities. High tourist visitation rates and associated sewage and poor anchoring practices are identified as a threat at some localities. Acute impacts such as ship groundings and associated spills would impact heavily on reef platform seagrasses.

Although deepwater seagrasses are the least understood seagrass community, they could be impacted by coastal runoff (and associated light reduction) and to some extent prawn/shrimp trawling activities[4, 63], although the scale of any impact is largely unknown and difficult to determine.

SEAGRASS PROTECTION

Seagrasses are habitat for juvenile fish and crustaceans that in many parts of the world form the basis of economically valuable subsistence and/or commercial fisheries. The need to manage fisheries in a sustainable way has itself become a motivating factor for the protection of seagrasses[64].

Approaches to coastal management decision-making are complex, and much of the information exists only in policy and legal documents that are not readily available. Local or regional government authorities have control over small jurisdictions with regulations and policies that may apply.

Approaches in eastern Australia to protecting seagrass tend to be location specific or at least state specific. The approach used depends to a large extent on the tools available in law and to the cultural approach of the community; in Australia these tools and approaches have their origin in British common law.

While there is no international legislation, there is a global acceptance through international conventions (e.g. the Ramsar Convention on Wetlands, the Convention on Conservation of Migratory Species of Wild Animals and the Convention on Biological Diversity) of the need for a set of standardized data on the location and values of seagrasses. Numerous studies worldwide have presented ideas for seagrass protection. Cappo *et al.*[65] summarized the main pressures on fish habitats and seagrasses in Australia. Leadbitter *et al.*[66], Lee Long *et al.*[10] and Coles and Fortes[64] expanded the implications for research and management, a discussion that has Australian as well as global relevance.

Protection by legislation

In the eastern Australian states of New South Wales and Queensland, marine plants cannot be damaged without a permit[67, 68]. In Queensland, the legislation directly protects marine plants. Marine plants are defined as "a plant (a tidal plant) that usually grows on, or adjacent to, tidal land, whether living, dead, standing, or fallen", a definition which includes living plants as well as seagrass plant material washed up on the beach. This definition recognizes the role of even dead plant material in the bacterial cycle that ultimately supports fisheries productivity.

The Queensland Fisheries Act[68] allows for destruction or damage of seagrass only when a permit

has been assessed and issued. All permit issue is directed by a policy that must be taken into account by the person delegated under the Act to make the decision. The policy requires that no reasonable alternative exists. In states such as Queensland, fines well in excess of US$0.5 million are applicable for damaging seagrasses, with the possibility of associated restoration orders.

All eastern Australian states have similar protections in either Fisheries Acts or in National Park or Marine Park Acts. Australia, in fact, has approximately 40 legislative instruments that directly influence marine plant and/or seagrass management[66], not including regulations and management plans that as subsidiary legislation may also be operationally vital to seagrass protection. An example of this would be fisheries legislation that limits areas where bottom trawling can take place.

Protection by marine protected areas (MPAs)

Overlying state and local approaches, Australia also has national legislation addressing international issues such as treaties and conventions including the Convention on International Trade in Endangered Species of Wild Fauna and Flora (CITES) and world heritage area declarations. The Great Barrier Reef World Heritage Area is protected in legislation by the world's largest MPA, the Great Barrier Reef Marine Park. This is unique in that it possibly has as much as 40 000 km^2 of seagrass[3], much of which is afforded a level of protection by the MPA. This can lead to confusingly high levels of regulation; a seagrass scientist working in east coast tropical Queensland requires permits and must meet conditions from national and state authorities.

However, the Great Barrier Reef Marine Park model would not be appropriate in many situations as the money to fund a large administrative authority, legislative support, ongoing research and long-term monitoring, and compliance is not available. More common are MPAs specific to a site and designed to protect an area identified as having important ecosystem functions.

The Queensland Fisheries Act[68] allows for the establishment of fish habitat areas (FHAs) that include areas of coastal seagrass. FHAs are usually small (up to several thousand hectares) MPAs designated specifically to protect fisheries habitat structure over areas considered especially important or critical to fisheries[69].

In recent years there has been a growing realization that we should identify and protect representative examples of the diversity of habitats and processes upon which species depend rather than just areas identified as having some especially important characteristic[70]. A representative area is an area that is typical of the surrounding habitats or ecosystem at a chosen scale. This approach would:

o maintain biological diversity at the ecosystem, habitat, species, population and genetic levels;
o allow species to evolve and function undisturbed;
o provide an ecological safety margin against human-induced and natural disasters;
o provide a solid ecological base from which threatened species or habitats can recover or repair themselves;
o maintain ecological processes or systems.

Typical of establishing a representative area approach to protecting seagrasses is the need for very detailed maps and quantitative data on species and biomass. Presently this information is inadequate for many of our seagrass areas. Compiling a global report card and synthesis of seagrass knowledge will provide a base for future protective decisions and implementation.

AUTHORS

Rob Coles, Len McKenzie and Stuart Campbell, Queensland Department of Primary Industries, Northern Fisheries Centre, P.O. Box 5396, Cairns, Queensland 4870, Australia. **Tel:** +61 (0)7 4035 0111. **Fax:** +61 (0)7 4035 4664. **E-mail:** rob.colesr@dpi.qld.gov.au

REFERENCES

1 Walker DI, Dennison WC, Edgar G [1999]. Status of Australian seagrass research and knowledge. In: Butler A, Jernakoff P (eds) *Seagrass in Australia: Strategic Review and Development of an R & D Plan*. CSIRO, Collingwood. pp 1-24.
2 Lee Long WJ, Mellors JE, Coles RG [1993]. Seagrasses between Cape York and Hervey Bay, Queensland, Australia. *Australian Journal of Marine and Freshwater Research* 44: 19-31.
3 Coles RG, Lee Long WJ, McKenzie LJ, Roelofs AJ, De'ath G [2000]. Stratification of seagrasses in the Great Barrier Reef World Heritage Area, Northeastern Australia, and the implications for management. *Biologia Marina Mediterranea* 7(2): 345-348.
4 Coles RG, Poiner IR, Kirkman H [1989]. Regional studies – seagrasses of North-eastern Australia. In: Larkum AWD, McComb AJ, Shepherd SA (eds) *Biology of Seagrasses: A Treatise on the Biology of Seagrasses with Special Reference to the Australian Region*. Elsevier, Amsterdam. pp 261-278.
5 Mukai H [1993]. Biogeography of the tropical seagrasses in the Western Pacific. *Australian Journal of Marine and Freshwater Research* 44: 1-17.
6 Coles RG, Lee Long WJ [1999]. Seagrasses. In: Maragos JE, Peterson MNA, Eldredge LG, Bardach JE, Takeuchi HF (eds) *Marine/Coastal Biodiversity in the Tropical Island Pacific Region: Vol. 2: Population, Development and Conservation Priorities*. Workshop proceedings, Pacific Science Association. East-West Centre, Honolulu. pp 21-46.
7 Larkum AWD, den Hartog C [1989]. Evolution and biogeography of seagrasses. In: Larkum AWD, McComb AJ, Shepherd SA (eds) *Biology of Seagrasses: A Treatise on the Biology of Seagrasses with Special Reference to the Australian Region*. Elsevier, Amsterdam. pp 112-156.
8 Larkum AWD, McComb AJ, Shepherd SA [1989]. *Biology of Seagrasses: A Treatise on the Biology of Seagrasses with Special Reference to the Australian Region*. Elsevier, Amsterdam.

9. Poiner IR, Staples DJ, Kenyon R [1987]. Seagrass communities of the Gulf of Carpentaria, Australia. *Australian Journal of Marine and Freshwater Research* 38: 121-131.
10. Lee Long WJ, Coles RG, McKenzie LJ [2000]. Issues for seagrass conservation management in Queensland. *Pacific Conservation Biology* 5: 321-328.
11. Carruthers TJB, Dennison WC, Longstaff B, Waycott M, McKenzie LJ, Lee Long WJ [2002 in press]. Seagrass habitats of north east Australia: Models of key processes and controls. *Bulletin of Marine Science*.
12. McKenzie LJ [1994]. Seasonal changes in biomass and shoot characteristic of a *Zostera capricorni* (Aschers.) dominant meadow in Cairns Harbour, Northern Queensland. *Australian Journal of Marine and Freshwater Research* 45: 1337-1352.
13. Bridges KW, Phillips RC, Young PC [1982]. Patterns of some seagrass distributions in the Torres Strait, Queensland. *Marine and Freshwater Research* 33: 273-283.
14. Preen AR, Lee Long WJ, Coles RG [1995]. Flood and cyclone related loss, and partial recovery, of more than 1,000 km^2 of seagrass in Hervey Bay, Queensland, Australia. *Aquatic Botany* 52: 3-17.
15. Inglis GJ [2000]. Variation in the recruitment behaviour of seagrass seeds: Implications for population dynamics and resource management. *Pacific Conservation Biology* 5: 251-259.
16. Lee Long WJ, Coles RG, McKenzie LJ [1996a]. Deepwater seagrasses in northeastern Australia – how deep, how meaningful? In: Kuo J, Phillips RC, Walker DI, Kirkman H (eds) *Seagrass Biology: Proceedings of an International Workshop, Rottnest Island, Western Australia 25-29 January, 1996*. Faculty of Sciences, University of Western Australia, Perth. pp 41-50.
17. Birch WR, Birch M [1984]. Succession and pattern of tropical intertidal seagrasses in Cockle Bay, Queensland, Australia: A decade of observations. *Aquatic Botany* 19: 343-367.
18. Kuo J, Lee Long WJ, Coles RG [1993]. Occurrence and fruit and seed biology of *Halophila tricostata* Greenway (Hydrocharitaceae). *Australian Journal of Marine and Freshwater Research* 44: 43-57.
19. Udy JW, Dennison WC, Lee Long WJ, McKenzie LJ [1999]. Responses of seagrass to nutrients in the Great Barrier Reef, Australia. *Marine Ecology Progress Series* 185: 257-271.
20. Short FT, Coles RG (eds) [2001]. *Global Seagrass Research Methods*. Elsevier Science BV, Amsterdam.
21. Kirkman H [1997]. Seagrasses of Australia. Australia: State of the Environment. Technical Paper Series (Estuaries and the Sea), Department of the Environment, Canberra.
22. Kirkman H, Kendrick G [1997]. Ecological significance and commercial harvesting of drifting and beach-cast macroalgae and seagrasses in Australia. *Journal of Applied Phycology* 9(4): 311-326.
23. Shepherd SA, McComb AJ, Bulthuis DA, Neverauskas V, Steffensen DA, West R [1989]. Decline of seagrasses. In: Larkum AWD, McComb AJ, Shepherd SA (eds) *Biology of Seagrasses: A Treatise on the Biology of Seagrasses with Special Reference to the Australian Region*. Elsevier, Amsterdam. pp 346-393.
24. King RJ, Hodgson BR [1986]. Aquatic angiosperms in coastal saline lagoons of New South Wales, IV. Long term changes. *Proceedings Linnean Society of New South Wales* 109: 51-60.
25. Larkum AWD, West RJ [1990]. Long-term changes of seagrass meadows in Botany Bay, Australia. *Aquatic Botany* 37: 55-70.
26. Seddon S, Connolly RM, Edyvane KS [2001]. Large seagrass dieback in northern Spencer Gulf, South Australia. *Aquatic Botany* 66(4): 297-310.
27. Bulthuis DA [1983a]. A Report of the Status of Seagrass in Westernport in May 1983. Internal Report Series No. 38, Marine Science Laboratories, Victoria.
28. Blake S, Ball D [2001]. Victorian Marine Habitat Database: Seagrass Mapping of Western Port. Marine and Freshwater Resources Institute, Department of Natural Resources and Environment, Report No. 29. Queenscliff, Victoria.
29. Campbell SJ, Miller CJ [2002]. Shoot and abundance characteristics of the seagrass *Heterozostera tasmanica* in Westernport estuary (south-eastern Australia). *Aquatic Botany* 73(1): 33-46.
30. Strong J, Malcolm J [1996]. Gippsland Lakes Fisheries Management Plan, Gippsland Region and Fisheries, Department of Natural Resources and Environment, October 1996. Victoria.
31. Roob R, Ball D [1997]. Victorian Marine Habitat Database: Gippsland Lakes Seagrass Mapping. Marine and Freshwater Resources Institute, Department of Natural Resources and Environment, Queenscliff, Victoria.
32. Bulthuis DA [1981]. Distribution of Seagrass in Port Phillip Victoria. Technical Report No. 5. Ministry for Conservation, Victoria.
33. Blake S, Ball D [2001]. Seagrass Mapping of Port Phillip Bay. Marine and Freshwater Resources Institute, Department of Natural Resources and Environment, Report No. 39. Queenscliff, Victoria.
34. Dennison WC, Abal EG [1999]. *Moreton Bay Study: A Scientific Basis for the Healthy Waterways Campaign*. South East Queensland Regional Water Quality Management Strategy, Brisbane. 246 pp.
35. Queensland Department of Primary Industries. Unpublished data.
36. Rasheed MA [1999]. Recovery of experimentally created gaps within a tropical *Zostera capricorni* (Aschers.) seagrass meadow, Queensland, Australia. *Journal of Experimental Marine Biology and Ecology* 235: 183-200.
37. Coles RG, McKenzie LJ, Yoshida RL [2001a]. Validation and GIS of seagrass surveys between Cape York and Tarrant Point – October/November 1986. DPI, QFS, Cairns. CD-ROM.
38. Long BG, Poiner IR [1997]. Seagrass Communities of Torres Strait, Northern Australia. Torres Strait Conservation Planning Final Report, December 1997. CSIRO, Cleveland.
39. McKenzie LJ, Lee Long WJ, Roelofs AJ, Roder CA, Coles RG [1998]. *Port of Mourilyan Seagrass Monitoring – First Four Years*. EcoPorts Monograph Series No. 15. Ports Corporation of Queensland, Brisbane.
40. Coles RG, McKenzie LJ, Yoshida RL [2001b]. Validation and GIS of seagrass surveys between Cape York and Cairns – November 1984. DPI, QFS, Cairns. CD-ROM.
41. Coles RG, McKenzie LJ, Mellors JE, Yoshida RL [2001c]. Validation and GIS of seagrass surveys between Cairns and Bowen – October/November 1987. DPI, QFS, Cairns. CD-ROM.
42. Coles RG, McKenzie LJ, Yoshida RL [2001d]. Validation and GIS of seagrass surveys between Bowen and Water Park Point – March/April 1987. DPI, QFS, Cairns. CD-ROM.
43. Coles RG, McKenzie LJ, Yoshida RL [2001e]. Validation and GIS of seagrass surveys between Water Park Point and Hervey Bay – October/November 1988. DPI, QFS, Cairns. CD-ROM.
44. Coles RG, Lee Long WJ, McKenzie LJ, Roder CA (eds) [2001f]. Seagrass and marine resources in the Dugong Protection Areas of Upstart Bay, Newry Region, Sand Bay, Ince Bay, Llewellyn Bay and Clairview Region, April/May 1999 and October 1999. Final report to the Great Barrier Reef Marine Park Authority. Queensland Department of Primary Industries, Cairns.
45. Hyland SJ, Courtney AJ, Butler CT [1989]. *Distribution of Seagrass in the Moreton Bay Region from Coolangatta to Noosa*. Queensland Department of Primary Industries Information Series QI89010. QFS, Brisbane.
46. Ayling AM, Roelofs AJ, McKenzie LJ, Lee Long WJ [1997]. *Port of Cape Flattery Benthic Monitoring Baseline Survey – Wet Season (February) 1996*. EcoPorts Monograph Series No. 5. Ports Corporation of Queensland, Brisbane.
47. McKenzie LJ, Lee Long WJ, Bradshaw EJ [1997]. Distribution of Seagrasses in the Lizard Island Group – A Reconnaissance Survey,

October 1995. CRC Reef Research Technical Report No. 14. NFC, QDPI, Cairns.
48 McKenzie LJ, Roder CA, Roelofs AJ, Lee Long WJ [2000]. *Post-flood Monitoring of Seagrasses in Hervey Bay and the Great Sandy Strait, 1999: Implications for Dugong, Turtle and Fisheries Management*. DPI Information Series QI00059. DPI, Cairns.
49 Lee Long WJ, McKenzie LJ, Coles RG [1997]. *Seagrass Communities in the Shoalwater Bay Region, Queensland – Spring (September) 1995 and Autumn (April) 1996*. Queensland Department of Primary Industries Information Series QI96042. QDPI, Brisbane.
50 Lee Long WJ, McKenzie LJ, Roelofs AJ, Makey LJ, Coles RG, Roder CA [1998]. *Baseline Survey of Hinchinbrook Region Seagrasses – October (Spring) 1996*. Research Publication No. 51. Great Barrier Reef Marine Park Authority, Townsville.
51 http://www.nre.vic.gov.au/fishing and aquaculture/bays and inlets (accessed July 2002).
52 Jordan AR, Lawler M, Halley V [2001]. Estuarine Habitat Mapping in the Derwent – Integrating Science and Management. Final report to the Natural Heritage Trust. 69 pp.
53 Barrett N, Sanderson J, Lawler M, Halley V, Jordan A [2001]. Mapping of Inshore Marine Habitats in South-Eastern Tasmania for Marine Protected Areas. Technical report of the Tasmanian Aquaculture and Fisheries Institute. 84 pp.
54 http://www.utas.edu.au/docs/tafi/TAFI_Homepage.html (accessed July 2002).
55 Shepherd SA, Robertson EL [1989]. Regional studies – seagrass of South Australia and Bass Strait. In: Larkum AWD, McComb AJ, Shepherd SA (eds) *Biology of Seagrasses: A Treatise on the Biology of Seagrasses with Special Reference to the Australian Region*. Elsevier, Amsterdam.
56 http://www.environment.sa.gov.au (accessed July 2002).
57 Coles RG, Lee Long WJ, Watson RA, Derbyshire KJ [1993]. Distribution of seagrasses, and their fish and penaeid prawn communities, in Cairns Harbour, a tropical estuary, northern Queensland, Australia. *Australian Journal of Marine and Freshwater Research* 44: 193-210.
58 Thomas M, Lavery P, Coles RG [1999]. Monitoring and assessment of seagrass. In: Butler A, Jernakoff P (eds) *Seagrass in Australia: Strategic Review and Development of an R & D Plan*. CSIRO, Collingwood. pp 116-139.
59 Haynes D, Michalek-Wagner K [2000a]. Water quality in the Great Barrier Reef World Heritage Area: Past perspectives, current issues and new research directions. *Marine Pollution Bulletin* 41: 428-434.
60 Haynes D, Ralph P, Prange J, Dennison B [2000b]. The impact of the herbicide Diuron on photosynthesis in three species of tropical seagrasses. *Marine Pollution Bulletin* 41: 288-293.
61 Rasheed MA, Coles RG, Thomas R [2001]. Fisheries habitat monitoring in Queensland tropical ports – a tool for port management. In: *Proceedings of the 15th Australasian Coastal and Ocean Engineering Conference and the 8th Australasian Port and Harbour Conference, 15th-18th September 2001*. pp 89-95.
62 Lee Long WJ, McKenzie LJ, Rasheed MA, Coles RG [1996b]. Monitoring seagrasses in tropical ports and harbours. In: Kuo J, Phillips RC, Walker DI, Kirkman H (eds) *Seagrass Biology: Proceedings of an International Workshop, Rottnest Island, Western Australia 25-29 January, 1996*. Faculty of Sciences, University of Western Australia, Perth. pp 345-350.
63 Coles RG, Lee Long WJ, Squire BA, Squire LC, Bibby JM [1987]. Distribution of seagrasses and associated juvenile commercial penaeid prawns in north-eastern Queensland waters. *Australia Journal of Marine and Freshwater Research* 38: 103-119.
64 Coles RG, Fortes M [2001]. Protecting seagrass – approaches and methods. In: Short FT, Coles RG (eds) *Global Seagrass Research Methods*. Elsevier Science BV, Amsterdam.
65 Cappo M, Alongi D, Williams D, Duke N [1998]. A Review and Synthesis of Australian Fisheries Habitat Research. FRDC Project No. 95/055. AIMS, Townsville.
66 Leadbitter D, Lee Long W, Dalmazzo P [1999]. Seagrasses and their management – implications for research. In: Butler A, Jernakoff P (eds) *Seagrass in Australia: Strategic Review and Development of an R & D Plan*. CSIRO, Collingwood. pp 140-171.
67 Smith A, Pollard D [1998]. Fish Habitat Protection Plan No. 2; Seagrasses. NSW Fisheries.
68 Queensland Fisheries Act [1994]. Queensland Government Printer, Brisbane, 137 pp. http://www.legislation.qld.gov.au/legislation.htm
69 http://chrisweb.dpi.qld.au/chris
70 Great Barrier Reef Marine Park Authority [1999]. *An Overview of the Great Barrier Reef Marine Park Authority Representative Areas Program*. GBRMPA, Townsville. 18 pp.
71 Fortes MD [1989]. *Seagrasses: A Resource Unknown in the ASEAN Region*. International Center for Living Aquatic Resources Management, Manila. 46 pp.
72 Hastie TJ, Tibshirani RJ [1990]. *Generalized Additive Models*. Chapman and Hall, London.
73 Derbyshire KJ, Willoughby SR, McColl AL, Hocroft DM [1995]. Small Prawn Habitat and Recruitment Study – Final Report to the Fisheries Research and Development Corporation and the Queensland Fisheries Management Authority. Queensland Department of Primary Industries, Cairns.
74 Bulthuis DA [1983b]. Effects of *in situ* light reduction on density and growth of the seagrass *Heterozostera tasmanica* (Martens ex Aschers.) den Hartog in Westernport, Victoria, Australia. *Journal of Experimental Marine Biology and Ecology* 67: 91-103.
75 Shapiro [1975]. *Westernport Bay Environmental Study 1973-74*. Ministry for Conservation, Victoria.
76 Bulthuis DA [1984]. Loss of Seagrass in Westernport, A Proposal for Further Research, 1984/85. Internal Report Series No. 72. Marine Science Laboratories, Victoria.
77 Edgar GJ, Hammond LS, Watson GF [1993]. Final Report to FRDC Committee on the Project 88/91: Consequences for Commercial Fisheries of Loss of Seagrass Beds in Southern Australia. Victorian Institute of Marine Sciences, Victoria.
78 Sargeant I [1977]. *A Review of the Extent and Environmental Effects of Erosion in the Westernport Catchment*. Environmental Studies Series Publication No. 174. Ministry for Conservation, Victoria.
79 Lee Long WJ, Coles RG, Helmke SA, Bennett RE [1989]. Seagrass Habitats in Coastal, Mid Shelf and Reef Waters from Lookout Point to Barrow Point in North-eastern Queensland. QDPI Report to the Great Barrier Reef Marine Park Authority. Queensland Department of Primary Industries, Cairns.
80 Baxter IN [1990]. Green Island Information Review. Report to the Great Barrier Reef Marine Park Authority – August 31. Great Barrier Reef Marine Park Authority, Townsville.
81 Morissette N [1992]. *Identifying Areas of Seagrasses within the Great Barrier Reef Region Threatened by Anthropogenic Activities*. Great Barrier Reef Marine Park Authority, Townsville.
82 Udy JW, Dennison WC [1997]. Growth and physiological responses of three seagrass species to elevated nutrients in Moreton Bay, Australia. *Journal of Experimental Marine Biology and Ecology* 217: 253-277.
83 Grice AM, Loneragan NR, Dennison WC [1996]. Light intensity and the interaction between physiology, morphology and stable isotope ratios in five species of seagrass. *Journal of Experimental Marine Biology and Ecology* 195: 91-110.

12 The seagrasses of NEW ZEALAND

G.J. Inglis

New Zealand (Aotearoa) is an isolated archipelago, consisting of two main islands and a number of smaller islands which lie in the southern Pacific Ocean. Until quite recently, the seagrass flora of New Zealand was thought to consist of two species of *Zostera*: *Zostera capricorni*, which also occurs in eastern Australia, and an endemic species, *Zostera novazelandica*. (Two species of *Ruppia* – *Ruppia polycarpa* and *Ruppia megacarpa* – also occur in brackish and freshwater wetlands in New Zealand[1], but are not considered further here.) *Zostera novazelandica* was originally described by Setchell in 1933 on the basis of morphological variation in vegetative characters, using a relatively limited sample of plants[2]. In fact, there is quite large morphological variation within natural stands of *Zostera* in New Zealand and reproductive structures occur infrequently in many populations[3-6]. This variation has caused considerable uncertainty in identification over the past century, and workers have variously referred to the New Zealand *Zostera* as *Zostera nana*, *Zostera muelleri*, *Zostera marina* and *Zostera tasmanica*[4, 7, 8]. A recent molecular phylogeny of the *Zostera* group, however, demonstrated that *Zostera capricorni* and *Zostera novazelandica* are, in fact, conspecific and that there is likely to be only a single species in New Zealand, hereafter referred to as *Zostera capricorni*[9].

DISTRIBUTION

Zostera capricorni occurs throughout the mainland coast of New Zealand, from Parengarenga Harbour in the north to Stewart Island in the south (Tables 12.1 and 12.2). It is found predominantly between mid and low tidal levels in estuaries and sheltered harbors[10]. On the eastern coastline of the two main islands, patchy stands of *Zostera capricorni* also occur on the tops of siltstone platform reefs in open coastal areas, where they are interspersed with algal beds and biotic assemblages more characteristic of rocky, intertidal assemblages[5, 11]. Stands vary in extent, biomass and stability, depending upon their location[5, 6, 12]. In large, shallow estuaries subject to wind fetch, and on platform reefs exposed to oceanic waves, stands of *Zostera capricorni* typically consist of a mosaic of patches that range in size from less than 1 m^2 to 15 m^2 and which exhibit large interannual fluctuations in extent[5, 12, 13]. The largest persistent stands appear to occur in estuaries and embayments with relatively clear, tidal waters that are situated away from major urban centers, such as Parengarenga Harbour, Farewell Spit, Whanganui Inlet and estuaries of the eastern Coromandel Peninsula.

Despite the wide geographic distribution of *Zostera capricorni* in New Zealand, there have been relatively few published studies of its extent, demography or ecology. Part of the reason for this may be its relative scarcity in many New Zealand estuaries. *Zostera capricorni* is absent from, or occurs in relatively small areas within, many of the shallow, turbid estuaries in New Zealand. Seagrass habitats have been mapped in only 22 of New Zealand's 300-plus estuaries (Table 12.1). The areas that have been mapped typically represent less than 3 percent of the total intertidal area of each estuary. Exceptions include tidally dominated embayments, such as Whanganui Inlet and Whangamata, where seagrass meadows cover up to 31 percent and 18 percent of the intertidal area, respectively. Just over half (54 percent) of New Zealand's estuaries are unsuitable for seagrass growth as they are shallow, barrier-formed estuaries, built around the mouths of rivers[14], so *Zostera capricorni* is likely to be a relatively uncommon benthic habitat in many estuarine environments.

ECOSYSTEM DESCRIPTION

Zostera capricorni stands in New Zealand, like those elsewhere in the world, support a diverse and abundant assemblage of invertebrates that is often richer than unvegetated habitats nearby[8, 13, 25-28]. The composition of the invertebrate assemblages varies with the size and stability of the seagrass stand and its position relative to other habitats[13]. Bullomorph and prosobranch gastropods are distinctive components of the epibenthic fauna[3]. Small crustaceans and polychaetes, which are particularly abundant within seagrass meadows, are important sources of food for wading birds, such as the South Island pied oystercatcher, bar-tailed godwit, pied stilt and royal spoonbill; and for fishes such as mullet, stargazers and juvenile flatfish[3, 29]. Seagrass fragments are also a common food of garfish (family Hemirhamphidae), which are popular with recreational fishermen[30].

Large densities of small cockles, *Austrovenus stutchburyi*, and other bivalves are common in seagrass habitats[3, 27, 28]. Many of these species are not restricted exclusively to seagrasses, but are often more abundant within them as juveniles. Several authors have also drawn a strong historical association between the distribution of seagrasses and beds of the New Zealand scallop, *Pecten novazelandiae*[3, 31]. However, the life cycle of the *Pecten novazelandiae* is not dependent on seagrass habitats and commercial stocks exist in areas where *Zostera capricorni* is not present[30].

A recent survey of more than 25 harbors in northern New Zealand suggests that seagrasses may be important nursery habitats for newly settled snapper (*Pagrus auratus*, family Sparidae)[32]. Snapper is arguably New Zealand's most sought-after marine fish and is the subject of a large commercial and recreational fishery. Adult snapper spawn in large bays, but juveniles are found predominantly in sheltered bays and shallow estuaries during their first summer, before they move to deeper coastal waters[30]. Snapper under one year old have been found in few other coastal habitats and appear to occur mostly in clear-water, sandy reaches of estuaries, the areas most favorable to seagrass growth. Juveniles of other estuarine and coastal fishes are also abundant in seagrass meadows[32].

The presence of *Zostera capricorni* on siltstone platform reefs allows some estuarine species to inhabit these open coastal environments. For example, the endemic burrowing crab *Macrophthalmus hirtipes* occurs exclusively in *Zostera capricorni* on siltstone reefs, where it feeds on seagrass detritus and associated invertebrates[33]. On estuarine mud flats, *Macrophthalmus hirtipes* is more widespread and not necessarily restricted to seagrasses.

HISTORICAL CHANGES IN DISTRIBUTION

The lack of detailed mapping and long-term study of seagrass habitats in New Zealand makes it difficult to determine how their distribution and extent have changed over time. Seagrass meadows undoubtedly supported elements of the economies of pre-European and early European life in New Zealand. The name given to *Zostera capricorni* by New Zealand's

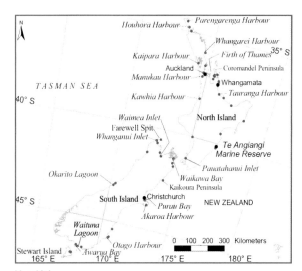

Map 12.1
New Zealand

Table 12.1
Area of seagrass in New Zealand estuaries where benthic habitats have been mapped

Estuary	Total area of *Zostera capricorni* (km²)
Mahurangi Harbour[15]	0.03
Whangateau Harbour[16]	0.33
Pahurehure Inlet (Manukau Harbour)[17]	0
Arm of Kaipara Harbour[18]	0
New River Estuary[18]	0.94
Matakana Harbour[19]	0
Manaia Estuary[20]	0.27
Whitianga Harbour[20]	0.05
Tairua Harbour[20]	1.25
Whangamata Estuary[20]	0.51
Wharekawa Estuary[20]	0.50
Otahu Estuary[20]	0.002
Te Kouma Estuary[20]	0.052
Firth of Thames[21]	0.30
Tauranga Harbour[22]	29.33
Ohiwa Estuary[18]	1.07
Waimea Estuary[18]	0.28
Havelock[18]	0.009
Whanganui Inlet[23]	8.59
Avon-Heathcote Estuary[18]	0.137
Kaikorai Estuary[18]	0
Harwood, Otago Harbour[24]	0.82

indigenous Maori – *rimurēhia* – suggests that they may have recognized the food value of its starchy underground rhizome. *Rimu* is the general term for seaweed or sea plant and *rēhia* was a type of jelly-like stew that was made by boiling marine plants (more usually algae) with tutu berries, the fruit of a wetland plant (*Coriaria* spp.)[34]. Seagrass leaves were also occasionally used by Maori to adorn items of clothing. Hamilton in 1901 described widows wearing mourning caps (*pōtae tauā*) that had veils made from seagrass[35].

Historical accounts by early European naturalists suggest that meadows were quite widespread at the end of the 19th century. Colenso in 1869 described them as "very plentiful" and occurring in "many places in the colony" from the top of the North Island to Stewart Island[36, 37]. Leonard Cockayne (1855-1934) described *Zostera* as "extremely common in shallow estuaries" where it "covers the muddy floor... for many square yards at a time"[38]. At that time, seagrass was apparently so abundant that at least two authors proposed harvesting it for export to London, where dried *Zostera* fetched between £7 10s and £10 per ton as a stuffing for mattresses and upholstered furniture[36, 39]. This suggestion does not appear to have been acted upon. Other accounts describe *Zostera* as "common in many of the lagoons and estuaries which occur along the coast"[40], and as covering "extensive areas of sheltered mud flats between the tides"[8]. Oliver in 1923 described extensive meadows of *Zostera* in Parengarenga Harbour, Tauranga Harbour and Golden Bay, Stewart Island[8]. According to him, "masses of *Zostera*" were occasionally torn up by

Table 12.2
List of locations where seagrasses have been recorded in New Zealand

Location	Description
Parengarenga Harbour[21]	Extensive tidal sand flats (42 km^2) mostly covered in seagrass. Important feeding grounds for large fish and bird populations
Muriwhenua Wetlands[21]	Includes Houhora (10.5 km^2) and Rangaunu Harbours (74 km^2). Extensive tidal sand flats mostly covered in dense beds of seagrass, supporting abundant mollusks, polychaetes, anemones, asteroids and crustaceans. Important feeding grounds for large fish and bird populations
Whangarei Harbour[14, 21]	Lush seagrass beds present until late 1960s. Some recent recovery
Whangapoua Wetlands[21]	Seagrass present on mud flats (ca 14% of the area). Significant site for shellfish gathering
Waitemata Harbour[52]	Seagrass meadows much reduced since 1960s, now in small abundance in a range of locations
Tairua Estuary[20]	Around 1.25 km^2 in 1995, covering ca 23% of the tidal flats
Whangamata Estuary[20]	Around 0.51 km^2 in 1995, covering ca 18% of the tidal flats
Wharekawa Estuary[20]	Around 0.50 km^2 in 1995, covering ca 32% of the tidal flats
Kaipara Harbour[21]	Extensive mud flats and sand flats, but limited area of seagrass
Manukau Harbour[12, 13, 21]	Extensive intertidal mud flats with large beds of seagrass in the 1960s. Current stands are patchy and temporally variable
Firth of Thames[21]	Internationally important feeding area for waterfowl (Ramsar Site). Around 0.3 km^2 of seagrass on tidal flats of ca 85 km^2 Traditional food gathering area. Important local fisheries for snapper and flounder
Kawhia Harbour[21]	Seagrass beds often present on tidal sand and mud flats
Tauranga Harbour[21, 22]	Around 29.3 km^2 of seagrass remaining (1996). Decline of 34% overall from 1959 and 90% in subtidal meadows. Important shellfishery, spawning and nursery areas for marine fishes
Maketu-Waihi Estuaries[21]	Intertidal mud flats and sand flats have local areas of seagrass
Ohiwa Harbour[18, 20, 21]	About 23.8 km^2 of intertidal flats with 1.1 km^2 of seagrass. Outstanding importance as an area for traditional shellfish collection
Ahuriri Estuary and Wetlands[21, 61]	Patches of seagrass in the marine reaches of the estuary, along with *Ruppia* and green algae. Important nursery for fish. High diversity and abundance of invertebrates, especially cockles
Te Angiangi Marine Reserve–East[11]	Patches of seagrass on coastal reefs. Marine reserve
Te Tapuwae O Rongokako	Patches of seagrass on coastal reefs. Marine reserve
Pauatahanui Inlet[56]	Large areas of *Zostera capricorni* on the banks of the inlet near deltas of Horokiwi and Pauatahanui streams
Farewell Spit[21, 58]	Extensive areas of sand and mud flats. Large areas of seagrass. Internationally important area for waterfowl (Ramsar site)

storms and washed onto beaches or swept out to sea in these areas.

In the major South Island city, Christchurch, mud flats of the Avon-Heathcote Estuary were reportedly "covered in great expanses of eelgrass (*Zostera*)" prior to European settlement[41]. Early photographs clearly show dense meadows lining the sand banks of the main channels[42, 43] and accounts described seeing eels feeding in "lush paddocks" of seagrass that grew in the deep channels[44]. Later records document the rapid disappearance and stuttering recovery of *Zostera* in the estuary. By 1929, the "lush paddocks" had been reduced to sparse, small patches[45]. Loss of the meadows was associated with the decline of small fisheries for shrimp and periwinkles in the estuary and caused a "severe and rapid degradation" of feeding grounds for wading birds, which were hunted extensively at the time for food and sport[29, 46]. At least ten families had made their living harvesting shrimp from what they referred to as "shrimp grass"[47]. By 1952, the seagrass had disappeared almost completely, with only a few, very small patches remaining in the northern channel of the Avon River[48]. Since then, patches have waxed and waned in abundance. By 1970 *Zostera* had almost completely disappeared[26, 49]. In 1981, small patches covered around 14 percent of the tidal flats, but these disappeared later in the same year with almost complete defoliation occurring in many areas[47]. The most recent surveys, in 1999, show a total area of around 0.137 km² that comprises around eight consolidated patches[18].

Seagrass losses have also been reported in other

Location	Description
Whanganui Inlet[21, 23]	Large seagrass beds (8.6 km²), especially in the northern part of the inlet. Important nursery for marine fishes. Marine reserve
Waimea Inlet[18, 21]	Extensive bar-built estuary. Around. 0.28 km² of seagrass
Parapara Inlet[62]	*Zostera capricorni* on silt deposits on the rock platform and on mud flats
Moutere Inlet[63]	Large mud flats with extensive beds of *Zostera capricorni* and shellfish. Important nursery for marine and freshwater fishes
Waikawa Bay (Queen Charlotte)[64]	Patches of *Zostera capricorni* present
Wairau Lagoons[21, 65]	Extensive areas of algae, *Ruppia megacarpa* and some *Zostera*. Nursery habitat for marine and freshwater fishes
Kaikoura Peninsula[5, 6, 33]	*Zostera capricorni* present on coastal siltstone reefs at Wairepo flats and Mudstone Bay
Karamea Estuary[21]	Mud flats with extensive areas of seagrass and large densities of invertebrates
Saltwater Lagoon[21]	Bare or sparsely vegetated tidal mud flats
Okarito Lagoon[21, 66]	Middle reaches of lagoon dominated by *Zostera*, upper reaches characterized by dense beds of *Ruppia*, *Lepidena* and *Nitella*
Avon-Heathcote Estuary, Christchurch[21, 47]	Patches of seagrass grow between low and mid tide close to the Avon Channel. Seagrass more abundant prior to 1929. Nationally important area for waterfowl. Among the most important food gathering sites for South Island Maori in pre-European times
Akaroa Harbour[67]	Extensive areas of seagrass on tidal flats at Duvaechelle and Takamatua Bays
Purau Bay, Lyttleton Harbour[67]	Patches of seagrass
Brooklands Lagoon[44, 68]	Scattered, large circular patches of *Zostera capricorni* prior to 1978. None recorded in 1991
Otago Harbour[24, 57]	Around 0.8 km² of seagrass on tidal flats at Harwood
New River Estuary[21]	Extensive mud flats with seagrass. Important source of *kaimoana*. Nationally important wildlife area
Awarua Bay[21]	Extensive mud flats with seagrass. Important source of *kaimoana*. Nationally important wildlife area
Toetoes Harbour[21]	Mud flats with extensive areas of seagrass and large densities of invertebrates. Nationally important wildlife area
Freshwater[21] Paterson Inlet, Stewart Island[21]	Mud flats beyond the river mouth support seagrass
Moeraki Beach[67]	Patches of seagrass on coastal reefs
Mahurangi Harbour[15]	Around 0.03 km² in 1999 in a single meadow.
Whangateau Harbour[16]	Around 0.33 km² in 1999 consisting of two main beds in the southern arm
Whitianga Harbour[20]	Around 0.5 km² in 1995, occupying ca 0.6% of the estuary area
Manaia Harbour[20]	Around 0.27 km² in 1995, covering ca 7.5% of the estuarine flats
Te Kouma Harbour[20]	Around 0.05 km² in 1995, covering ca 2% of the tidal flats
Otahu Estuary[20]	Around 0.002 km² in 1995, covering ca 0.4% of the tidal flats

parts of the country. *Zostera* was reputedly once very abundant in Waitemata Harbour, the location of New Zealand's largest city, Auckland (population ca 1 million). Before 1921, seagrass dominated large areas of Hobson Bay and Stanley Bay, but by 1931, it had all but disappeared[50, 51]. Powell[51] associated this loss with marked reductions in catches of snapper and other carnivorous fishes. At the time, he speculated that "in respect to depletion of harbour fishing grounds generally [loss of seagrass] may be a more important factor than either over-fishing or assumed harbour pollution". This hypothesis has been given greater weight by research that suggests an important nursery role of seagrasses for juvenile snapper[32].

Extensive meadows in the Tamaki Estuary, Howick Beach, Okahu Bay, Kawakawa Bay, Torpedo Bay and Cheltenham in the Auckland district that were present during the early 1960s disappeared by the 1980s[31, 52]. Well-developed stands of seagrass also occurred on Te Tau Banks and along the northern tidal flats of Manukau Harbour in the early 1960s[31]. Descriptions at the time referred to "splendid *Zostera* fields of the Manukau Harbour... in some places up to a mile across"[3]. Most of these areas had also disappeared or were severely reduced in size by the early 1980s[27].

Further north, "lush beds" of seagrass on mud flats in Whangarei Harbour disappeared in the early 1960s[14, 21]. In Tauranga Harbour, Park recorded a decline of around 15 km^2 (about a third of the total area of seagrass) between 1959 and 1996[22]. Subtidal meadows were most affected, with just 0.46 km^2 remaining out of the 4.79 km^2 present in 1959 (a 90 percent reduction).

The causes of these declines are generally unclear. They have variously been attributed to a range of different human activities and natural events. In the Avon-Heathcote Estuary, the loss was linked to the practice of "river sweeping" which began in 1925 to clear silt and plant growth that had accumulated in the two rivers which feed into the estuary[44]. Large quantities of sediment were released during the 25 years that the sweeper operated, producing a muddy sediment layer in the estuary up to 25 cm deep. Untreated sewerage effluent and industrial waste from the rapidly growing city of Christchurch were also discharged into the estuary at this time and may have contributed to the decline[44]. In Waitemata Harbour, the disappearance of seagrass was attributed to waterfront construction, channelization of tidal streams and runoff of fine sediments from surrounding land development[50]. In Whangarei Harbour, a major cement works discharged around 106 000 metric tons of limestone washings each year into the surrounding waters. The discharge significantly reduced water clarity and has been implicated in the disappearance of extensive areas of seagrass[14, 21].

Armiger reported the widespread die-back of seagrass throughout New Zealand during the 1960s[31]. In 1964, she isolated a slime mold from some of the affected populations that resembled *Labyrinthula zosterae*[4], the pathogen responsible for the infamous wasting disease epidemic in North Atlantic *Zostera* during the 1930s[53, 54]. Subsequent collections and observations showed that the mold and symptoms of die-off were present throughout both the North and South Islands[31]. Other studies have reported sporadic outbreaks in some populations[5, 33, 47]. Curiously, the first recorded disappearance of seagrass meadows in New Zealand, from Waitemata Harbour and the Avon-Heathcote Estuary, occurred at much the same time as the northern hemisphere epidemic and corresponded with reports of the large-scale disappearance of *Zostera* in South Australia[31].

PRESENT THREATS

There has been no recent assessment of the condition of New Zealand's estuaries and, therefore, of contemporary threats to seagrass habitats. New Zealand is relatively sparsely populated (ca 3.8 million people in a total land area of 268 021 km^2) so that, although most of its estuaries have settlements nearby, only six are located within urban environments that contain more than 80 000 people[14]. Estuarine habitats have, however, been progressively modified since the times of Polynesian (ca 800 years ago) and European (ca 200 years ago) settlement. Land clearance, shoreline reclamation, harbor development, flood mitigation works and discharge of pollutants have had direct impacts. Less than 23 percent of the land area of the country now remains in native forest with significant areas converted to agricultural production (51 percent) or plantation forestry (6 percent)[14].

Sedimentation is the most widespread problem in New Zealand's estuaries. New Zealand is a predominantly mountainous and hilly country, with nearly half of the land mass at slopes steeper than 28 degrees. Its rivers carry a particularly high load of suspended sediments as a result of the steep terrain and relatively high annual rainfall[55]. Deforestation and rural land management have exacerbated the delivery of suspended sediments to coastal areas and many of New Zealand's larger estuaries are very turbid (light attenuation coefficients up to 0.75/m^2), with comparatively high rates of sediment accretion. In some northern estuaries, this has meant that the area of intertidal habitat has slowly been reduced by increases in the area of mangroves and supratidal salt marsh. Losses of seagrass habitat have been attributed to increased sedimentation and turbidity in a number of

estuaries[14, 22, 44, 52, 56] and it remains the biggest challenge for restoration of submerged aquatic vegetation.

Large areas of plantation forest are now coming into production in New Zealand and harvests are expected to double within the next ten years to more than 600 km² per year. The largest increases are likely to occur in regional areas of the North Island (Northland, Coromandel, East Cape, Hawkes Bay, and southern North Island), and to include areas bordering some of the most significant remaining areas of seagrass (e.g. Parengarenga, Houhora and Coromandel Harbours). It will be important for industry and regional authorities to manage sediment and nutrient runoff from this activity to avoid additional impacts on the ecology of these estuaries.

Nutrient enrichment from land-based sources is a significant problem in some urban estuaries. Recurrent blooms of macroalgae in Tauranga Harbour and the Avon-Heathcote Estuary in Christchurch have been attributed to nutrient loads from wastewater discharge and urban runoff. No direct studies have been done of the effects of nutrient loading from these sources on seagrass growth, although both estuaries have had significant seagrass meadows in the past[22, 44, 47]. Less information is available on nutrient loads to estuaries outside the major urban centers. The most widespread sources of nitrates entering New Zealand rivers are likely to be associated with effluent and runoff from agricultural production[14]. It is unclear what impacts these diffuse sources have had on seagrass habitats.

In some areas, recreational activities have had localized impacts on seagrasses[33, 57]. In Otago Harbour, for example, horse riding and four-wheel drive bikes occasionally rip up rhizomes and roots leading to the formation of large bare patches that can take longer than one year to regrow. Heavy trampling (more than ten passes in one area) across the seagrass flats has also been shown to cause trench formation and lasting damage, but it is unclear how widespread this is[57].

Occasional, recurrent outbreaks of wasting disease appear likely in New Zealand seagrass populations. Further study is required to understand the epidemiology of these outbreaks and, in particular, if they are exacerbated by human activities. In the first instance, this requires an understanding of the resilience of different meadow types (e.g. large versus patchy, persistent versus ephemeral) to outbreaks of the disease and the method of transmission of the pathogen from one location to another.

Nevertheless, there are positive signs that seagrass meadows are slowly returning to some areas from which they had been lost. In Whangarei Harbour, improvements in water quality over the past two decades have led to the re-establishment of seagrasses in areas from which they had disappeared. Discharges of limestone washings into the harbor ceased in 1983 and, since then, improvements in sewerage wastewater and other discharges have greatly increased water quality[14]. This pattern has been repeated in other estuaries as point sources of pollution have been removed or have been better managed over the past 20-30 years. Regrowth of seagrasses in the Avon-Heathcote Estuary is no doubt attributable, in part, to improvements in water quality that have been made through upgrading treatment of

Figure 12.1
An example of changes in the historical distribution of seagrasses in New Zealand

Moncks Bay in the Avon-Heathcote Estuary, Christchurch at low tide in 1885 (left) and 2003 (right). The channel morphology has changed considerably since 1885 and the once extensive intertidal sand banks have all but disappeared. Seagrass meadows, which can be seen clearly as dark bands lining the sand banks in 1885, are no longer present. It is unclear whether the change in channel morphology preceded the loss of seagrass or resulted from it, as the root and rhizomes of seagrass meadows trap and hold soft sediments in place.

wastewater and urban runoff and ending disturbance of river habitat[44, 47]. Non-point sources of pollution and urban stormwater, however, remain significant problems for many estuaries.

POLICY AND MANAGEMENT

Seagrasses and other aquatic macrophytes are not specifically protected by legislation in New Zealand, but are provided for under a variety of resource management and conservation legislation. Responsibility for the protection and management of coastal habitats is split among several national and regional authorities.

The Resource Management Act 1991 (RMA) is an overarching piece of legislation that governs the use of most natural and physical resources (excluding fisheries) in New Zealand. Under the RMA, regional authorities have principal responsibility for managing the use of coastal environments and are required to prepare regional coastal plans as the strategic basis for guiding decisions about resource use in these areas. Development activities within the coastal marine area require approval ("resource consent") from the local authorities under the RMA and must be consistent with the provisions of the coastal plan. Priorities for coastal management were set by the Minister of Conservation in the New Zealand Coastal Policy Statement and these serve as a guide for development of the regional coastal plans. Wetlands are specifically identified as a "matter of national importance" in the RMA that must be taken into account when decisions are being made about resource use. Because of this, many freshwater and estuarine wetlands are specifically listed as "areas of significant conservation value" in existing and proposed coastal plans, and are subject to relatively strict development controls. In some instances, regional authorities have used regulatory measures such as estuarine protection zones to exclude damaging activities from sensitive environments. Regional authorities are also responsible for maintaining coastal water quality under the RMA and regulate land-based activities that can detrimentally affect water quality.

The New Zealand Fisheries Act 1996 provides for the "utilization of fisheries resources while ensuring sustainability". This includes managing the current and potential production of fisheries in New Zealand and their impact on the habitats that support them. Although marine vegetation is not specifically mentioned in the Act, it establishes environmental principles to guide the utilization of fisheries resources. These include the "maintenance of biological diversity" and the "protection of habitat of particular significance for fisheries management". Provisions allow for the protection of specific areas that are important for local and

Case Study 12.1
A SEAGRASS SPECIALIST

An unusual seagrass specialist in New Zealand is the small endemic limpet, *Notoacmea helmsi* (*scapha*) (see drawing, right). This species appears to occupy an almost identical niche to the North Atlantic species *Lottia alveus*, which reputedly became extinct during the wasting disease epidemic of the 1930s[59]. Like *Lottia alveus*, *Notoacmea helmsi* (*scapha*) is a small, elongate limpet (ca 4 mm long x 1.75 mm wide) that fits perfectly onto the narrow leaves of *Zostera*[60].

Unfortunately, there have been no studies of its life history, so it is unclear if it is as specialized as its North American counterpart and, although it was reportedly once widespread in New Zealand[60], there is no contemporary information on its distribution and abundance. The absence of detailed study also means that there is some uncertainty about whether *Notoacmea helmsi* (*scapha*) is a true species or simply a morphological variant of the larger estuarine limpet *Notoacmea helmsi helmsi*[3, 60].

The New Zealand limpet provides a unique opportunity to study the causes of rarity and extinction in marine environments and to determine what impact (if any) loss of the North Atlantic limpet may have had on other species that live and feed in seagrass meadows.

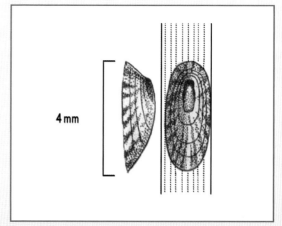

Morphology and habit of the New Zealand seagrass limpet *Notoacmea helmsi* (*scapha*).

Source: Redrawn from Morton and Miller[3].

customary fisheries (*taiapure*), traditional fishing (*mataitai*) and for the protection of specific stocks or their habitat.

Legal protection of coastal waters is mostly administered by the Department of Conservation under the Marine Reserves Act 1971. Marine reserves contain the highest level of protection for natural marine environments in New Zealand; all species and habitats are protected from exploitation. There are currently 16 marine reserves in New Zealand that encompass around 7 633.5 km². However, only two of these contain significant areas of seagrass. Whanganui (Westhaven) Inlet contains around 8.59 km² of seagrass[23] which are protected through a combination of a marine reserve and a wildlife management reserve that cover a total area of 26.48 km². Te Angiangi Marine Reserve, on the east coast of the North Island, also encompasses extensive stands of seagrass on intertidal platform reefs[11]. The exact area of seagrass in the reserve is not known, but in this open coast environment it is likely to be highly variable[6].

The Wildlife Act 1953 and Reserves Act 1977, also administered by the Department of Conservation, have been used to protect intertidal habitats in some estuaries where there are important wildlife, scenic, scientific, recreational or natural values.

Five wetlands in New Zealand are registered under the Ramsar Convention as of special importance to wading birds[21]. Three of these contain coastal or marine environments that include areas of seagrass. Farewell Spit, on the northwest of the South Island, contains an extensive area of intertidal sand and mud flats with *Zostera capricorni* meadows[21]. It has been protected as a Nature Reserve since 1938 and is a significant area for a variety of wading birds and waterfowl. In particular, it is the site of the major molting congregation of the native black swan, *Cygnus atratus*. More than 13 000 swans have been recorded in the area, at densities of up to 1 000 birds per km² [58]. During these congregations, *Zostera capricorni* is the largest component of their diet. The two other coastal Ramsar sites are in the Firth of Thames in the North Island and Waituna Lagoon at the southern tip of the South Island.

CONCLUSION

The collage of historical and contemporary information assembled in this review suggests strongly that seagrass habitats were once much more widespread in New Zealand's estuaries. Their demise appears to be the result of a combination of disease and human activities that have reduced the quality of estuarine waters. Despite relatively limited information on the ecological functions of these habitats in New Zealand, historical information suggests that loss of seagrass

Photographers negotiating the tidal channels near New Brighton in the Avon-Heathcote Estuary in the early 1900s. The elevated intertidal banks are clearly vegetated with extensive stands of *Zostera*.

has had similarly dramatic effects on the distribution and abundance of invertebrates, fishes and other estuarine wildlife that depend upon them, including some species of commercial significance. The high turbidity of many New Zealand estuaries – caused by a combination of natural topography and changes in land use – means that restoration efforts are likely to be long term and broad based, necessitating changes in land and catchment management. Immediate conservation is, therefore, best focused on the relatively few areas where there are large, persistent meadows. There are, however, promising signs of improving water quality in a number of estuaries and of the recent expansion of seagrass habitats in some areas.

ACKNOWLEDGMENTS

Preparation of the review and attendance at the Global Seagrass Workshop was supported by a Technical Participatory Programme grant from the International Science and Technology Linkages Fund. Thanks are due to Diane Gardiner (Ministry of Research, Science and Technology), Megan Linwood (Ministry for the Environment) and Rick Pridmore (National Institute of Water and Atmospheric Research Ltd) for facilitating this. Valuable information from, and discussions with, Mark Morrison, Anne-Maree Schwarz (NIWA), Paul Gillespie (Cawthron Institute), Stephanie Turner (Environment Waikato) and Don Les (University of Connecticut) improved the content of the manuscript.

AUTHOR

Graeme Inglis, National Institute of Water and Atmospheric Research Ltd, P.O. Box 8602, Christchurch, New Zealand. **Tel:** +64 (0)3 348 8987. **Fax:** +64 (0)3 348 5548. **E-mail:** g.inglis@niwa.cri.nz

REFERENCES

1. Mason R [1967]. The species of *Ruppia* in New Zealand. *New Zealand Journal of Botany* 5: 519-531.
2. Setchell WA [1933]. A preliminary survey of the species of *Zostera*. *Proceedings National Academy of Sciences* 19: 810-817.
3. Morton J, Miller M [1968]. *The New Zealand Sea Shore*. Collins, London.
4. Armiger LC [1964]. An occurrence of *Labyrinthula* in New Zealand *Zostera*. *New Zealand Journal of Botany* 2: 3-9.
5. Ramage DL [1995]. The Patch Dynamics and Demography of *Zostera novazelandica* Setchell on the Intertidal Platforms of the Kaikoura Peninsula. Unpublished MSc thesis, University of Canterbury, New Zealand.
6. Ramage DL, Schiel DR [1998]. Reproduction in the seagrass *Zostera novazelandica* on intertidal platforms in southern New Zealand. *Marine Biology* 130: 479-489.
7. Cheeseman TF [1925]. *Manual of the New Zealand Flora*. 2nd edn. Government Printer, Wellington, New Zealand.
8. Oliver WRB [1923]. Marine littoral plant and animal communities in New Zealand. *Transactions & Proceedings of the Royal New Zealand Institute* 54: 496-545.
9. Les DH, Moody ML, Jacobs SWL, Bayer RJ [2002]. Systematics of seagrasses (Zosteraceae) in Australia and New Zealand. *J Sys Botany* 27: 468-484.
10. Webb C, Johnson P, Sykes B [1990]. *Flowering Plants of New Zealand*. Caxton Press, Christchurch, New Zealand.
11. Cresswell PD, Warren EJ [1990]. The Flora and Fauna of the Southern Hawkes Bay Coast. Unpublished report prepared for Department of Conservation, Napier, New Zealand.
12. Turner SJ, Thrush SF, Morrisey DJ, Wilkinson MR, Hewitt JE, Cummings VJ, Schwarz A-M, Hawes I [1996]. Patch dynamics of the seagrass *Zostera novazelandica* (?) at three sites in New Zealand. In: Kuo J, Walker DI, Kirkman H (eds) *Seagrass Biology: Scientific Discussion from an International Workshop*. University of Western Australia, Perth, Western Australia. pp 21-31.
13. Turner SJ, Hewitt JE, Wilkinson MR, Morrisey DJ, Thrush SF, Cummings VJ, Funnell G [1999]. Seagrass patches and landscapes: The influence of wind-wave dynamics and hierarchical arrangements of spatial structure on macrofaunal seagrass communities. *Estuaries* 22: 1016-1032.
14. Ministry for the Environment [1997]. The state of our waters. In: *The State of New Zealand's Environment*. Ministry for the Environment and GP Publications, Wellington, New Zealand. pp 7.1-7.28.
15. Morrison M, Shankar U, Hartill B, Drury J [2000]. Mahurangi Harbour habitat map. NIWA Information Series No. 13.
16. Hartill B, Morrison M, Shankar U, Drury J [2000]. Whangateau Harbour habitat map. NIWA Information Series No. 10.
17. Morrison M, Hartill B, Shankar U, Drury J [2000]. Pahurehure Inlet, Manukau Harbour habitat map. NIWA Information Series No. 12.
18. Cawthron Institute [1999]. Unpublished data.
19. Shankar U, Morrison M, Hartill B, Drury J [2000]. Matakana Harbour habitat map. NIWA Information Series No. 11.
20. Turner SJ, Riddle B [in review]. Estuarine Sedimentation and Vegetation – Management Issues and Monitoring Priorities. Environment Waikato Internal Series 2001/05, Hamilton, New Zealand.
21. Cromarty P, Scott DA [1996]. *A Directory of Wetlands in New Zealand*. Department of Conservation, Wellington, New Zealand.
22. Park SG [1999]. Changes in the Abundance of Seagrass (*Zostera* spp.) in Tauranga Harbour from 1959-96. Environmental Report 99/30, Environment Bay of Plenty, Tauranga, New Zealand.
23. Davidson RJ [1990]. A Report on the Ecology of Whanganui Inlet, North-west Nelson. Department of Conservation, Nelson/Marlborough Conservancy. Occasional Publication No. 2.
24. Fyfe J, Israel SA, Chong A, Ismael N, Hurd CL, Probert K [1999]. Mapping marine habitats in Otago, southern New Zealand. *Geocarto International* 14: 15-26.
25. Wood DH [1962]. An Ecological Study of a Sandy Beach (Howick). Unpublished MSc thesis, University of Auckland, New Zealand.
26. Thompson DJB [1969]. A Quantitative Study of the Macrofauna of *Zostera novazelandica* Setchell. Unpublished BSc (honours) project, Department of Zoology, University of Canterbury, New Zealand.
27. Henriques PR [1980]. Faunal community structure of eight soft shore, intertidal habitats in the Manukau Harbour. *New Zealand Journal of Ecology* 3: 97-103.
28. Ball A [1997]. Seasonal Changes and Demography of *Zostera novazelandica* Setchell in the Avon-Heathcote Estuary. Unpublished MSc thesis, Zoology Department, Canterbury University, New Zealand.
29. Crossland AC [1993]. Birdlife of the Avon-Heathcote Estuary and Rivers, and their Margins. Canterbury Conservancy Technical Report Series No. 6, August 1993. Department of Conservation.
30. Paul L [2000]. *New Zealand Fishes: Identification, Natural History and Fisheries*. Reed Books, Auckland, New Zealand.
31. Armiger LC [1965]. A Contribution to the Autecology of *Zostera*. Unpublished MSc thesis, University of Auckland, New Zealand.
32. Morrison M, Francis M [2001]. 25-harbour fish survey. *Water & Atmosphere* 9(1): 7.
33. Woods CMC, Schiel DR [1997]. Use of seagrass *Zostera novazelandica* (Setchell, 1933) as habitat and food by the crab *Macrophthalmus hirtipes* (Heller, 1862) (Brachyura: Ocypodidae) on rocky intertidal platforms in southern New Zealand. *Journal of Experimental Marine Biology and Ecology* 214: 49-65.
34. Herries Beattie J [1920]. *Traditional Lifeways of the Southern Maori: The Otago Museum Ethnological Project*. Anderson A (ed), 1994. University of Otago Press, Dunedin, New Zealand.
35. Hamilton A [1901]. *Maori Art*. The New Zealand Institute, Wellington, New Zealand.
36. Colenso W [1869]. *Transactions of the New Zealand Institute* 1: 233.
37. Kirk T [1878]. Notice of the occurrence of a variety of *Zostera nana*, Roth, in New Zealand. *Transactions & Proceedings of the Royal New Zealand Institute* X: 392-393.
38. Cockayne L [1967]. *New Zealand Plants and their Story*. 4th edn. RE Owen Government Publisher, Wellington, New Zealand.
39. Smith JA [1878]. On two indigenous productions – manganese and *Zostera marina* – which might be made fair articles of export. *Transactions & Proceedings of the Royal New Zealand Institute* X: 568-569.
40. Thompson GM [1909]. *A New Zealand Naturalist's Calendar and Notes by the Wayside*. RJ Stark & Company, Dunedin, New Zealand.
41. Ogilvie G [1978]. *The Port Hills of Christchurch*. AH & AW Reed, Christchurch, New Zealand.
42. Findlay RH, Kirk RM [1988]. Post-1847 changes in the Avon-Heathcote Estuary, Christchurch: A study of the effect of urban development around a tidal estuary. *New Zealand Journal of Marine and Freshwater Research* 22: 101-127.
43. Knox GA, Kilner AR [1973]. The Ecology of the Avon-Heathcote Estuary. Report to the Christchurch Drainage Board, Christchurch, New Zealand.
44. Deely J [1992]. The last 150 years – the effects of urbanisation. In: Owen SJ (ed) *The Estuary – Where Our Rivers Meet the Sea. Christchurch's Avon-Heathcote Estuary and Brooklands Lagoon*. Parks Unit, Christchurch City Council, Christchurch, New Zealand. pp 108-121.

45. Thompson EF [1930]. An Introduction to the Natural History of the Heathcote Estuary and Brighton Beach. Unpublished MSc thesis, Canterbury College, Christchurch, New Zealand.
46. de Their W [1976]. *Sumner to Ferrymead: A Christchurch History*. Pegasus Press, Christchurch, New Zealand.
47. Knox GA [1992]. The Ecology of the Avon-Heathcote Estuary. Report for the Christchurch City Council and Canterbury Regional Council.
48. Bruce A [1953]. Report on a Biological and Chemical Investigation of the Waters of the Estuary of the Avon and Heathcote Rivers. Christchurch Drainage Board Report, Christchurch, New Zealand.
49. Cameron J [1970]. Biological aspects of pollution in the Heathcote River, Christchurch, New Zealand. *New Zealand Journal of Marine and Freshwater Research* 4: 431-444.
50. Hounsell WK [1935]. Hydrographical observations in Auckland Harbour. *Transactions of the Royal Society of New Zealand* 64: 257-274.
51. Powell AWB [1937]. Animal communities of the sea-bottom in Auckland and Manukau Harbours. *Transactions of the Royal Society of New Zealand* 66: 354-401.
52. Dromgoole FI, Foster BA [1983]. Changes to the marine biota of the Auckland Harbour. *Tane* 29: 79-96.
53. Muehltstein LK, Porter D, Short FT [1988]. *Labyrinthula* sp., a marine slime mold producing the symptoms of wasting disease in eelgrass, *Zostera marina*. *Marine Biology* 99: 465-472.
54. Muehltstein LK [1992]. The host-pathogen interaction in the wasting disease of eelgrass, *Zostera marina*. *Canadian Journal of Botany* 70: 2081-2088.
55. Griffiths GA, Glasby GP [1985]. Input of river-derived sediment to the New Zealand continental shelf: I Mass. *Estuarine, Coastal and Shelf Science* 21: 773-787.
56. Healy WB [1980]. *Pauatahanui Inlet: An Environmental Study*. DSIR Information Series 141. New Zealand Department of Scientific and Industrial Research, Wellington, New Zealand.
57. Miller S [1998]. Effects of Disturbance on Eelgrass, *Zostera novazelandica*, and the Associated Benthic Macrofauna at Harwood, Otago Harbour, New Zealand. Unpublished MSc thesis, University of Otago, New Zealand.
58. Sagar PM, Schwarz A-M, Howard-Williams C [1995]. *Review of the Ecological Role of Black Swan* (Cygnus atratus). NIWA Science and Technology Series No. 25, NIWA Christchurch.
59. Carlton JT, Vermeij GJ, Lindberg DR, Carlton DA, Dudley EC [1991]. The first historical extinction of a marine invertebrate in an ocean basin: The demise of the eelgrass limpet *Lottia alveus*. *Biological Bulletin* 180: 72-80.
60. Powell AWB [1978]. *New Zealand Mollusca*. Collins, Auckland, New Zealand.
61. Knox GA, Bolton LA, Sagar P [1978]. The Ecology of Westshore Lagoon, Ahuriri Estuary, Hawkes Bay. Estuarine Research Unit Report No. 15. Estuarine Research Unit, University of Canterbury, New Zealand.
62. Knox GA, Bolton LA, Hackwell K [1977]. Report on the Ecology of the Parapara Inlet, Golden Bay. Estuarine Research Unit Report No. 11. Estuarine Research Unit, University of Canterbury, New Zealand.
63. Davis SF [1987]. *Wetlands of National Importance to Fisheries*. Freshwater Fisheries Centre, MAFFish, Christchurch. New Zealand.
64. Stephenson RL [1977]. Waikawa Bay (Queen Charlotte Sound): An Intertidal Biological Survey. Estuarine Research Unit Report No. 9. Estuarine Research Unit, University of Canterbury, New Zealand.
65. Knox GA [1983]. An Ecological Survey of the Wairau River Estuary. Estuarine Research Unit Report No. 27. Estuarine Research Unit, University of Canterbury, New Zealand.
66. Knox GA, Fenwick GD, Sagar P [1976]. A Preliminary Investigation of Okarito Lagoon, Westland. Estuarine Research Unit Report No. 5. Estuarine Research Unit, University of Canterbury, New Zealand.
67. Inglis G. Personal observations.
68. Knox GA, Bolton LA [1978]. The Ecology of the Benthic Macroflora and Fauna of Brooklands Lagoon, Waimakariri River Estuary. Estuarine Research Unit Report No. 16. Estuarine Research Unit, University of Canterbury, New Zealand.

13 The seagrasses of THAILAND

C. Supanwanid
K. Lewmanomont

The coastline of Thailand is 2583 km along the Gulf of Thailand and the Andaman Sea, and coastal habitats support abundant populations of commercial fish and associated nearshore fisheries. Seagrasses occur in many locations along the Thai shoreline. The occurrence, community structure and biomass of seagrasses have been studied at different locations in 19 provinces along the coastal areas of the Gulf of Thailand and the Andaman Sea. Among the 12 species of seagrasses found in Thailand, *Halophila ovalis* is the most widely distributed, because of its ability to grow in different habitats. *Enhalus acoroides*, the largest species, is also common in the major seagrass areas. Seagrasses are more abundant in the Andaman Sea than in the Gulf of Thailand.

BIOGEOGRAPHY

Most of the seagrass beds are multispecies beds located in enclosed or semi-enclosed embayments from the intertidal area to 5 m in depth depending on seagrass species, chemical and physical factors. Distribution and habitat of the 12 seagrass species in Thailand is summarized in Table 13.1. Seven species are widespread in both the Gulf of Thailand and the Andaman Sea. *Enhalus acoroides* occurs in brackish water canals down to the lower intertidal and subtidal zones on mud, muddy sand and sandy coral substrates; *Thalassia hemprichii* grows on muddy sand or fragmented dead coral substrates in the upper littoral zone or coral sand substrate in subtidal areas; *Halophila beccarii* grows on mud or muddy sand substrates in estuarine and coastal areas in the intertidal zone; *Halophila decipiens* was previously thought only to occur in waters 9-36 m in depth but has been found in the intertidal areas where it is exposed during low tides; *Halophila ovalis* is found growing on various substrates such as mud, muddy sand and dead coral fragments in the upper littoral to subtidal areas; *Halodule pinifolia* and *Halodule uninervis* both grow in sandy or muddy sand substrates from the upper littoral to subtidal areas. Two species occur only in the Gulf of Thailand: *Halophila minor* which grows on muddy sand in the intertidal zone and *Ruppia maritima* in mangrove areas or brackish water ponds. *Cymodocea serrulata*, which grows on muddy sand, fine sand or sand with coral rubble substrates in the intertidal zone, occurs in both regions but is mainly distributed along the Andaman Sea coastline. Two species are found in the Andaman Sea and not the Gulf: *Cymodocea rotundata*, which occupies the lower littoral zone on muddy sand area or sandy bottom mixed with dead coral fragments and *Syringodium isoetifolium* which occurs densely in subtidal areas on fine sediment.

A total of 68.5 km^2 along the coast of Thailand is known to be covered by seagrasses, but actual coverage must be much greater given the lack of measurements in 11 of the 24 locations in Table 13.1. Seagrass distribution is more extensive in the Andaman Sea than in the Gulf of Thailand.

The four most important seagrass beds in Thailand are Haad Chao Mai National Park, in Trang province on the southern coast of the Andaman Sea and just north of Malaysia, Ko Talibong (Talibong Island), also in Trang province, Kung Krabane Bay, in Chanthaburi province on the eastern coast of the Gulf of Thailand near to Cambodia, and Ko Samui (Samui Island), in Surat Thani province, and part of the southern coast of the Gulf of Thailand.

The seagrass beds at Haad Chao Mai National Park, Trang province are the largest of these seagrass beds and cover 18 km^2, with the highest species diversity for a single area in Thailand[13]. The beds cover a small area around a peninsula called Khao Bae Na and a larger area between the islands of Ko Muk and Laem Yong Lum on the mainland. There are nine species in this area: *Enhalus acoroides, Thalassia hemprichii, Halophila decipiens, Halophila ovalis, Halodule pinifolia, Halodule uninervis, Cymodocea*

rotundata, *Cymodocea serrulata* and *Syringodium isoetifolium*[14]. *Halophila decipiens* is considered to be a deepwater seagrass species in Thailand. However this species occurs in the intertidal zone at Khao Bae Na and pure stands of *Halophila decipiens* are therefore exposed during low tide down to the depths of 5 m[1, 14]. Until recently the only available information on the seagrass beds at Haad Chao Mai National Park was qualitative and restricted to the intertidal zone, but in 2000 the distribution and biomass of seagrasses over the entire subtidal and intertidal bed was investigated[13]. The biomass was highest at shallower depths (<2 m) all along the coastline. *Enhalus acoroides* was the most abundant species, followed by *Halophila ovalis* and *Thalassia hemprichii*. Both *Halophila ovalis* and *Thalassia hemprichii* were dominant at the upper intertidal area and formed monospecific patches in sand dunes and tide pools respectively. Average above-ground biomass of seagrasses in the intertidal area (15 g/m^2) was 1.5 times greater than the biomass of subtidal seagrass beds (10 g/m^2). *Enhalus acoroides* was the most dominant species in the subtidal and lower intertidal zones[13]. The sedimentation rate inside the *Enhalus acoroides* beds was greater than those inside the *Thalassia hemprichii* and *Halophila ovalis* beds because of the shape and size of the *Enhalus acoroides* plants[15]. It has been suggested that distribution of seagrass beds in this area is primarily controlled by the physical conditions of the local environment, principally the roughness of weather during the monsoon season and the amount of shelter available at different locations[16]. This is also true for the seagrass beds at Ko Talibong. The strong southwest waves during the monsoon season (May-October) induces instability of bottom sediments and high turbidity preventing seagrass settlement and growth in the area directly facing offshore waters[16]. Consequently seagrasses only flourish in areas sheltered by the offshore islands.

At the muddy flat of Ko Talibong, 15 km from the southern end of the Haad Chao Mai National Park bed, 7.0 km^2 of nine seagrass species are distributed along the northern, eastern and southeastern coasts of this island. This bed is very important as a feeding ground for the dugong (*Dugong dugon*)[17]. One hundred and twenty-three dugongs were found in Haad Chao Mai National Park and Ko Talibong seagrass beds in 2001, with the largest herd size being 53 dugongs in the Ko Talibong seagrass bed[18].

Compared to the seagrass bed at Haad Chao Mai National Park, the seagrass bed at Ko Talibong is highly affected by siltation from the Trang River. These seagrasses grow in a highly turbid environment with a transparency of about 1-2 m on mud and muddy sand substrates. As a result the maximum depth of seagrass

Map 13.1
Thailand

is limited to 2.5 m[11]. At the eastern end of the island, seagrasses grow on muddy flats and are exposed to the air during low tide. Nine seagrass species were found: *Enhalus acoroides*, *Thalassia hemprichii*, *Halophila beccarii*, *Halophila ovalis*, *Halodule pinifolia*, *Halodule uninervis*, *Cymodocea rotundata*, *Cymodocea serrulata* and *Syringodium isoetifolium*[10]. *Enhalus acoroides* and *Halophila ovalis* were the dominant species in intertidal flats while *Halophila ovalis* was widely distributed in the subtidal area to the southeast of the island.

In the Gulf of Thailand, two major seagrass beds are located in the almost enclosed Kung Krabane Bay in Chanthaburi province and Ko Samui in Surat Thani province[1, 4, 5, 7]. Kung Krabane Bay has a small narrow opening to the sea and an area of approximately 15 km^2 which is surrounded by mangroves and shrimp ponds. Five species of seagrasses grow here: *Enhalus acoroides*, *Halophila decipiens*, *Halophila minor*, *Halophila ovalis* and *Halodule pinifolia*, and cover 7.0 km^2 [1, 12-14]. The deepest part of this bay does not exceed 6 m. *Enhalus acoroides* and *Halodule pinifolia* were the two dominant species among the five[4].

Ko Samui is the largest island on the west coast of the Gulf of Thailand and a major destination for foreign tourists. Five species of seagrasses grow in beds that almost completely surround the island: *Halodule uninervis*, *Halophila minor*, *Halophila ovalis*, *Halophila decipiens* and *Enhalus acoroides* cover a total area of 7.7 km^2 and grow in association with corals, mainly *Acropora* spp. and massive species of coral, scattered around the island. Most of the seagrass

Table 13.1
Occurrence of seagrass species in Thailand

Province/major seagrass area	Ea	Th	Hb	Hd	Hm	Ho	Hp	Hu	Cr	Cs	Si	Rm	No. of species	Area (km²)
Chon Buri	✓			✓	✓	✓				✓		✓	6	id
Rayong				✓	✓	✓	✓	✓					5	id
Makampom Bay							✓						1	2.5
Chanthaburi	✓			✓	✓	✓	✓	✓					6	id
Kung Krabane Bay	✓			✓	✓	✓	✓						5	7.0
Trat	✓		✓	✓		✓	✓	✓		✓			7	id
Phetchaburi												✓	1	id
Prachuab Khiri Khan							✓	✓					2	id
Chumphon	✓												1	id
Surat Thani	✓	✓	✓	✓	✓	✓		✓					7	id
Ko Samui	✓			✓	✓	✓		✓					5	7.7
Nakhon Si Thammarat	✓	✓				✓		✓					4	id
Songkhla		✓				✓	✓					✓	4	id
Pattani		✓				✓	✓	✓				✓	5	4.2
Ranong	✓	✓		✓		✓		✓		✓	✓		7	1.2
Phangnga	✓	✓	✓	✓		✓	✓	✓	✓	✓	✓		10	4.0
Krabi	✓	✓		✓		✓	✓	✓	✓	✓	✓		9	10.0
Phuket	✓	✓	✓	✓		✓	✓	✓	✓	✓	✓		10	4.7
Trang	✓	✓	✓	✓		✓	✓	✓	✓	✓	✓		10	27.1
Haad Chao Mai National Park	✓	✓		✓		✓	✓	✓	✓	✓	✓		9	18.0
Ko Talibong	✓	✓	✓			✓	✓	✓	✓	✓	✓		9	7.0
Satun	✓		✓			✓			✓	✓			5	0.06
Phatthalung			✓				✓						2	id
Narathiwat			✓				✓						2	0.04

Notes: Ea *Enhalus acoroides*; Th *Thalassia hemprichii*; Hb *Halophila beccarii*; Hd *Halophila decipiens*; Hm *Halophila minor*; Ho *Halophila ovalis*; Hp *Halodule pinifolia*; Hu *Halodule uninervis*; Cr *Cymodocea rotundata*; Cs *Cymodocea serrulata*; Si *Syringodium isoetifolium*; Rm *Ruppia maritima*.

id insufficient data.

Source: Various sources[1-14].

areas were formed outside the area of living corals or on reef flats inside the coral reef. *Enhalus acoroides* grows on coarse substrates ranging from medium and coarse sand to coral rubbles at a depth of 0.5-1.0 m. *Halodule uninervis*, *Halophila ovalis*, *Halophila minor* and *Halophila decipiens* are distributed on fine to medium sand at 2.5-7.0 m in depth.[7]

HISTORICAL PERSPECTIVES

The first report of *Halophila ovalis* and *Halodule uninervis* in Thai water was made in 1902 when *Halophila decipiens* was also described as a new species. There were no further reports until 1970 when den Hartog found five species in Thailand: *Cymodocea rotundata*, *Thalassia hemprichii*, *Halophila ovalis*, *Halophila ovata* and *Halophila decipiens*[19]. In 1976, Lewmanomont reported the occurrence of seagrasses belonging to *Halophila*, *Enhalus* and *Cymodocea* in the mangrove areas[20]. Christensen and Anderson found two seagrass species in Surin Island in 1977[21]. Two species were recorded in Koh Kram in Chon Buri province[2]. After this, many reports were published on the occurrence, community structure, biomass and area of seagrasses. Many studies on the ecology and biology of seagrasses have been initiated under the ASEAN-Australia Marine Science project since 1988[7, 22-25].

For Thai people, the main importance of seagrasses is their role as fishing grounds and as habitats for many commercially important species and endangered marine mammals, but the value of seagrasses to provincial and national economies has not been quantified. Indirect uses of seagrasses in

Thailand include their role in coastal protection and as nursery grounds for marine species.

Before 1999 there was no information on the importance of seagrasses in coastal protection in Thailand. Then studies on the water flow and hydrological factors in seagrass beds at Haad Chao Mai National Park were conducted. The studies showed that the intensity of bottom water movement in seagrass beds at lower depths was less than that at the upper depths. This study demonstrated the effectiveness with which *Enhalus acoroides* beds retard the intensity of water motion: current speed inside the *Enhalus acoroides* beds was 15 cm/s on the seafloor and 25 cm/s at 0.5 m in depth. This was a slower movement of water than inside the other seagrass beds, and over bare sand where currents speeds were 22.5 cm/s and 35 cm/s on the seafloor and at 0.5 m depth, respectively. The width and length of *Enhalus acoroides* blades is the greatest among the seagrass species of Thailand, and the blades not only greatly reduce the rate of water flow under and over the meadow but also induce a higher sedimentation rate as a result. In this way, the seagrass beds at Haad Chao Mai National Park create and maintain a unique physical environment in terms of water motion and sedimentation which protects the coastline from the adverse effects of high wave action during the monsoon season[15].

Thai seagrass beds are a nursery ground for juvenile fishes and other marine animals. At Haad Chao Mai National Park, 30 families of fish larvae have been recorded in the nearshore seagrass bed. The abundance of fish larvae in the seagrass bed, at 2064 individuals/1000 m^3, was higher than in open sandy areas, with 1217 individuals/1000 m^3. Economically important fish larvae found in this area were Carangidae, Nemipteridae, Engraulidae, Mullidae and Callionymidae[26]. At Haad Chao Mai National Park seagrass bed, juveniles of the Malabar grouper, *Epinephelus malabaricus*, were collected by small fish traps and cultured in net cages in the canals near the seagrass bed[27]. Twenty-two species of juvenile fishes were reported in the seagrass bed at Kung Krabane Bay, Chanthaburi province. Among these, Serranidae are the most abundant and are also the most important species for fisheries. From October to December, fishermen collect juveniles of Serranidae species

---

Case Study 13.1
THE DUGONG – A FLAGSHIP SPECIES

In Thailand, most fishermen and local people know that seagrass is an important food for the dugongs (*Dugong dugon*). The dugong in Thailand is an endangered species and is protected under the Thai Fishery Act 1947.

Before the first aerial survey for dugong in 1992, not many Thais knew what dugongs and seagrasses were. During the first survey in 1993, dugongs were found near the seagrass bed in Trang Province and the Royal Forestry Department announced that this was the last herd of dugong in Thailand[32]. However, dugongs may still exist on the eastern coast of the Gulf of Thailand[33]. Fishermen in Rayong province have seen dugongs and their feeding trails on small seagrass species[34].

More dugong feeding trails on *Halophila ovalis* at Haad Chao Mai National Park were reported in 1996[35]. At that time, Thai people believed that dugongs preferred feeding on small seagrass species. In 1998, the study on dugong grazing on *Halophila ovalis* beds at Haad Chao Mai National Park was carried out. It was reported that in a 100 x 100 m quadrat, one dugong could produce 14.9 feeding trails (5.1 m^2/day). The estimated grazing rate of *Halophila ovalis* by a dugong was 1.1 kg dry weight, 13.0 kg wet weight/day[36]. Recently, other seagrass species were found in the stomach content of dugongs in Trang province. The species included *Halodule pinifolia, Halodule uninervis, Halophila ovalis, Cymodocea rotundata, Cymodocea serrulata, Syringodium isoetifolium, Thalassia hemprichii* and *Enhalus acoroides*[37].

The dugongs in the Andaman Sea are a flagship species based on their specialized relationship with seagrasses and they are further evidence of the value and importance of the seagrass ecosystem. Recent surveys have shown that more than 60 percent of the people along the Andaman Sea appreciate the importance of the dugongs and seagrasses[38].

Dugongs and seagrass on a Thai stamp.

(approximately 2.5 cm in length) in the morning using scoop nets, and culture them in net cages until they grow to marketable size, when each individual weighs more than 0.8-1.5 kg[28].

Seagrass beds in Thailand are very important areas for fisheries, over and above their role as nursery areas, with both demersal and highly mobile species of fish being harvested from seagrass areas throughout the country. At least 318 species representing 51 families have been identified in seagrass beds in ASEAN countries. They have economic value mainly as food and aquarium specimens[29]. In Thailand the diversity of fish is lower in seagrass beds in the Gulf of Thailand (where 38 species of fishes from 29 families have been recorded from six seagrass beds[30]) than in the Andaman Sea (where 78 species of fishes from 46 families have been recorded from the seagrass beds at Haad Chao Mai National Park). Many species are very important in terms of economic value such as *Epinephelus malabaricus*, orange-spotted grouper (*Epinephelus coioides*), great barracuda (*Sphyraena barracuda*), squaretail mullet (*Liza vaigiensis*), brown-stripe red snapper (*Lutjanus vitta*), Russell's snapper (*Lutjanus russelli*), mangrove red snapper (*Lutjanus argentimaculatus*), oriental sweetlips (*Plectorhinchus orientalis*), silver sillago (*Sillago sihama*) and Indian mackerel (*Rastrelliger kanagurta*)[27].

In addition to the fishes in the seagrass area, crabs and sea cucumbers are also important to fisheries. Since 1998, local fishermen have been collecting sea cucumbers from many seagrass beds in summer, during low tide. After drying the sea cucumbers, the fishermen sell them to Malaysian buyers. At present, three species of sea cucumber have been harvested, namely, *Holothuria scabra*, *Holothuria atra* and *Bohadschia marmorata*. Fresh sea cucumber costs US$12-15 (500-600 Baht) per kg while the dried ones cost US$25 per kg[30]. Eighty percent of the crabs exported from Thailand are portunids, mainly *Portunus pelagicus*, coming mostly from seagrass areas.[31]

Direct use of seagrass is less apparent in Thailand although the seeds of *Enhalus acoroides* are eaten by Thai fishermen. They believe that someone who has a chance to eat the seeds of *Enhalus acoroides* will be lucky. However, they do not like to harvest the fruits of *Enhalus acoroides* for food because of the time necessary to collect enough seeds. Local people in some areas in Thailand use dry seagrass leaves and rhizomes for the treatment of diarrhea. At present, extracts from many species are being screened for biological properties. For example a group of researchers from Kasetsart University has been testing crude seagrass extracts and conducting five bioassays (antibacterial, antifungal, cytotoxicity, antialgal and toxicity tests) on these extracts.

HISTORICAL LOSSES

It is very difficult to estimate the seagrass loss in Thailand because there are no reports on historical coverage or loss. Most of the studies on seagrasses in Thailand were conducted recently and over very short periods of one to two years. There has been no long-term monitoring in the country. Even the present seagrass coverage cannot be completely estimated. However there is evidence showing that a small seagrass bed at Khao Bae Na in Haad Chao Mai National Park has been covered by sand.

Khao Bae Na is a small embayment of flat sand which had a dense *Halophila ovalis* meadow extending over approximately 30 000 m^2 and served in the past as a feeding ground for dugongs. The feeding trails of dugongs were clearly seen during low tide. Some *Cymodocea rotundata*, *Halophila decipiens* and small patches of *Enhalus acoroides* occurred in this area. Tidal level of the meadow was about 1.8 m above mean lower low water[36]. Since the monsoon season in 2000, this seagrass bed has been covered with a high level of sediment. Only small patches of *Halophila ovalis* and *Cymodocea rotundata* have survived and their distributions have been limited by the high sedimentation rate. It is thought that the dugongs have moved their feeding grounds to Ko Muk and Talibong Island seagrass beds.

On 20 January 2002, damage to the seagrasses at Baan Pak Krok in Phuket by the use of mechanized push seines was reported in the press, but the area affected was not estimated. The Natural Resource Conservation Group of Baan Pak Krok requested the government to strengthen law enforcement. There is other anecdotal evidence of damage to seagrass areas in Thailand but it would be impossible to determine the actual loss.

THREATS

Seagrasses in Thailand are threatened by a combination of illegal fisheries and fishing practices, and land-based activities, especially mining. The destruction of seagrass beds is caused by fishing gear such as small-mesh beach seines and mechanized push seines.

Before 1992, the local fishermen in five villages near Haad Chao Mai National Park used mechanized push seines that decreased the number of marine animals and seagrass area. Paradoxically the fishermen's income also decreased while the use of these illegal fishing gears increased. They started to fish by using dynamite and cyanide in the seagrass bed. After 1992, the Royal Forestry Department announced the occurrence of dugong in Haad Chao Mai National Park, and a mass media campaign helped to spread awareness of dugong and seagrass conservation in

Thailand. Local organizations implemented dugong and seagrass conservation projects to persuade local fishermen to stop using beach and push seines in seagrass areas. They can now only use traps for fishing. One year later, the seagrass bed at Haad Chao Mai National Park had increased in size and the fishermen's income had increased because of larger catches from within the protected seagrass areas. However, the Royal Forestry Department still found mechanized push seine trails in other seagrass areas[38].

In Thailand, tin mining is centered in Phuket, Phangnga and Ranong provinces. It has been suggested that sediments from tin mining in Phuket cause chronic problems for seagrass beds in Phuket and Phangnga provinces. Mining activities have now decreased drastically in most areas, but the seagrasses are still affected by other activities, such as land development resulting in landfill, open topsoil on roads and construction on hill slopes[11].

A major threat to seagrasses in Thailand is reduced water clarity in many areas resulting from upland clearing, development along rivers and destruction of mangrove forests.

LEGAL AND POLICY INITIATIVES

In Thailand, there are only two seagrass protected areas. These are Haad Chao Mai National Park and Libong Island Non-hunting Area. Haad Chao Mai National Park is administered by the Royal Forestry Department under the auspices of the Marine National Park Division. Haad Chao Mai National Park was established in 1981 and encompasses 230.9 km^2 – 59 percent of the area is an aquatic zone. Hunting and collecting are forbidden since this is the largest seagrass bed with the highest diversity in Thailand. Libong Island Non-Hunting Area (Ko Talibong Non-Hunting Area) was established in 1960. The only activity restricted here is hunting. Seven square kilometers of seagrass bed distributed in this area serves as a feeding ground for more than 53 dugongs. Most of the officers of Libong Island Non-hunting Area are the local people of the island. They not only protect the area from hunting but also help other local people understand the importance of seagrasses to the marine environment.

There have been several other policy initiatives designed, in part, to conserve seagrasses. In 1972, the Ministry of Agriculture and Co-operatives declared that all mechanized fishing gears were prohibited within 3 000 meters of the coastline in all coastal provinces. In 1993, Trang Provincial Notification was empowered under Fisheries Act B.E. 2490 (Fisheries Act 1947) Section 32[2] to declare that trawlers, mechanized push seines, beach seines and gill nets were prohibited in Haad Chao Mai National Park seagrass bed and at Ko Talibong. In 1997, the Ministry of Agriculture and Co-

Dugong feeding trails on *Halophila ovalis* at Haad Chao Mai National Park.

Seeds of *Enhalus acoroides*.

operatives declared the prohibition of trawlers, mechanized push seines, purse-seines and nets in the area along Phangnga Bay which includes Phuket, Phangnga and Krabi coastlines.

In 1998, the Office of Environmental Policy and Planning proposed policies for the management of seagrass resources including:
o   accelerated management and control of water pollution;
o   increasing efficiency in management of seagrass conservation through landuse planning;

- support for studies on seagrass research and conservation;
- campaigns to heighten and improve public awareness of the importance of conserving seagrasses, at all levels of the community;
- review and adjustment of laws, regulations and enforcement concerning seagrasses so that they work more efficiently by recognizing the important roles of local authorities and communities;
- the monitoring of the status and problems of the seagrass beds, with the cooperation of central government, local authorities and local people[39].

So far seagrass monitoring, restoration and conservation in Thailand has not been widely successful in the long term because of a lack of funding and a suitable methodology. Law enforcement alone has not led to the successful protection of the seagrass ecosystem. It is necessary to involve local people through information and education. A non-governmental organization in Thailand, the Yad Fon Association, has been deeply involved with local people, spreading knowledge and enabling them to understand the importance and usefulness of seagrass beds. At this stage, seagrass and dugong conservation are mostly concentrated in the Andaman Sea.

ACKNOWLEDGMENTS

We are grateful to Dr F.T. Short and Dr R.G. Coles for their great support. Sincere thanks go to Assistant Professor Dr C. Meksampan, P. Wisespongpand, S. Putchakarn, S. Wongworalak, S. Pitaksintorn, K. Adulyanukosol, Assistant Professor Dr S. Satumanatpan and N. Suksunthon for their information and help.

AUTHORS

Chatcharee Supanwanid and Khanjanapaj Lewmanomont, Department of Fishery Biology, Faculty of Fisheries, Kasetsart University, Chatujak, BKK 10900, Thailand. **Tel:** +66 (0)2 579 5575. **Fax:** +66 (0)2 940 5016. **E-mail:** ffischs@ku.ac.th

REFERENCES

1. Lewmanomont K, Deethae S, Srimanopas V [1991]. Taxonomy and Ecology of Seagrass in Thailand. Final report submitted to the National Research Council of Thailand, Bangkok (in Thai).
2. Srimanobhas V [1980]. Some marine plants of Koh Kram, Chon Buri. *Marine Fisheries Laboratory Technical Paper* MFL/22/3.
3. Phuket Marine Biological Center [1998]. Seagrass Management and Dugong Conservation. 1998 progress report submitted to Fishery Department, Bangkok (in Thai).
4. Sudara S, Nateekarnjanalarp S, Plathong S [1992]. Survival and growth rate of transplanted seagrasses *Halodule pinifolia* and *Enhalus acoroides* in different types of substrata. In: Chou LM, Wilkinson CR (eds) *Third ASEAN Science and Technology Week Conference Proceedings, Vol.6, Marine Science: Living Coastal Resources*. National University, Singapore. pp 261-266.
5. Aryuthaka C, Sangthong S, Areeyanon K [1992]. Seagrass Beds in Kung Krabane Bay. Report to Fishery Department Research Conference, 16-28 September 1992. Fishery Department, Bangkok (in Thai).
6. Lewmanomont K, Deetae S, Srimanobhas V [1996]. Seagrass of Thailand. In: Kuo J, Phillips RC, Walker DI, Kirkman H (eds) *Seagrass Biology: Proceedings of an International Workshop, 25-29 January 1996, Rottnest Island, Western Australia*. University of Western Australia, Nedlands, Western Australia. pp 21-26.
7. Nateekarnchanalarp S, Sudara S [1992]. Species composition and distribution of seagrasses at Koh Samui, Thailand. In: Chou LM, Wilkinson CR (eds) *Third ASEAN Science and Technology Week Conference Proceedings, Vol. 6, Marine Science: Living Coastal Resources*. National University, Singapore. pp 251-260.
8. Chumphon Fishery Center [1998]. Coral Reef Observation Report. Cited by: Office of Environmental Policy and Planning. Final Report of the Study of Seagrass Management in Thailand. Office of Environmental Policy and Planning, Bangkok (in Thai).
9. Phuket Marine Biological Center [1996]. Seagrass Management and Dugong Conservation. 1996 progress report submitted to Fishery Department, Bangkok (in Thai).
10. Supanwanid C, Nimsanticharoen S, Chirapart A [1998]. The Biodiversity of Seagrass in Ranong Research Station Coastal Area, 1997-1998. Report submitted to Kasetsart University Research and Development Institute, Kasetsart University, Bangkok (in Thai).
11. Chansang H, Poovachiranon S [1994]. The distribution and species composition of seagrass beds along the Andaman Sea coast of Thailand. *Phuket Marine Biology Center Bulletin* 59: 43-52.
12. Poovachiranon S, Puangprasarn S, Yamarunpattana C [2001]. A Survey of Seagrass Beds at Krabi Bay. Abstract paper presented at the Seminar on Fisheries 2001, 18-20 September 2001. Department of Fisheries, Bangkok (in Thai).
13. Nakaoka M, Supanwanid C [2000]. Quantitative estimation of the distribution and biomass of seagrasses at Haad Chao Mai National Park, Trang Province, Thailand. *Kasetsart University Fishery Research Bulletin* 22: 10-22.
14. Lewmanomont K, Supanwanid C [2000]. Species composition of seagrasses at Haad Chao Mai National Park, Trang Province, Thailand. *Kasetsart University Fishery Research Bulletin* 22: 1-9.
15. Komatsu T [1999]. Water flow and several environmental factors in seagrass beds at Haad Chao Mai National Park in Trang Province, Thailand. In: Koike I (ed) *Effects of Grazing and Disturbance by Dugongs and Turtles on Tropical Seagrass Ecosystem*. University of Tokyo, Tokyo. pp 1-16.
16. Koike I, Nakaoka M, Iizumi H, Umezawa Y, Kuramoto T, Komatsu T, Yamanuro M, Kogure K, Supanwanid C, Lewmanomont K [1999]. Environmental factors controlling biomass of a seagrass bed at Haad Chao Mai National Park, Trang, Thailand. In: Koike I (ed) *Effects of Grazing and Disturbance by Dugongs and Turtles on Tropical Seagrass Ecosystem*. University of Tokyo, Tokyo. pp 66-81.
17. Ostenfeld CH [1902]. Hydrocharitaceae, Lemnaceae, Potederiaceae, Potamagetonaceae, Gentianaceae (Limnanthemum), Nymphaeaceae. In: Schmidt J (ed) *1900-1916 Flora of Koh Chang*. Bianco Luno, Copenhagen. pp 363-366.
18. Adulyanukosal K, Hines E [2001]. Current Research on the Status, Distribution and Biology of the Dugong (*Dugong dugon*) in Thailand. Abstract paper presented at the Seminar on Fisheries 2001, 18-20 September 2001. Department of Fisheries, Bangkok.
19. den Hartog C [1970]. *Seagrasses of the World*. North-Holland Publishing, Amsterdam.
20. Lewmanomont K [1976]. Algal flora of the mangrove areas. In: *Proceedings of the First National Seminar of Ecology of Mangrove*, Vol. 2, Part 2. National Research Council of Thailand, Bangkok. pp 202-213 (in Thai).

21. Christensen B, Anderson W [1977]. Mangrove plants, seagrasses and benthic algae at Surin Islands, west coast of Thailand. *Phuket Marine Biological Center Research Bulletin* 14: 1-15.
22. Poovachiranon S [1989]. Survey on Seagrass in the Andaman Sea from Phuket to Satun Provinces. Final report of ASEAN-Australia Coastal Living Resources Project, submitted to Office of the National Environmental Board. 34 pp.
23. Sudara S, Nateekarnjanalarp S [1989]. Seagrass Community in the Gulf of Thailand. Final report of ASEAN-Australia Coastal Living Resources Project, submitted to Office of the National Environmental Board. 68 pp.
24. Fortes MD [1990]. Taxonomy and Distribution of Seagrasses in the ASEAN Region. Paper presented during the SEAGRAM 2 Advanced Training Course/Workshop on Seagrass Resources Research and Management, 8-26 January 1990, Quezon City, Philippines.
25. Nateekanjanalarp S, Sudara S, Chidonnirat W [1991]. Observation on the spatial distribution of coral reef and seagrass beds in the Gulf of Thailand. In: Alcala AC (ed) *Proceedings of the Regional Symposium on Living Resources in Coastal Areas*. Marine Science Institute, University of the Philippines, Quezon City, Philippines. pp 363-366.
26. Duangdee T [1995]. Identification and Distribution of Fish Larvae in Seagrass Bed at Haad Chao Mai National Park, Changwat Trang. MSc thesis, Kasetsart University, Bangkok (in Thai).
27. Janekitkarn S [1995]. Some Ecological Aspects of Fishes in the Seagrass Bed at Haad Chao Mai National Park, Changwat Trang. MSc thesis, Kasetsart University, Bangkok (in Thai).
28. Sudara S, Nateekanjanalarp S, Thamrongnawasawat T, Satumanatpan S, Chindonwiwat W [1991]. Survey of fauna associated with the seagrass community in Aow Khung Krabane, Chantaburi, Thailand. In: *Proceedings of the Regional Symposium on Living Resources in Coastal Areas*. University of the Philippines, Manila, Philippines. pp 347-352.
29. Poovachiranon S, Fortes MD, Sudara S, Kiswara W, Satumanaptan S [1994]. Status of ASEAN seagrass fisheries. In: Wilkinson CR, Sudara S, Ming CL (eds) *Third ASEAN-Australia Symposium on Living Coastal Resources, 16-20 May 1994*. Department of Marine Science, Chulalongkorn University, Bangkok. pp 251-257.
30. Sudara S, Satumanatpan S, Nateekanjanalarp S [1992]. Seagrass fish fauna in the Gulf of Thailand. In: Chou LM, Wilkinson CR (eds) *Third ASEAN Science and Technology Week Conference Proceedings, Vol. 6, Marine Science: Living Coastal Resources*. National University of Singapore and National Science and Technology Board, Singapore. pp 301-306.
31. Putchakarn S. Personal communication.
32. Sea Aueng S, Witayasak W, Lukanawakulra R, Pearkwisak W, O'Sullivan P [1993]. A survey of dugong in seagrass bed at Changwat Trang. In: *The 31st Kasetsart University Annual Conference*, Kasetsart University, Bangkok. pp 363-368.
33. Nateekanjanalarp S, Sudara S [1994]. Dugong protection awareness: An approach for coastal conservation. In: Sudara S, Wilkinson CR, Ming CL (eds) *Third ASEAN-Australia Symposium on Living Coastal Resources, 16-20 May 1994*. Department of Marine Science, Chulalongkorn University, Bangkok. pp 515-525.
34. Pitaksintorn S. Personal communication.
35. Supanwanid C [1996]. Recovery of the seagrass *Halophila ovalis* after grazing by dugong. In: Kuo J, Phillips RC, Walker DI, Kirkman H (eds) *Seagrass Biology: Proceedings of an International Workshop, 25-29 January 1996, Rottnest Island, Western Australia*. University of Western Australia, Nedlands, Western Australia. pp 315-318.
36. Mukai H, Aioi K, Lewmanomont K, Matsumasa M, Nakaoka M, Nojima S, Supanwanid C, Suzuki T, Toyohara T [1999]. Dugong grazing on *Halophila* beds in Haad Chao Mai National Park, Trang Province, Thailand: How many dugongs can survive? In: Koike I (ed) *Effects of Grazing and Disturbance by Dugongs and Turtles on Tropical Seagrass Ecosystem*. University of Tokyo, Tokyo. pp 239-254.
37. Adulyanukosol K, Poovachiranon S, Natakuathung P [2001]. Analysis of stomach contents of dugongs (*Dugong dugon*) from Trang Province. *Fishery Gazette* 54(2): 129-137 (in Thai).
38. Hines E [2000]. Population and Habitat Assessment of the Dugong (*Dugong dugon*) off the Andaman Coast of Thailand. Final report submitted to the Ocean Park Conservation Foundation, Hong Kong.
39. Satumanatpan S, Sudara S, Navanugraha C [2000]. State of the seagrass beds in Thailand. *Biologia Marina Mediterranea* 7(2): 417-420.

14 The seagrasses of MALAYSIA

J.S. Bujang
M.H. Zakaria

Malaysia's coastline is around 4800 km long, stretching along the Malay Peninsula, Sabah and Sarawak, bounding much of the southern part of the South China Sea. In and adjacent to this coastline are three major coastal ecosystems – mangroves, coral reefs and, less well known, seagrasses. Corals are found on the outer edge of the coastal zone while mangroves are on the inner edge. In general, coastal areas between mangroves and corals, from low-tide level to the coral reef fringe, form the habitats for seagrasses in Malaysia. Seagrasses are also found around offshore islands with fringing corals. Here they are usually found in the outer region between the corals and the semi-open sea. The earliest account of seagrasses in the shallow bays all around the coast of Peninsular Malaysia dates back to 1924[1]. Information on seagrasses is scattered and appears in a number of books, scientific publications and monographs[2-10]. These have been largely taxonomic in nature and list habitats of at least seven species of seagrasses: *Enhalus acoroides* (then referred to as *Enhalus koenigii* by Ridley and Holttum), *Halophila ovalis, Halophila minor* (referred to as *Halophila ovata* by Henderson), *Halophila spinulosa, Halodule uninervis* (then referred to as *Diplanthera uninervis*), *Thalassia hemprichii* and *Ruppia maritima*.

In recent years more research has been carried out on seagrasses in Malaysia. Consequently there are now a number of reports in the literature that describe the extent and richness of flora[11-26] and fauna[12, 27-30] in Malaysian seagrass beds. Unlike other terrestrial communities that can be lived in, managed or exploited, seagrasses offer only a few direct uses. The ecological role and importance of seagrasses has not been fully understood. Much more effort has been spent on quantifying and managing mangroves and corals. Mangrove reserves have been established and coral reefs are protected and conserved in marine parks and marine protected areas. There are guidelines and policies governing the conservation and management of mangroves by the National Mangrove Committee[31] and corals under the Fisheries Act 1985. However the importance of seagrasses at local and national levels, and from the standpoint of conservation, has received far less attention. There are no specific reserves or legislation for seagrasses. Given the importance of seagrass as fisheries habitat, nursery and feeding grounds in Malaysia, this neglected and relatively lesser known resource must be afforded the same priority and be as well managed as mangroves and corals to provide for future renewable resource utilization, education and training, science and research, conservation and protection.

ECOSYSTEM DESCRIPTION

The majority of seagrasses in Malaysia are restricted to sheltered situations in the shallow intertidal associated ecosystem, semi-enclosed lagoons and also in subtidal zones. In these areas they sometimes form diverse extensive communities. The overview of the seagrass distribution and description in this section is given separately for Peninsular and East Malaysia (Sabah). We include specific examples to illustrate the types of seagrass bed found in Malaysia.

Peninsular Malaysia

Along the west coast, patches of mixed species seagrass communities usually occur on substrates from the sandy mud to sand-covered corals in the extreme northern region along the coast of Langkawi Island, Kedah, to the central region of Port Dickson, Negri Sembilan[15], extending as far as Pulau Serimbun, Malacca[32]. The Port Dickson area, at Teluk Kemang, is the only area in mainland Peninsular Malaysia that has intertidal seagrass on reef platform. In the southern region, around the Sungai Pulai area, Johore, mixed species seagrass beds exist at depths of 2-3 m on both sandy mud banks of the mangrove estuary[33] and

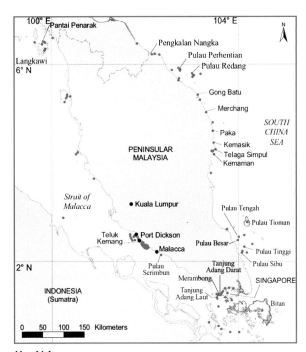

Map 14.1
Peninsular Malaysia

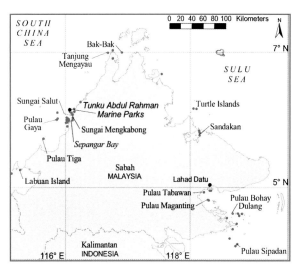

Map 14.2
Sabah

calcareous sandy mud subtidal shoals of Merambong[15], Tanjung Adang Darat[16] and Tanjung Adang Laut[22]. These subtidal shoals, at depths of 2-2.7 m, support nine species (*Enhalus acoroides, Halophila ovalis, Halophila minor, Halophila spinulosa, Thalassia hemprichii, Cymodocea serrulata, Halodule pinifolia, Halodule uninervis, Syringodium isoetifolium*) of seagrasses, the highest species number for any locality in Peninsular Malaysia or East Malaysia[22]. These beds measure 1-1.2 km in length and 100-200 m in width according to estimates based on the visible portion exposed several times in a year by low tides. This is therefore probably the largest single seagrass bed in Peninsular Malaysia. The south has a greater diversity of seagrasses than the northern region with just three species (*Halophila ovalis, Cymodocea serrulata* and *Halodule uninervis*) in Tanjung Rhu and Pantai Penarak in the north.

Intertidal areas of the eastern coastline are devoid of seagrasses. Beds of two species, *Halophila beccarii* and *Halodule pinifolia*, inhabit the fine sand substrate of the shallow inland coastal lagoons from Pengkalan Nangka, Kelantan, to Paka, Terengganu, while *Halodule pinifolia* and *Halophila ovalis* inhabit a similar substrate type at Gong Batu and Merchang. Monospecific beds of *Halodule pinifolia* were found at Kemasik, Terengganu, and pure stands of *Halophila beccarii* grew on the mud flat of the mangroves in Kemaman, Terengganu. Monospecific beds of *Halodule pinifolia*, *Halophila decipiens* and mixed species seagrass beds occur in the waters of the offshore islands with fringing coral reefs such as Pulau Sibu, Pulau Tengah, Pulau Besar and Pulau Tinggi[13, 15], Pulau Redang and Pulau Perhentian[28] and Pulau Tioman[34]. Seagrasses are usually found in the outer region between the corals and the semi-open sea.

East Malaysia
The west and southeastern coasts of Sabah harbor mixed species seagrass beds in the intertidal zone down to a depth of 2.5 m. Seagrasses grow on substrates ranging from sand and muddy sand to coral rubble. There are six areas of intertidal mixed associations of seagrass and coral reef along the west coast at Bak-Bak, Tanjung Mengayau, Sungai Salut, Sungai Mengkabong, Sepangar Bay and Pulau Gaya. The four isolated offshore islands of Pulau Maganting, Pulau Tabawan, Pulau Bohay Dulang and Pulau Sipadan along the southeastern coast have subtidal seagrasses growing on coral rubble[14, 17-20, 35].

In Sarawak, other than records of herbarium specimens of *Halophila beccarii*, collected by Beccari in Sungai Bintulu[2, 10], and *Halophila decipiens* collected at Pulau Talang Talang, Semantan[32], nothing much is known about the seagrass habitats, distribution and species composition.

BIOGEOGRAPHY
Peninsular Malaysia
The distribution of seagrasses in Peninsular Malaysia has been detailed in various publications[10, 13, 15, 16, 22]. A very broad distinction can be made between the seagrass distribution of the west and east coasts. Differences in the available habitats and prevailing environmental characteristics along the east and west coasts probably explain these distributions. On the west coast seagrasses occur in the sandy mud sediments of

shallow coastal waters while on the east coast the coastline is fringed with sandy to rocky areas which are not suitable for the growth of seagrasses. On the east coast seagrasses inhabit sandy mud lagoons, behind the sand ridges in areas sheltered from the open sea. Seagrasses are also found around relatively calm offshore eastern islands with fringing reefs such as Pulau Redang, Pulau Perhentian, Pulau Tengah, Pulau Sibu, Pulau Tinggi and Pulau Besar. The west coast of Peninsular Malaysia does not generally experience strong wave action, whereas the east coast is exposed annually to the northeast monsoon from November to January[34].

Clarity of water and sufficient light irradiance play a significant role in the depth distribution of the seagrasses. Coastal waters are often turbid or high in suspended solids that limit the depth at which most seagrasses grow, more so on the western coast of Peninsular Malaysia than the east. This is reflected in seagrass communities along the west coast which are generally found inhabiting the shallow waters at depths of less than 4.0 m. Seagrasses on the east coast, however, extend to deeper areas, 5.0-7.0 m. Seagrasses will colonize greater depth if the water is clear. By way of comparison, in the clear water of the east coast the depth limit for *Halophila decipiens* ranged from 6 to 24 m in the Sungai Redang Estuary and Cagar Hutang of Pulau Redang, Terengganu, respectively[39] while in the turbid water of the west coast, at Teluk Kemang, Port Dickson, it grows at 1.5-3.1 m[22].

Sabah, East Malaysia

Seagrass distribution along the west and southeastern coasts of Sabah was described by Ismail in 1993[14] and in the Tunku Abdul Rahman Marine Parks (Pulau Gaya, Pulau Mamutik, Pulau Sulug, Pulau Manukan and Pulau Sapi) by Josephine in 1997[35]. Almost all seagrasses are associated with degraded coral reef, although a few are associated with mangroves and habitats damaged through illegal fishing by explosive. There were no broad differences regionally with respect to species distribution and composition.

HISTORICAL PERSPECTIVES

In Peninsular Malaysia seagrasses (*Enhalus acoroides, Halophila ovalis*) were apparently locally common all around the coast on muddy shores and areas exposed at low tide[1,3,4,5,7]. Historical accounts of the distribution of seagrass species at three places in Sabah, Labuan Island, Sandakan and Lahad Datu, were given by den Hartog in 1970[10]. Information on their abundance was not given. Ismail[14] described seagrass habitats that were already degraded by human activities in Sabah, East Malaysia, in 1993. Since the early reports, which indicated extensive seagrass beds, many of the habitats (e.g. the west coast of Peninsular Malaysia, East Malaysia, Sabah) have been exploited or have deteriorated to a greater or lesser extent as a result of coastal development, especially in the last 15 years[11,15]. Such phenomena would explain the present seagrass distribution, which is no longer extensive, and its patchy distribution along the Malaysian coastline[15,22,28].

Known uses of seagrasses were few. Burkill[3] in his book, *A Dictionary of the Economic Products of the Malay Peninsula*, mentioned that Ridley recorded in 1924 that the leaves of *Enhalus acoroides* were one of the chief foods of the dugong, *Dugong dugon*, which was then common in Malaysia. Later the dugong became rare because it was hunted for meat and hide[6]. Presently dugongs are found in areas with abundant seagrasses such as Pulau Sibu, Pulau Tengah, Pulau Besar and Pulau Tinggi on the east coast and around Merambong, Tanjung Adang Darat and Tanjung Adang Laut shoals of Sungai Pulai, Johore. *Enhalus acoroides* fruits are edible[1,3] and the coastal communities of Sungai Pulai, Johore, still collect them for consumption. In addition the softer parts of *Enhalus acoroides* form fibers that are made into fishing nets. *Ruppia maritima* plants are used in fish ponds to aid in the aeration of the water, and the milk fish (*Chanos* spp.) feeds on it. This functional role, though mentioned, has not been observed in Peninsular Malaysia, and is probably based on observations made in the fishponds of Java, Indonesia[3]. *Ruppia maritima* is rare in Peninsular Malaysia[15].

Other forms of utilization include using seagrass areas for fish (*Lates calcarifer* and *Epinephelus sexfasciatus*) cage farming, for example at Pengkalan Nangka, Kelantan, and Gong Batu, Terengganu, which started in 1991, or oyster (*Saccostrea cacullata*) farming as at Merchang from 1998[19]. Seagrass areas at Pengkalan Nangka, Kelantan, Paka shoal, Terengganu, and Tanjung Adang Laut shoal, Johore, are used as collection and gleaning sites for food including fishes, gastropods (*Lambis lambis, Strombus canarium*), bivalves (*Gafrarium* sp., *Meretrix* sp., *Modiolus* sp.) and echinoderms (sea cucumber e.g. *Pentacta quadrangularis, Mensamaria intercedens*). Gleaning for food in seagrass areas associated with coral reefs is widespread in Sabah, East Malaysia.

ESTIMATE OF HISTORICAL LOSSES

There is no information in the form of historical maps or aerial photographs that can be used to determine the loss of seagrass beds over time. The losses reported here have been observed during repeated visits to the various seagrass sites. On the west coast of Peninsular Malaysia, at Port Dickson, localized depletion of seagrass (narrow-leaved *Halodule*

uninervis and *Enhalus acoroides*) began in 1994, representing about 50 percent of the area originally present. This area was heavily utilized as a public recreational area. At Teluk Kemang in 1997 there was intensive sand mining for reclamation activities in mangrove swamps as part of the construction of a condominium. This caused the loss of *Halophila ovalis* and *Halodule pinifolia* in the subtidal seagrass bed of Teluk Kemang. Suspended particles in the water settled on the leaves of the seagrasses, blocking light for photosynthesis and causing considerable stress and mortality through burial. The presence of an oil refinery, intense shipping activity and frequent oil spills in the adjacent waters have also been suggested as potential causes for the decline or loss of seagrasses along the coastline of Port Dickson. Tar balls in significant quantities, frequently washed ashore, were evidence of oil spills. In addition, petrogenic hydrocarbons were detected in the water and sediments at Teluk Kemang[28].

The Sungai Pulai seagrass beds Tanjung Adang Laut and Tanjung Adang Darat are diverse and extensive, and were only discovered in 1991 and 1994 respectively, yet by 1998 they were at risk from port development involving dredging of shallow passageways and land reclamation for new facilities, both causing an increase in the suspended solids in the water column. Localized losses were observed with the death of sand-smothered *Halophila ovalis* clearly visible. In addition dense overgrowth of the macroalgae *Gracilaria coronopifolia* and *Amphiroa fragilissima* caused the seagrasses in the area to die back. However, recovery occurred with regrowth of seagrasses and the disappearance of the macroalgae.

On the east coast at Pengkalan Nangka, Kelantan, the decline was the result of human activities such as

Case Study 14.1
THE SEAGRASS MACROALGAE COMMUNITY OF TELUK KEMANG

At Teluk Kemang, Port Dickson, Negri Sembilan, the intertidal community consists of non-uniform patches of mixed seagrasses and macroalgae on a coral reef platform 1.0-1.5 m deep. Seagrasses grow in various substrates, from sand-covered coral to a combination of silt, coarse sand and coral rubbles (see photograph). *Halophila ovalis* is dominant and widespread, interspersed with *Thalassia hemprichii*, *Cymodocea serrulata*, *Enhalus acoroides* and *Halodule pinifolia*. *Syringodium isoetifolium*, a rare species here, occurs in patches in the sand-filled spaces amongst coral rubble areas. Macroalgae coexist with these seagrasses.

The most common, and seasonal, macroalgae species are (Chlorophytae) *Caulerpa sertularioides*, *Caulerpa prolifera*, *Caulerpa racemosa*, *Caulerpa lentillifera*; (Phaeophytae) *Sargassum polycystum*, *Sargassum cristaefolium*, *Sargassum ilicifolium* and *Padina tetrastomatica*; and (Rhodophytae) *Laurencia corymbosa* and *Jania decussato-dichotoma*[36]. This intertidal community extends into the subtidal zone to depths of 3.5 m with a clear zonation of seagrass species that are confined to sandy mud and silty substrates. Pure stands of *Halophila ovalis* and *Halodule pinifolia* with isolated individuals of *Enhalus acoroides* occur at a depth of 1.5 m. *Halophila decipiens* grows in small patches at a depth of 1.5-2.0 m in association with *Halophila ovalis* and *Halodule pinifolia*. Slightly deeper, at 2.0-3.0 m, *Halophila decipiens* forms a continuous meadow. Occasionally patches of pure *Halophila ovalis* occur at depths of 3.2 to 3.5 m. Morphological differences are observed in *Halophila ovalis* in these two communities. Subtidal *Halophila ovalis* plants possess much bigger leaf blades and more cross-veins[37] than plants of the same species growing in the intertidal zone.

Another conspicuous seagrass is *Halophila decipiens* which occurs at shallow depths of 1.5-3.0 m[38]. *Halophila decipiens* was previously thought to be a deepwater species growing at depths between 10 m and 30 m[10,17,39].

The Teluk Kemang seagrass macroalgae community on coral reef platform. Seagrasses occupy the sand-filled spaces of the coral reef platform, and macroalgae dominated by *Sargassum* spp. inhabit the boulders and coral rubbles.

Photo: J.S. Bujang

the dredging of sand for landfills which have totally removed two shoals of *Halophila beccarii* and *Halodule pinifolia*, representing 30 percent of the total seagrass area. At Merchang and Kemasik, Terengganu, the effect of wind and resulting wave action on lagoon seagrass is reduced by the sheltering presence of the sand ridges. Despite this protection, 50-70 percent of *Halodule pinifolia* and *Halophila ovalis* seagrass beds were severely damaged by intense winds, waves and sediment movement during the northeast monsoon storms of October 1998 to January 1999. No recovery to the original areal extent has been observed yet. Mining of sand at Telaga Simpul, Terengganu, in March 1997, for the shoreline stabilization and protection of Kuala Kemaman village, resulted in high total suspended solids in the water column and sedimentation smothered the dense *Halophila beccarii* bed there. The bed was transformed to sparse and scattered patches

---

Case Study 14.2
THE SUBTIDAL SHOAL SEAGRASS COMMUNITY OF TANJUNG ADANG LAUT

The subtidal shoal of Tanjung Adang Laut in the Sungai Pulai estuary, Johore, is 1.5-2.7 m below mean sea level and is vegetated with seagrasses (see photograph)[16]. This shoal is one of the feeding grounds for dugongs around Sungai Pulai, Johore, and their feeding trails can be seen clearly at low tides. The shoal is made up of calcareous sandy mud substrate and supports a mixed species community dominated by *Enhalus acoroides*, *Halophila ovalis* and *Halophila spinulosa*. This association occupies the middle zone (1.5-1.8 m) and is exposed during extreme low spring tides. *Cymodocea serrulata*, *Syringodium isoetifolium* and *Halodule uninervis* inhabit the deeper, narrow edge zones (1.8-2.1 m) which remain unexposed.

The edge zone is bare at some places, while at others isolated patches of *Cymodocea rotundata*, *Halophila spinulosa*, *Halophila minor* or *Halodule pinifolia* occur. In the deeper zone (2.1-2.7 m) sparse, isolated patches of *Enhalus acoroides* and *Halophila ovalis* are found. *Enhalus acoroides* and *Halophila ovalis* occur at depths of 1.5-1.8 m, and are also exposed during low spring tides, but are able to withstand short periods of desiccating conditions. *Cymodocea serrulata* and *Syringodium isoetifolium* are less resistant and therefore tend to occur in the unexposed edge zone (1.8-2.1 m).

This seagrass bed also supports a total of 25 species of macroalgae. Rhizophytic macroalgae such as *Avrainvillea erecta*, *Caulerpa* spp. and *Udotea occidentalis* are set into the sandy or sandy mud substrates whereas epiphytes such as *Bryopsis plumosa*, *Ceramium affine*, *Chaetomorpha spiralis*, *Cladophora spatentiramea*, *Cladophora fascicularis*, *Cladophora fuliginosa*, *Dictyota dichtoma*, *Hypnea cervicornis*, *Gracilaria coronopifolia*, *Gracilaria fisherii* and *Gracilaria salicornia* are attached directly to seagrasses. Species such as *Enteromorpha calthrata* and *Gracilaria textorii* attach to mollusk shells or polycheate tubes. Drift macroalgae, such as *Acanthophora spicifera*, *Amphiroa rigida*, *Amphiroa fragilissima*, *Hypnea esperi* and *Ulva* spp. lie loosely amongst the seagrasses. Attached (e.g. *Gracilaria coronopifolia*) and drift macroalgae (e.g. *Amphiroa fragilissima*) form important components of this shoal community and seasonally, from April to July and in November, the seagrass bed is overgrown with them.

The waters around Tanjung Adang Laut as well as those of Tanjung Adang Darat and Merambong shoals support the fisheries which feed the inhabitants of coastal communities. Seventy-six species of fishes (including the Indian anchovy *Stolephorus indicus*, barramundi *Lates calcarifer* and Spanish flag snapper *Lutjanus carponotatus*) and others including prawn (e.g. *Penaeus indicus*) and crabs (*Portunus pelagicus* and *Scylla serrata*) have been reported in the area[12,29,30]. The locals also used the shoal as a gleaning site for collection of gastropods such as *Strombus canarium* and *Lambis lambis* and bivalves such as *Gafrarium* spp. and *Modiolus* spp.

Tanjung Adang Laut subtidal shoal with mixed species seagrass community. Nine species of seagrass inhabit the calcareous sandy mud substrate of the shoal.

and *Halophila beccarii* has been largely replaced by the more aggressive *Halodule pinifolia* which now forms a monospecific bed. Standing biomass of *Halophila beccarii* has been dramatically reduced from 0.89-4.34 g dry weight/m^2 (shoot density of 2078-6798/m^2) before the mining in 1996 to 0.58-0.59 g dry weight/m^2 (shoot density of 758-1386/m^2) from April 1997 until January 1999. *Halodule pinifolia* biomass and shoot density fluctuated from 10.1 to 56.6 g dry weight/m^2 and 2145.3 to 8946/m^2 respectively during that period.

In Sabah, no information on decline or loss of seagrasses has been reported. However, symptoms of a declining seagrass bed were visible at Sepangar Bay. The middle sublittoral belt of *Halodule uninervis* and *Cymodocea rotundata* was eroded by wave action. Edge plants have exposed rhizomes and roots. Sediment erosion and instability appear to be implicated in the progressive decline of these seagrasses in the shallow water.

PRESENT COVERAGE

Information on the total area, extent or size of seagrass beds in Malaysia is incomplete. The individual and total estimated areas presented (Table 14.1) are for the known seagrass areas in Peninsular Malaysia. This is an underestimate as seagrass areas in the offshore islands are not included. Although Ismail[14] has reported that seagrass beds in Sabah occur in patches ranging in size from 10 m to 150 m in diameter, no further data are available, though it is known that, compared with Peninsular Malaysia, seagrasses are common in Sabah. An approximate estimate for seagrass areas in Sabah would be many times that of the known seagrass areas in Peninsular Malaysia.

PRESENT THREATS

The Malaysian coastal zone is being subjected to a high degree of resource exploitation as well as pollution. Seagrass beds grow in shallow, coastal zone waters and this renders them susceptible to unplanned and unmanaged urban and industrial development. These problems are compounded by a lack of environmental assessment procedures for developments and lack of awareness about the importance of seagrasses. In the past, and even at present, losses of seagrass communities in the coastal areas of Malaysia caused either by natural causes or human activities generally pass unnoticed or unrecorded. States such as Kedah and Malacca are undertaking land reclamation and expansion programs. Land reclamation and expansion in Johore is occurring for the development of new port facilities. With more expansion planned, the future intention is to completely reclaim the stretch of seagrass beds of Merambong-Tanjung Adang shoals, the feeding ground of dugongs. Sourcing for sand on the east coast is a common activity for landfill and shoreline stabilization projects. Dredging is being carried out in the *Halophila beccarii* and *Halodule pinifolia* beds of Pengkalan Nangka, Paka shoal and Telaga Simpul. This dredging will lead inevitably to increased sedimentation and smothering of seagrasses. More bed removal will eventually occur if dredging is to be continued to supply the increasing demand for sand.

Small-scale destructive fishing by pull net at Pengkalan Nangka, Kelantan, and Paka shoal, Terengganu, dislodges the seagrasses and reduces the seagrass cover. Harvesting of bivalves, *Hiatula solida*, *Meretrix meretrix* and *Geloina coaxans* at Pengkalan Nangka, Kelantan, has been shown to cause mechanical damage, reduce seagrass cover and retard the spread and colonization of seagrasses. Other threats include the increasing public use of natural seagrass areas, such as for recreational boating, fishing and swimming in Port Dickson, Negri Sembilan, and as avenues for transportation such as in the narrow channels in the Paka Lagoon, Terengganu, and Sungai Pulai-Merambong-Tanjung Adang shoals, Johore.

In Sabah, seagrass and coral reef associated ecosystems are areas of gleaning and collection for food resources. Uncontrolled collection of flora such as

Table 14.1
Estimate of known seagrass areas in Peninsular Malaysia

State and location	Area (ha)
Kelantan	
Pengkalan Nangka Lagoon	40.0
Kampung Baru Nelayan-Kampung Sungai Tanjung	27.0
Pantai Baru Lagoon	20.0
Terengganu	
Sungai Kemaman	17.0
Chukai, Kemaman	3.3
Telaga Simpul	28.0
Sungai Paka Lagoon	4.7
Sungai Paka shoal	43.0
River bank of Sungai Paka	1.5
Merchang	3.0
Gong Batu	5.0
Negri Sembilan	
Teluk Kemang	11.0
Johore	
Tanjung Adang Laut shoal	40.0
Tanjung Adang Darat shoal	42.0
Merambong shoal	30.0
Total estimated area in Peninsular Malaysia	**315.5**

> **Case Study 14.3**
> **COASTAL LAGOON SEAGRASS COMMUNITY AT PENGKALAN NANGKA, KELANTAN**
>
> The intertidal area and two shoals in the lagoon all harbor a mixed *Halodule pinifolia* and *Halophila beccarii* community. *Halodule pinifolia* grows in pure and extensive subtidal meadows on soft muddy substrates at depths of 1.6 to 2.0 m. *Halophila beccarii* grows in shallower parts, at depths of 0.9 to 1.5 m in monospecific and very dense meadows on sandy substrates. The two species are able to withstand a wide fluctuation of salinity from 0 to 18 psu. The meadow is a site for the collection of bivalves (e.g. *Hiatula solida* and *Geloina coaxans*) and artisanal fishing. Digging for bivalves has caused a lot of damage to the meadow (see photograph). Since 1991 the lagoon has also been used for fish cage farming of *Lates calcarifer* and *Epinephelus sexfasciatus*. Seasonally, from June to July, the migrant wader, *Egretta garzetta*, used the shoals as a feeding ground on its migrations until two shoals were completely destroyed by sand dredging in early 1999.
>
>
> *Halophila beccarii* meadow is a harvesting site for *Hiatula solida* and *Geloina coaxans*. Digging has caused damage to the bed.

Caulerpa spp. and fauna such as sea cucumbers, gastropods and bivalves, and illegal fishing with explosives are among the major causes of damage to coral reefs and associated seagrasses. Such activities not only cause loss of flora and fauna but also create an imbalance within the ecosystem from which seagrass beds are unlikely to recover quickly.

POLICY RESPONSES
In the earlier part of this chapter, it was mentioned that seagrass beds are the least protected of the three main marine ecosystems in Malaysia. It is strongly recommended that seagrass beds, especially those around offshore islands that have been gazetted as marine parks (Pulau Redang; Pulau Perhentian, Terengganu; Pulau Tioman, Phang; Pulau Tengah; Pulau Besar; Pulau Sibu; Pulau Tinggi, Johore) be given protection as marine parks or reserves under the Fisheries Act 1985. Under Part IX, Section 41(1) and (2) of the Fisheries Act 1985 the Minister of Agriculture may order in the Gazette the establishment of any area or part of an area in Malaysian fisheries waters as a marine park or marine reserve in order to:

"(a) afford special protection to the aquatic flora and fauna of such area or part thereof and to protect, preserve and manage the natural breeding grounds and habitat of aquatic life, with particular regard to species of rare or endangered flora and fauna;
(b) allow for the natural regeneration of aquatic life in such area or part thereof where such life has been depleted;
(c) promote scientific study and research in respect of such area or part thereof;
(d) preserve and enhance the pristine state and productivity of such area or part thereof; and
(e) regulate recreational and other activities in such area or part thereof to avoid irreversible damage to its environment."

Furthermore:

"(2) The limits of any area or part of an area established as a marine park or marine reserve under subsection (1) may be altered by the Minister by order in the Gazette and such order may also provide for the area or part of the area to cease to be a marine park or marine reserve."

The question of affording comprehensive protection to marine ecosystems gazetted under the present Fisheries Act 1985 has been the subject of intense scrutiny by marine scientists, government officials and conservationists. The bone of contention has been the separation of the land on islands gazetted as marine parks and reserves from the waters surrounding the islands. Under these circumstances, while the authorities vested with the powers to manage and enforce the marine park laws can do so at sea, they have no jurisdiction whatsoever over what happens on land.

This could be resolved based on practices adopted by Sabah Parks and the present trend of promulgating state parks enactment for the protection of ecosystems. At present, Sabah Parks has under its auspices three marine protected areas: Tunku Abdul Rahman Marine Parks, Pulau Tiga Parks and Turtle Islands Parks. All harbor seagrasses and were gazetted as state parks under the State Parks Enactment 1984. Marine areas gazetted as state parks in Sabah are afforded more comprehensive protection under the enactment than marine parks or reserves in Peninsular Malaysia. These parks are protected in their entirety without separating the marine and terrestrial components.

Several states in Peninsular Malaysia have promulgated enactments for the gazettement of state parks. Johore has gazetted the National Parks (Johore) Corporation Enactment 1991. Terengganu has a Terengganu State Parks Enactment.

Can the above policies be applied for the management of marine protected areas in Peninsular Malaysia? The answer lies in encouraging concurrent gazettement of marine protected areas under both federal and state legislation using the Fisheries Act 1985 to gazette the protection of the waters surrounding the islands as marine parks or reserves, and state park enactments to gazette the terrestrial component of the marine protected areas as state parks.

Boat unable to move because of thick *Enhalus acoroides* bed.

ACKNOWLEDGMENTS

Support from the United Kingdom Department for International Development (DfID) is gratefully acknowledged.

AUTHORS

Japar Sidik Bujang, Department of Biology, Faculty of Science and Environmental Studies, Universiti Putra Malaysia, 43400 UPM, Serdang, Selangor Darul Ehsan, Malaysia. **Tel:** +603 (0)8946 6626. **Fax:** +603 (0)8656 7454. **E-mail:** japar@fsas.upm.edu.my

Muta Harah Zakaria, Faculty of Agricultural Sciences and Food, Universiti Putra Malaysia Bintulu Campus, Jalan Nyabau, P.O. Box 396, 97008 Bintulu, Sarawak, Malaysia.

REFERENCES

1. Ridley HN [1924]. *The Flora of the Malay Peninsula, Vol. IV, Monocotyledons*. A. Asher & Co., Amsterdam and L. Reeve and Co., Brook, near Ashford, UK.
2. Beccari O [1904]. *Wanderings in the Great Forests of Borneo*. Oxford University Press, Oxford and New York.
3. Burkill IH [1935]. *A Dictionary of the Economic Products of the Malay Peninsula*. Ministry of Agriculture and Co-operatives, Kuala Lumpur.
4. Henderson MR [1954]. *Malayan Wild Flowers: Monocotyledon*. The Malayan Nature Society, Kuala Lumpur.
5. Holttum RE [1954]. *Plant Life in Malaya*. Longman, London.
6. Tweedie MWF, Harrison JI [1954]. *Malayan Animal Life*. Longman, London.
7. Sinclair J [1956]. Additions to the flora of Singapore II. *Gardens Bulletin Singapore* 15: 22-30.
8. Hsuan Keng [1969]. *Orders and Families of Malayan Seed Plants. Synopsis of orders and families of Malayan gymnosperms, dicotyledons and monocotyledons*. Singapore University Press, Singapore.
9. den Hartog C [1964]. An approach to the taxonomy of the sea-grass genus *Halodule* Endll. (Potamogetonaceae). *Blumea* 12: 19-312.
10. den Hartog C [1970]. *Seagrasses of the World*. North-Holland Publishing, Amsterdam.
11. Phang SM [1989]. Seagrasses – a neglected natural resource in Malaysia. In: Phang SM, Sasekumar A, Vickineswary S (eds) *Proceedings 12th Annual Seminar of the Malaysian Society of Marine Sciences. Research Priorities for Marine Sciences in the 90's*. Malaysian Society of Marine Science and Institute for Advanced Studies, Universiti Malaya, Kuala Lumpur. pp 269-278.
12. Sasekumar A, Charles Leh MU, Chong VC, D'Cruz R, Audery MI [1989]. The Sungai Pulai (Johore): A unique mangrove estuary. In: Phang SM, Sasekumar A, Vickineswary S (eds) *Proceedings 12th Annual Seminar of the Malaysian Society of Marine Sciences. Research Priorities for Marine Sciences in the 90's*. Malaysian Society of Marine Science and Institute for Advanced Studies, Universiti Malaya, Kuala Lumpur. pp 191-211.
13. Mohd Rajuddin MK [1992]. The Areas and Species Distribution of Seagrasses in Peninsular Malaysia. Paper presented at the First National Symposium on Natural Resources, 23-26 July 1992, FSSA, UKM, Kota Kinabalu, Sabah.
14. Ismail N [1993]. Preliminary study of seagrass flora of Sabah, Malaysia. *Pertanika Journal of Tropical Agricultural Science* 16(2): 111-118.

15. Japar Sidik B [1994]. Status of seagrass resources in Malaysia. In: Wilkinson CR, Sudara S, Chou LM (eds) *Proceedings Third ASEAN Australia Symposium on Living Coastal Resources, Status Reviews 1.* Chulalongkorn University, Bangkok. pp 283-290.
16. Japar Sidik B, Arshad A, Hishamuddin O, Muta Harah Z, Misni S [1996]. Seagrass and macroalgal communities of Sungai Pulai estuary, south-west Johore, Peninsular Malaysia. In: Kuo J, Walker DI, Kirkman H (eds) *Seagrass Biology: Scientific discussion from an international workshop, Rottnest Island, Western Australia.* University of Western Australia, Nedlands, Western Australia. pp 3-12.
17. Japar Sidik B, Lamri A, Muta Harah Z, Muhamad Saini S, Mansoruddin A, Josephine G, Fazrullah Rizally AR [1997]. *Halophila decipiens* (Hydrocharitaceae), a new seagrass for Sabah. *Sandakania* 9: 67-75.
18. Japar Sidik B, Muta Harah Z, Lamri A, Francis L, Josephine G, Fazrullah Rizally AR [1999]. *Halophila spinulosa* (R. Br.) Aschers.: An unreported seagrass in Sabah, Malaysia. *Sabah Parks Nature Journal* 2: 1-9.
19. Japar Sidik B, Muta Harah Z, Mohd Pauzi A, Suleika M [1999]. *Halodule* species from Malaysia – distribution and morphological variation. *Aquatic Botany* 65: 33-46.
20. Japar Sidik B, Muta Harah Z, Fazrullah Rizally AR, Kamin B [2000]. New observations on *Halophila spinulosa* (R. Br.) Aschers. in Neumayer, Malaysia. *Biologia Marina Mediterranea* 7(2): 75-78.
21. Japar Sidik B, Muta Harah Z, Misri K, Hishamuddin O [2000]. The seagrass, *Halophila decipiens* (Hydrocharitaceae) from Teluk Kemang, Negeri Sembilan: An update. In: Shariff M, Yusoff FM, Gopinath N, Ibrahim HM, Nik Mustapha RA (eds) *Towards Sustainable Management of the Straits of Malacca.* Malacca Straits Research and Development Centre (MASDEC), Universiti Putra Malaysia, Serdang. pp 233-238.
22. Japar Sidik B, Muta Harah Z, Kanamoto Z, Mohd Pauzi A [2001]. Seagrass communities of the Straits of Malacca. In: Japar Sidik B, Arshad A, Tan SG, Daud SK, Jambari HA, Sugiyama S (eds) *Aquatic Resource and Environmental Studies of the Straits of Malacca: Current Research and Reviews.* Malacca Straits Research and Development Centre (MASDEC), Universiti Putra Malaysia, Serdang. pp 81-98.
23. Mashoor M, Yassin Z, Abdul Razak L, Ganesan M [1994]. Abstract: The Distribution of Phytobenthos along the Coastal Water of Pulau Besar, Johore, Malaysia. Presented at The International Conference on Applied Ecology and Biology of Benthic Organisms. The Centre for Extension and Continuing Education, Universiti Pertanian Malaysia.
24. Anisah A, Zulfigar Y [1998]. Seagrass of Pulai Sipadan, Sabah, Malaysia: A preliminary survey. *Malayan Nature Journal* 52(3, 4): 223-235.
25. Muta Harah Z, Japar Sidik B, Hishamuddin O [1999]. Flowering, fruiting and seeding of *Halophila beccarii* Aschers. (Hydrocharitaceae) from Malaysia. *Aquatic Botany* 65: 199-207.
26. Muta Harah Z, Japar Sidik B, Law AT, Hishamuddin O [2000]. Seeding of *Halophila beccarii* Aschers. in Peninsular Malaysia. *Biologia Marina Mediterranea* 7(2): 99-102.
27. Mohd Rajuddin MK [1992]. Species composition and size of fish in seagrass communities of Peninsular Malaysia. In: Chou LM, Wilkinson CR (eds) *Third ASEAN Science and Technology Week Conference Proceedings, Vol. 6, Marine Science: Living Coastal Resources.* Department of Zoology, National University of Singapore and National Science and Technology Board, Singapore. pp 309-313.
28. Japar Sidik B, Arshad A, Law AT [1995]. Inventory for seagrass beds in Malaysia. In: Japar Sidik B (ed) *UNEP: EAS-35: Malaysian Inventory of Coastal Watersheds, Coastal Wetlands, Seagrasses and Coral Reefs.* Department of Environment, Ministry of Science, Technology and Environment, Kuala Lumpur. pp 48-79.
29. Arshad A, Siti Sarah MY, Japar Sidik B [1994]. A comparative qualitative survey on the invertebrate fauna in seagrass and non seagrass beds in Merambong shoal Johore. In: Sudara S, Wilkinson CR, Chou LM (eds) *Proceedings Third ASEAN Australia Symposium on Living Coastal Resources, Chulalongkorn University, Bangkok.* Department of Marine Science, Chulalongkorn University. *Bangkok Research Papers* 2: 337-348.
30. Arshad A, Japar Sidik B, Muta Harah Z [2001]. Fishes associated with seagrass habitat. In: Japar Sidik B, Arshad A, Tan SG, Daud SK, Jambari HA, Sugiyama S (eds) *Aquatic Resource and Environmental Studies of the Straits of Malacca: Current Research and Reviews.* Malacca Straits Research and Development Centre (MASDEC), Universiti Putra Malaysia, Serdang. pp 151-162.
31. National Mangrove Committee [1989]. Guidelines on the Use of the Mangrove Ecosystem for Brackish Water Aquaculture in Malaysia. The Working Group to the Malaysian National Mangrove Committee and the National Council for Scientific Research and Development, Ministry of Science, Technology and the Environment, Kuala Lumpur.
32. Phang SM [2000]. *Seagrasses of Malaysia.* Universiti Malaya, Botanical Monographs No. 2. Institute of Biological Sciences, Universiti Malaya, Kuala Lumpur.
33. Audery ML, Sasekumar A, Charles Leh MU, Chong VC, Rebecca D [1989]. A Survey of the Living Aquatic Resources of the Sungai Pulai Estuary, Johore.
34. Zelina ZI, Arshad A, Lee SC, Japar Sidik B, Law AT, Nik Mustapha RA, Maged Mahmoud M [2000]. East coast of Peninsular Malaysia. In: Sheppard C (ed) *Seas at the Millennium.* Vol. 2. Pergamon, Amsterdam. pp 345-359.
35. Josephine G [1997]. A Study on Seagrass Biodiversity, Distribution and Biomass in Tunku Abdul Rahman Park, Sabah, Malaysia. Final year thesis, Universiti Kolej Terengganu, Universiti Putra Malaysia.
36. Lim LH, Hishamuddin O, Japar Sidik B, Noro T [2001]. Seaweed community in degraded coral reef area at Teluk Kemang, Port Dickson, Negeri Sembilan. In: Japar Sidik B, Arshad A, Tan SG, Daud SK, Jambari HA, Sugiyama S (eds) *Aquatic Resource and Environmental Studies of the Straits of Malacca: Current Research and Reviews.* Malacca Straits Research and Development Centre (MASDEC), Universiti Putra Malaysia, Serdang. pp 99-110.
37. Lee CN [1998]. Seagrass Communities in the Coastal Waters of Port Dickson. BSc thesis, Department of Biology, Universiti Putra Malaysia, Serdang.
38. Japar Sidik B, Arshad A, Hishamuddin O, Shamsul Bahar A [1995]. *Halophila decipiens* Ostenfeld (Hydrocharitaceae): A new record of seagrass for Malaysia. *Aquatic Botany* 52: 151-154.
39. Muta Harah Z, Japar Sidik B, Fazrullah Rizally AR, Mansoruddin A, Sujak S [2002, in press]. Occurrence and systematic account of seagrasses from Pulau Redang, Terengganu, Malaysia. In: *Proceedings National Symposium on Marine Park and Islands of Terengganu: Towards Sustainable Usage and Management of Islands, Grand Continental Hotel, Kuala Terengganu.* Department of Fisheries Malaysia, Ministry of Agriculture Malaysia, Kuala Lumpur.

15 The seagrasses of
THE WESTERN PACIFIC ISLANDS

R. Coles
L. McKenzie
S. Campbell
M. Fortes
F. Short

The western Pacific island region includes the countries and island states of Micronesia, Melanesia and Polynesia. These countries are located in the tropical Pacific Ocean; almost all the islands are in a zone spanning the equator from the Tropic of Cancer in the north to the Tropic of Capricorn in the south.

Most islands, with the exception of Papua New Guinea and Fiji, are small by continental standards and are separated by expanses of deep ocean waters. It is no easy task estimating even the number of islands in the western Pacific region. For example there are in excess of 2000 islands in Micronesia alone, some of which may not be permanent and can be swamped by high tides. There are two main types of islands – the high islands such as Fiji, Papua New Guinea and most of the Solomon Islands, and the low islands and coral reef atolls such as Majuro and Kiribati. In the Pacific, as in the rest of the world, most of the cities and towns are located in the coastal region. Only in Papua New Guinea are there large towns located away from the coast. There has been a marked change away from mostly subsistence living. As a consequence Pacific islanders are no longer totally rural, and urban growth is outstripping total growth. Human populations are increasing throughout the region and can be as high as 23 000 people per km^2 (e.g. Marshall Islands)[1].

Most countries in the Pacific list human waste disposal as a significant issue and this is likely to affect seagrass meadows. Only larger towns have sewage systems, but most of the effluent discharges into the sea[1]. Along with septic systems and village latrines, the eventual nutrient loads of sewage systems to inshore and reef platform seagrasses may be significant. Custom ownership of land (inherited ownership of land and nearshore regions by indigenous villages or families) gives the owners the right to do as they wish with the land even if that leads to environmental damage. While these issues are recognized and are being addressed by planning legislation, enforcement is difficult or impossible in many of the islands. This dilemma of land tenure may be an obstacle to the environment planning needed to ensure a sustainable habitat for seagrass.

There are 24 species of seagrasses, including *Ruppia*[2], found throughout the tropical Indo-Pacific[3]. Our best estimate is that 13 of these are found in the western Pacific islands. These include the genera of *Cymodocea*, *Enhalus*, *Halodule*, *Halophila*, *Syringodium*, *Thalassia* and *Thalassodendron*. It is possible that new species remain to be described from the western Pacific, as collections from this region are relatively few. Seagrass species distribution across the western Pacific is believed to be influenced by the equatorial counter-current in the northern hemisphere and the equatorial current in the southern hemisphere[4], with the number of species declining with easterly distance. The reduced bottom area available and the effect of past changes in sea level would also reduce species numbers along an easterly gradient[5]. The numbers are greatest near the biggest land mass, with 13 species in Papua New Guinea[6], and least in the easternmost islands; only one species is known from Tahiti[7].

Seagrasses across the region are also often closely linked, with complex interactions, to mangrove communities and coral reef systems. Dense seagrass communities of *Enhalus* and *Cymodocea* are often present on the intertidal banks adjacent to mangroves and fringing reefs.

BIOGEOGRAPHY
Western Pacific seagrass communities grow on fringing reefs, in protected bays and on the protected side of barrier reefs and islands. Habitats most suited to tropical seagrasses are reef platforms and lagoons with mainly fine sand or muddy sediments enclosed

by outer coral reefs. These habitats are influenced by pulses of sediment-laden, nutrient-rich freshwater, resulting from seasonally high summer rainfall. Cyclones and severe storms or wind waves also influence seagrass distribution to varying degrees. On reef platforms and in lagoons the presence of water pooling at low tide prevents drying out and enables seagrass to survive tropical summer temperatures.

Enhalus acoroides is the only species that releases pollen to the surface of the water when reproducing sexually. This feature restricts its distribution to intertidal and shallow subtidal areas. It is a slow-growing, persistent species with a poor resistance to perturbation[8], suggesting that areas where it is found are quite stable over time. *Cymodocea* is an intermediate genus that can survive a moderate level of disturbance, while *Halophila* and *Halodule* are described as being ephemeral genera with rapid turnover and high seed set, well adapted to high levels of disturbance[8].

Thalassia hemprichii is the dominant seagrass found throughout Micronesia and Melanesia, although it is absent from Polynesia and Fiji. *Thalassia hemprichii* is often associated with coral reefs and is common on reef platforms where it may form dense meadows. It is able to grow on hard coral substrates with little sediment cover. It can also be found colonizing muddy substrates, particularly where water pools at low tide. In the Indo-Pacific region, *Thalassia hemprichii* is commonly the climax seagrass species. Species of *Halodule*, *Cymodocea* and *Syringodium* may at times also be found in dense meadows associated with reefs and on reef platforms. *Enhalus acoroides* and *Cymodocea rotundata* are also widespread throughout the region but absent from Polynesia and Fiji. *Halodule uninervis* is abundant throughout Melanesia and Polynesia, but is only found in Guam and Palau in Micronesia. Both *Cymodocea serrulata* and *Cymodocea rotundata* were recorded in intertidal regions of Micronesia[9] and in Papua New Guinea, eastern Micronesia and Vanuatu[10].

Syringodium isoetifolium has only been recorded in the most westerly islands of Micronesia (e.g. Palau and Yap), in Tonga and Samoa in Polynesia, in Papua New Guinea, and in Vanuatu and Fiji in Melanesia. In Fiji *Syringodium isoetifolium* occurs as a widespread and dominant seagrass species.

Halophila species are widespread through the Pacific islands with the exception of the eastern Micronesia islands. In the western islands of the western Pacific, *Halophila ovalis* is found in intertidal habitat mixed with larger seagrass species like *Enhalus acoroides* in Palau or *Thalassia hemprichii* in Yap. *Halophila ovalis* is also commonly found in deep water at the offshore edge of mixed seagrass meadows. The only *Halophila* species present in Fiji is the subspecies *Halophila ovalis bullosa* identified by den Hartog[7].

Thalassodendron ciliatum has been recorded from Palau, Papua New Guinea and Vanuatu[10]. It is unusual in being restricted almost exclusively to rocky or reef substrates. It is often found on reef edges exposed to wave action, protected from damage by its flexible woody stem and strong root system. It can be difficult to locate because of its exposed reef edge habit and is uncommon in records from most Pacific island countries.

Generally low nutrient availability[11] is a likely determining factor in seagrass extent on reef habitats across the western Pacific islands. Seagrasses frequently grow more abundantly on intertidal reef platforms and mud flats adjacent to populated areas where they can utilize the available nutrients. Seagrass communities in the western Pacific islands must tolerate fluctuating and extreme temperatures, fluctuating salinity during rainfall seasons, and exposure to storm-driven waves and erosion. Often the sediments are unstable and their depth on the reef platforms can be very shallow, restricting seagrass growth and distribution.

Most tropical species in the western Pacific are found in waters less than 10 m deep. There is a complex depth range for seagrasses as the availability of bottom substrate and shelter for seagrass growth is controlled by the topography of coral reef communities which often protect the seagrass habitats from wave action. The location of the seaward edge may be determined by the depth or location at which coral cover becomes consistent or by the edge of a platform that drops rapidly into deeper water. This distribution and the topographic features controlling it differ from many temperate regions where availability of light for photosynthesis controls the depth penetration of seagrasses.

Exposure at low tide, wave action and low salinity from freshwater inflow determine seagrass species survival at the shallow edge. Seagrasses survive in the intertidal zone especially at sites sheltered from wave action or where there is entrapment of water at low tide (e.g. reef platforms and tide pools) protecting the seagrasses from exposure (to excessive heat or drying) at low tide. At the deeper edge, light, wave action and the availability of suitable bottom substrate limit distribution.

The stresses and limitations to seagrasses in the tropics are generally different from those in temperate or subarctic regions. They include thermal impacts from high water temperatures; desiccation from overexposure to warm air; osmotic impacts from hypersalinity due to evaporation or hyposalinity from wet season rain; radiation impacts from high irradiance and UV exposure. Both *Halophila ovalis* and *Thalassia*

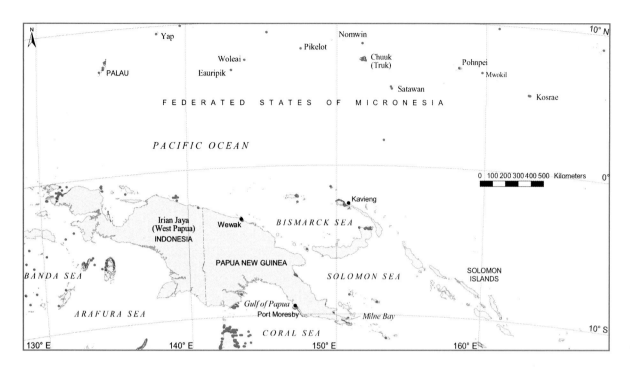

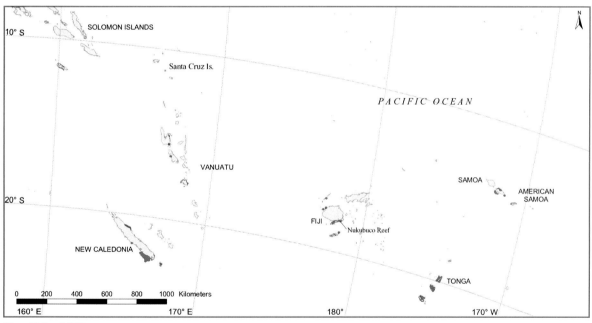

Maps 15.1 and 15.2
Western Pacific islands (west) (top) and Western Pacific islands (east)

hemprichii were found in intertidal regions in Yap, Micronesia[12], where tolerance to 40°C temperatures and low salinity allow these species to colonize. Other species present in Yap, *Syringodium isoetifolium* and *Cymodocea serrulata*, were restricted to deeper water by these conditions.

Reef platform seagrass meadows support a wide range of mollusks, fish, holothurians and decapods. The available literature does not focus on the ecological role of seagrasses and information on complex community interactions presented for reef flat species may not necessarily refer to areas with seagrass. Munro[13] lists 75 species of mollusks collected by subsistence gleaners in the Solomon Islands, Papua New Guinea and Fiji from mangroves, reefs, seagrass meadows and sand flats. Other mollusks such as the trochus shell (*Trochus niloticus*) found in seagrass meadows are collected as a source of cash income. Similarly the holothurians have been a valuable source of cash income, although now heavily overfished[14]. We

have found lower value species such as *Holothuria atra* to be still common in seagrass meadows in parts of Micronesia.

Pyle[15] lists at least 3 392 reef and shore fish from the Pacific islands but it is not possible to distinguish which species are from seagrass meadows. Klumpp *et al.*[16] refer to 154 species of tropical invertebrates and fish that feed directly on seagrasses and Coles *et al.*[17] list and classify 134 taxa of fish and 20 shrimp species found in tropical Australian seagrass meadows giving some indication of the likely use of tropical Pacific seagrass meadows.

Seagrasses are also food for the green turtle (*Chelonia mydas*), found throughout the Pacific island region, and for the dugong (*Dugong dugon*), found in small numbers feeding on seagrasses in the western islands – Palau, Vanuatu and the Solomon Islands.

HISTORICAL PERSPECTIVES

The major changes in Pacific island seagrass meadows have occurred mostly in the post-Second World War period and are related to transport infrastructure, tourist development and population growth. Some islands have seagrass maps available but most do not

Case Study 15.1
KOSRAE

The Federated States of Micronesia is made up of four states: Kosrae, Pohnpei, Chuuk and Yap. Kosrae is the easternmost state and consists of two islands: a large mountainous island approximately 20 km long and 12 km wide, and a smaller 70 ha island, Lelu, approximately 1 km off the northeast coast of Kosrae.

A detailed assessment of Kosrae reef environments in 1989 (carried out by the US Army Corps of Engineers, Coastal Engineering Research Center) mapped approximately 3.5 km² of seagrass meadows around the islands. Seagrass meadows were restricted to reef tops. Large dense meadows were mapped adjacent to Okat and Lelu Harbours.

Species of seagrass found were *Enhalus acoroides*, *Thalassia hemprichii* and *Cymodocea rotundata*.

Over the last three to four decades there has been considerable coastal construction activity on the islands to build modern transportation facilities, and the seagrass meadows and reef flats at those locations have been severely impacted. Two aircraft runways and associated causeways have been constructed on the only available flat area on the island – the reef flat.

The first runway was constructed on the shallow flat between Kosrae and Lelu Islands in the late 1960s and early 1970s. Maragos[19] reported that the causeway connecting to this runway construction had adverse effects on Lelu Harbour. The original causeway blocks the water circulation and fish runs into inner Lelu Harbour,

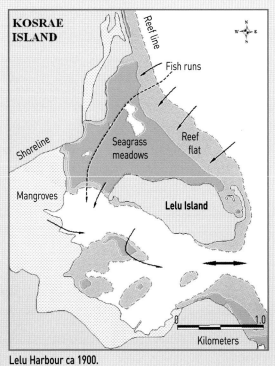

Lelu Harbour ca 1900.

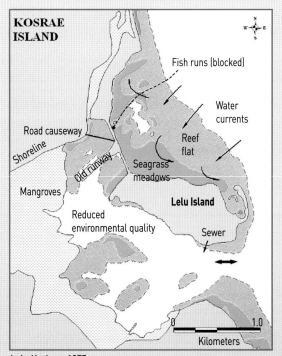

Lelu Harbour 1975.

have information recorded with the precision required to identify any historical change. It is likely that some information exists in unpublished reports and environmental assessments for areas subject to development but, where it exists, this information is not readily available.

Human population growth and the need to provide tourist accommodation have led to filling in some coastal areas to provide new land. Certainly port developments and small boat marinas have been constructed in locations without taking the presence of seagrass meadows into account[18]. Nutrient inputs from expanding coastal urban development may have increased the biomass of seagrass on nearby reef platforms. In general, though, there is not sufficient historical written information from which to draw direct conclusions on historic trends. Munro[13] does report that 2000-year-old mollusk shell middens in Papua New Guinea have essentially the same species composition as present-day harvests, suggesting indirectly that the habitats, including seagrass habitats and their faunal communities, are stable and any changes occurring are either short term or the result of localized impacts.

leading to a decline in seagrasses and fish catches and increased pollution problems. Fill for the runway expansion further reduced water circulation, fish yields, water quality and seagrasses in the harbor.

In the mid-1980s, a new airport and dock were constructed in Okat Harbour on the north of Kosrae Island. Construction buried a large area of the offshore reef flat seagrass meadows (see sketch maps below). Also, during dredging activities, the rate of slurry discharged into a retention basin exceeded the basin's capacity, causing the slurry to overflow and burying an adjacent 10 ha of seagrass and coral habitat under 0.25-0.5 m of fine mud.

The construction also changed the water circulation, and the strong currents caused shoreline erosion. These impacts are reported to have reduced Okat reef's fish harvest to half that of pre-construction levels.

The unintended environmental effects of these constructions are continuing with shore erosion and restoration by revetment still occurring at Lelu Harbour and adjacent to villages near the new airport. While it is easy to criticize a decision to build infrastructure on top of coral reef platforms, it is hard to suggest a feasible land-based solution on such a mountainous island. Flat areas available are either inhabited or mangrove covered. It would be hoped that if these projects or similar were undertaken today, better environment management systems in place would at least reduce the unintended effects and slurry overflow that occurred.

Source: Maragos[19].

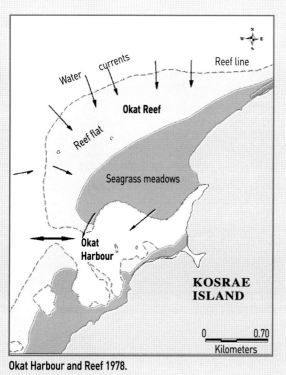
Okat Harbour and Reef 1978.

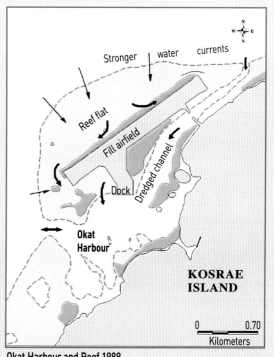
Okat Harbour and Reef 1988.

Banded sea snake swimming over *Syringodium isoetifolium* and *Halodule uninervis* meadow, Nukubuco Reef, Fiji.

AN ESTIMATE OF HISTORICAL LOSSES

In the western Pacific, local coastal developments for tourism or transport infrastructure are the major cause of seagrass loss. In Kosrae and other members of the Federated States of Micronesia the development of local airports has contributed to a loss of seagrass on reef platforms. The Kosrae airport, for instance, is placed on landfill covering a reef platform and seagrasses[19]. In Palau, the building of causeways without sufficient consideration of the need for culverts to maintain water flow has caused localized seagrass loss[18]. In Fiji, eutrophication and coastal development are the primary causes of seagrass loss. Little information is available on the loss of seagrass habitats in Papua New Guinea, but away from major population centers losses are likely to be small and again associated with transport infrastructures.

Maragos[19] details the loss of mainly coral reef flat habitat, but including seagrasses and mangroves, in the Federated States of Micronesia from construction activities associated with plantations, transportation, military activity, urban development, aquaculture development and resort development. Coastal road construction around the islands of Pohnpei and Kosrae resulted in the dredging of many hectares of seagrass and mangrove habitat.

Losses of seagrasses such as these are likely to be widespread across the Pacific islands as there has been little attention paid to protecting seagrasses. Modern mapping and monitoring techniques should in the near future enable some baseline estimation of the total areas of the seagrass resources of the region.

AN ESTIMATE OF PRESENT COVERAGE

Species lists are available for the western Pacific region[3] but they are not available for many of the individual islands. Coles and Kuo[10] list seagrass species from 26 islands (including the Hawaiian Islands and Papua New Guinea) based on published records, examination of herbarium specimens and/or site visits by the authors. Species numbers ranged from 11 on Vanuatu to a single species in the Marshall Islands. The numbers in Coles and Kuo[10] are conservative in some cases because they do not include unpublished reports or records. Maps of seagrass are not readily available or are of relatively poor quality and/or reliability. Some estimation might be possible based on the high likelihood of almost all shallow (<2 m below mean sea level) reef flats having at least a sparse seagrass cover, but no numerical estimation of seagrass cover in the western Pacific has been made to date.

Geographic information system (GIS) initiatives in the Federated States of Micronesia by the South Pacific Regional Environment Program should improve map coverage. Simple GIS maps are already available for Kosrae although they are based on earlier aerial mapping and would not be precise enough for detailed management purposes. Project assistance to update and validate these maps would accelerate the process of providing a publicly available set of maps for these islands. Partial maps are available for other western Pacific islands although their validity is uncertain and likely to be variable.

CSIRO (Australia's Commonwealth Scientific and Industrial Research Organisation) has recently surveyed Milne Bay Province in Papua New Guinea. Seagrass was seen at 103 locations out of a total of 1 126. Seagrass was found at several areas throughout the province, mostly on shallow areas adjacent to the larger islands such as the Trobriand, Woodlark and Sudest Islands. Cover was up to 95 percent in these areas. The dominant species were *Thalassia hemprichii*, *Enhalus acoroides* and *Halophila ovalis* with some *Cymodocea serrulata*, *Halodule uninervis* and *Syringodium isoetifolium*[20]. To the best of our knowledge no other broadscale surveys have been conducted for Papua New Guinea outside individual published site descriptions.

USES AND THREATS

Traditional uses of seagrass by communities in the western Pacific include manufacture of baskets; burning for salt, soda or warmth; bedding; roof thatch; upholstery and packing material; fertilizer; insulation for sound and temperature; fiber substitutes; piles to build dikes; and for cigars and children's toys[21]. *Enhalus acoroides* fiber is also reported to be used on Yap, Micronesia, in the construction of nets[22]. *Enhalus acoroides* fruit is

Regional map: The Pacific

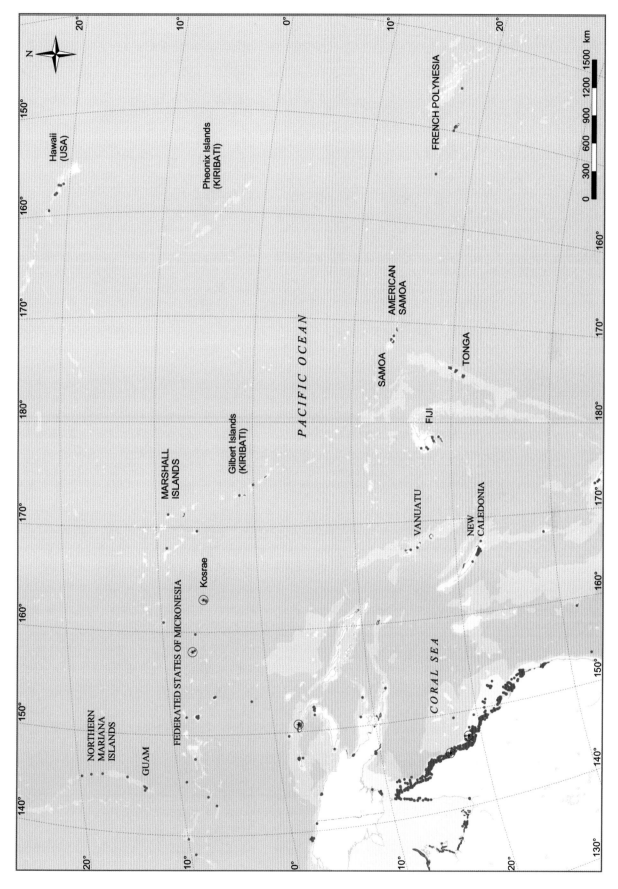

SEAGRASSES AND PEOPLE

Recreational fisher standing in a seagrass bed, Bali, Indonesia.

Food (bivalves) collected from seagrass beds, Mozambique.

Harvesting abalone from *Phyllospadix* spp. ca 1914, Pacific coast of USA.

Harvesting *Zostera marina* for transplanting, Maine, USA.

Beach seine is pulled over a seagrass meadow in the Philippines to catch sardines or anchovies that pass by.

Snorkeler over a long-bladed *Zostera marina* bed in New Hampshire, USA.

Trap fisherman Anibal Amade in Montepuez bay, Mozambique, with a *marema* fish trap and holding a seagrass parrotfish *Leptoscarus vaigiensis*.

A fisher family on Quirimba Island, Mozambique with the catch from a trap fishing trip in the seagrass beds.

eaten in some Australian traditional communities and in some parts of the western Pacific.

Coastal development, dredging and marina developments are generic threats to seagrass in the tropical tourist regions but areas lost are generally small. Causeway development in Palau[18] without culverts to allow water flow has led to large seagrass losses as water stagnates and sediment builds up. Coastal agriculture may add to sediment loads in catchments in Papua New Guinea and Fiji. Shipping management influences seagrass survival adjacent to shipping lanes and port locations.

Climate change and associated increase in storm activity, water temperature and/or sea-level rise have the potential to damage seagrasses in the region and to influence their distribution causing widespread loss. Reef platform seagrasses are already exposed to water temperatures at low tide greater than 40°C and an increase in temperature may restrict the growth of the inner shallow edge of reef platform seagrass. Sea-level rise and associated increased storm activity could lead to large seagrass losses through increased water movement over seagrass beds and erosion of sediments. It is possible that with a rise in sea level areas that are now seagrass habitat may be colonized by coral.

SEAGRASS PROTECTION

Many western Pacific island communities have complex and at times unwritten approaches to land ownership, custom rights and coastal sea rights. These are partially overlaid by arrangements put in place by colonizing powers during and after the Second World War, leaving the nature and strength of protective arrangements open for debate[18]. In implementing any protective arrangements for seagrasses the challenge will be to develop an approach that will suit all parties and that will respect traditional ownership rights. This must also be achieved in an area where enforcement, at least in the sense it is used in North America and Europe, is absent or ineffective and more of a consensus approach will be required.

We are not able to find any legislation or protective reserve systems that are specifically designed to protect seagrasses. Existing reserves, however, often include seagrasses and legislation to protect mangroves or marine animals such as trochus shell may indirectly protect seagrass meadows.

The South Pacific Regional Environment Programme Action Strategy for Nature Conservation in the Pacific Islands region lists 232 established protected areas and community-based conservation areas in the Pacific islands. Some, such as the Okat trochus sanctuary in Kosrae and the Ngerukewid Islands reserve in Palau, provide some level of indirect protection.

Under the Law of the Sea Treaty, coastal nations are bound to protect the marine environments under their control. There are some 13 other international conventions and treaties which could have some bearing on seagrass management although in reality it is hard to measure any quantifiable outcomes that protect seagrasses whether the programs are ratified or not.

At a regional level, laws relating to impact assessment and town planning have an indirect ability to protect seagrass from loss. In Fiji the Town Planning Act deals with environmental impact assessments.

Halophila ovalis, a species commonly found in the western Pacific.

Land below high water is administered by the Ministry for Land and Mineral Resources through the Department of Lands and Surveys. If a mangrove area (and presumably also seagrasses) is be reclaimed, the application is referred to the Department of Town and Country Planning for comment, recommendation and suggested conditions. It may also be referred to the Department of Fisheries and the Native Fisheries Commission for arbitration of compensation[1].

In Papua New Guinea the Environmental Planning Act requires a plan for a development project to be submitted to the Department of Environment and Conservation for approval[1]. Palau's conservation laws are cited in the Palau National Code Annotated and are described by the Palau Conservation Society[23] in an easy-to-understand form.

Two trends are emerging from the Pacific islands. One is the recognition of the need for sanctuaries and protected areas and the other the concept of traditional or community management of these areas[24]. The role being played by non-governmental organizations, focused on conservation and environment protection integrated with traditional leadership and government

Case Study 15.2
SEAGRASSNET – A WESTERN PACIFIC PILOT STUDY

SeagrassNet is a global monitoring program that investigates and documents the status of seagrass resources worldwide and the threats to this important ecosystem. Seagrasses, which grow at the interface of the land margin and the world's oceans, are threatened by numerous anthropogenic impacts. There is a lack of information on the status and health of seagrasses, particularly in the less economically developed countries. SeagrassNet's efforts to monitor known seagrass areas and to map and record uncharted seagrasses in the western Pacific are important first steps in understanding and maintaining seagrass resources worldwide. Synchronous and repeated global sampling of selected environment and plant parameters is critical to comprehending seagrass status and trends; monitoring these ecosystems will reveal both human impacts and natural fluctuations in coastal environments throughout the world.

SeagrassNet was developed with two components. Research-oriented monitoring methods are based on recently compiled seagrass research techniques for global application[2], while community-based seagrass monitoring effort is modeled after Seagrass-Watch[25] – an Australian seagrass community (citizen) monitoring program that is conducted simultaneously with research-based monitoring so that comparisons of the resulting data are possible.

An important part of the communication strategy for SeagrassNet is an interactive system established on a website, with data entry, archiving, display and retrieval of seagrass habitat-monitoring data, ranging from plant species distribution to animal abundance and records of localized die-offs. SeagrassNet both acquires and provides monitoring data in a format for information sharing.

A PILOT STUDY

Before the program can become fully established, a pilot study is being conducted to develop a globally applicable seagrass monitoring protocol, to compare science-based with community-based monitoring efforts and to test the feasibility and usefulness of this publicly available database retrieval network. The western Pacific was chosen for the pilot because it has extensive and diverse seagrass habitats and a myriad of coastal issues with the potential to threaten seagrass growth and survival. Challenges to seagrasses in the western Pacific are numerous and, similar to those in most parts of the world, range from human population increase, fisheries practices, pollution and onshore development to global climate change and sea-level rise. The combination of these factors and the remoteness of many locations provide a complex set of circumstances that challenges our scientific ability to monitor seagrass habitat and to test the diversity of habitat impacts. The western Pacific region includes underdeveloped countries that have extensive seagrass habitat linked to important economic activities such as fishing, tourism and sports diving. The constraints of resources and the relatively small number of seagrass scientists in the western Pacific have to date precluded extensive surveys and monitoring of the kind common in Europe and parts of the US and Australian coast.

With funding assistance from the David and Lucile Packard Foundation, eight locations, five of which are western Pacific islands, were identified as suitable. In mid-2001, long-term monitoring sites were established in Kosrae, Pohnpei, Palau, Kavieng (Papua New Guinea) and Fiji. Scientists were identified at each location to take part in the direct field monitoring aspect of the research. Quarterly monitoring is now being conducted at designated sites in each country. Sites chosen were representative of the dominant seagrass habitat existing in each location.

In Kosrae, a monitoring site was established in a trochus sanctuary adjacent to Okat Harbour on the north of Kosrae Island. The site is on an intertidal fringing reef borded by mangroves landward and the reef edge seaward. Seagrass meadows cover much of the fringing reef where coral is absent. It is predominately an *Enhalus acoroides* meadow inshore, which changes to a meadow dominated by *Thalassia hemprichii* and *Cymodocea rotundata* seaward. The Fisheries Development Division is monitoring the site with some assistance from the Kosrae Development Review Commission.

On the island of Pohnpei, the largest island and location of the capital of the Federal States of Micronesia, a monitoring site was established on a relatively remote fringing reef at the southernmost point of the main island in an area free from physical disturbance by human activity. The site is on a reef flat where water pools at low tide, and is similar to the site monitored on Kosrae, including the species *Enhalus acoroides* and *Thalassia hemprichii*. Scientists from the College of Micronesia are monitoring the site.

In the Republic of Palau a monitoring site was established on a fringing reef at the edge of the shipping channel on Koror. The meadow extends across the intertidal reef flat from the mangrove-lined shore to the reef crest. Inshore, the meadow is predominately *Enhalus acoroides*, becoming interspersed with *Thalassia hemprichii*, which increases in presence along with *Halodule uninervis* and *Halophila ovalis* seaward. The site is adjacent to coastal development and receives stormwater and agricultural runoff. Scientists of the Palau International Coral Reef Centre and Coral Reef Research Foundation are monitoring this site.

In Papua New Guinea, seagrass monitoring was conducted near Kavieng in New Ireland, an island province in the northeast. With the permission of the village leader a monitoring site was established on the fringing reef flat of a small island, Nusa Lik. The site is intertidal with a mixture of *Halodule uninervis*, *Enhalus acoroides*, *Thalassia hemprichii*, *Cymodocea serrulata* and *Halophila ovalis*. The outer edge of the seagrass was determined by the edge of the coral reef. Staff attached to the Fisheries Research Laboratory and local fisheries college are monitoring the site.

Fiji has environmental issues similar to the other western Pacific island countries, such as deforestation, soil erosion and sewage effluent. A monitoring site was established on Nukubuco Reef in Laucala Bay. This monitoring site is different from sites at other localities, as it is on a barrier reef. No suitable fringing reef sites similar to other participating countries could be found. The site was chosen because the seagrass distribution and abundance of Nukubuco Reef have been mapped as part of a University of the South Pacific postgraduate project and the site was easily accessible from Suva. The monitoring site is adjacent to a sand cay at the northwestern edge of the reef. It is an intertidal site with a mixture of *Halodule uninervis*, *Halodule pinifolia* and *Halophila ovalis* subsp. *bullosa* close to the cay, becoming a monospecific *Syringodium isoetifolium* meadow seaward. The outer edge of the meadow was determined by the edge of the channel. Scientists from the University of the South Pacific are monitoring the site.

The community-based seagrass monitoring program that forms the second stage of the project was initiated in the western Pacific islands in April 2002 in New Ireland, Papua New Guinea. In June and July 2002 local citizens also began monitoring sites in Kosrae, Palau and Fiji. Community participants were mostly school students and local villagers. Community monitoring sites were established on intertidal fringing reefs and local scientists, government and non-governmental organizations are providing support. The program is using the existing Australian Seagrass-Watch program[25] protocols and data entry systems.

Thalassia hemprichii and *Cymodocea rotundata* meadow on intertidal fringing reef, Kosrae, Federated States of Micronesia.

ENCOURAGING RESULTS

Preliminary results from the scientific monitoring indicate that the sampling protocols appear suitable, although adjustments and refinements may occur from time to time as the program develops. Data entry via the website (www.seagrassnet.org) was successful, although access to the Internet is limited in some countries. Quality control and data validation are being completed at the University of New Hampshire's Jackson Estuarine Laboratory. Photographic collections are being cataloged and archived by the Queensland Department of Primary Industry Marine Plant Ecology Group. Herbarium samples were also verified at the University of New Hampshire and sent to the International Seagrass Herbarium at the Smithsonian Institution, Washington, DC, USA.

The initial success of the pilot study has encouraged scientists and coastal resource managers in Africa, South America, Asia, Europe, Australia and North America to participate. The goal is to expand SeagrassNet to other areas of the globe and, ultimately, to establish a network of monitoring sites linked through the Internet by an interactive database. The ultimate aim is to preserve the seagrass ecosystem by increasing scientific knowledge and public awareness of this threatened coastal resource.

agencies, suggests that conservation measures and the acceptance of enforcement will continue to improve.

There is a growing understanding that community types such as seagrasses are vital to the health of the reef environment and that they are threatened by climate change as well as direct human impacts. There is clearly a need in the Pacific island nations to quantify the risks to seagrass of present management practices and to quantify the extent and value of seagrass protection afforded by the present reserves and legislative approach.

REFERENCES

1. Bryant-Tokalalu JJ [1999]. The impact of human settlements on marine/coastal biodiversity in the tropical Pacific. In: Maragos JE, Peterson MNA, Eldredge LG, Bardach JE, Takeuchi HF (eds) *Marine and Coastal Biodiversity in the Tropical Island Pacific Region: Vol 2: Population, development and conservation priorities*. Workshop proceedings. Pacific Science Association, East-West Center, Honolulu. pp 215-235.
2. Short FT, Coles RG, Pergent-Martini C [2001]. Global seagrass distribution. In: Short FT, Coles RG (eds) *Global Seagrass Research Methods*. Elsevier Science BV, Amsterdam. pp 5-30.
3. Short FT, Coles RG (eds) [2001]. *Global Seagrass Research Methods*. Elsevier Science BV, Amsterdam.
4. Mukai H [1993]. Biogeography of the tropical seagrasses in the western Pacific. *Australian Journal of Marine and Freshwater Research* 44: 1-17.
5. Coles RG, Lee Long WJ [1999]. Seagrasses. In: Maragos JE, Peterson MNA, Eldredge LG, Bardach JE, Takeuchi HF (eds) *Marine and Coastal Biodiversity in the Tropical Island Pacific Region: Vol 2: Population, development and conservation priorities*. Workshop proceedings. Pacific Science Association, East-West Center, Honolulu. pp 21-46.
6. Fortes MD [1998]. Indo-West Pacific affinities of Philippine seagrasses. *Botanica Marina* 31: 237-242.
7. den Hartog C [1970]. *The Seagrasses of the World*. North Holland Publishing, Amsterdam.
8. Walker DI, Dennison WC, Edgar G [1999]. Status of Australian seagrass research and knowledge. In: Butler A, Jernakoff P (eds) *Seagrass in Australia: Strategic review and development of an R&D plan*. CSIRO, Collingwood, Australia. pp 1-24.
9. Tsuda RT, Kamura S [1990]. Comparative review on the floristics, phytogeography, seasonal aspects and assemblage patterns of the seagrass flora in Micronesia and the Ryukyu Islands. *Galaxea* 9: 77-93.
10. Coles RG, Kuo J [1995]. Seagrasses. In: Maragos JE, Peterson MNA, Eldredge LG, Bardach JE, Takeuchi HF (eds) *Marine and Coastal Biodiversity in the Tropical Island Pacific Region: Vol 1: Species systematics and information management priorities*. Workshop proceedings. Pacific Science Association, East-West Center, Honolulu. pp 39-57.
11. Stapel J, Manuntun R, Hemminga MA [1997]. Biomass loss and nutrient redistribution in an Indonesian *Thalassia hemprichii* seagrass bed following seasonal low tide exposure during daylight. *Marine Ecology Progress Series* 148: 251-262.
12. Bridges KW, McMillan C [1986]. The distribution of seagrasses of Yap, Micronesia, with relation to tide conditions. *Aquatic Botany* 24: 403-407.
13. Munro JL [1999]. Utilization of coastal molluscan resources in the tropical insular Pacific and its impacts on biodiversity. In: Maragos JE, Peterson MNA, Eldredge LG, Bardach JE, Takeuchi HF (eds) *Marine and Coastal Biodiversity in the Tropical Island Pacific Region: Vol 2: Population, development and conservation priorities*. Workshop proceedings. Pacific Science Association, East-West Center, Honolulu. pp 127-144.
14. Richmond RH [1999]. Sea cucumbers. In: Maragos JE, Peterson MNA, Eldredge LG, Bardach JE, Takeuchi HF (eds) *Marine and Coastal Biodiversity in the Tropical Island Pacific Region: Vol 2: Population, development and conservation priorities*. Workshop proceedings. Pacific Science Association, East-West Center, Honolulu. pp 145-156.
15. Pyle RL [1999]. Patterns of Pacific reef and shore fish biodiversity. In: Maragos JE, Peterson MNA, Eldredge LG, Bardach JE, Takeuchi HF (eds) *Marine and Coastal Biodiversity in the Tropical Island Pacific Region: Vol 2: Population, development and conservation priorities*. Workshop proceedings. Pacific Science Association. East-West Center, Honolulu. pp 157-176.
16. Klumpp DW, Howard RK, Pollard DA [1989]. Trophodynamics and nutritional ecology of seagrass communities. In: Larkum AWD, McComb AJ, Shepherd SA (eds) *Biology of Seagrasses: A treatise on the biology of seagrasses with special reference to the Australian region*. Elsevier, Amsterdam. pp 394-457.
17. Coles RG, Lee Long WJ, Watson RA, Derbyshire KJ [1993]. Distribution of seagrasses, their fish and penaid prawn communities in Cairns Harbour, a tropical estuary, northern Queensland, Australia. *Australian Journal of Marine and Freshwater Research* 44: 193-210.
18. Coles RG [1996]. Coastal Management and Community Coastal Resource Planning in the Asia Pacific Region. Final report to the Churchill Fellowship Foundation. 51 pp.
19. Maragos JE [1993]. Impact of coastal construction on coral reefs in the US affiliated Pacific islands. *Coastal Management* 21: 235-269.
20. Skewes. Personal communication.
21. Fortes MD [1990]. *Seagrasses: A resource unknown in the ASEAN region*. United States Coastal Resources Management Project. International Center for Living Aquatic Resources Management, Manila, Philippines. ICLARM Education Series No. 6. 46 pp.
22. Falanruw MC [1992]. Seagrass nets. *Atoll Research Bulletin* (364): 1-12.
23. Palau Conservation Society [1996]. *A Guide to the Conservation Laws and Regulations of the Republic of Palau*. Palau Conservation Society, Koror. 19 pp.
24. Federated States of Micronesia [2001]. Preliminary Report to the Conference of the Parties of the Convention on Biological Diversity. 24 pp.
25. McKenzie LJ, Campbell SJ, Roder CA [2001]. *Seagrass-Watch: Manual for mapping & monitoring seagrass resources by community (citizen) volunteers*. Northern Fisheries Service, Cairns, Australia. 94 pp.

AUTHORS

Rob Coles, Len McKenzie and Stuart Campbell, DPIQ, Northern Fisheries Centre, P.O. Box 5396, Cairns, Queensland 4870, Australia. **Tel:** +61 (0)7 4035 0111. **Fax:** +61 (0)7 4035 4664. **E-mail:** rob.coles@dpi.qld.gov.au

Miguel Fortes, Marine Science Institute CS, University of the Philippines, Diliman, Quezon City 1101, Philippines.

Fred Short, Department of Natural Resources, University of New Hampshire, Jackson Estuarine Laboratory, 85 Adams Point Road, Durham, NH 03824, United States.

16 The seagrasses of INDONESIA

T.E. Kuriandewa
W. Kiswara
M. Hutomo
S. Soemodihardjo

Although seagrasses cover at least 30 000 km² throughout the Indonesian Archipelago, from Pulau Weh in Aceh to Merauke, Papua, they have only been studied in relatively small areas and information is therefore rather limited. Nonetheless an encouraging and increased understanding of the importance, ecology and biology of Indonesian seagrasses has developed in recent years[1]. Vast areas of the archipelago (e.g. the north coast of Papua, the southwest coast of Indonesia, the south and west coasts of Kalimantan) are yet to be studied, however.

The diversity of marine habitats in Indonesia is among the highest in the world and Indonesian seagrass diversity is comparable to other countries in the region. Seven genera and 12 species of seagrasses currently occur in Indonesian waters[2-5]. Two species, *Halophila spinulosa* and *Halophila decipiens*, have been recorded in just a few locations: *Halophila spinulosa* in Sorong (Papua), Lombok, East Java, Sunda Strait and Riau, and *Halophila decipiens* in Aru, Kotania Bay, Lembata, Sumbawa and Jakarta Bay. Two further species, *Halophila beccarii* and *Ruppia maritima*, are known only through specimens at the Bogor Herbarium and have not recently been found in the field.

Indonesian seagrasses either form dense monospecific meadows or mixed stands of up to eight species. *Thalassia hemprichii*, *Enhalus acoroides*, *Halophila ovalis*, *Halodule uninervis*, *Cymodocea serrulata* and *Thalassodendron ciliatum* usually grow in monospecific beds[6], and muddy substrates on the seaward edges of mangroves often have meadows of high biomass. Mixed-species meadows occur in the lower intertidal and shallow subtidal zones, growing best in well-sheltered, sandy (not muddy), stable and low-relief sediments[6]. These beds are typically dominated by pioneer species such as *Halophila ovalis*, *Cymodocea rotundata* and *Halodule pinifolia*. *Thalassodendron ciliatum* dominates the lower subtidal zone – this species can grow in silt as well as in medium-to-coarse sand and coral rubble. High bioturbation by, for example, burrowing shrimps tends to decrease seagrass density and favor the pioneer species. Seagrasses growing in terrigenous sediment are more influenced by the turbidity, seasonality, fluctuating nutrient and salinity concentrations, and subsequent light limitation, of land runoff than those in reef-derived carbonate sediments with less variable seasonal dynamics.

Monospecific beds of *Thalassia hemprichii* are the most widespread throughout Indonesia and occur over a large vertical range from the intertidal zone down to the lower subtidal zone[7]. *Halophila ovalis* also has a wide vertical range, from the intertidal zone down to more than 20 m depth, and grows especially well on disturbed sediments such as the mounds of burrowing invertebrates. *Enhalus acoroides*, too, grows in a variety of different sediment types, from silt to coarse sand, in subtidal areas or localities with heavy bioturbation. *Halodule uninervis* is a pioneer species, usually forming monospecific beds on the inner reef flat or on steep sediment slopes in both the intertidal and subtidal zones.

ECOLOGY

The majority of detritus produced by Indonesian seagrasses is believed to settle within the beds, with an estimate of only 10 percent exported to other ecosystems[6]. Most of the nutrients lost by leaf fragmentation through decomposition or harvesting by alpheid shrimps are translocated to the sediment and about 80 percent of the nitrogen content is denitrified there[8]. This retention of nutrients within the beds may explain why seagrass beds in Indonesia maintain a high level of productivity despite low nutrient availability.

Detailed studies of the nutrient concentrations at six different locations in the Spermonde Archipelago of South Sulawesi have indicated that there are structural and functional differences between coastal beds growing on the sand and mud deposited by rivers, and

Table 16.1
Average biomass of seagrasses (g dry weight/m^2) at various locations throughout Indonesia

Species	Sunda Strait	Banten Bay	Jakarta Bay	Flores Sea	Lombok
Enhalus acoroides	1 976	353-560	250-663	155-546	393-2 479
Cymodocea rotundata	37-106	139	18-23	34-113	39-243
Cymodocea serrulata	48-104	15-35	240	45-174	111
Halodule pinifolia	-	-	-	29-126	47
Halodule uninervis	10-36	6-80	64	13-516	29-128
Halophila ovalis	2-4	8	1-8	1-3	4-46
Syringodium isoetifolium	74	102-372	25-90	33-127	85-262
Thalassia hemprichii	87-193	120-257	90-278	115-322	53-263
Thalassodendron ciliatum	-	-	-	231-444	-

Source: Kiswara[1].

Table 16.2
Average density of seagrasses (shoots/m^2) at various locations throughout the Indonesian Archipelago

Species	Sunda Strait	Banten Bay	Jakarta Bay	Flores Sea	Lombok
Enhalus acoroides	160	40-80	36-96	60-146	50-90
Cymodocea rotundata	38-756	690	26-1 136	220-1 800	253-1 400
Cymodocea serrulata	48-1 120	60-190	1 056	115-1 600	362
Halodule pinifolia	-	-	-	430-2 260	7 120
Halodule uninervis	10-335	40-1 160	604	360-5 600	80-160
Halophila ovalis	15-240	820	18-115	100-2 160	400-1 855
Syringodium isoetifolium	630	124-3 920	144-536	360-3 740	1 160-2 520
Thalassia hemprichii	30-315	220-464	68-560	160-1 820	200-865
Thalassodendron ciliatum	-	-	-	400-840	-

therefore of terrestrial origin, and those growing offshore on sediments derived from coral reefs. Concentrations of dissolved reactive phosphate, ammonium and nitrate+nitrite were low (<2 µM) in the water column at all sites, often below detectable limits, but considerably higher in sediment porewater[8]. Porewater phosphate concentrations (3-13 µM) were comparable between the two sediment types, but exchangeable phosphorus contents were two to five times higher in carbonate sediment (18.2-23.6 mg phosphorus/100 g versus 4.4-10.9 mg phosphorus/100 g) than in terrigenous sediments. Carbonate sediments were extremely low in organic matter compared with terrigenous sediments.

The more vigorous growth of coastal seagrasses is attributed to a higher level of nutrients in the sediment than offshore. Leaf size of Enhalus acoroides is significantly larger in coastal than offshore beds[9], biomass and shoot densities are higher and epiphyte cover lower, factors attributed to the less severe environmental fluctuations of offshore beds[8].

Biomass

The below-ground rhizome biomass of Enhalus acoroides is six to ten times larger than that of above-ground biomass[6]. Cymodocea rotundata, Cymodocea serrulata and Halodule uninervis have higher below-ground biomass when growing in established mixed vegetation beds than in monospecific pioneer beds[6]. In general, species characteristic of climax Indonesian seagrass meadows (Thalassia hemprichii, Enhalus acoroides and Thalassodendron ciliatum) invest two to four times more energy into below-ground biomass growth than the colonizing species (Halodule uninervis, Cymodocea rotundata and Cymodocea serrulata)[6]. Biomass values show high variability (Table 16.1) due to habitat differences, species composition, plant densities between locations and sampling techniques[10].

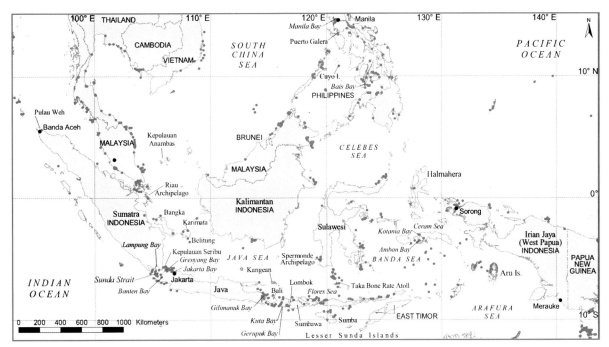

Map 16.1
Indonesia

Seagrass density also varies considerably between locations (Table 16.2). Kiswara found that the density of *Halodule uninervis* depends on the phenotype (normal shoots or thin shoots). In Gerupuk Bay, southern Lombok, *Halodule uninervis* densities ranged from 870 normal shoots/m^2 to 6 560 thin shoots/m^2 within the same seagrass bed[1]. Nienhuis reported that *Halodule uninervis* had the highest density of all seagrass species in mixed as well as in monospecific seagrass beds (Table 16.3). In seagrass beds where foliage covers more than 70 percent of substrate, the density of seagrasses frequently depends on the species composition of the community and the relative age of the seagrasses. In some species, such as *Thalassia hemprichii*, biomass is frequently a function of shoot density and total leaf area per leaf cluster[11].

Seasonal studies of seagrass biomass and shoot densities in Indonesian waters are scarce but significant seasonal fluctuations are known to occur[8].

Productivity

Growth studies have been carried out in Indonesia using several techniques[7, 8, 12-17]. Using the oxygen evolution (photosynthesis) technique, Lindeboom and Sandee[12] demonstrated that gross primary production rates of various seagrass communities in the Flores Sea vary from 1 230 mg carbon/m^2/day to 4 700 mg carbon/m^2/day. Seagrass respiration consumption rates were between 860 mg carbon/m^2/day and 3 900 mg carbon/m^2/day. They concluded that net primary production rates of seagrass communities in the Flores Sea vary between 60 mg carbon/m^2/day and 1 060 mg carbon/m^2/day which, assuming the same rates of production throughout the year as during the study period (October), translates to a maximum annual net primary production of about 387 g carbon/m^2. Epiphyte production alone accounted for a maximum annual net primary production of about 84 mg carbon/m^2 of leaf surface area[12], or 36 percent of the net primary production rate of the seagrass communities studied.

Table 16.3

Average shoot density of seagrass species in mixed and monospecific seagrass meadows in the Flores Sea

Species	Mixed seagrass meadow (number of shoots/m^2)	Monospecific seagrass meadow (number of shoots/m^2)
Cymodocea rotundata	324 (276)	–
Cymodocea serrulata	696 (767)	533 (543)
Enhalus acoroides	54 (86)	136 (58)
Halodule uninervis	2 847 (5 689)	14 762 (6 076)
Halophila ovalis	69 (117)	–
Syringodium isoetifolium	2 504 (1 736)	–
Thalassia hemprichii	754 (748)	1 459 (811)
Thalassodendron ciliatum	–	692 (272)

Note: In all sampling locations foliage cover is >70 percent, except for *Thalassodendron ciliatum* (>50 percent) (SD in parentheses).

A comparative study of two different seagrass environments in the Spermonde Archipelago obtained very similar results using the same techniques[8]. Gross primary production rates ranged from 900 mg carbon/m^2/day to 4400 mg carbon/m^2/day. Interestingly, the bell-jar technique used in the Spermonde Archipelago did not reveal any significant difference in seagrass production rates between coastal and reef environments. Net primary production was slightly negative in a number of stations and was generally below 500 mg carbon/m^2/day. Low net primary rates were attributed to high community oxygen consumption rates. Higher net primary production rates were obtained from monospecific stands of *Thalassia hemprichii*, where combined seagrass and epiphyte net production rates reached 1.5 mg carbon/m^2/day to 1.9 mg carbon/m^2/day, equivalent to a maximum of 694 mg carbon/m^2/year.

Nienhuis has suggested that Indonesian seagrass communities are self-sustaining systems and export very little of their photosynthetically fixed carbon to adjacent ecosystems such as coral reefs[6]. The results obtained from the Flores Sea and Spermonde Archipelago seem to support this general hypothesis. Erftemeijer points out that many seagrass communities (58 percent of his study sites) seem to use more energy than is actually produced by the autotrophic seagrass community. This suggests that, while recycling of nutrients and organic carbon is high, seagrass beds may not be self-sustaining. Filter and suspension-feeding macroinvertebrates constitute a significant consumer component of the Indonesian seagrass community.

Marking methods have been used to measure leaf production[18] in seagrass meadows at Taka Bone Rate Atoll[7], Kepulauan Seribu[13, 14], Banten Bay[15] and, most recently, in Lombok[16] and the Spermonde Archipelago[8]. Production rates obtained from these studies are summarized in Table 16.4.

Erftemeijer[8] demonstrated that, while there was no difference in primary production rates of coastal and offshore seagrass beds in Sulawesi, the leaf growth rate (3.1 mm/day) of *Enhalus acoroides* was significantly higher in muddy coastal habitats than in offshore reef habitats (1.6 mm/day). Similar results were obtained for *Thalassia hemprichii*.

ASSOCIATED BIOTA

Seagrass-associated flora and fauna remain one of the most open and exciting fields of research for Indonesian scientists. Recent studies have focused on establishing species lists and measuring abundance and biomass of various seagrass-associated taxa. With a few exceptions[19, 20] the majority of seagrass-associated faunal studies have dealt with infauna, macrofauna, motile epifauna and epibenthic fauna (Table 16.5).

Algae

Fishermen at Benoa, Bali and West Lombok have recorded seven economically important species of seaweeds growing in the mixed seagrass meadow of *Cymodocea serrulata*, *Halodule uninervis*, *Thalassia hemprichii* and *Thalassodendron ciliatum*[33]. In South Sulawesi 117 species of macroalgae are associated with seagrasses, composed of 50 species of Chlorophyta, 17 species of Phaeophyta and 50 species of Rhodophyta. Thirteen species were exclusively associated with seagrass vegetation[9].

Meiofauna

The meiofauna associated with monospecific *Enhalus acoroides* seagrass beds on the south coast of Lombok consisted of nematodes, foraminiferans, cumaceans, copepods, ostracods, turbelarians and polychaetes[20],

Table 16.4
Average growth rates (mm/day) of seagrass leaves using leaf-marking techniques

Species	West Java Sea	Spermonde Archipelago	Lombok	Flores Sea
Cymodocea rotundata	-	-	5.5 (6.8)	-
Cymodocea serrulata	5.0 (0.6)	-	-	-
Enhalus acoroides	7.3 (3.6)	2.4 (2.3*)	6.5 (1.5)	-
Syringodium isoetifolium	4.1 (6.8)	-	-	-
Thalassia hemprichii	4.9 (1.5)	1.6 (3.5*)	3.8 (8.1)	-
Thalassodendron ciliatum	-	-	-	2.7 (4.7)

Notes:
Production rates in parentheses (g dry weight/m^2/day).
* In mg ash-free dry weight/m^2/day.

Table 16.5
Indonesian seagrass-associated flora and fauna: number of species

Taxon	Banten Bay	Jakarta Bay	Lombok	Ambon Bay	Kotania Bay	South Sulawesi
Algae			37		34	117
Meiofauna			6 groups			
Mollusks	15		55		143 (hermit crabs)	
Crustaceans	25	32	84		30	
Echinoderms	3		45			10
Fishes	180	78	85	168	205	
Fish larvae			53			

Source: Various sources[21-32].

many of which were actively emergent. A high abundance of nematodes was indicative of nutrient enrichment. Benthic foraminifera are an important component of Indonesian seagrass communities, but have received only rudimentary attention[19]. In the Kepulauan Seribu patch reef complex, seagrass beds are abundant and frequently dominated by associations of *Enhalus acoroides* and *Thalassia hemprichii*[32]. Benthic foraminifera in this location are dominated by the suborders Miliolina and Rotaliina[19]. The most abundant rotaliinids were *Ammonia beccarii*, *Ammonia umbonata*, *Calcarina calcar*, *Elpidium advenum*, *Elpidium crispum*, *Elpidium craticulatum* and *Rosalina bradyi*. The genus *Ammonia* is a euryhaline group, common in shallow-water tropical environments, and *Calcarina calcar* is indicative of coral reef habitats. The abundance of *Elpidium* spp. is interesting, since this euryhaline, shallow-water species is extremely tolerant of low salinities and can be found far up estuaries. The miliolinids are represented by *Adolesina semistriata*, *Milionella sublineata*, *Quinqueloculina granulocostata*, *Quinqueloculina parkery*, *Quinqueloculina* sp., *Spiroloculina communis*, *Spirolina cilindrica* and *Triloculina tricarinata*. Both *Quinquiloculina* and *Triloculina* are characteristic of shallow tropical waters.

Crustaceans

Crustaceans are a key component of seagrass food webs. Recent gut analyses from the south coast of Lombok[34] demonstrated that crustaceans are the dominant food source for seagrass-associated fish. Aswandy and Hutomo[21] recorded 28 species of crustaceans in Banten Bay seagrass beds. The tanaidacean *Apseudes chilkensis* and an unknown species of melitidae amphipod are the most abundant crustaceans in *Enhalus acoroides* meadows in Grenyang Bay[27]. Moosa and Aswandy[30] recorded 70 crustacean species from seagrass meadows in Kuta and Gerupuk Bays but many specimens were apparently collected from coral rubble areas adjacent to the seagrass meadows. One hipollitid shrimp there, *Tozeuma* spp., has special morphological adaptations to live specifically in seagrass meadows. Its lancelet body shape and coloration, green mottled with small white spots, provides almost perfect camouflage when it adheres to seagrass leaves. Many stomatopods are found in Indonesian seagrass beds with *Pseudosquilla ciliata*, an obligate seagrass-associated species[30]. Other stomatopods, such as *Odontodactylus scyllarus*, leave the reefs to forage for mollusks in adjacent seagrass beds[35]. Rahayu collected 30 species of hermit crabs from Kotania Bay seagrass bed. Three were species of *Diogenes*, one was a species of *Pagurus* and four were undescribed species. It is believed that crustaceans in the seagrass beds of Kotania Bay are much more diverse than those of other locations.

Mollusks

The mollusks are one of the best-known groups of seagrass-associated macroinvertebrates and perhaps the most overexploited. Mudjiono et al. recorded 11 gastropods and four bivalves from the seagrass meadows in Banten Bay[26]. This rather impoverished mollusk fauna was collected from monospecific *Enhalus acoroides* beds, in mixed beds of *Enhalus acoroides*, *Cymodocea serrulata* and *Syringodium isoetifolium*, and mixed beds of *Enhalus acoroides*, *Cymodocea rotundata*, *Cymodocea serrulata*, *Halodule uninervis*, *Halophila ovalis*, *Syringodium isoetifolium* and *Thalassia hemprichii*. The entire bay is heavily exploited and only two gastropods were common to all locations, *Pyrene versicolor* and *Cerithium tenellum*. Just four juvenile (3-5 mm diameter) *Trochus niloticus* were collected[26].

Seventy species were collected from less disturbed sites in Lombok[31], many of which are economically valuable. Gastropod families included Bullidae, Conidae, Castellariidae, Cypraeidae, Olividae, Pyrenidae, Strombidae, Trochidae and Volutidae;

bivalve families were Arcidae, Cardiidae, Glycymeridae, Isognomonidae, Lucinidae, Mesodesnatidae, Mytilidae, Pinnidae, Pteridae, Tellinidae and Veneridae. *Pyrene versicolor*, *Strombus labiatus*, *Strombus luhuanus* and *Cymbiola verspertilio* were the most abundant gastropod species and *Anadara scapha*, *Trachycardium flavum*, *Trachycardium subrugosum*, *Peryglypta crispata*, *Mactra* spp. and *Pinna bicolor* were the most common bivalve species[31]. A number of *Conus* species were found.

A high diversity of mollusks, 142 species from 43 families, has also been reported from seagrass beds in Kotania Bay[36].

Echinoderms

The most significant echinoderm species is a sea star, *Protoreaster nodosus*, which feeds on seagrass detritus and the surface of broken seagrass leaves. Forty-five species of Echinodea, Holothuridae, Ophiuroidae and Crinoidae have been recorded in the seagrass beds of Kuta and Gerupuk Bays. Several economically important species of *Holothuria* and *Actinopyga*, and the sea urchin *Tripneustes gratilla*, have declined in abundance[28].

Similar depletions in echinoderm populations have been reported from Kotania Bay on west Ceram Island, Moluccas, where seagrass meadows formerly supported a high abundance of economically important holothuroids. In 1983, the extensive seagrass meadows in Kotania Bay supported high population densities (i.e. 1-2 individuals/m^2) of nine economically important sea cucumber species, namely *Bohadschia marmorata*, *Bohadschia argus*, *Holothuria (Metrialyta) scabra*, *Holothuria nobilis*, *Holothuria vagabunda*, *Holothuria impatiens*, *Holothuria edulis*, *Thelenota ananas* and *Actinopyga miliaris*. In a 1993 inventory of the same area, only three sea cucumbers were recorded within a distance of 500 m. The average body size of sea cucumbers decreased from around 22 cm in 1983 to less than 15 cm in 1993. The decline of the stock and size are attributed to intensive collections by local people to supply the lucrative *teripang* (*bêche de mer*) trade. Another heavily overexploited echinoderm species whose population has declined sharply during the past ten years is the edible sea urchin *Tripneustes gratilla*.

Fish

In 1977 one of the first studies of seagrass-associated fish in Indonesia collected 78 species from *Thalassia hemprichii* and *Enhalus acoroides* meadows amongst lagoonal patch reefs in Pari Island, in the Kepulauan Seribu complex[37]. Only six (Apogonidae, Atherinidae,

---

Case Study 16.1
BANTEN BAY, WEST JAVA

Banten Bay covers 120 km^2, and harbors several coral islands. The biggest inhabited island is Pulau Panjang; the other islands are small and uninhabited. The rivers Domas, Soge, Kemayung, Banten, Pelabuhan, Wadas, Baros and Ciujung discharge into the bay. Seagrass is found along the mainland Java coast in the western part of the bay, on the reef flat of the coral islands (Pulau Panjang, Pulau Tarahan, Pulau Lima, Pulau Kambing and Pulau Pamujan Besar) and on submerged coral reefs in the intertidal area down to a depth of 6 m. The total area of seagrass beds at Banten Bay is about 330 ha, consisting of 168 ha on the mainland and 162 ha on the coral islands.

The depth of the bay is not more than 10 m. Its sediment consists of mud and sand[24, 25], and the salinity varies between 28.23 and 35.34 psu[38]. The rainy season is from November to March. Mangrove is found at Grenyang in the eastern part of the bay up to Tanjung Pontang in the west part, and in the southern part of Pulau Panjang. Eight species of seagrasses occur here: *Cymodocea rotundata*, *Cymodocea serrulata*, *Enhalus acoroides*, *Halodule uninervis*, *Halophila ovalis*, *Halophila minor*, *Syringodium isoetifolium* and *Thalassia hemprichii*. Beds between 25 and 300 m in length[43] are continuous along the coast of Banten Bay, from the beach to the reef edge.

They are nursery grounds for 165 species of fish[38] which feed either directly on algae and seagrass or on seagrass-associated invertebrates[44], including six juveniles of grouper (*Epinephelus bleekeri*, *Epinephelus fuscoguttatus*, *Epinephelus merra*, *Epinephelus septemfasciatus*, *Epinephelus coioides* and *Plectropomus* spp.[45]). Dugongs also occur here[46]. The cultivation of seaweeds in Banten Bay has increased enormously in recent years along the coastline of all the islands, on the coral reef and lately also outside the reef flat area. Approximately 35 ha, including 25 ha or 10 percent of the reef flat area and 10 ha outside the coral reef flat, are now used for the cultivation of seaweeds and have been cleared of seagrass[47]. Transplantation studies using *Enhalus acoroides* were conducted in Banten Bay in 1998[17]. Only rhizomes transplanted to muddy substrate survived more than five months – these new seagrass beds are now used by local fishermen to collect fishes and prawns[48].

Labridae, Gerridae, Siganidae and Monacanthidae) of the families recorded, however, could be considered as important seagrass residents. The Pari Island study was followed in 1985 by a long-term study of seagrass fish assemblages in Banten Bay, southwest Java Sea. The results from the Banten Bay study[35] supported earlier views that only small numbers of fish species permanently reside in seagrass beds. However, it was also reconfirmed that seagrass beds act as nursery grounds for many economically valuable fish species. Beds with higher densities of seagrass supported higher abundance of fish, and *Enhalus acoroides* meadows supported higher fish abundance than *Thalassia hemprichii*.

Studies on seagrass fish in Indonesia have been gradually increasing since the late 1980s[34, 38, 39]. Indonesian seagrass fish communities are commonly dominated by Siganidae (rabbitfishes), such as *Siganus*

Case Study 16.2
KUTA AND GERUPUK BAYS, LOMBOK

Kuta and Gerupuk Bays are covered by gravel, small pebbles, fine sand and mud in the river mouth, where *Enhalus acoroides* grows. The tidal range in the bays is about 2 m, tidal velocity and direction are 2.8-10.8 cm/min and 315°-350° at high tide and 4.5-10.0 cm/min and 270°-310° at low tide. During the wet season, December to April, salinity varies from 28 to 29 psu and surface water temperature from 18 to 24°C. In the dry season, May to November, these measurements are approximately 34 psu and 27°C.

The most diverse seagrass beds in Indonesia occur here, with 11 of the 12 species present in Gerupuk (*Cymodocea rotundata, Cymodocea serrulata, Enhalus acoroides, Halodule pinifolia, Halodule uninervis, Halophila minor, Halophila ovalis, Halophila spinulosa, Syringodium isoetifolium, Thalassia hemprichii, Thalassodendron ciliatum*). *Halophila spinulosa* is absent from Kuta. *Enhalus acoroides* and *Thalassodendron ciliatum* form monospecific beds in both bays, and *Halophila spinulosa* in Gerupuk Bay. Mixed beds of *Cymodocea rotundata, Cymodocea serrulata, Halodule pinifolia, Halodule uninervis, Halophila minor, Halophila ovalis, Syringodium isoetifolium* and *Thalassia hemprichii* occur at both locations.

Coverage area of habitat types at Kuta and Gerupuk Bays

	Coverage area (ha)	
	Kuta Bay	Gerupuk Bay
Enhalus acoroides	7.68	29.40
Thalassodendron ciliatum	10.50	id
Halophila spinulosa	–	11.07
Habitat types		
Mixed vegetation	96.37	76.86
Sandy bar	12.03	42.97
Lagoon	55.19	–
Dead coral	70.97	27.36
Live coral	38.08	–
Volcanic stone	2.93	–

Large amounts of seagrass detritus wash up and accumulate on the beach during the strong winds of the east monsoon. Interestingly *Thalassodendron ciliatum* at Kuta Bay is able to grow on volcanic stone. During low tide the local community collects milkfish, sea cucumbers, octopus, shellfish, sea urchins and seaweeds (*Caulerpa* spp., *Gracilaria* spp. and *Hypnea cervicornis*) from the seagrass beds. The commercial alga, *Kappaphycum alvarezi*, is cultivated here.

Associated flora and fauna of the seagrass bed of Lombok

Taxon group	Number of species
Algae	37
Meiofauna[20]	6 (higher taxa)
Mollusk[31]	55
Echinoderm[28]	45
Crustacean[30]	71
Fish[29]	85
Fish larvae	53

Source: Various sources – see references by groups.

Only four of the fish found here – *Syngnathoides biaculeatus, Novaculichthys* spp., *Pervagor* spp. and *Centrogenys valgiensis* – are typical seagrass fishes. *Halichoeres argus* and *Cheilio enermis* are abundant not only in seagrass but also in algal beds. The dominance of *Syngnathoides biaculeatus* and *Cheilio enermis* is unusual because, more commonly, the fish populations of Indonesian seagrass beds are characterized by abundant rabbitfish, especially *Siganus canaliculatus*.

The main threat to the seagrass of Kuta and Gerupuk Bays is the intensive collecting of intertidal organisms during low tide, often involving digging with sharp iron sticks which disturb the substrate, cut the leaves of seagrasses and uproot their rhizomes. Future threats may include hotel construction and operation as the area has been earmarked for development by the local government.

Case Study 16.3
KOTANIA BAY

Kotania Bay, Ceram Island, contains five small islands: Buntal, Burung, Marsegu, Tatumbu and Osi. Only Osi has freshwater and is inhabited, along with two villages at Pelita Jaya and Kotania on Ceram Island. The water around Pelita Jaya (40 m) is deeper than that at Kotania village (20 m). The intertidal area in the northern part of Kotania Bay is very narrow (4 to 10 m) but wider in the east and south (50 to 250 m). Seagrasses are found along the whole coast area of the bay, except in the north. On Buntal and Osi Islands the sediment trapped by these seagrass beds has, over time, created "cliffs" which have served as substrate for the development and seaward expansion of mangrove communities.

The seagrass beds have been mapped using remote-sensing techniques which estimated a total area of 11.2 km². The pattern of seagrass distribution depends on the type of substrate.

Muddy substrate is mostly dominated by monospecific beds of *Enhalus acoroides*. Mixtures of mud, sand and coral rubble are usually covered by *Thalassia hemprichii*. The highest density of seagrass is found in the area between Osi and Burung. The eastern part of the bay, called Wai Tosu, has two kinds of substrates. The sediment at the mouth of a small creek is deep and muddy, and is covered only by *Enhalus acoroides* (10-20 percent coverage) while there is a thin layer of mud, sand and coral rubble about 100 m in front of the mangroves. Underneath this thin substrate is a hard layer of coral rock. *Thalassia hemprichii, Cymodocea rotundata, Halodule uninervis, Halophila ovalis* and *Enhalus acoroides* grow sporadically here to less than 35 percent coverage. Local people have set a fish trap around the seagrass area and built a large cage to rear sea cucumbers.

In the southern part of Kotania Bay the intertidal zone is very flat and almost all is exposed during the lowest tides. The substrate near to the mangrove area is mixed mud and sand dominated by *Thalassia hemprichii* and *Enhalus acoroides*. Along

Table 16.6
Present coverage of seagrasses in Indonesia

	North Sulawesi	Derawan Islands	Karimata Island	Anambas/Natuna Island	Lampung Bay	Bangka/Belitung	Banten Bay	Taka Bone Rate	Gilimanuk Bay	Benoa Bay	Kuta Bay	Gerupuk Bay
Coverage (ha)	200-300	50-150	<2	5-150	2-5	0.5-18	336	20-80	30	1	73	36
Cover (%)	20-80	15-80	12-25	10-15	5-15	5-15	25-45	30-70	15-50	5-10	30-70	20-50
Recorded species (number)	9	8	5	7	8	9	8	10	5	9	10	11
Hydrocharitaceae												
Enhalus acoroides	C	A	R	R	R	C	VA	-	R	R	VA	VA
Halophila decipiens	-	-	-	-	-	-	-	R	-	-	-	-
Halophila minor	-	-	-	VR	-	-	R	R	-	-	R	R
Halophila ovalis	R	R	R	VR	C	R	R	R	R	R	R	R
Halophila spinulosa	-	-	-	-	-	-	-	-	-	-	-	R
Thalassia hemprichii	C	VA	C	R	R	R	VA	VA	R	R	VA	VA
Cymodoceaceae												
Cymodocea rotundata	R	C	R	VR	R	R	R	R	R	R	R	R
Cymodocea serrulata	R	R	-	-	VR	VR	VR	R	-	R	R	R
Halodule pinifolia	R	R	-	VR	R	R	-	R	-	R	R	R
Halodule uninervis	R	R	-	-	R	R	R	R	R	R	R	R
Syringodium isoetifolium	R	C	VR	VR	R	R	R	R	-	R	R	R
Thalassodendron ciliatum	C	-	-	-	-	C	-	A	-	R	C	R

Notes: C common; A abundant; R rare; VA very abundant; VR very rare.

Distribution of seagrass in Kotania Bay

Species of seagrass	% cover	Substrate type	Depth (m)	PJ	TL	OI	BRI	BTI	TI	MI
Cymodocea rotundata	10-40	Sand	+0.2-2.0	✓	✓	✓	✓	✓	✓	✓
Cymodocea serrulata	<5	Sand	0.5-2.0	✓			✓			✓
Enhalus acoroides	20-60	Mud, sand	0.5-2.5	✓	✓	✓	✓	✓	✓	✓
Halodule pinifolia	<5	Sand	+0.2-1.5	✓	✓	✓	✓	✓	✓	✓
Halodule uninervis	<5	Sand	0.5-2.0	✓	✓	✓	✓	✓	✓	✓
Halophila decipiens	40-100	Coral rubble					✓			✓
Halophila ovalis	<5	Sand	+0.2-1.5	✓	✓	✓	✓	✓	✓	✓
Halophila minor				✓						
Syringodium isoetifolium	<5	Mud, sand	0.5-2.0	✓	✓	✓				✓
Thalassia hemprichii	<5	Mud, sand	+0.2-2.5	✓	✓	✓	✓	✓	✓	✓

Notes: PJ Pelita Jaya; TL Tanjung Lalansoi; OI Osi Island; BRI Burung Island; BTI Buntal Island; TI Tatumbu Island; MI Marsegu Island.

the southern part of the bay up to Tanjung Lalansoi the substrate in the deeper areas is a mixture of sand and coral rubble. The most common seagrasses there are *Thalassia hemprichii, Cymodocea rotundata, Syringodium isoetifolium, Cymodocea serrulata, Halodule pinifolia, Halodule uninervis, Halophila ovalis* and *Enhalus acoroides*. The seagrass density is quite high and varies seasonally. Percent coverage ranges from 40 to 70 percent with the highest values always close to the mangrove areas.

	Sermata Islands	Ambon Bay	Baguala Bay	Kotania Bay	Central Maluku Island	Aru Islands	Yamdena Islands	West Timor Island	Saleh Bay	Wakatobi Island	Makassar Strait	Malaka Strait
Coverage (ha)	10-50	0.3-1	4-5	212	25-75	100-1000	5-50	10-100	25-75	100-1000	5-50	10-100
Cover (%)	30-60	5-20	15-30	30-80	30-60	50-99	30-70	30-50	30-60	50-99	30-70	30-50
Recorded species (number)	9	7	8	10	8	8	8	8	8	8	8	8
Hydrocharitaceae												
Enhalus acoroides	VA	R	R	A	VA	A	VA	VA	VA	A	VA	VA
Halophila decipiens	-	-	-	R	-	-	-	-	-	-	-	-
Halophila minor	R	VR	-	R	-	-	-	-	-	-	-	-
Halophila ovalis	R	R	R	R	R	R	R	R	R	R	R	R
Halophila spinulosa	-	-	-	-	-	-	-	-	-	-	-	-
Thalassia hemprichii	VA	C	R	VA	VA	R	VA	VA	VA	R	VA	VA
Cymodoceaceae												
Cymodocea rotundata	R	C	C	C	R	R	R	R	R	R	R	R
Cymodocea serrulata	R	-	R	R	R	R	R	R	R	R	R	R
Halodule pinifolia	R	VR	R	R	VR	R	R	R	VR	R	R	R
Halodule uninervis	R	R	R	R	R	R	R	R	R	R	R	R
Syringodium isoetifolium	R	-	R	C	R	R	R	R	R	R	R	R
Thalassodendron ciliatum	-	-	-	-	-	-	-	-	-	-	-	-

Notes: C common; A abundant; R rare; VA very abundant; VR very rare.

canaliculatus in Jakarta Bay[37], except in Lombok (see Case Study 16.2). Indonesian seagrass fish have been classified into four principal species assemblages:

1. permanent residents which spend most of their lives in seagrass beds (e.g. the chequered cardinalfish, *Apogon margaritophorus*);
2. residents which live in seagrass throughout their life cycle but which spawn outside the seagrass beds (e.g. *Halichoeres argus*, *Atherinomorus duodecimalis*, *Cheilodipterus quinquelineatus*, *Gerres macrosoma*, *Stephanolepis hispidus*, *Acreichthys hajam*, *Hemiglyphidodon plagiometopon*, *Syngnathoides biaculeatus*);
3. temporary residents which occur in seagrass beds only during their juvenile stage (e.g. *Siganus canaliculatus*, *Siganus virgatus*, *Siganus punctatus*, *Lethrinus* spp., *Scarus* spp., *Abudefduf* spp., *Monacanthus chinensis*, *Mulloidichthys flavolineatus*, *Pelates quadrilineatus*, *Upeneus tragula*);
4. occasional residents or transients that visit seagrass beds to seek shelter or food.

Measuring the primary productivity of seagrass meadows in Sulawesi using enclosures equipped with oxygen electrodes.

The first study on seagrass fish larvae and juveniles took place in Kuta Bay and recorded 53 species belonging mainly to four families: Channidae, Ambassidae, Engraulidae and Gobiidae. High numbers of species and individuals were found in unvegetated areas full of broken seagrass leaves, and in the *Enhalus acoroides* beds.

HISTORICAL PERSPECTIVE

Herbarium collections of seagrasses from Indonesian waters were made by Zollenger in 1847 and Kostermans in 1962 and include both *Ruppia maritima* from Ancol-Jakarta Bay and Pasir Putih, East Java, and one specimen of *Halophila beccarii* from an unknown location. The development of Jakarta has destroyed the mangrove swamp in Ancol, the only place that *Ruppia maritima* had been reported, and this is thought to have caused the disappearance of this species from Indonesia.

For 15 years the Ancol Oceanorium in Jakarta kept two male dugongs in captivity, feeding them with seagrasses (*Syringodium isoetifolium* and *Halodule uninervis*) harvested from Banten Bay. Unfortunately they died in November 1991[40]. There is one female dugong in Surabaya Zoo, which has been in captivity since 1985. Its food is harvested from Celengan-Muncar, East Java, about 340 km from Surabaya. It feeds mostly on *Syringodium isoetifolium*, which forms 95 percent of the dugong's dietary intake. The consumption rate of the captive dugong is approximately 30 kg wet weight/day[41]. Recently Sea World of Indonesia in Ancol-Jakarta has acquired two male dugongs. One of them was caught in seine nets in Banten Bay in 1998 and the other one was trapped in a *sero* (fish trap) on the seagrass bed at Miskam Bay in 2001.

The degradation of seagrass beds in Indonesian waters has been poorly documented from only limited areas. The decline of seagrass beds at Banten Bay was caused by converting agricultural areas and fish ponds into an industrial estate, with a total loss of about 116 ha or 26 percent of seagrass mainly in the western part of the bay[42]. The decline of other seagrass beds has been caused by reclamation activities. Less damaging than the reclamation was the uprooting of seagrasses by fishing boats using seine nets to catch shrimp and fish[22]. In Kuta and Gerupuk the decline of seagrass was caused by people collecting dead coral for building material in the seagrass beds.

PRESENT COVERAGE

It is difficult to present accurate information about the present coverage of Indonesian seagrass, since observations on seagrass ecosystems in Indonesia vary considerably in duration, location, method of sampling and object of study, and many places in Indonesia have not been studied yet. Table 16.6 summarizes existing knowledge about the present coverage in Indonesia. Based on this available information, and to the best of our knowledge, we estimate that seagrass covers at least 30 000 km² throughout the Indonesian Archipelago.

Seagrasses in Indonesia are presently threatened mainly by physical degradation such as mangrove cutting and coral reef damage, and by marine pollution from both land- and marine-based sources, and by overexploitation of living marine resources such as fish, mollusks and sea cucumbers. The alarming amount of land reclamation is an increasing cause of seagrass habitat loss in Indonesia.

POLICY

No specific regulation relating to seagrass is currently available and so management is implemented through general regulations pertaining to marine affairs, environmental protection and management of living resources. Of primary importance is the Act of the Republic of Indonesia (RI) No. 5 1990, concerning the conservation of living resources and their ecosystems, together with Act of the RI No. 5 1994 on the ratification of the Convention on Biodiversity and Act of the RI No. 23 1997 concerning the management of the living environment. Apart from acts and statutes, there are three other types of regulation which are hierarchically lower than the former: they are government regulation, presidential decree and ministerial decree.

To have a proper management system for coastal ecosystems, appropriate laws and regulations must be established. The Indonesian Seagrass Committee (ISC) has therefore prepared a draft Seagrass Policy, Strategy and Action Plan to guide the management of the seagrass ecosystem in Indonesia. It forms an integral part of the activities of the South China Sea Project, financed by UNEP-GEF (the United Nations Environment Programme section of the Global Environment Facility), and seeks to address the main issues concerning the management of seagrasses. The draft, scheduled to be completed in 2004, is expected to become a reference document in the formulation of official regulations by the government.

Enhalus acoroides growing among living coral in Komodo National Park, Indonesia.

AUTHORS

T.E. Kuriandewa, W. Kiswara, M. Hutomo, S. Soemodihardjo, Research Centre for Oceanography – Indonesian Institute of Sciences (LIPI), Jalan Pasir Putih 1, Ancol Timur, Jakarta Utara 12190, Indonesia. **Tel:** +62 (0)21 68 3850. **Fax:** +62 (0)21 68 1948. **E-mail:** indo-seagrass@centrin.net.id

REFERENCES

1. Kiswara W [1994]. A review: Seagrass ecosystem studies in Indonesian waters. In: Wilkinson CR, Sudara S, Chou LM (eds) *Proceedings Third ASEAN-Australia Symposium on Living Coastal Resources.* Volume 1. Culalongkorn University, Bangkok. pp 259-281.
2. Larkum AWD, den Hartog C [1989]. Evolution and biogeography of seagrasses. In: Larkum AWD, McComb AJ, Shepherd SA (eds) *Biology of Seagrasses.* Elsevier, Amsterdam. pp 112-156.
3. den Hartog C [1970]. *The Seagrasses of the World.* North-Holland, Amsterdam.
4. Soegiarto A, Polunin N [1981]. Marine Ecosystems of Indonesia: A Basis for Conservation. Report IUCN/WWF Indonesia Programme, Bogor. 254 pp.
5. Kiswara W, Hutomo M [1985]. Seagrass, its habitat and geographical distribution. *Oseana* X(1): 21-30.
6. Nienhuis PH, Coosen J, Kiswara W [1989]. Community structure and biomass distribution of seagrass and macrofauna in the Flores Sea, Indonesia. *Netherlands Journal of Sea Research* 23(3): 197-214.
7. Brouns JJWM [1985]. A preliminary study of the seagrass *Thalassodendron ciliatum* from eastern Indonesia: Biological results of the Snellius II Expedition. *Aquatic Botany* 23(3): 249-260.
8. Erftemeijer PLA [1993]. Factors Limiting Growth and Production of Tropical Seagrasses: Nutrient Dynamics in Indonesian Seagrass Beds. PhD thesis, Nijmegen Catholic University, Nijmegen, Netherlands.
9. Verheij E, Erftemeijer PLA [1993]. Distribution of seagrasses and associated macroalgae in South Sulawesi, Indonesia. *Blumea* 38: 45-64.
10. Zieman JC, Wetzel RG [1980]. Productivity in seagrasses: Methods and rates. In: Phillips RC, McRoy CP (eds) *Handbook of Seagrass Biology: An Ecosystem Perspective.* Garland STPM, New York.
11. Brouns JJWM, Heijs FML [1991]. Seagrasses ecosystem in the tropical west Pacific. In: Mathieson AC, Nienhuis PH (eds) *Intertidal and Littoral Ecosystems. Ecosystems of the World.* Volume 24. Elsevier, Amsterdam. pp 371-390.
12. Lindeboom HJ, Sandee AJJ [1989]. Production and consumption of tropical seagrass fields in eastern Indonesia measured with bell jars and micro-electrodes. *Netherlands Journal of Sea Research* 23(2): 181-190.
13. Azkab MH [1988a]. Pertumbuhan dan produksi lamun, *Thalassia hemprichii* (Ehrenb.) di rataan terumbu Pulau Pari, Kepulauan Seribu. In: Moosa MK, Praseno DP, Sukarno (eds) *Teluk Jakarta, Biologi, Budidaya, Geology, dan Kondisi Perairan.* Lembaga Ilmu Pengetahuan Indonesia, Jakarta. pp 60-66.
14. Azkab MH [1988b]. Pertumbuhan dan produksi lamun, *Enhalus acoroides.* In: Moosa MK, Praseno DP, Sukarno (eds) *Teluk Jakarta, Biologi, Budidaya, Geology, dan Kondisi Perairan.* Lembaga Ilmu Pengetahuan Indonesia, Jakarta. pp 55-59.
15. Moro DS [1988]. Pertumbuhan dan Produksi Jenis Lamun di Pulau Panjang, Teluk Banten. Master's thesis, Universitas Nasional, Jakarta.
16. Azkab MH, Kiswara W [1994]. Pertumbuhan dan produksi lamun di Teluk Kuta, Lombok Selatan. In: Kiswara WK, Moosa MK, Hutomo M (eds) *Struktur Komunitas Biologi Padang Lamun di Pantai selatan Lombok dan Kondisi Lingkungannya.* Lembaga Ilmu Pengetahuan Indonesia, Jakarta. pp 33-41.
17. Kiswara W [2002]. Tehnik transplantasi tunas tunggal dengan perbedaan panjang rimpang *Enhalus acoroides* LF Royle di Teluk

Banten. Makalah disampaikan pada Seminar Nasional Biologi XVI Padang, 22-24 Juli 2002.
18. Dennison WC [1990]. Leaf production. In: Phillips RC, McRoy CP (eds) *Seagrass Research Methods*. UNESCO, Paris. pp 77-79.
19. Suhartati M [1994]. Benthic foraminifera in the seagrass beds of Pari Island, Seribu Island, Jakarta. In: Sudara S, Wilkinson CR, Chou LM (eds) *Proceedings Third ASEAN-Australian Symposium on Living Coastal Resources, Bangkok*. Volume 2. Culalongkorn University, Bangkok. pp 323-329.
20. Susetiono [1993]. Struktur dan kelimpahan meiofauna di antara vegetasi lamun *Enhalus acoroides* di pantai Kuta, Lombok Selatan. In: Kiswara WK, Moosa MK, Hutomo M (eds) *Struktur Komunitas Biologi Padang Llamun di Pantai Selatan Lombok dan Kondisi Lingkungannya*. Lembaga Ilmu Pengetahuan Indonesia, Jakarta. pp 79-95.
21. Aswandy I, Hutomo M [1988]. Komunitas fauna benthic pada padang lamun (seagrass) di Teluk Banten. In: Moosa MK, Praseno JP, Sukarno (eds) *Perairan Indonesia: Biologi, Budidaya, Kualitas Airdan Oseanografi*. Lembaga Ilmu Pengetahuan Indonesia. pp 45-47.
22. Kiswara W [1991]. Sebaran jenis, kerapatan dan biomas lamun (seagrass) di Teluk Lampung. Makalah disampaikan pada Seminar Ilmiah dan Kongres Nasional Biologi X, Bogor, 24-26 September 1991.
23. Kiswara W [1992a]. Vegetasi lamun (seagrass) di rataan terumbu Pulau Pari, Pulau-pulau Seribu, Jakarta. *Oseanologi di Indonesia* 25: 31-49.
24. Aswandy I, Kiswara W [1992]. Studies of crustacean communities living on the seagrass beds of Banten Bay, West Java. In: Chou CL, Wilkinson CR (eds) *Third ASEAN Science and Technology Week Conference Proceedings, Vol 6, Marine Science: Living Coastal Resources, 21-23 September 1992*. National University of Singapore, Singapore. pp 21-23.
25. Kiswara W [1992b]. Community structure and biomass distribution of seagrasses at Banten Bay, West Java-Indonesia. In: Wilkinson C, Ming CL (eds) *Proceedings of the Third ASEAN Science & Technology Week, Vol. 6, Marine Science: Living Coastal Resources, Singapore 21-23 September 1992*. pp 241-250.
26. Mudjiono, Kastoro W, Kiswara W [1992]. Molluscan community of seagrass beds of Banten Bay, West Java. In: Chou CL, Wilkinson CR (eds) *Third ASEAN Science and Technology Week Conference Proceedings, Vol. 6, Marine Science: Living Coastal Resources, 21-23 September 1992*. National University of Singapore, Singapore. pp 241-250.
27. Azis A [1994]. Aktivitas "grazing" bulu babi jenis *Tripneutes gratilla* pada padang lamun di pantai Lombok Selatan. In: Kiswara WK, Moosa MK, Hutomo M (eds) *Struktur Komunitas Biologi Padang Lamun di Pantai selatan Lombok dan Kondisi Lingkungannya*. Lembaga Ilmu Pengetahuan Indonesia, Jakarta. pp 64-70.
28. Azis A, Soegiarto H [1994]. Fauna ekhninodermata padang lamun di pantai Lombok Selatan. In: Kiswara WK, Moosa MK, Hutomo M (eds) *Struktur Komunitas Biologi Padang Lamun di Pantai selatan Lombok dan Kondisi Lingkungannya*. Lembaga Ilmu Pengetahuan Indonesia, Jakarta. pp 52-63.
29. Hutomo M, Parino [1994]. Fauna ikan padang lamun di Lombok Selatan. In: Kiswara WK, Moosa MK, Hutomo M (eds) *Struktur Komunitas Biologi Padang Lamun di Pantai selatan Lombok dan Kondisi Lingkungannya*. Lembaga Ilmu Pengetahuan Indonesia, Jakarta. pp 96-110.
30. Moosa MK, Aswandy I [1994]. Krustasea dari padang lamun di pantai Selatan Lombok dan kondisi lingkunggannya. In: Kiswara WK, Moosa MK, Hutomo M (eds) *Struktur Komunitas Biologi Padang Lamun di Pantai selatan Lombok dan Kondisi Lingkungannya*. Lembaga Ilmu Pengetahuan Indonesia, Jakarta. pp 42-51.
31. Mudjiono, Sudjoko B [1994]. Fauna moluska padang lamun dari pantai Lombok Selatan. In: Kiswara WK, Moosa MK, Hutomo M (eds) *Struktur Komunitas Biologi Padang Lamun di Pantai selatan Lombok dan Kondisi Lingkungannya*. Lembaga Ilmu Pengetahuan Indonesia, Jakarta. pp 71-78.
32. Azkab MH [1991]. Study on seagrass community structure and biomass in the southern part of Seribu Islands. In: Alcala AC (ed) *Proceedings of the Regional Symposium on Living Resources in Coastal Areas. ASEAN-Australia Cooperative Program in Marine Sciences*, Australian Institute of Marine Science and University of the Philippines. pp 353-362.
33. Atmadja. Personal communication.
34. Peristiwadi T [1994a]. Makanan ikan-ikan utama di padang lamun Lombok Selatan. In: Kiswara WK, Moosa MK, Hutomo M (eds) *Struktur Komunitas Biologi Padang Lamun di Pantai selatan Lombok dan Kondisi Lingkungannya*. Lembaga Ilmu Pengetahuan Indonesia, Jakarta. pp 111-125.
35. Tomascik. Personal observation.
36. Cappenberg HAW [1995]. Komunitas Moluska di Padang Lamun Teluk Kotania, Seram Barat. Perairan Maluku dan Sekirtarnya. Balitbang Sumberdaya Laut, Puslitbang Oseanologi –LIPI Ambon. pp 19-26.
37. Hutomo M, Martosewojo S [1997]. The fishes of seagrass community on the west side of Burung Island (Pari Islands, Seribu Island) and their variation in abundance. *Marine Research in Indonesia* 17: 147-172.
38. Hutomo M [1985]. Telaah Ekologik Komunitas Ikan Pada Padang Lamun (Seagrass, Anthophyta) di Perairan Teluk Banten. Thesis Doktor, Fakultas Pasca Sarjana, Institut Pertanian Bogor, Bogor.
39. Peristiwady T [1994b]. Ikan-ikan di padang lamun P. Osi dan P. Marsegu, Seram Barat: I. Struktur komunitas. *Perairan Maluku dan Sekitarnya*. 7: 35-52.
40. Widodo [1991]. Personal communication.
41. Hernowo [1993]. Personal communication.
42. Douven WJAM, Buurman J, Kiswara W [In preparation]. Spatial tools to support coastal research and management: The example of seagrass beds in Banten Bay, Indonesia.
43. Kiswara W, Djamali A [1995]. Seagrass community and associated biota in Banten Bay, West Java-Indonesia: Problems and research priorities. In: Heun JC, Lindeboom HJ, Tiwi DA, Garno YS, Burhannuddin, Praseno D, Makarim H, Mukaryanti, Anyuta, Nugroho S (eds) *Proceedings of the Workshop on Marine and Coastal Research, BPP Teknologi and Netherlands Organization. for Scientific Research (NOW) 1-4 May 1995, Jakarta*. pp 93-107.
44. Kiswara W, Genisa AS, Purnomo LH [1991]. Preliminary Study: Species Composition, Abundance and Distribution of Fishes in the Seagrass Beds of Banten Bay. Mangrove and Seagrass Fisheries Connection Workshop, 26-30 August 1991, Ipoh-Perak, Malaysia.
45. Sugama K, Eda H [1986]. Feasibility Survey on Collection of Natural Seeds of some Promising Fish for Floating Net-Cage Culture in Banten Bay. Scientific Report.
46. Hendrokusumo S, Sumitro S, Tas' an [1979]. The distribution of the dugong in Indonesian waters. In: Marsh H (ed) *The Dugong. Proceedings of a Seminar/Workshop held at James Cook University 8-13 May 1979, North Queensland, Australia*. pp 5-10.
47. Douven WJAM [1999]. Human Pressure on Marine Ecosystems in the Teluk Banten Coastal Zone: Recent Situation and Future Prospects. Teluk Banten Research Program. Report Series No. 3. IHE-Delft, WOTRO.
48. Zainudin [2002]. Personal communication.

Regional map: Asia IX

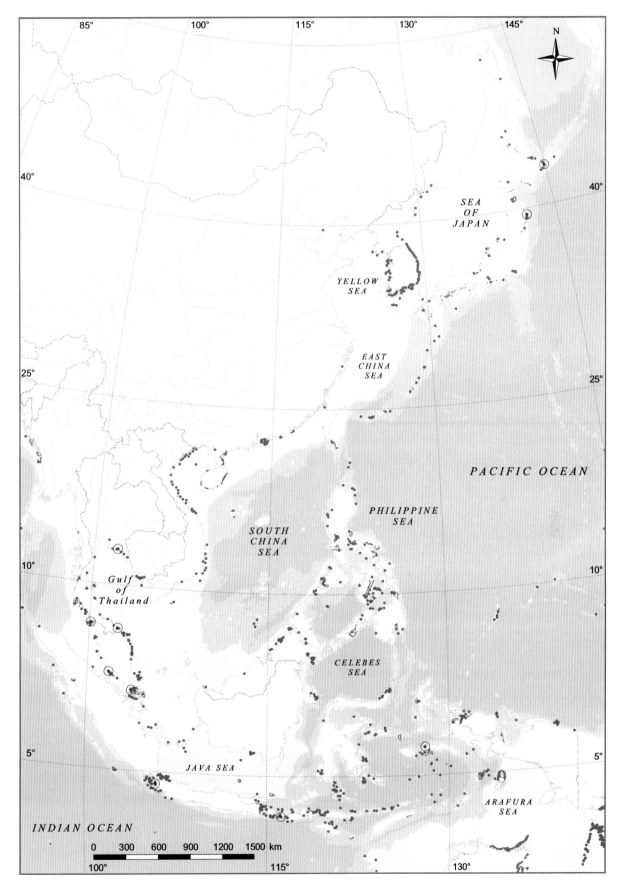

THE SEX LIFE OF SEAGRASSES

Posidonia oceanica inflorescence with fruits, western Mediterranean.

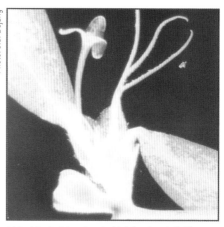

Halophila decipiens female (right) and male (left) flowers, Malaysia.

Enhalus acoroides in Malaysia and Indonesia: female flowers (above); pollen on the water surface (right, top); fruit (right, middle); seed dispersal (right, bottom).

Spadix of eelgrass, *Zostera marina*, with male flowers releasing pollen.

Posidonia australis fruit meadow, Western Australia.

Thalassia hemprichii fruit.

The seagrasses of
The Philippines and Viet Nam

Seagrasses are found extensively throughout the Philippine Archipelago. There are documented sizeable beds offshore from western, northwestern and southern islands covering 978 km^2 at 96 well-studied sites. Approximately one third of this area has been mapped in detail using a combination of remote sensing and field survey techniques. The remainder is estimated. With many other areas not surveyed for seagrasses, the total seagrass area is likely to be many times greater.

The Philippines is reported to have 15 species of seagrass[1-6]. In addition to *Ruppia maritima* and *Halophila beccarii*, Fortes lists a new variety of *Halophila minor*[7]. Calumpong and Meñez[8] consider *Halophila beccarii* to have been extirpated from Philippine waters, because the only specimens to be collected were in 1912 from Manila Bay[9], now heavily impacted by the growth of metropolitan Manila. Fortes disagrees, believing this species still occurs in Manila Bay[5] and to be common in Lingayen Gulf, northwestern Philippines.

Many plants and animals live in the seagrass beds of the Philippines and Viet Nam, supporting fisheries with their rich nutrient pool and the diversity of physical structures protecting juveniles from predators. Major commercial fisheries occur immediately adjacent to seagrass beds[10]. Fish and shrimp are the most important elements of the commercial fishery, although coastal villages derive their sustenance from other components of the grass beds. The major invertebrates found in the beds are shrimps, sea cucumbers, sea urchins, crabs, scallops, mussels and snails. Some endangered species of sea turtles reported in seagrass beds include the green sea turtle, the olive Ridley, the loggerhead and the flatback. In the Philippines and Viet Nam the sea cow (*Dugong dugon*), which is almost completely dependent on seagrass, is an endangered species.

Coral reefs and their associated seagrasses potentially supply more than 20 percent of the fish catch in the Philippines. A total of 1 384 individuals and 55 species from 25 fish families were identified from five seagrass sites[1].

Calumpong and Meñez[8] describe two mixed species associations, one of *Syringodium* (or sometimes *Thalassia*) with *Cymodocea* and *Halodule* spp., growing primarily on sandy sediment, the other of *Enhalus* and *Thalassia* spp. on muddy substrates. Monospecific seagrass beds are less common than mixed populations in the Philippines and tend to occur under certain conditions: *Enhalus acoroides* colonizes turbid, quiet, protected areas such as bays and estuaries and *Thalassia hemprichii* occurs as pure stands in the tidepools of the most northerly islands in the Philippines[11]. *Halophila decipiens* grows primarily at depths of 11-23 m[3]. *Thalassodendron ciliatum* and *Halophila spinulosa* are found in deep, clear water[5]. *Thalassodendron ciliatum* also grows in shallow waters but only in conditions of low turbidity on coarse or rocky substrates[2]. Cuyo Island is the northernmost limit of *Thalassodendron ciliatum* in the Pacific[5].

The vital role of seagrasses as nursery grounds and food for fish and invertebrates in the Philippines has been appreciated for some time[12]. The rabbitfish, *Siganus canaliculatus*, is a voracious herbivore and particularly important as a food species. In Bais Bay, Negros Oriental, the population of *Siganus canaliculatus* consumes 0.64 metric tons per day from a 52-ha *Enhalus acoroides* meadow[13]. However, this represents less than 1 percent of the daily organic production of *Enhalus acoroides*. Rabbitfish are often caught in seagrass beds using bamboo traps[7], representing a direct link between seagrass habitat and human subsistence.

Seagrass beds in the Philippines are threatened by eutrophication, siltation, pollution, dredging and unsustainable fishing methods. Many thousands of hectares of seagrass have been lost as a result of land reclamation for housing, airports and shipping

Spear-fisher over a seagrass bed in the Philippines.

facilities[8]. Some attempts have been made at rehabilitating damaged seagrass beds using transplanting techniques[14].

Puerto Galera, a quiet ecotourist destination south of Manila, is the site of one of the first SeagrassNet global monitoring locations. Quarterly sampling of the seagrass habitat has been conducted at reference and impacted seagrass sites monitored by graduate students from the University of the Philippines. Even in the early stages of monitoring the SeagrassNet team has clearly shown the impacts of eutrophication at the site adjacent to a coastal town. Seagrass-Watch is now established in Puerto Galera. This community-based seagrass monitoring program is coordinating with SeagrassNet to provide a second data stream, generated by volunteer members.

VIET NAM

There are 11 species of seagrass in Viet Nam distributed along the coastline but mostly from the middle to the southern sections. Their status is unknown though in general the Viet Nam coastal zone has been heavily impacted by sedimentation and domestic and agricultural pollution. Viet Nam has at least 440 km^2 of seagrasses determined from remote sensing and ground-truth surveys. Viet Nam is at the overlap of temperate and tropical seagrass species with *Zostera japonica* growing intertidally in the north and mixing with *Halophila ovalis*, while in the south the species composition is similar to the Philippines and Malaysia.

AUTHORS

Miguel Fortes, Marine Science Institute CS, University of the Philippines, Diliman, Quezon City 1101, Philippines. **Tel:** + 63 (0)2 9205301. **Fax:** +63 (0)2 9247678. **E-mail:** mdfortes138@yahoo.com

Ed Green, UNEP World Conservation Monitoring Centre, 219 Huntingdon Road, Cambridge CB3 0DL, UK.

Fred Short, University of New Hampshire, Jackson Estuarine Laboratory, 85 Adams Point Road, Durham, NH 03824, USA.

REFERENCES

1. Meñez EG, Phillips RC, Calumpong HP [1983]. *Seagrasses from the Philippines*. Smithsonian Contributions to the Marine Sciences No. 21. pp 1-40.
2. Meñez EG, Calumpong HP [1983]. *Thalassodendron ciliatum*: An unreported seagrass from the Philippines. *Micronesia* 18: 103-111.
3. Meñez EG, Calumpong HP [1985]. *Halophila decipiens*: An unreported seagrass from the Philippines. *Proceedings of the Biological Society of Washington* 98(1): 232-236.
4. Fortes MD [1990]. Seagrass resources in East Asia: Research status, environmental issues and management perspectives. In: *ASEAMS/UNEP Proceedings of First ASEAMS Symposium on Southeast Asian Marine and Environmental Protection*. UNEP Regional Seas Reports and Studies No. 116. pp 135-144.
5. Fortes MD [1986]. Taxonomy and Ecology of Philippine Seagrasses. PhD dissertation, University of the Philippines, Diliman, Quezon City. 245 pp.
6. Short FT, Coles RG (eds) [2001] *Global Seagrass Research Methods*. Elsevier Science, Amsterdam.
7. Fortes MD [1989]. *Seagrasses: A Resource unknown in the ASEAN Region*. ICLARM Education Series 5. 46 pp.
8. Calumpong HP, Meñez EG [1997]. *Field Guide to the Common Mangroves, Seagrasses and Algae of the Philippines*. Bookmark, Makati City, Philippines. 197 pp.
9. Merrill ED [1912]. *A Flora of Manila*. Philippine Islands Bureau of Science Publication Number 5. Bureau of Printing, Manila. 490 pp.
10. Phillips RC, Meñez EG [1988]. *Seagrasses*. Smithsonian Contributions to the Marine Sciences No. 34. 104 pp.
11. Calumpong HP, Meñez EG, Phillips RC [1986]. Seagrasses in Batanes Province, northern Philippines. *Silliman Journal* 33(1-4): 148-154.
12. Dolar MLL [1991]. A survey on the fish and crustacean fauna of the seagrass beds in North Bais Bay, Negros Oriental, Philippines. In: *Proceedings of the Regional Symposium on Living Resources in Coastal Areas*. University of the Philippines Marine Science Institute, Quezon City. pp 367-377.
13. Leptein MV [1992]. The gut passage rate and daily food consumption of the rabbitfish *Siganus canaliculatus* (Park). In: *Third ASEAN Science and Technology Week Conference Proceedings Vol. 6*. National University of Singapore and National Science Technology Board, Singapore. pp. 327-336.
14. Calumpong HP, Phillips RC, Meñez EG, Estacion JS, de Leon ROD, Alava MNR [1993]. Performance of seagrass transplants in Negros Island, central Philippines and its implications in mitigating degraded shallow coastal areas. In: *Proceedings of the 2nd RP-USA Phycology Symposium/Workshop*. Philippine Council for Aquatic and Marine Research and Development, Los Baños, Laguna. pp 295-313.

17 The seagrasses of JAPAN

K. Aioi

M. Nakaoka

Sixteen seagrass species, including seven temperate (Zosteraceae) species and nine tropical species (Hydrocharitaceae and Cymodoceaceae), occur on the coasts of Japan, about a quarter of the total number of seagrass species in the world (Table 17.1). Species diversity is high not only for seagrasses but also for algal flora, with about 1500 species of algae occurring around Japan. Such a high species diversity in Japanese marine flora is probably related to complex hydrodynamic properties around the Japanese coasts that are affected by several major ocean currents such as the Oyashio cold current, and the Kuroshio and Tsushima warm currents.

Among the 16 species of seagrasses in Japan, nine belong to the families Hydrocharitaceae and Cymodoceaceae and are tropical species commonly found in tropical and subtropical areas of the Indo-West Pacific region[1]. In Japan, their distribution is restricted to the southwestern islands (Ryukyu and Amami Islands) except for *Halophila ovalis*. In contrast, distribution of all the species of Zosteraceae is limited to the main island areas, except for *Zostera japonica* which also occurs in the Ryukyu Islands. Thus, the seagrass flora in Japan differ distinctly between the subtropical southwestern islands and the temperate coasts of the main islands. The southern limit of the temperate species of Zosteraceae is determined by the summer high seawater temperature of 28°C around Kyushu, while the distribution of tropical seagrass species is restricted by the winter seawater temperature of 15°C[2].

Along the temperate coasts of China, the Korean Peninsula and the islands of Japan, species diversity of Zosteraceae is high. In addition to *Zostera marina*, a cosmopolitan species widespread in the northern hemisphere in both the Pacific and Atlantic Oceans, six species of the family Zosteraceae are present that are considered to be endemic to the northwestern Pacific (Japanese, Korean, Chinese and southeast Russian waters), namely *Zostera asiatica*, *Zostera caespitosa*, *Zostera caulescens*, *Zostera japonica*, *Phyllospadix iwatensis* and *Phyllospadix japonicus*[3,4]. The region can be regarded as a "hotspot" of seagrass floral diversity within the temperate waters of the northern hemisphere. Most of these species have limited distribution in some localities along the northern part of Japan (see below). The hemispheric distributions in the western Pacific may reflect the speciation process of Zosteraceae from its possible ancestral origins in equatorial regions[5].

Despite the high species diversity of Japanese seagrasses, there are relatively few ecological studies of these species with the exception of *Zostera marina*. After pioneer studies by Tomitaro Makino and Shigeru Miki who described these species in the late 19th and early 20th centuries, few seagrass studies were conducted in Japan until the early 1970s. This is especially true for the endemic species of Zosteraceae for which information on distribution and ecology was not available until recently, partly due to their occurrence in deep water (see below).

Eelgrass, *Zostera marina*, is a cosmopolitan species commonly found in temperate to subarctic coasts in the northern hemisphere[1]. In Japan, *Zostera marina* occurs in numerous localities along the coastlines of the main islands, i.e. Honshu, Hokkaido, Kyushu and Shikoku[4]. The northernmost population of *Zostera marina* in Japan is found near Soya Cape, Hokkaido (45°30'N)[6] and the southernmost population in Satsuma Peninsula, Kyushu (31°10'N)[7]. Most eelgrass populations in Japan are perennial and extend their distribution both by clonal propagation of rhizomes and by seed production, although an annual form of *Zostera marina* is found in some localities such as Hamana-ko, Okayama and Kagoshima[8,9].

Zostera japonica is a small seagrass that generally inhabits intertidal and shallow subtidal bottoms along the coast of East Asia, from Viet Nam to

Sakhalin and Kamchatka, Russia[1, 10]. In Japan, *Zostera japonica* is found in various localities, such as Notsuke Bay in the northeastern part of Hokkaido[11], Toyama Bay in the Sea of Japan[12-14], Sagami Bay on the Pacific coast of central Honshu[15] and the Ryukyu Islands, in the southwestern part of Japan[16-18].

Zostera asiatica was originally recorded by Miki from southern Sakhalin (Russia), the northeastern and southern parts of Hokkaido, in the central part of Honshu facing the Sea of Japan, and on the eastern coast of the Korean Peninsula[19]. In Japan, populations of *Zostera asiatica* are currently known only in Hamanaka and Akkeshi Bay, Hokkaido[3], and in Funakoshi Bay, on the northeastern coast of Honshu[20]. Additionally, the stranded dead plants have been collected at several beaches in Hokkaido and one site in Toyama Bay[3, 14].

Zostera caulescens was known from limited localities along the central to northern coast of Honshu and the southern coast of the Korean Peninsula when Miki first described this species[19, 21]. Some recent papers report the existence of populations in Mutsu Bay, northern Honshu[3], along the Sanriku coast, northeastern Honshu[3, 22, 23], in Tokyo Bay and Sagami Bay, on the Pacific coast of central Honshu[2, 15, 24-26], and in Toyama Bay, on the Sea of Japan[13].

Zostera caespitosa was reported to occur in Hokkaido, in the northern half of Honshu and on the east of the Korean Peninsula[19]. Populations of this species were recently reported from Notoro Lake and Notsuke Bay in Hokkaido[10, 27], Mutsu Bay, northern Honshu[3], Yamada Bay and Otsuchi Bay, northeastern Honshu[28, 29], and Toyama Bay, in the Sea of Japan[14].

Two *Phyllospadix* species, *Phyllospadix iwatensis* and *Phyllospadix japonicus*, inhabit the intertidal and subtidal rocky bottoms of temperate regions of Japan. Distribution of *Phyllospadix iwatensis* ranges from Hokkaido to the northern part of Honshu. *Phyllospadix japonicus* occurs in the central part of the Pacific coast of Honshu and the western part of the Sea of Japan coast of Honshu[30].

Among the nine tropical seagrass species belonging to the families Hydrocharitaceae and Cymodoceaceae, *Halophila ovalis* has the widest distribution, occurring from the Yaeyama Islands to Chiba Prefecture (Odawa Bay) and to Toyama Bay, on the Sea of Japan[13]. The distribution of the other eight species is restricted to the Amami and Ryukyu Islands (Table 17.1). Detailed information on island-by-island distribution of each species has been given[17, 18, 31].

Geographical distribution of these species overlaps widely, with multispecific seagrass habitats commonly observed both in the temperate and tropical regions of Japan. In Hokkaido, three species coexist in a single bed in Notoro-ko (*Zostera marina*, *Zostera japonica* and *Zostera caespitosa*) and in Akkeshi Bay (*Zostera marina*, *Zostera asiatica* and *Phyllospadix iwatensis*). In Honshu, four seagrass species co-occur in Odawa Bay (*Zostera marina*, *Zostera japonica*, *Zostera caulescens* and *Halophila ovalis*)[14] and in Otsuchi Bay (*Zostera marina*, *Zostera caespitosa*, *Zostera caulescens* and *Phyllospadix iwatensis*), and three species co-occur in Funakoshi Bay (*Zostera marina*, *Zostera asiatica* and *Zostera caulescens*) and in Iida Bay (*Zostera marina*, *Zostera japonica* and *Zostera caespitosa*). In the Ryukyu Islands, nine species were found in a single seagrass bed in Iriomote Island (*Enhalus acoroides*, *Thalassia hemprichii*, *Halophila ovalis*, *Cymodocea rotundata*, *Cymodocea serrulata*, *Halodule pinifolia*, *Halodule uninervis*, *Syringodium isoetifolium* and *Zostera japonica*)[17]; eight species were found in some beds at Ishigaki Island, Miyako Island and Okinawa Island[17, 32, 33].

BIOGEOGRAPHY

The depth range of seagrasses in Japan is reported for some multispecific seagrass beds where two or more species coexist in a single bed[34]. Generally, each species in the mixed beds shows a different depth

Table 17.1
Seagrasses recorded in Japan

Species	Distribution
Hydrocharitaceae	
Enhalus acoroides	Ryukyu Islands
Thalassia hemprichii	Ryukyu Islands
Halophila decipiens	Ryukyu Islands
Halophila ovalis	From Ryukyu Islands to central Honshu
Cymodoceaceae	
Cymodocea rotundata	Ryukyu Islands
Cymodocea serrulata	Ryukyu Islands
Halodule pinifolia	Ryukyu Islands
Halodule uninervis	Ryukyu Islands
Syringodium isoetifolium	Ryukyu Islands
Zosteraceae	
Phyllospadix iwatensis	North Honshu and Hokkaido
Phyllospadix japonicus	South Honshu
Zostera asiatica	Hokkaido and north Honshu
Zostera caespitosa	Hokkaido and north Honshu
Zostera caulescens	Central and north Honshu
Zostera japonica	From Ryukyu Islands to Hokkaido
Zostera marina	From Kyushu to Hokkaido

distribution, forming some specific patterns of zonation along depth gradients.

Among *Zostera* spp. in temperate multispecific seagrass meadows, *Zostera japonica* is always found in the uppermost parts of the bed, as its main habitat is intertidal flats. *Zostera marina* occurs in the shallowest parts of subtidal beds, mostly between 1 and 5 m deep, but in some places down to 10 m. *Zostera asiatica* occurs between the intertidal zone and a depth of 5 m. Two other species generally occur in deeper habitats than *Zostera marina*: *Zostera caespitosa*, between 1 and 20 m, and *Zostera caulescens*, between 3 and 17 m. In most of the mixed seagrass beds, the plants' depth ranges overlap to some degree with *Zostera marina*. Depth zonation in *Zostera asiatica*, *Zostera caespitosa* and *Zostera caulescens* cannot be described since these species do not generally co-occur.

In multispecific seagrass beds in the Ryukyu Islands, *Halodule pinifolia*, *Cymodocea rotundata* and *Thalassia hemprichii* are dominant in the intertidal to upper subtidal zone, while *Cymodocea serrulata* and *Enhalus acoroides* are more abundant in the deeper subtidal zone[32,33]. The observed depth distribution of these species generally agrees with those reported in other parts of the tropical Indo-West Pacific region[35-38].

Quantitative studies on biomass, shoot density, shoot size and productivity have been conducted in about 30 seagrass beds in Japan[34]. Biomass, shoot density and shoot size of *Zostera marina* vary greatly among and within populations. Within populations, biomass, shoot density and shoot size have sometimes varied more than twofold, with greater biomass generally observed in shallower parts of seagrass beds[39,40]. Biomass as high as 500 g dry weight/m^2 was recorded in some areas such as Notsuke Bay, Otsuchi Bay, Ushimado and Maizuru Bay, whereas maximum biomass was less than 200 g dry weight/m^2 for populations in Odawa Bay, Yanai Bay and Toyama Bay[34]. Estimates for above-ground net production were available for several *Zostera marina* populations, and varied between less than 1 g to 13 g dry weight/m^2/day at different sites, depths and seasons. Between-site variation in these parameters did not appear to be correlated with variations in latitude or geographical distances.

Quantitative information on abundance and productivity is very sparse for other species of *Zostera*, and for the tropical seagrass species, in Japan. Biomass of *Zostera japonica* varies greatly with season; a maximum biomass of 270 g dry weight/m^2 was recorded in July and a minimum of 30 g dry weight/m^2 from December to January at Mikawa Bay, the Pacific side of central Honshu[41]. For *Zostera caespitosa*, maximum above-ground biomass of 60 g dry weight/m^2 was recorded for the population in Iida Bay, Noto Peninsula (the Sea of Japan coast)[42]. For *Zostera asiatica* at Akkeshi Bay, Hokkaido, biomass (427 g dry weight/m^2) was twice that of *Zostera marina*, whereas the shoot density (134 shoots/m^2) was about half of *Zostera marina* when comparing monospecific stands at the same depth, reflecting the larger shoot size of the former[43]. The above-ground net production of *Zostera asiatica* was estimated to be 3-5 g dry weight/m^2/day.

Information on the flowering and fruiting seasons of *Zostera marina* and other *Zostera* species is available from some localities in Japan[34]. In *Zostera marina*, seasons for flowering and fruiting vary by 2-4 months across the region, with early flowering and fruiting observed at lower latitudes. Seed germination of *Zostera marina* was generally observed during the winter in southern populations and during spring in northern populations. For other *Zostera* species, flowering and fruiting seasons have been reported only from limited localities, and vary greatly among

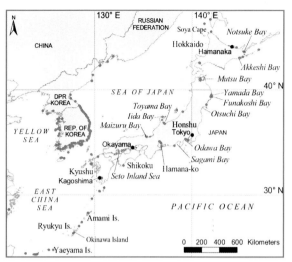

Map 17.1
Japan

Meadow of *Zostera caespitosa* at Yamada Bay, Japan.

localities. For *Zostera asiatica* and *Zostera caulescens*, the flowering and fruiting seasons are generally the same as those of coexisting *Zostera marina*, whereas *Zostera caespitosa* flowers and fruits about one month earlier than sympatric *Zostera marina*[34].

HISTORICAL PERSPECTIVES

Although seagrasses have been very familiar to, and traditionally utilized by, Japanese people living in coastal areas, very little information is available about the historical distribution of seagrass beds. In Tokyo Bay, an old chart issued in 1908 shows the distribution of *Zostera marina* and *Zostera japonica* in the early 20th century. Extensive seagrass meadows of approximately 3-5 km^2 in area were located in shallow waters (<3 m) in some localities such as Yokohama, Tokyo, Funabashi and Chiba[44]. Unfortunately, all these seagrass beds were destroyed in land reclamation (filling and hardening of the shoreline) projects during industrialization in the mid-20th century.

Traditional uses of seagrasses in Japan include direct use as fiber for rope or padding (e.g. cushions and tatami mats) or use as agricultural compost (Table 17.2). Besides those listed, there may well be further traditional uses. For example, eelgrass was named *moshiogusa* in Japanese, which means salt grass, and might have been used to produce salt.

Table 17.2
Traditional uses of seagrasses in Japan

Traditional use	Species	Locality
Rope for gill net	*Phyllospadix iwatensis*	Hokkaido
Cushions for horse saddles	*Zostera marina*	Sanriku (Miyagi Pre.)
Fishermen's skirts	*Phyllospadix iwatensis* and *Zostera marina*	Sanriku (Iwate Pre.)
Cushions for train seats	*Zostera marina*	Tokyo Bay area
Tatami mats	*Zostera marina*	Seto Inland Sea
Agricultural compost	Algae and *Zostera marina*	Miura Peninsula (Kanagawa Pre.)
Agricultural compost	Algae and *Zostera marina*	Hamana-ko (Shizuoka Pre.)
Agricultural compost	*Zostera marina*	Mikawa Bay
Agricultural compost	*Zostera marina*	Seto Inland Sea (Okayama Pre.)
Agricultural compost	*Zostera marina* and freshwater plants	Nakaumi (Shimane Pre.)

Note: Pre. = Prefecture.

ESTIMATES OF HISTORICAL LOSS AND PRESENT COVERAGE

The area of seagrass beds, especially those consisting of eelgrass *Zostera marina*, has declined since the 1960s, mainly because of land reclamation. During these last decades, the Japanese economy has developed rapidly. The Environment Agency of Japan surveyed the status of algal and seagrass beds along most of the coastal areas of Japan in 1978 and again in 1991, from which the loss during this period was estimated (Table 17.3). The total area of algal and seagrass beds together was 3262 km^2 in 1978 and 3159 km^2 in 1991. For seagrass beds, the total area declined from 515 km^2 to 495 km^2 during the period, i.e. about 4 percent of Japan's total seagrass resource was lost in 13 years. In particular, more than 30 percent of *Zostera marina* beds disappeared in localities such as Ariake Bay, Kagoshima Bay and Hyuga-nada in Kyushu during this period[7]. In the Seto Inland Sea, more than 70 percent of *Zostera marina* beds have been lost since 1977, a loss which has seriously affected coastal fisheries[4, 45].

For regionally endemic species of *Zostera*, the situation may be more serious than for *Zostera marina*, because populations are now known to exist in only a few localities around Japan. In fact, *Zostera asiatica* and *Zostera caulescens* are now ranked as VU (vulnerable stage) in the Red Data Book of threatened Japanese plant species[46]. Among tropical seagrass species, *Enhalus acoroides* and *Halophila decipiens* are found only in limited localities in the Ryukyu Islands, and they are also listed as VU in the Red Data Book.

Table 17.3
Estimates of total areas of algal and seagrass beds in Japan in 1978 and 1991, and the percent area lost during the period

	Area of macrophyte beds (km^2)		Area lost (km^2)	% lost
	1978	1991		
Algal beds*	2748	2664	83	3.0
Seagrass beds	515	495	21	4.0
Total	3263	3159	104	3.2

Note: *Algal beds consisted of *Ulva*, *Enteromorpha*, *Sargassum*, *Laminaria*, *Eisenia* and *Gelidium*..

Source: As reported by the Japanese Environment Agency in 1994.

PRESENT THREATS

As described above, seagrasses have been disappearing rapidly due to industrial development in the coastal regions of Japan. Major threats for further

Case Study 17.1
AKKESHI, EASTERN HOKKAIDO

In Hokkaido, the northernmost island of Japan, some healthy *Zostera marina* beds remain. At Akkeshi (42°50'N, 144°50'E), located in eastern Hokkaido, extensive seagrass meadows occur in Akkeshi-ko (a brackish lagoon) adjacent to Akkeshi Bay. In Akkeshi-ko, the dominant seagrass is *Zostera marina* with minor amounts of *Zostera japonica*. A large-scale study was recently initiated here to examine the interactions between terrestrial and coastal ecosystems[48]. It was found that considerable amounts of nutrients of terrestrial origin flow into this lagoon and these are important for the productivity of *Zostera marina* and associated communities. Studies on food web dynamics in the *Zostera marina* bed have revealed that the major consumer of *Zostera marina* was the whooper swan (*Cygnus cygnus*) which overwinters in the lagoon, and that mysids are the most dominant herbivores grazing on epiphytic algae on eelgrass[49, 50]. Both the biomass and the diversity of mysids are high, and this supports high productivity of commercially important fish and shellfish species such as epifaunal shrimps and several species of fish[50].

Zostera marina leaves coated with epiphytes.

decline in present seagrass coverage include land reclamation, environmental deterioration such as reduced water quality, and rise in water temperature and water level due to global warming.

The loss of seagrass vegetation over the last two decades along the Japanese coast is mostly attributed to land reclamation[44]. Many land reclamation projects are still ongoing or at the planning stage, and will probably further accelerate the loss of seagrass beds. For example, the coastline has been damaged by land reclamation and port construction in the Ryukyu Islands where large economic investments have been made toward rapid modernization. The natural ecosystems of coral reefs and lagoons were greatly impacted, especially along the coasts of Okinawa Island. Dugongs inhabit several seagrass beds in the northeastern coast of Okinawa Island, which is the northern limit of global distribution of this threatened marine mammal. Nevertheless, a large-scale land reclamation project is now planned in the center of the seagrass beds (Henoko coral lagoon) to build an offshore runway for the US air base. Such construction would almost certainly be fatal to the lagoon ecosystem and directly destroy seagrass habitats for dugongs. The Environment Agency of Japan decided to make a general survey of the northernmost dugongs and their habitats in February 2002. Scientists and non-governmental organizations in Japan must support and collaborate in these surveys to save the dugongs and conserve their habitats.

Some seagrass beds have been declining rapidly even in areas where no major land reclamation has occurred, such as the Seto Inland Sea. In these areas, water pollution and disturbance of habitats by fish trawling are major causes of decline in seagrass beds.

In the case of multispecific seagrass meadows, changes in environmental conditions due to human activities have effects not only on overall seagrass distribution and abundance but also on the species composition of the seagrass beds. In Odawa Bay near the Tokyo metropolitan area, for example, reduced light condition due to eutrophication over the past 20 years caused a decrease in areas of *Zostera marina* in shallow habitats, but possibly favored the deeper-living *Zostera caulescens* to expand its populations into shallower depths[15]. However, due to lack of species-by-species data in past literature, it is not possible to determine whether *Zostera caulescens* truly increased in recent years. Long-term field surveys of seagrass beds using a unified approach are necessary in order to monitor future changes in seagrasses in relation to changes in environmental conditions.

Case Study 17.2
RIAS COAST IN IWATE PREFECTURE, NORTHEASTERN HONSHU

Five temperate seagrass species, *Zostera marina*, *Zostera caulescens*, *Zostera caespitosa*, *Zostera japonica* and *Phyllospadix iwatensis*, occur in the three bays along the Rias Coast facing the northeastern Pacific, namely Yamada Bay, Funakoshi Bay and Otsuchi Bay, in Iwate Prefecture. The species composition of seagrasses varies among the bays. In Yamada Bay, *Zostera caespitosa* is the most abundant, with *Zostera marina* co-occurring. The dominant species in the seagrass bed in Funakoshi Bay is *Zostera caulescens* with small patches of *Zostera marina* and *Zostera asiatica* occurring at the shallower part of the bed. *Zostera caulescens*, *Zostera marina* and *Zostera caespitosa* are found to coexist in several seagrass beds in Otsuchi Bay.

A large-scale census of these seagrass beds has been undertaken using an acoustic sounding technique to estimate overall distribution and abundance of seagrasses[51]. The survey in Funakoshi Bay has shown that the areal extent of the seagrass bed was approximately 0.5 km² with the depth distribution extending from 2 to 17 m. Variation in canopy height of *Zostera caulescens* by depth was also analyzed from the echo-trace of the sounder. The same technique has been utilized to estimate the abundance of *Zostera caespitosa* in Yamada Bay and to monitor long-term changes in patch dynamics of a seagrass bed at a river mouth on Otsuchi Bay.

In the seagrass bed at Funakoshi Bay, *Zostera caulescens* develops a high canopy at the deeper parts of the bed (>10 m) by extending long flowering shoots. A maximum shoot height of 6.8 m was recorded in 1998, known as the world's record longest among all seagrasses[22]. In July 2000, an even longer shoot (7.8 m) was found at the same site. Studies of the dynamics and production of the *Zostera caulescens* population revealed that most of the flowering shoots emerge in winter and grow rapidly, reaching an average height of 5 m in late summer[52]. Annual above-ground net production per area was estimated to be 426 g dry weight/m²/year, similar to estimates for other *Zostera* species that live in intertidal and shallow subtidal beds (<1 m deep). Thus, the productivity of *Zostera caulescens* is quite high despite its distribution in deep water (4-6 m) with poor light conditions. Comparative morphological and phenological studies of *Zostera caulescens* between Iwate and Sangami Bay (near Tokyo) showed that the large differences in shoot height and seasonal dynamics are probably related to differences in environmental factors such as temperature[23].

In these seagrass beds, the abundance and dynamics of associated communities have been investigated for epiphytic algae[53], sessile epifauna[54], mobile epifauna[53,55,56] and benthic infauna[57]. The dynamics of these organisms are greatly influenced by spatial and temporal variations in seagrass abundance. Most interestingly, a species of tanaid crustacean (*Zeuxo* sp.) was found to feed on predispersal seeds of *Zostera marina* and *Zostera caulescens*[58]. The crustacean consumes up to 30 percent of the seeds, which may have a large negative impact on the seed abundance of the seagrasses.

Photo: K. Aioi

The world's longest seagrass, *Zostera caulescens*, at 10 m deep in Funakoshi Bay.

The global circulation of ocean currents is important not only for land vegetation but also for marine plants, as distributions of temperate and tropical seagrass species are restricted by seawater temperature in summer and winter, respectively, along the Japanese archipelago. Most physicists and meteorologists believe that the seas have warmed from 2°C to 5°C over the past 50 years due to global warming. Global warming is predicted to affect the photosynthetic activities of marine plants in Japan. A further warming of 2 or 3°C in the seawater temperature may prove fatal to seagrass beds in shallower areas[47]. Shallow water vegetation such as seagrass and algae along the coasts of Japan is also at risk of accelerated loss due to water level rise caused by global warming.

ACKNOWLEDGMENTS
We thank Y. Tanaka and M. Watanabe for providing information and photographs of Japanese seagrasses, and N. Kouchi and F.T. Short for reviewing the manuscript.

AUTHORS
Keiko Aioi, Aoyama Gakuin Women's Junior College, Shibuya 4-4-25, Shibuya, Tokyo 150-8366, Japan. **Tel/fax:** +81 (0)3 3313 1296. **E-mail:** aioi357@galaxy.ocn.ne.jp

Masahiro Nakaoka, Graduate School of Science and Technology, Chiba University, Inage, Chiba 263-8522, Japan.

REFERENCES
1 den Hartog C [1970]. *The Seagrasses of the World*. North Holland Publishing, Amsterdam.
2 Nozawa R [1974]. Aquatic plants in the sea. *The Heredity* 28: 43-49 (in Japanese).
3 Omori Y [1993]. Zosteraceous species endemic to Japan. *Bull Water Plant Soc Japan* 51: 19-25 (in Japanese).
4 Aioi K [1998]. On the red list of Japanese seagrasses. *Aquabiology* 114: 7-12 (in Japanese with English abstract).
5 Kuo J, Aioi K [1997]. Australian seagrasses and marine balls. *Bull Water Plant Soc Japan* 62: 1-7 (in Japanese with English abstract).
6 Omori Y [1992]. Geographical variation of the size of spadix and spathe and the number of flowers among the four species of the subgenus *Zostera* (Zosteraceae). *Scientific Report of Yokosuka City Museum* 40: 69-74 (in Japanese with English abstract).
7 Environment Agency of Japan [1994]. The Report of the Marine Biotic Environment Survey in the 4th National Survey on the Natural Environment. Vol. 2: Algal and Seagrass Beds. Nature Conservation Bureau. Environment Agency of Japan, Tokyo (in Japanese).
8. Imao K, Fushimi H [1985]. Ecology of the eelgrass (*Zostera marina* L.), especially environmental factors determining the occurrence of annual eelgrass in Lake Hamana-ko. *Japanese Journal of Phycology* 33: 320-327 (in Japanese with English abstract).
9 Fukuda T, Tsuchiya Y [1987]. Relation between shoot and seed distributions of eelgrass bed. *Nippon Suisan Gakkaishi* 53: 1755-1758.
10 Shin H, Choi HK [1998]. Taxonomy and distribution of *Zostera* (Zosteraceae) in eastern Asia, with special reference to Korea. *Aquatic Botany* 60: 49-66.
11 Mizushima T [1985]. Seasonal changes in standing crop and production of eelgrass (*Zostera marina* Linné) in Notsuke Bay, eastern Hokkaido. *Scientific Reports of Hokkaido Fisheries Experimental Station* 27: 111-118 (in Japanese with English abstract).
12 Tsutusi I, Sano O [1996]. The marine plants observed at seven localities along the coast of Noto Peninsula. *Report Noto Marine Center* 2: 81-84 (in Japanese with English abstract).
13 Fujita D, Takayama S [1999]. Records of seagrasses *Halophila ovalis* and *Zostera japonica* off Uozu City, Toyama Prefecture. *Bulletin Toyama Prefectural Fisheries Research Institute* 11: 67-70 (in Japanese with English abstract).
14 Higashide Y, Tsutsui I, Sakai K [1999]. Seaweed and seagrass specimens deposited in the Noto Marine Center, Ishikawa Prefecture. *Report Noto Marine Center* 5: 35-48 (in Japanese).
15 Kudo T [1999]. Distribution of seagrasses in Odawa Bay, Central Japan. *Bulletin of Kanagawa Prefectural Fisheries Experimental Station* 4: 51-60 (in Japanese with English abstract).
16 Kanamoto Z, Watanabe T [1981]. Ecological study of the seagrass meadows in Nagura Bay, Ishigaki Island, Okinawa. I. Ecological distribution of seagrass. *Benthos Research* 21/22: 1-14 (in Japanese with English abstract).
17 Tsuda RT, Kamura S [1990]. Comparative review of the floristics, phytogeography, seasonal aspects and assemblage patterns of the seagrass flora in Micronesia and the Ryukyu Islands. *Galaxea* 9: 77-93.
18 Toma T [1999]. Seagrasses from the Ryukyu Islands. I. Species and distribution. *Biological Magazine Okinawa* 37: 75-92 (in Japanese with English abstract).
19 Miki S [1993]. On the seagrasses in Japan. (I) *Zostera* and *Phyllospadix*, with special reference to morphological and ecological characters. *Botanical Magazine* 47: 842-862.
20 Aioi K, Nakaoka M, Kouchi N, Omori Y [2000]. A new record of *Zostera asiatica* Miki (Zosteraceae) in Funakoshi Bay, Iwate Prefecture. *Otsuchi Marine Science* 25: 23-26.
21 Miki S [1932]. On seagrasses new to Japan. *Botanical Magazine* 46: 774-788.
22 Aioi K, Komatsu T, Morita K [1998]. The world's longest seagrass, *Zostera caulescens* from northern Japan. *Aquatic Botany* 61: 87-93.
23 Omori Y, Aioi K [2000]. Seasonal changes of the erect shoot of *Zostera caulescens* Miki (Zosteraceae) in Sanriku Kaigan, northern Honshu. *Scientific Report of Yokosuka City Museum* 47: 67-72 (in Japanese with English abstract).
24 Omori Y [1991]. Peculiarity of the flowering shoot of *Zostera caulescens* (Zosteraceae). *Scientific Report of Yokosuka City Museum* 39: 45-50 (in Japanese with English abstract).
25 Omori Y [1994]. Seasonal changes of the reproductive shoot of *Zostera caulescens* (Zosteraceae) in Sagami Bay, central Japan. *Scientific Report of Yokosuka City Museum* 42: 65-69 (in Japanese with English abstract).
26 Nakase K [2000]. Quantitative estimation of eelgrass distribution from the viewpoint of external forces in Takeoka Beach, Tokyo Bay, Japan. *Biologia Marina Mediterranea* 7: 397-400.
27 Hokkaido Abashiri Fisheries Experimental Station [1997]. Research on Utilization of Coastal Area with Conservation of Environment. Hokkaido Abashiri Fisheries Experimental Station, Abashiri (in Japanese).

28 Omori Y, Aioi K, Morita K [1996]. A new record of *Zostera caespitosa* Miki (Zosteraceae): Its distribution in Yamada Bay, Iwate Prefecture, Japan. *Otsuchi Marine Research Center Report* 21: 32-37 (in Japanese).

29 Omori Y, Aioi K [1998]. Rhizome morphology and branching pattern in *Zostera caespitosa* Miki (Zosteraceae). *Otsuchi Marine Research Center Report* 23: 49-55 (in Japanese).

30 Omori Y [2000]. Japanese seagrass – distribution and morphology. *Aquabiology* 22: 524-532 (in Japanese with English abstract).

31 Nozawa K [1981]. Distribution of Japanese seagrasses. *Plant and Nature* 15(13): 15-19 (in Japanese).

32 Tanaka Y [1999]. Distribution of Seagrasses in Ishigaki Island and its Regulating Factors. MSc thesis, Faculty of Science, University of Tokyo (in Japanese with English abstract).

33 Kanamoto Z [2001]. Spatial distributions of seagrass and their seasonal change in Nagura Bay, Ishigaki Island, Okinawa. *Otsuchi Marine Science* 26: 28-39.

34 Nakaoka M, Aioi K [2001]. Ecology of seagrasses *Zostera* spp. (Zosteraceae) in Japanese waters: A review. *Otsuchi Marine Science* 26: 7-22.

35 Mukai H, Nojima S, Nishihira M [1987]. Seagrass coverage and distribution in Loloata seagrass bed. In: Hattori A (ed) *Studies on Dynamics of the Biological Community in Tropical Seagrass Ecosystems in Papua New Guinea: The Second Report*. Ocean Research Institute, University of Tokyo, Tokyo. pp 18-27.

36 Bach SS, Borum J, Fortes MD, Duarte CM [1998]. Species composition and plant performance of mixed seagrass beds along a siltation gradient at Cape Bolinao, Philippines. *Marine Ecology Progress Series* 174: 247-256.

37 Björk M, Uku J, Weil A, Beer S [1999]. Photosynthetic tolerances to desiccation of tropical intertidal seagrasses. *Marine Ecology Progress Series* 191: 121-126.

38 Nakaoka M, Spanwanid C [2000]. Quantitative estimation of the distribution and biomass of seagrasses at Haad Chao Mai National Park, Trang province, Thailand. *Kasetsart University Fishery Research Bulletin* 22: 10-22.

39 Aioi K [1980]. Seasonal change in the standing crop of eelgrass (*Zostera marina* L.) in Odawa Bay, central Japan. *Aquatic Botany* 8: 343-354.

40 Iizumi H [1996]. Temporal and spatial variability of leaf production of *Zostera marina* L. at Otsuchi Bay, northern Japan. In: Kuo J, Phillips RC, Walker DI, Kirkman H (eds) *Seagrass Biology: Proceedings of an International Workshop, Rottenest Island, Western Australia, 25-29 January 1996*. Faculty of Science, University of Western Australia, Nedlands, Perth, Western Australia. pp 143-148.

41 Arasaki S [1950]. Studies on the ecology of *Zostera marina* and *Zostera nana* (a). *Bull Jap Soc Sci Fish* 15: 567-572 (in Japanese with English abstract).

42 Taniguchi K, Yamada Y [1979]. Vertical distribution and natural life history of *Zostera marina* Linné and some other species of seagrass in Iida Bay of the Noto Peninsula on the Honshu, Japan Sea coast. *Bull Jap Sea Reg Fish Res Lab* 30: 111-112 (in Japanese with English abstract).

43 Watanabe M, Nakaoka M, Mukai H [2000]. Growth and productivity of *Zostera asiatica* and *Zostera marina* in Akkeshi, northern Japan. *Biologia Marina Mediterranea* 7: 156-159.

44 Aioi K [2000]. A daybreak in the studies on Japanese *Zostera* beds. *Aquabiology* 22(6): 516-523 (in Japanese with English abstract).

45 Komatsu T [1997]. Long term changes in *Zostera* bed area in the Seto Inland Sea (Japan), especially along the coast of the Okayama Prefecture. *Oceanologica Acta* 20: 209-216.

46 Environment Agency of Japan [2000]. *Threatened Wildlife of Japan: Red Data Book*. 2nd edn. Japan Wildlife Research Center, Tokyo (in Japanese).

47 Aioi K, Omori Y [2000]. Warming and Japanese seagrasses. In: Domoto A, Iwatsuki K, Kawamichi T, Mcneely J (eds) *A Threat to Life*. Tsukiji-Shokan Publishing Company Limited, Japan, and IUCN, Gland, Switzerland. pp 57-60.

48 Mukai H, Iizumi H, Kishi M [2002]. Comparison of terrestrial input to Akkeshi estuarine system between stable and unstable conditions. *Gekkan Kaiyo* 34: 449-457 (in Japanese).

49 Supanwanid C, Albertsen JO, Mukai H [2001]. Methods for assessing the grazing effects of large herbivores on seagrasses. In: Short FT, Coles RG (eds) *Global Seagrass Research Methods*. Elsevier, Amsterdam. pp 293-312.

50 Takahashi K, Vallet C, Kawamura H, Taguchi S [2000]. Abundance, species composition, and distribution of mysids in subarctic eelgrass beds at Akkeshi-ko Lagoon, Hokkaido, northern Japan. *Biologia Marina Mediterranea* 7: 286-289.

51 Taksukawa K, Komatsu T, Aioi K, Morita K [1996]. Distribution of seagrasses off Kirikiri in Funakoshi Bay, Iwate Prefecture, Japan. *Otsuchi Marine Research Center Report* 21: 38-47 (in Japanese).

52 Nakaoka M, Kouchi N, Aioi K [2000]. Growth and shoot dynamics of *Zostera caulescens* Miki in Funakoshi Bay, Japan: How does it maintain high canopy structure? *Biologia Marina Mediterranea* 7: 103-106.

53 Nakaoka M, Toyohara T, Matsumasa M [2001]. Seasonal and between-substrate variation in mobile epifaunal community in a multispecific seagrass bed of Otsuchi Bay, Japan. *PSZN Marine Ecology* 22: 379-395.

54 Kouchi N, Nakaoka M [2000]. Distribution of encrusting bryozoa on *Zostera caulescens* in Funakoshi Bay, Japan: Effects of seagrass vertical structure on epifauna. *Biologia Marina Mediterranea* 7: 247-250.

55 Toyohara T, Nakaoka M, Aioi K [1999]. Population dynamics and reproductive traits of phytal gastropod in seagrass bed in Otsuchi Bay, north-eastern Japan. *PSZN Marine Ecology* 20: 273-289.

56 Tohoyara T, Nakaoka M, Tsuchida E [2001]. Population dynamics and life history traits of *Siphonacmea oblongata* (Yokoyama) on seagrass leaves in Otsuchi Bay, north-eastern Japan (Siphonariidae, Pulumonata). *Venus* 60: 27-36.

57 Nojima S [1996]. Biodiversity and stabilization mechanisms in seagrass communities. *Japanese Journal of Ecology* 46: 327-337 (in Japanese).

58 Nakaoka M [2002]. Predation on seeds of seagrasses *Zostera marina* and *Zostera caulescens* by a tanaid crustacean *Zeuxo* sp. *Aquatic Botany* 72: 99-106.

18 The seagrasses of
THE REPUBLIC OF KOREA

K.-S. Lee

S.Y. Lee

The Korean peninsula, located at the eastern end of the Eurasian continent, lies between 33°N and 43°N. The total coastline of the peninsula, including the coastlines of the islands, reaches 17 000 km. About 3 400 islands are distributed along the coasts of the Republic of Korea. Since each coast shows very distinct characteristics, seagrass habitat properties also vary. The west and the south coasts have highly complex and indented coastlines, while the east coast has a simple and linear one. Sand dunes are well developed, and several lagoons are formed on the east coast of the peninsula. Tidal flats are located at several places on the south coast. Tidal range is 1-4 m on the south coast and higher along the west part of the coastline. About 2 000 islands are distributed in the western part of the south coast. Although the linear distance of the west coast is some 650 km, the actual length of the coastline is about 4 700 km. Tidal range is extremely high on the west coast of the Korean peninsula; maximum tidal range is about 10 m on this coast. Very large tidal flats are formed because of the flat sea bottom and high tidal range.

Eight temperate seagrass species, five *Zostera*, two *Phyllospadix* and *Ruppia maritima*, are distributed on the coasts of the Korean peninsula[1-4]. Although seagrasses are relatively abundant, few studies have reported on their physiology and ecology in this area and most of these reports were written in the Korean language. In this paper, we review the status, habitat characteristics and ecology of seagrasses on the coasts of the Republic of Korea.

The Korean peninsula is enclosed by the Yellow Sea (to the west of Korea), the South Sea and the East Sea, which have considerably different characteristics (Table 18.1). The coastline of the Yellow Sea (the west coast) shows a heavily indented coast with maximum tidal range of about 10 m. The hydrographic properties and circulation characteristics of the Yellow Sea are strongly influenced by climatic conditions. The South Sea is connected to the East China Sea and the Tsushima current, a branch of Kuroshio, flows towards the East Sea through the South Sea. The coastline of the South Sea is also heavily indented. Tidal ranges on the south coast vary from about 1.0 m in the east part of the coast to about 4.0 m in the west part. The East Sea is deeper than the Yellow Sea or the South Sea, and the eastern coastline is very simple and linear. Tidal range is usually less than 0.3 m.

PRESENT SEAGRASS DISTRIBUTION

Eight temperate seagrass species are distributed on the coasts of the Korean peninsula. *Zostera marina* is the most abundant seagrass species, widely distributed throughout all coastal areas (Table 18.2) in relatively large meadows. *Zostera asiatica* is mostly distributed in the cold and temperate coasts of northeastern Asia. In the Republic of Korea *Zostera asiatica* occurs on the east coast; the distribution of this species on the west and the south coasts of the Korean peninsula is not clear. *Zostera caespitosa*, *Zostera caulescens* and *Zostera japonica* are found on all the Republic of Korea's coasts (Table 18.2).

Two *Phyllospadix* species, *Phyllospadix iwatensis* and *Phyllospadix japonicus*, are found on Korean coasts[2]. *Phyllospadix japonicus* occurs on all coasts of the peninsula, while *Phyllospadix iwatensis* occurs on the east and west coasts. On the east coast, *Phyllospadix iwatensis* usually appears in the northern parts of the coast, while *Phyllospadix japonicus* is distributed in the southern parts. Distribution of *Ruppia maritima* in the Republic of Korea has been reported from limited areas on the west and south coasts[1, 5].

BIOGEOGRAPHY OF THE REGION

Seagrasses are distributed in numerous locations along the coast of the Korean peninsula with habitat types varying among the different coasts (Table 18.3). Seagrasses are widely distributed throughout the south

Table 18.1
Physical characteristics of seagrass beds on the west, south and east coasts of the Republic of Korea

Characteristics	West coast	South coast	East coast
Wave energy	Low	Low	High
Sediment	Muddy sand	Muddy sand	Sand
Tidal range (m)	3-10	1-4	0.1-1
Coastline	Heavily indented	Heavily indented	Simple linear
Seagrass habitat	Bays, islands	Bays, islands	Lagoons, bays

coast, while on the east coast, where the wave energy is high, the distribution of seagrasses is limited to lagoons, ports and barrier reefs. On the west coast, seagrasses are mainly distributed in the intertidal and subtidal zones of islands. Seagrasses usually form small patches on the east coast, while large seagrass meadows occasionally occur on the west and the south coasts of the Republic of Korea.

Zostera marina appears at the intertidal and subtidal zones, where the water depth is usually less than 5 m, and forms relatively large meadows. *Zostera marina* can be observed in both muddy and sandy sediments[6]. *Zostera asiatica* is distributed in relatively deep water (9-15 m) in bays or along open shores, and forms small patches. *Zostera asiatica* is usually observed in sandy sediments along the east coast. *Zostera caespitosa* also usually forms small patches and is distributed in deeper water (3-8 m) than *Zostera marina*. *Zostera caespitosa* occurs on all coasts of the Korean peninsula, but is limited to a few areas with mixed sediments of sand and gravel. *Zostera caulescens* is distributed on both sandy and muddy bottoms. As a result of its height, *Zostera caulescens* usually occurs in deep water (6-12 m). *Zostera japonica* is mainly distributed in the intertidal zone around islands.

Both *Phyllospadix* species, *Phyllospadix iwatensis* and *Phyllospadix japonicus*, occur mainly on rocky substrata along the east coast of the Korean peninsula. Although *Phyllospadix* species are observed in both sheltered and open shores, they usually grow in high-energy environments. Distribution of *Ruppia maritima* has been reported from a few estuaries on the west and the south coasts. However, the *Ruppia maritima* habitats have been severely disturbed recently, so the present distribution of this species should be investigated. Although a few mixed beds with *Zostera marina* and *Zostera japonica* occur on the west coast, different seagrass species do not usually coexist in the Republic of Korea.

The vegetative shoot height of *Zostera marina* in the Republic of Korea ranges from 30 cm to 210 cm (Table 18.4), and varies significantly among habitats. Reproductive shoots, which are usually taller than vegetative shoots, range from 50 cm to 350 cm. The leaf width of *Zostera marina* also shows significant variation according to environment. Some plants from eelgrass beds on the east and the south coasts have

Table 18.2
Seagrass species distributed on the coasts of the Republic of Korea

Species	Distribution		
	West coast	South coast	East coast
Genus *Zostera*			
Zostera marina	✓	✓	✓
Zostera asiatica	?	?	✓
Zostera caespitosa	✓	✓	✓
Zostera caulescens	✓	✓	✓
Zostera japonica	✓	✓	✓
Genus *Phyllospadix*			
Phyllospadix iwatensis	✓	?	✓
Phyllospadix japonicus	✓	✓	✓
Genus *Ruppia*			
Ruppia maritima	✓	✓	?

very wide leaves (about 15 mm). *Zostera marina* in the Republic of Korea has 5-11 leaf veins.

The shoot heights of *Zostera asiatica* are 50-90 cm for vegetative shoots and 60-80 cm for reproductive shoots. Leaf widths range from 11 to 15 mm, and the leaves have 9-11 veins (Table 18.4). The shoot height of *Zostera caespitosa* for both vegetative and reproductive shoots ranges from 50 cm to 170 cm. *Zostera caespitosa* has relatively narrow leaves (5-8 mm wide), and usually 5 leaf veins. *Zostera caulescens* is a very tall seagrass species; the height of its reproductive shoot reaches 7-8 m. *Zostera caulescens* has wide leaves (10-16 mm) and 9-11 leaf veins. *Zostera japonica* is a very small seagrass species, usually less than 30 cm tall, with 3 leaf veins. The leaf width of *Zostera japonica* is 1-2 mm.

Phyllospadix iwatensis and *Phyllospadix japonicus* show several morphological differences. The shoot heights of both *Phyllospadix* species range from 20 cm to 100 cm. *Phyllospadix iwatensis* has 5 veins in the lower portion of the leaf, but only 3 veins in the apical portion[2]. *Phyllospadix japonicus* has 3 leaf veins. The leaf width of *Phyllospadix japonicus* ranges from 1.5 mm to 2.5 mm, while the width of *Phyllospadix iwatensis* ranges from 2.0 mm to 4.5 mm.

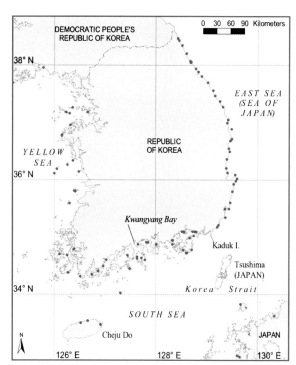

Map 18.1
Republic of Korea

Table 18.3
Habitat characteristics of seagrass species in the Republic of Korea

Species	Sediment type	Wave energy	Water depth (m)	Location
Zostera marina	Muddy, sandy	Low	0-5	Bay, lagoon
Zostera asiatica	Sandy	Intermediate	9-15	Bay, open shore
Zostera caespitosa	Gravelly	Low	3-8	Bay
Zostera caulescens	Muddy, sandy	Low	6-12	Bay
Zostera japonica	Muddy, sandy	Low	Intertidal zone	Bay
Phyllospadix iwatensis	Rocky	High	0-3	Open shore
Phyllospadix japonicus	Rocky	High	0-3	Open shore
Ruppia maritima	Muddy, sandy	Low	0-2	Estuary

Table 18.4
Morphological characteristics of seagrasses distributed in the Republic of Korea

Species	Shoot height (cm)		Leaf width (mm)	Number of leaf veins
	Vegetative	Reproductive		
Zostera marina	30-210	50-350	5-15	5-11
Zostera asiatica	50-90	60-80	11-15	9-11
Zostera caespitosa	50-170	50-170	5-8	5-7
Zostera caulescens	90-200	150-800	10-16	9-11
Zostera japonica	15-40	10-20	1-2	3
Phyllospadix iwatensis	20-100	-	2-4.5	5
Phyllospadix japonicus	20-100	-	1.5-2.5	3

> **Case Study 18.1**
> **RECENT RESEARCH ON SEAGRASSES**
>
> Few biological and ecological studies have been conducted on seagrasses of the Korean peninsula. Recently, the Government of the Republic of Korea began to realize the ecological and economic importance of seagrasses in coastal and estuarine ecosystems. There is now an effort to preserve seagrass habitats and to restore disturbed and destroyed habitats. Since there is little physiological and ecological information on seagrasses in the Republic of Korea, basic research on seagrass biology and ecology is necessary for efficient management and restoration of seagrass habitats on the Korean coast.
>
> Most seagrass research in the Republic of Korea, except for taxonomic studies, has been conducted during the past few years. Current seagrass research projects are as follows: Dr Choi (Ajou University), Dr Shin (Soonchunhyang University), Dr Oh (Kyungsang National University) and Drs Choi and Lee (Hanyang University) are conducting taxonomic studies. Drs Choi and Lee have conducted a molecular phylogenetic study to examine the phylogeny of the *Zostera* species in the Republic of Korea. The sequences of the internal transcribed spacer (ITS) regions in nuclear ribosomal DNA have been determined for five *Zostera* species in the Republic of Korea. From the study, *Zostera marina* and *Zostera caespitosa* were the most closely related species, and *Zostera japonica* was the most distinctive *Zostera* species.
>
> Dr Hong (Inha University) is investigating benthos in seagrass beds on the south and west coasts of the Korean peninsula. Dr Huh (Pukyung National University) and Dr Kim (Chonnam National University) are studying fish populations and phytoplankton communities in seagrass beds. Dr K. Lee (Pusan National University) is conducting basic physiological and ecological studies on seagrass in the Republic of Korea, and research into carbon and nutrient dynamics in seagrass beds.

Few studies on seagrass ecology have been reported from the Korean peninsula. Therefore, few data exist concerning seagrass biomass or productivity. The density, biomass and productivity of seagrasses change significantly with environmental conditions such as water temperature, underwater irradiance and nutrient concentration. Since water temperature along the coasts of the Korean peninsula shows obvious seasonal variation, being less than 10°C during winter and about 25°C during summer, seagrass biomass and productivity also show significant seasonal variations. The shoot density of *Zostera marina* varies from about 50 shoots/m^2 to 300 shoots/m^2 depending on environmental conditions. Shoot densities of *Zostera caulescens* and *Zostera japonica* are about 120 shoots/m^2 and about 8000 shoots/m^2, respectively, during the summer months on the south coast[7]. Biomass of both *Zostera marina* and *Zostera caulescens* on the south coast is about 500 g dry weight/m^2 during summer months, while *Zostera japonica* has about 200 g dry weight/m^2. Leaf productivity is about 2 g dry weight/m^2/day for *Zostera marina* and *Zostera caulescens*, and about 5 g dry weight/m^2/day for *Zostera japonica* during summer months. The small seagrass species *Zostera japonica* has higher leaf productivity because of its much higher shoot density.

More than 60 fish species can be observed in seagrass beds in the Republic of Korea; fish collected in seagrass beds are primarily small fish species or juveniles of big fish species. Fish species abundant in seagrass beds include *Sgnathynus schlegeli*, *Pholis* sp., *Pseudoblennius cottoides*, *Sebastes inermis* and *Acanthogobius flavimanus*[8,9]. Juveniles of economically valuable fish species such as *Sebastes inermis*, *Platycephalus indicus* and *Limanda yokohamae*, *Acanthopagrus schlegeli* and *Lateolabrax japonicus* grow in seagrass beds. In seagrass beds on the south coast, *Acanthopagrus schlegeli* were not observed during winter and spring. However, many small-size juveniles (less than 3 cm body length) of *Acanthopagrus schlegeli* were found in July; then they were seen infrequently in October when their body length reaches about 6 cm[10]. *Sebastes inermis*, which is a very valuable fish species in the Republic of Korea, also migrates into seagrass beds when its body size is about 2 cm and spends its juvenile stage in seagrass beds.

There are seasonal variations in the species composition and abundance of fish populations in seagrass meadows. A peak of fish abundance occurs during spring, with a secondary peak during the fall[8]. The peaks are probably caused by increased larval recruitment. Seasonal variation in fish species composition is closely related to standing crops of seagrasses[11].

A peak of shrimp abundance occurs in the late winter and spring[12]. Shrimp species in seagrass beds were most diverse during the late summer, and least diverse during the late fall. Crab species in seagrass beds were most abundant during summer, and most

diverse during spring and summer[13]. The dominant group of benthic macrofauna in seagrass beds was polychaete worms[14]. About 15 epiphytic algae species on seagrass leaf tissues have been reported from seagrass beds in the Republic of Korea. The dominant epiphyte species are *Callophyllis rhynchocarpa* and *Champia* sp. during spring and summer, and *Polysiphonia japonica* and *Lomentaria hakodatensis* during fall and winter. Epiphyte biomass is lowest during summer, and highest during winter. Epiphytic algae account for approximately 15 to 20 percent of total plant standing crop of the seagrass beds in Kwangyang Bay[15].

USES OF SEAGRASSES

Seagrasses are rarely used either directly or indirectly in the Republic of Korea. Seagrasses have been considered as useless weeds around ports and boat channels and in shallow fishing grounds; fishermen cut off seagrasses to have a better waterway. However, fishermen and the Government now realize the ecological and economical importance of seagrasses for fisheries and coastal ecosystems in the Republic of Korea. We are trying to use seagrass for conservation and restoration of coastal ecosystems. The coasts of the Korean peninsula have been highly disturbed and polluted as a consequence of industrial development since the 1970s. Additionally, an expanse of tidal flats and seagrass habitats has been reclaimed for factory sites or residential districts. These coastal disturbances have led to a reduction in spawning grounds and nursery areas for economically valuable fish, and consequently led to decreases in coastal fish production. Concrete constructs have been added to coastal waters in the Republic of Korea for artificial fish-breeding reefs. Most artificial fish reefs were constructed in deep water, so few types of seaweed can grow on the construct, and the construct provides habitat for only adult fish. However, seagrass beds can provide a good fish spawning ground and nursery area for juvenile fish, so we are now trying to restore seagrass habitats on Korean coasts. Seagrasses will also be used as a nutrient filter to reduce algal blooms, especially red tide, which damage coastal fisheries in the Republic of Korea almost every year.

Seagrass leaf detritus piled up on the beach is collected in some coastal areas to make compost, which is used for fertilizing agricultural land. In earlier times, children in the coastal zones chewed seagrass rhizomes for their sweet taste.

ESTIMATES OF HISTORICAL LOSSES AND PRESENT DISTRIBUTION

There are no studies of seagrass areas in the Republic of Korea. Therefore, we must estimate historical losses and present seagrass coverage using personal observation and verbal information obtained from fishermen. Most large seagrass beds in the Republic of Korea were located in the bays of the south coast. Many of these bay areas are now urbanized, so bay water is highly over-enriched by anthropogenic nutrients. Most seagrass has disappeared from these eutrophic bays. Numerous former tidal flats and seagrass habitats on the south and west coasts of the Republic of Korea have been reclaimed for factory sites, residential districts or agricultural land. Large areas of seagrass have been lost due to reclamation, particularly from the west and south coasts. For example, a large *Zostera marina* bed (13.6 km^2) existed in front of Kaduk Island, on the south coast of the Republic of Korea, until the late 1980s[1]. However, this eelgrass bed disappeared after reclamation of the adjacent mud flats during the early 1990s.

Many oyster and seaweed farms are located in shallow coastal waters, where seagrasses exist. Lots of seagrass beds were destroyed by the construction and maintenance of these farms. Seagrass areas in the Republic of Korea also have been lost to boat traffic, trawling and clamming. From the estimates of

Table 18.5
The estimated areas of seagrasses distributed on the coasts of the Republic of Korea

Species	Area (km^2)
Zostera marina	50-60
Zostera asiatica	<1.0
Zostera caespitosa	1.0
Zostera caulescens	1.0-5.0
Zostera japonica	1.0
Phyllospadix iwatensis	<1.0
Phyllospadix japonicus	1.0-2.0
Ruppia maritima	<1.0
Total	55-70

potential seagrass area and present seagrass coverage, and verbal information from fishermen, we believe that more than 50 percent (and maybe as much as 70 or 80 percent) of the seagrass area in the Republic of Korea has been lost since the beginning of industrial development during the 1970s.

Most of the seagrass area in the Republic of Korea is located on the south coast, and *Zostera marina* beds account for about 90 percent of total seagrass coverage in the Republic of Korea. Our estimated area of *Zostera marina* on the coasts of the Korean peninsula is about 50 to 60 km^2 (Table 18.5). The estimated area for *Zostera caulescens* is 1 to 5 km^2. Most *Zostera asiatica* and *Phyllospadix iwatensis* beds

are found on the east coast; the estimated area for both species is less than 1 km^2. Most *Phyllospadix japonicus* beds are also found on the east coast, and the bed area is estimated at 1 to 2 km^2. The estimated area for *Zostera caespitosa* and *Zostera japonica* is about 1 km^2 (Table 18.5). There is no information on the area of *Ruppia maritima* in the Republic of Korea, but it is probably less than 1 km^2.

PRESENT THREATS

Seagrasses in the Republic of Korea have been severely impacted by coastal eutrophication, land reclamation, aquaculture and fishing activities, and these threats still exist. Estuaries and coastal ecosystems in the Republic of Korea are receiving extraordinary amounts of nutrients as a consequence of anthropogenic loading, as well as through industrial pollutants. Nutrient over-enrichment and pollutant discharge widely affect estuarine and coastal ecosystems and damage seagrass habitats. On the west and south coasts of the Korean peninsula land reclamation is a major contributor to loss of seagrass habitats. Since 1945, more than 62 km^2 of tidal flats have been reclaimed[16]. Reclamation of tidal flats caused loss of adjacent seagrass beds, and many seagrass beds have disappeared due to reclamation over the last decades. Many land reclamation projects, which did not consider the ecological and economical values of tidal flats and seagrass beds, are still being constructed by the Government of the Republic of Korea. Oyster and seaweed farming, and fishing activities such as clamming and trawling, are also serious threats to Korean seagrasses.

POLICY

There is no policy which directly serves to protect seagrasses in the Republic of Korea. However, several coastal areas are protected as Environmental Conservation Areas or Special Coastal Management Areas, and many seagrass beds are located in the protected areas. There are critical problems associated with the management of estuaries and designated protected areas. There is no long-term management plan for effective management and protection of estuarine and coastal ecosystems in the Republic of Korea. Punishment for illegal activities in the protected areas, such as illegal fishing activities or illegal dumping, is very weak, so protection of Korean coastal ecosystems by the law is not effective. Additionally, economic advantages are usually given priority over ecological conservation or environmental preservation. Therefore, even protected areas can be developed or reclaimed for industrial facilities in the Republic of Korea based on economic considerations.

AUTHORS

Kun-Seop Lee and Sang Yong Lee, Department of Biology, Pusan National University, Pusan 609-735, Republic of Korea. **Tel:** +82 (0)51 510 2255. **Fax:** +82 (0)51 581 2962. **E-mail:** klee@hyowon.cc.pusan.ac.kr

REFERENCES

1. Chung YH, Choi HK [1985]. Distributional abundance and standing crop of the hydrophytes in the estuary of the Nadong River. *Nature Conservation* 49: 37-42 (in Korean).
2. Shin H, Choi HK, Oh YS [1993]. Taxonomic examination of Korean seagrasses: Morphology and distribution of the genus *Phyllospadix* (Zosteraceae). *Korean Journal of Plant Taxonomy* 23: 189-199.
3. Shin H, Choi HK [1998]. Taxonomy and distribution of *Zostera* (Zosteraceae) in eastern Asia, with special reference to Korea. *Aquatic Botany* 60: 49-66.
4. Lee SY [2001]. A Study on the Ecological and Taxonomical Characteristics of *Zostera* (Zosteraceae) in Korea. PhD thesis, Hanyang University, Seoul. 165 pp (in Korean).
5. Choi HK [2000]. *Aquatic Vascular Plants.* Korea Research Institute of Bioscience and Biotechnology.
6. Lee SY, Kwon CJ, Choi CI [2000]. Sediment characteristics from the beds of *Zostera marina* and *Z. asiatica*. *Journal of Natural Science & Technology, Hangyang University* 2: 25-29 (in Korean).
7. Lee K-S, Lee SY [2001]. Status and restoration of seagrass habitat on the south coast of the Korean peninsula. *Nature Conservation* 116: 15-20 (in Korean).
8. Huh SH [1986]. Species composition and seasonal variations in abundance of fishes in eelgrass meadows. *Bulletin of the Korean Fisheries Society* 19: 509-517 (in Korean).
9. Huh SH, Kwak SN [1997]. Species composition and seasonal variations of fishes in eelgrass (*Zostera marina*) bed in Kwangyang Bay. *Korean Journal of Ichthyology* 9: 202-220 (in Korean).
10. Huh SH, Kwak SN [1998]. Feeding habits of juvenile *Acanthopagrus schlegeli* in the eelgrass (*Zostera marina*) bed in Kwangyang Bay. *Korean Journal of Ichthyology* 10: 168-175 (in Korean).
11. Lee TW, Moon HT, Hwang HB, Huh SH, Kim DJ [2000]. Seasonal variation in species composition of fishes in the eelgrass beds in Angol Bay of the southern coast of Korea. *Journal of the Korean Fisheries Society* 33: 439-447 (in Korean).
12. Huh SH, An Y-R [1997]. Seasonal variation of shrimp (Crustacea: Decapoda) community in the eelgrass (*Zostera marina*) bed in Kwangyang Bay, Korea. *Journal of the Korean Fisheries Society* 30: 532-542 (in Korean).
13. Huh SH, An Y-R [1998]. Seasonal variation of crab (Crustacea: Decapoda) community in the eelgrass (*Zostera marina*) bed in Kwangyang Bay, Korea. *Journal of the Korean Fisheries Society* 31: 535-544 (in Korean).
14. Yun SG, Huh SH, Kwak SN [1997]. Species composition and seasonal variations of benthic macrofauna in eelgrass, *Zostera marina*, bed. *Journal of the Korean Fisheries Society* 30: 744-752.
15. Huh SH, Kwak SN, Nam KW [1998]. Seasonal variations of eelgrass (*Zostera marina*) and epiphytic algae in eelgrass beds in Kwangyang Bay. *Journal of the Korean Fisheries Society* 31: 56-62 (in Korean).
16. Koh C-H (ed) [2001]. *The Korean Tidal Flat: Environment, Biology and Human.* Seoul National University Press, Seoul.

19 The seagrasses of THE PACIFIC COAST OF NORTH AMERICA

S. Wyllie-Echeverria

J.D. Ackerman

The region along the Pacific coast of North America extending from the Baja Peninsula in Mexico through Alaska includes a wide variety of ecosystems ranging from subtropical through arctic in a northerly transect. Given the nature of the leading edge coast and the resultant paucity of large regions where soft sediments can accumulate, one would expect a rather limited diversity of marine angiosperms or seagrasses. However, a reasonably large number of species exist in this region for a number of reasons, related in part to the ability of members of the genus *Phyllospadix* to colonize rocky shores. Eight seagrass species are recognized: *Halodule wrightii*, *Ruppia maritima*, *Zostera marina*, *Zostera japonica*, *Zostera asiatica*, *Phyllospadix scouleri*, *Phyllospadix serrulatus* and *Phyllospadix torreyi*[1-4]. Four of them, *Zostera marina*, *Phyllospadix scouleri*, *Phyllospadix serrulatus* and *Phyllospadix torreyi*, have probably been growing in the region since the Pliocene[5]; one, *Zostera japonica*, is a recent addition to the northeast Pacific flora, being introduced as a result of oyster enhancement programs[6]; and little is known about the phytogeographic history of the three other species (*Zostera asiatica*, *Ruppia maritima* and *Halodule wrightii*).

In terms of ecosystems, members of the genus *Phyllospadix* (the surfgrasses) dominate the rocky subtidal and intertidal zones, where their condensed rhizomes allow them to colonize hard substrates. The three species in the genus *Phyllospadix* (*Phyllospadix serrulatus*, *Phyllospadix scouleri* and *Phyllospadix torreyi*) are endemic to the Northeast Pacific[2]. Both *Phyllospadix torreyi* and *Phyllospadix scouleri* were widely used in the region by indigenous people before European contact[7]. For example the flowers of *Phyllospadix torreyi* were sucked for sweetness by children of the Makah people who live on the Olympic Peninsula of Washington state, and leaves of the same plant were woven into pouches by the coastal Chumash in the Channel Islands of California.

In contrast, soft-bottom habitats in the subtidal and intertidal zones and estuaries are more commonly associated with plants in the genus *Zostera*, which can form large monotypic stands in the Northeast Pacific estuaries, and mixed stand populations of *Zostera marina* and *Zostera japonica*, and sometimes *Ruppia maritima*, in estuaries from southern British Columbia, Canada, to Coos Bay, Oregon[8, 9]. *Zostera marina* provides important habitat for migrating waterfowl, juvenile salmon, resident forage fish, invertebrates and wading birds[8], and *Zostera japonica* is commonly eaten by resident and migratory waterfowl[10]. Noteworthy is the use of *Zostera marina* as substrate for the laying of Pacific herring (*Clupea harengus pallasi*) roe; the roe is also used by humans[8].

Whereas little is known about the primary production rates for *Zostera japonica*, *Zostera marina* productivity can be quite high on an annual basis (84-480 g carbon/m^2/year) and standing stocks may cover many hectares of seafloor[8, 11]. For example, the large populations at Izembek Lagoon, Alaska, United States (160 ha) and Laguna Ojo de Liebre, Baja California, Mexico (175 ha), which are the primary staging grounds for migratory waterfowl, may be the largest *Zostera marina* ecosystems in the world[12, 30]. In addition, pre-contact First Nations peoples recognized *Zostera marina* and its ecosystems as valuable cultural and food resources. In British Columbia a number of First Nations people (Nuu-chah-nulth, Haida and Kwakwaka'wakw) ate fresh rhizomes and leaf bases or dried them into cakes for winter food[13]. Moreover, the Seri Indians living on the Gulf of California in Sonora, Mexico, used the *Zostera marina* seeds to make flour[14].

Ruppia maritima grows in many of the brackish water coastal lagoons from Alaska south to Mexico[4, 14]. Interestingly, *Ruppia maritima* was recognized as a separate species (from *Zostera marina*) by the Seri elders but was not used by them[14].

The last two species, namely *Zostera asiatica* and *Halodule wrightii*, have rarely been the focus of biogeographic investigation within this region. *Zostera asiatica* was found recently at three sites in southern California[3] and, leaving aside some regional studies documenting the presence of *Halodule wrightii* in the Gulf of California[1], there are no studies that discuss the habitat value or autecology of these plants in the Northeast Pacific region.

BIOGEOGRAPHY

Zostera marina (or eelgrass) is the dominant species in terms of biomass and habitats on the Pacific coast of North America, where it grows in:
o   the shallow waters of the continental shelf;
o   the Gulf of California (Sea of Cortez);
o   coastal lagoons such as San Quintin, Baja California, Mexico, and Izembek Lagoon, Alaska[11];
o   estuaries formed by tectonic processes like San Francisco Bay;

Case Study 19.1
THE LINK BETWEEN SEAGRASS AND MIGRATING BLACK BRANT ALONG THE PACIFIC FLYWAY

Black brant (or sea goose, *Branta bernicla nigricans*) forage on seagrass flats (primarily *Zostera marina*) from Alaska to Mexico. In late August, after raising young in the Yukon-Kuskokwim Delta (61°N, 165°W), flocks gather at Izembek Lagoon (55°N, 163°W) to graze on one of the largest intertidal stands of *Zostera marina* in the world (see photograph below left). In the fall, most of the population moves on a non-stop, three-day transoceanic flight to *Zostera marina* and *Ruppia maritima* beds in Baja California at Bahía San Quintin (30°N, 116°W; see photograph below right), Laguna Ojo de Liebre (27°N, 114°W) and Laguna San Ignacio (26°N, 113°W).

Spring migration coincides with midday maximum low water events, which allow brant daylight opportunities to graze on the extensive seagrass resources growing on the tide flats at locations like Morro Bay and Humboldt Bay, California; South Slough and Yaquina Bay, Oregon; Willapa Bay and Padilla Bay, Washington; and Boundary Bay, British Columbia, Canada. International conservation efforts by the United States, Canada and Mexico are under way at wintering and migration stopover sites along the eastern Pacific Flyway to protect seagrass habitats in coastal embayments and estuaries.

In collaboration with David Ward (US Geological Survey, Anchorage) and Dr Silvia Ibarra-Obando (Centro de Investigaciones Cientifica y de Educación Superior de Ensenada, Baja California, Mexico).

Black brant grazing on the *Zostera marina* bed in Izembek Lagoon, United States.

Black brant on the *Zostera marina* and *Ruppia maritima* beds in Bahía San Quintin, Mexico.

o coastal fjords similar to Puget Sound, Washington[8].

It is found along the coast of British Columbia including the coasts of Vancouver Island and Queen Charlotte Islands (Haida Gwaii) in sheltered bays and coves including Bamfield Harbour and Sooke Basin[4]. The species also extends well into Alaskan waters to the Arctic Circle[15]. In the intertidal zone *Zostera marina* can co-mingle with *Zostera japonica* in the Pacific Northwest and *Ruppia maritima* in Baja California[8, 14].

Whereas the majority of *Zostera marina* populations are perennial, annual populations (e.g. Bahía Kino, Gulf of California, Mexico; Yaquina Bay, Oregon, United States) in which 100 percent of the population are generative shoots that recruit from seeds each year have been reported[4, 16]. The appearance of branched reproductive shoots, a dimorphic expression quite distinct from the ribbon-shaped leaves of the vegetative shoots, begins to occur as water temperatures warm in the spring. In the Northeast Pacific, reproductive shoots are visible in February at southern sites such as Baja California, Mexico and southern California; in late March or early April in Puget Sound, Washington; and as late as June in northern sites like Izembek Lagoon, Alaska. Flowering phenology is protogynous and the emergence of stigmas and then anthers effect the release, transport and capture of pollen, which rotate in the shear around stigmas[17]. *Zostera marina* is monoecious; however the release of pollen and its stigmatic capture is separated in time to promote an outcrossing breeding system[17, 18].

In a region-wide analysis of population structure[19], Alberte and colleagues found that:
o there was high genetic diversity among *Zostera marina* populations in the region;
o gene flow restriction existed for populations that were near each other;
o intertidal plants in disturbed environments were less diverse genetically than those in undisturbed sites.

In a subsequent study, focused on San Diego, California, and Baja California, Mexico, Williams and Davis discovered that transplanted *Zostera marina* populations were less diverse genetically than naturally occurring populations[20].

Although *Zostera japonica* is typically smaller than *Zostera marina*, it can be confused with the intertidal growing habit of *Zostera marina* (var. *typica*)[21]. However *Zostera japonica* commonly grows higher in the intertidal zone and has an open (as opposed to tubular) leaf sheath characteristic of its subgenus *Zosterella*[6]. It is a possible invader to the region coming by way of the oyster trade with Japan in

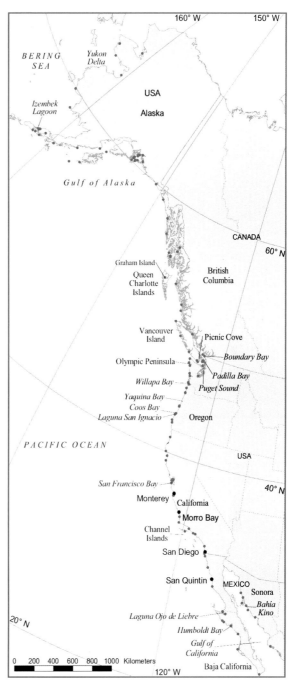

Map 19.1
The Pacific coast of North America

the early part of the 1900s[6]. At several sites from southern British Columbia, Canada, to southern Oregon, *Zostera japonica* co-mingles with *Zostera marina* (and occasionally *Ruppia maritima*[9]) in the intertidal region of many estuaries[6]. *Zostera japonica* is restricted in its northerly extent to the region near the city of Vancouver including Boundary Bay and Tsawwassen, where it is found primarily in the upper intertidal zones in muddy or silty areas[4]. Some unconfirmed reports also exist of the species further

Case Study 19.2
THE LINK BETWEEN THE SEAGRASS *ZOSTERA MARINA* (*TS'ÁTS'AYEM*) AND THE KWAKWAKA'WAKW NATION, VANCOUVER ISLAND, CANADA

Photo: K. Recalma-Clutesi

Chief Adam Dick twisting the seagrass (*Zostera marina*) for dipping in eulachon (*Thaleichthys pacificus*) grease. Plants were harvested at Deep Bay on the east coast of Vancouver Island on 28 July 2002.

Chief Adam Dick (Kwaxsistala) and Kim Recalma-Clutesi (OqwiloGwa) are members of the Kwakwaka'wakw Nation on the northeast coast of Vancouver Island, British Columbia, Canada. Both are keenly aware of the value of *Zostera marina* or *ts'áts'ayem* from oral tradition of their nation.

They recall that at Grassy Point or *wáwasalth*, *ts'áts'ayem* is collected with a long thin pole or *k'elpawi* that is stuck into the substrate, rotated to entwine the leaves of *ts'áts'ayem*, and pulled from the bottom to reveal leaves, rhizomes and roots. On removal the plants are peeled exposing the tender soft tissue of the leaf base. The leaves are then wrapped around the rhizome, dipped in *klina* (eulachon (*Thaleichthys pacificus*) grease) and eaten as a ceremonial food.

Whereas Grassy Point has both cultural and ecological value, Chief Dick, Kim Recalma-Clutesi and others of the Kwakwaka'wakw Nation have concern about regional and global practices that threaten the survival of *ts'áts'ayem*.

In collaboration with Chief Adam Dick and Kim Recalma-Clutesi (Kwakwaka'wakw Nation) and Dr Nancy J. Turner (University of Victoria, Victoria, Canada).

---

north in the Strait of Georgia[4]. Large stands are present in Boundary Bay, southern British Columbia, Canada, and Padilla Bay and Willapa Bay, Washington[10]. The sediments and fauna within *Zostera japonica* beds were found to be largely similar to those found in *Zostera marina* beds in Oregon, although some differences in sediment grain size and organic constituents were observed[22]. *Zostera japonica* has been shown to be important to resident and migratory waterfowl in Boundary Bay[10], and is used as habitat by epibenthic crustaceans in Padilla Bay. To the best of our knowledge, there are no other studies linking *Zostera japonica* to secondary consumers either as a food or habitat in North America.

The three species of the genus *Phyllospadix* are found on exposed rocky coasts in the surf zone and in tide pools in the intertidal zone where their condensed rhizome allows them to attach to hard substrates. Three of the five species in this North Pacific genus are found on the west coast of North America. Turner and Lucas[23]

and Phillips and Menez[2] describe the habitat and regional distribution of the three species. *Phyllospadix serrulatus* grows in the upper intertidal zone (+1.5 m to mean lower low water) on the outer coasts of Alaska, British Columbia, Washington and Oregon. Its distribution is often confused with *Phyllospadix scouleri*, which inhabits the lower intertidal and shallow subtidal zone. It can be locally quite common, as on Graham Island (Haida Gwaii)[4], and has a distribution that extends from southeast Alaska to Baja California, although it is reported to be more abundant north of Monterey, California[2, 4]. *Phyllospadix torreyi* grows at greater depths and is generally more abundant on the exposed parts of the coast and even in tidal pools with sandy bottoms, which are typically devoid of the other two *Phyllospadix* species. The distribution of *Phyllospadix torreyi* nearly overlaps *Phyllospadix scouleri*, but it is more abundant south of Monterey, California. The lack of information on its distribution may be related to the difficulty of making collections in

energetic habitats where *Phyllospadix torreyi* is found[4]. Whereas little is known about the biogeography of these species, studies have revealed aspects of their autecology and life history such as the adaptations of seeds and roots to cling to surfaces in the rocky intertidal zone[23, 24]. *Phyllospadix* spp. form patches of various sizes in the surf zones, except *Phyllospadix serrulatus*, which is often found in more protected environments[2]. Plants in the genus *Phyllospadix* are largely clonal, and can be of a single sex due to the dioecious nature of the genus in this region.

Few studies have documented the habitat value of *Phyllospadix*; however, infaunal polychaetes are known to live in the rhizome mats of *Phyllospadix scouleri* and *Phyllospadix torreyi*. Surfgrass wrack, identified as *Phyllospadix torreyi*, has also been found in the macrophyte detritus layers in submarine canyons, in southern and central California[25]. This decomposing vegetation provides food and habitat for deep-sea benthic fauna[25]. In terms of commercially important species, researchers in southern California found that in their larval pelagic stage, spiny lobsters, *Panulirus interruptus*, were attracted to experimental treatments containing *Phyllospadix torreyi*.

Ruppia maritima is a variable plant with a number of named varieties with characteristic features, and is found in both freshwater and marine habitats. *Ruppia maritima* var. *spiralis* occurs along the southern coast of Alaska including the Alaska Peninsula, in British Columbia and in California[4]. Varieties *longipes* and *maritima* occur in coastal lagoons and estuaries throughout British Columbia including the Haida Gwaii[4]. Further south *Ruppia maritima* occurs at many sites influenced by saline water in Washington, Oregon, California and northern Mexico[1, 14]. The leaves, rhizomes and seeds of this plant are eaten by resident and migratory waterfowl.

Halodule wrightii is a subtropical species that occurs in the Gulf of California off the coast of mainland Mexico[1, 2], and *Zostera asiatica* is known to occur in three subtidal regions in central California, United States, where it forms underwater forests ca 3 m tall[3]. More work is necessary to elucidate the life history traits and habitat value of these species in this region.

HISTORICAL PERSPECTIVES

Potential changes in the standing crop and areal extent of *Zostera marina* have concerned natural resource managers in the Northeast Pacific for more than two decades[8, 26]. This concern is primarily a function of the habitat value provided by the large ecosystems created by these plants. Changes in the distribution of *Zostera japonica* have received less attention. This species is an exotic but provides valuable waterfowl habitat[10]. Given this, and the fact that *Zostera japonica* has not yet been shown to negatively impact the indigenous *Zostera marina*, prompts some to argue for detailed resource inventories. Information about changes in the local or

---

Case Study 19.3
THE LINK BETWEEN SEAGRASSES AND HUMANS IN PICNIC COVE, SHAW ISLAND, WASHINGTON, UNITED STATES

Picnic Cove is a sheltered embayment on the southeast corner of Shaw Island, which is centrally located in the San Juan Archipelago in the Pacific Northwest (see photograph; 48°35'N 122°57'W).

It contains a *Zostera marina* meadow of ca 0.05 km² [33] and a very small patch of *Zostera japonica*, in addition to multi-layered shell middens on the low bank at the head of the cove, which indicate historical use by coastal Salish people.

After European contact, Picnic Cove became a favorite picnic spot.

It is now the site of a long-term monitoring station for Washington State Department of Natural Resources' Submerged Vegetation Monitoring Project, and is the location of quadrat-based investigations by S. Wyllie-Echeverria, University of Washington.

The San Juan Archipelago with insert of Picnic Cove on Shaw Island, Washington, United States.

Photos: S. Wyllie-Echeverria

Table 19.1
Zostera marina and *Zostera japonica* basal area cover in the Northeast Pacific

Country	Region	Area (km²)
USA	Port Clarence, AK[28, 29]	4.2
USA	Safety Lagoon, AK[28, 29]	9.1
USA	Izembek Lagoon, AK[30]	159.5
USA	Kinzarof Lagoon, AK[28, 29]	8.7
USA	East Prince William Sound, AK[31]	4.4
Canada	Roberts Bank, BC[32]	4*
Canada	Boundary Bay, BC[10]	56*
USA	Puget Sound, WA[34]	200
	Sites within Puget Sound, WA	
	Padilla Bay[29]	32*
	King County[35]	2
USA	Grays Harbor, WA[8]	47.3
USA	Willapa Bay, WA[36]	159*
USA	Netarts Bay, OR[37]	3.4
USA	Yaquina Bay, OR[38]	0.9*
USA	Tillamook Bay, OR[39]	3.6
USA	Coos Bay, OR[9]	0.01*
USA	Humboldt Bay, CA[40]	12.2
USA	Tomales Bay[41]	3.9
USA	San Francisco Bay[42]	1.9
USA	San Diego Bay[43]	4.4
Mexico	Bahía San Quintin[30]	20
Mexico	Laguna Ojo de Liebre[30]	175
Mexico	Laguna San Ignacio[30]	53

Note: * Includes both *Zostera marina* and *Zostera japonica*.

Source: Various sources – see individual references by regions.

regional abundance of the other seagrass species in the Northeast Pacific remains largely unknown[26].

Direct use of seagrasses by humans for food, technology and medicine in the Northeast Pacific was widespread before European contact[7, 13, 14]. However, use now is quite localized and involves the weaving of *Phyllospadix* spp. as a decorative element in small personal baskets and the collection of *Zostera marina* plants for green mulch or the protection of culturally significant sites (see Case Study 19.2). The United States Department of Agriculture investigated the potential use of *Zostera marina* as a cultivar in coastal desert ecosystems during the 1980s, but we are unaware of any projects to further this goal.

It is difficult to ascertain the extent of seagrass losses due to coastal development and population expansion since the beginning of the 20th century, as no baseline data exist prior to the onset of these changes. Any attempt to do so would be conjecture. However, there is anecdotal information to suggest that losses of *Zostera marina* have occurred in the two largest estuaries on the west coast of the continental United States – Puget Sound and San Francisco Bay[27] – and we suspect losses have occurred at other sites as well. Widespread efforts to monitor and map this species in this region should help to determine if local and regional losses continue in the 21st century. More effort is needed to develop a programmatic response for comprehensive resource inventories of the other seven seagrass species[26].

SEAGRASS COVERAGE

Whereas an estimate of the seagrass cover is marginally possible for *Zostera marina*, it is not possible for the other seven species. The conservative estimate for *Zostera marina* is approximately 1 000 km² and is based on studies cited in Table 19.1 and our personal knowledge of sites not yet mapped.

THREATS

The following discussion of the present and potential threats to seagrasses is based on some information documented by Phillips[8], Wyllie-Echeverria and Thom[26] and studies therein, as well as a degree of unpublished observation and conjecture.

Coastal modifications and overwater structures in the form of ferry terminals, commercial docks, and smaller residential docks and floats threaten the survival of species that could be shaded in either soft-bottom or rocky littoral zones. Shoreline armoring, which can alter the trajectory of reflected wave energy, may also displace seagrasses. The direct removal of seagrasses through maintenance dredging is a rare occurrence in the Northeast Pacific, but resuspension of sediment associated with activity outside the seagrass zone may reduce transmission of light and/or bury plant populations. The deposition of upland soils into the littoral zone as a result of industrial, commercial and residential development may smother and/or kill seagrass. Moreover, modifications to the coastline projection may alter longshore current patterns resulting in changes to water clarity.

Recreational watercraft (powerboats, jet skis, etc.) may scar seagrasses in soft-bottom environments resulting in the fragmentation of populations and the subsequent loss of wildlife habitat. Whereas larger vessels (ferries, freighters, tankers, etc.) rarely venture into shallow waters, accelerated currents associated with propeller wash connected with landing and getting under way may displace seagrasses and affect current flow during pollen and seed release. The swing of anchor chains, and chains and lines connected to permanent buoys, can uproot plants and leave permanent scars in populations.

Rack and rope culture techniques used in commercial shellfish culture may shade the bottom and alter nutrient regimes and current flow, which may result in the loss of seagrass cover in localized areas. Human trampling associated with the harvest of market-sized oysters from stakes used to set and grow oyster spat can also result in reductions in seagrass cover. Moreover, recreational clam removal using shovels can destroy meters of seagrass cover.

Spills associated with oil production from offshore oil platforms such as those located on the continental shelf of southern California or the transport by oceanic tankers from Alaska to southern ports in Washington and California may result in the death of seagrasses in the littoral zone depending on the intensity and duration of the spill. Proposals for offshore oil development in Canada, the United States and Mexico are most problematic in this regard.

Episodic events such as ENSO (El Niño Southern Oscillation) and interdecadinal variation have the potential to alter ocean temperature and rainfall regimes, which may affect local populations and may also operate on the regional scale. However, a preliminary investigation in Puget Sound found that both biomass and productivity of a subtidal *Zostera marina* population increased during an El Niño year (1991-92) demonstrating the need for time-series data collection to evaluate the status and trend of seagrass ecosystems. Subduction associated with the nearshore plate tectonic activity may alter the shape and size of the littoral zone, reducing or eliminating the seagrass cover, as in the case of the 1964 Alaska earthquake.

The fragmentation of populations caused by natural or anthropogenic disturbance can also provide habitat for introduced species such as the mussel, *Musculista senhousia*, which can in turn prevent regrowth into fragmented areas, potentially leading to a more widespread decline in seagrass cover.

POLICY OPTIONS AND SEAGRASS PROTECTION

It is not clear that federal, provincial or state, or local administrative laws and ordinances recognize the eight seagrass species in the Northeast Pacific. However, in the United States, Canada and Mexico, protection is afforded to *Zostera marina* because the ecosystems provided by this plant are valuable habitat for commercially and recreationally important species such as Pacific salmon (*Oncorhynchus* spp.), Pacific herring (*Clupea harengus pallasi*) and black brant (*Branta bernicla nigricans*)[8]. In Washington state, *Zostera japonica* is also protected, but to the best of our knowledge no other seagrasses are protected by administrative code in the Northeast Pacific.

Small net bag or flexible basket woven from the leaves of *Phyllospadix torreyi*, found at Santa Rosa Island, California, and dated 1100-1500 in the Common Era.

Zostera marina prairie adjacent to the ferry terminal on Whidbey Island in central Puget Sound. Prairie density is influenced by both the overwater structure and ferry propwash.

AUTHORS

Sandy Wyllie-Echeverria, School of Marine Affairs, Box 355685, University of Washington, Seattle, WA, USA, 98105-6715. **Tel:** +1 360 468 4619; 293 0939. **Fax:** +1 206 543 1417. **E-mail:** zmseed@u.washington.edu

Josef Daniel Ackerman, Physical Ecology Laboratory, University of Northern British Columbia, Prince George, BC, Canada, V2N 4Z9.

REFERENCES

1. Aguilar-Rosas R, Lopez-Ruelas J [1985]. *Halodule wrightii* Aschers. (Potamogetonales: Cymodoceae) in Topolobampo Sinaloa, Mexico. *Ciencias Marinas* 11(2): 87-91.
2. Phillips RC, Menez EG [1988]. *Seagrasses*. Smithsonian Contributions to the Marine Sciences No. 34.
3. Phillips RC, Wyllie-Echeverria S [1990]. *Zostera asiatica* Miki on the Pacific Coast of North America. *Pacific Science* 44(2): 130-134.
4. Brayshaw TC [2000]. *Pondweeds, Bur-reeds and their Relatives of British Columbia*. Royal British Columbia Museum, Victoria. 250 pp.
5. Domning DP [1976]. An ecological model for late Tertiary sirenian evolution in the North Pacific Ocean. *Systematic Zoology* 25: 352-362.
6. Harrison PG, Bigley RE [1982]. The recent introduction of the seagrass *Zostera japonica* Aschers. and Graebn. to the Pacific Coast of North America. *Canadian Journal of Fisheries and Aquatic Sciences* 39: 1642-1648.
7. Wyllie-Echeverria S, Cox PA [2000]. Cultural saliency as a tool for seagrass conservation. *Biologia Marina Mediterranea* 7(2): 421-424.
8. Phillips RC [1984]. The Ecology of Eelgrass Meadows in the Pacific Northwest: A Community Profile. FWS/OBS-84/24.
9. Rumrill S [2001]. South Slough National Estuarine Research Reserve, Charleston, OR (SSNERR). Personal communication.
10. Baldwin JR, Lovvorn JR [1994]. Expansion of seagrass habitat by the exotic *Zostera japonica*, and its use by dabbling ducks and brant in Boundary Bay, British Columbia. *Marine Ecology Progress Series* 103: 119-127.
11. Ibarra-Obando SE, Boudouresque CF, Roux M [1997]. Leaf dynamics and production of a *Zostera marina* bed near its southern distributional limit. *Aquatic Botany* 58(2): 99-112.
12. Ward DH, Markon CJ, Douglas DC [1997]. Distribution and stability of eelgrass beds at Izembek Lagoon, Alaska. *Aquatic Botany* 58: 229-240.
13. Kuhnlein HV, Turner NJ [1991]. *Traditional Plant Foods of Canadian Indigenous Peoples: Nutrition, Botany and Use*. Gordon and Breach Science Publishers, Philadelphia.
14. Felger RS, Moser MB [1985]. *People of the Desert and Sea: Ethnobotany of the Seri Indians*. University of Arizona Press, Tucson, AR.
15. McRoy CP [1968]. The distribution and biogeography of *Zostera marina* (eelgrass) in Alaska. *Pacific Science* 22: 507-513.
16. Meling-Lopez AE, Ibarra-Obando SE [1999]. Annual life cycles of two *Zostera marina* L. populations in the Gulf of California: Contrasts in seasonality and reproductive effort. *Aquatic Botany* 65(1-4): 59-69.
17. Ackerman JD [2000]. Abiotic pollen and pollination: Ecological, functional, and evolutionary perspectives. *Plant Systematics and Evolution 222*: 167-185.
18. Ruckelshaus MH [1995]. Estimates of outcrossing rates and of inbreeding depression in a population of the marine angiosperm *Zostera marina*. *Marine Biology* 123(3): 583-593.
19. Alberte RS, Suba, GK, Procaccini G, Zimmerman RC, Fain SR [1994]. Assessment of genetic diversity of seagrass populations using DNA fingerprinting: Implications for population stability. *Proceedings of the National Academy of Sciences* 91: 1049-1053.
20. Williams SL, Davis CA [1996]. Population genetic analyses of transplanted eelgrass (*Zostera marina*) beds reveal reduced genetic diversity in southern California. *Restoration Ecology* 4(2): 163-180.
21. Backman TWH [1991]. Phenotypic and genotypic variability of *Zostera marina* on the west coast of North America. *Canadian Journal of Botany* 69: 1361-1371.
22. Posey MH [1988]. Community changes associated with the spread of an introduced seagrass, *Zostera japonica*. *Ecology* 69(4): 974-983.
23. Turner T, Lucas J [1985]. Differences and similarities in the community roles of three rocky intertidal surfgrasses. *Journal of Experimental Marine Biology and Ecology* 89(2-3): 175-189.
24. Ramirez-Garcia P, Lot A, Duarte CM, Terrados J, Agawin NSR [1998]. Bathymetric distribution, biomass and growth dynamics of intertidal *Phyllospadix scouleri* and *Phyllospadix torreyi* in Baja California (Mexico). *Marine Ecology Progress Series* 173: 13-23.
25. Vetter EW, Dayton PK [1999]. Organic enrichment by macrophyte detritus and abundance patterns of megafaunal populations in submarine canyons. *Marine Ecology Progress Series* 186: 137-148.
26. Wyllie-Echeverria S, Thom RM [1994]. Managing Seagrass Systems in Western North America: Research Gaps and Needs. Alaska Sea Grant Program Report No. 94-01.
27. Zimmerman RC, Reguzzoni JR, Wyllie-Echeverria S, Alberte RS, Josselyn MN [1991]. Assessment of environmental suitability for growth of *Zostera marina* in San Francisco Bay. *Aquatic Botany* 39: 353-366.
28. McRoy CP [1970]. Standing stocks and other features of eelgrass (*Zostera marina*) populations on the coast of Alaska. *Journal of the Fisheries Board of Canada* 27: 1811-1821.
29. Bulthuis DA [1995]. Distribution of seagrasses in a North Puget Sound estuary: Padilla Bay, Washington, USA. *Aquatic Botany* 50: 99-105.
30. Ward DH, Tibbitts TL, Morton A, Carrera-Gonzéles E, Kempka R [in press]. Use of airborne remote sensing techniques to assess seagrass distribution in Bahia San Quintin, Baja California, Mexico. *Ciencias Marinas*.
31. McRoy CP, Bridges KW [1998]. Eelgrass Survey of Eastern Prince William Sound. Report to the US Army Corps of Engineers. Anchorage, Alaska.
32. Harrison PG [1987]. Natural expansion and experimental manipulation of seagrass (*Zostera* spp.) abundance and the response of infaunal invertebrates. *Estuarine Coastal and Shelf Science* 24: 799-812.
33. Norris JG, Wyllie-Echeverria S, Mumford T, Bailey A, Turner T [1997]. Estimating basal area coverage of subtidal seagrass beds using underwater videography. *Aquatic Botany* 58: 269-287.
34. Submerged Vegetation Monitoring Project [2003]. Nearshore Habitat Program. Washington State Department of Natural Resources, Olympia, Washington.
35. Woodruff DL, Farley PJ, Borde AB, Southard JS, Thom RM [2001]. King County Nearshore Habitat Mapping Data Report: Picnic Point to Shilsole Marina. Report to King County Department of Natural Resources, King County, Washington.
36. C-CAP Coastal Change Analysis Program [1995]. Matthew Van Ess, Director Columbia River Estuary Study Taskforce (CREST), 750 Commercial Street, Room 205 Astoria, OR 9710 (tel: 503 325 0435).
37. Kentula ME, McIntire CD [1986]. The autecology and production dynamics of eelgrass (*Zostera marina* L.) in Netarts Bay, Oregon. *Estuaries* 9(3): 188-199.
38. Clinton PJ, Young DR, Specht DT [2002]. A hybrid high-resolution image classification method for mapping eelgrass distribution in Yaquina Bay, Oregon. (Submitted to 7th ERIM Conference for Remote Sensing in Marine and Coastal Environments, Miami, Florida, 20-22 May 2002.) www.veridian.com/conferences
39. Strittholt JR, Frost PA [1996]. Determining Abundance and Distribution of Eelgrass (*Zostera* spp.) in the Tillamook Bay Estuary, Oregon Using Multispectral Airborne Imagery. www.cwrc.org/html/reportse-h.htm
40. Harding LW Jr, Butler JH [1979]. The standing stock and production of eelgrass, *Zostera marina*, in Humboldt Bay, *California*. *California Fish and Game* 65(3): 151-158.
41. Spratt JD [1989]. The distribution and density of eelgrass *Zostera marina* in Tomales Bay, California. *California Fish and Game* 75(4): 204-212.
42. Wyllie-Echeverria S [1990]. Distribution and geographic range of *Zostera marina* eelgrass in San Francisco Bay. In: Merkel KW, Hoffman RS (eds) *Proceedings of the California Eelgrass Symposium*. Sweetwater River Press, National City, CA. pp 65-69.
43. US Navy [2000]. Eelgrass Survey. SWDIV Naval Facilities Engineering Command, Port of San Diego, San Diego Bay.

20 The seagrasses of
THE WESTERN NORTH ATLANTIC

F.T. Short
C.A. Short

Eelgrass (*Zostera marina*) is the overwhelmingly dominant seagrass in coastal and estuarine areas of the western North Atlantic, a region considered here as the Atlantic coast from Quebec (Canada) at approximately 60°N to New Jersey (United States) at 39°N[1, 2]. Eelgrass meadows provide a wide array of ecological functions important for maintaining healthy estuarine and coastal ecosystems[3], creating essential habitat and forming a basis of primary production that supports ecologically and economically important species in the region[4-6]. The importance of eelgrass to estuarine and coastal productivity was highlighted in the 1930s, when a large-scale die-off of eelgrass occurred on both sides of the Atlantic due to wasting disease[7]. The wasting disease has since been shown to be caused by a pathogenic slime mold, *Labyrinthula zosterae*[8] and has been reported in several species of *Zostera*[9, 10]. The disease resulted in the loss of over 90 percent of the North Atlantic eelgrass population, and this loss had a catastrophic effect on estuarine productivity including the disappearance of the scallop (*Argopecten irradians*) fishery and drastic reduction in brant (*Branta bernicla*) populations[11]. After 30-40 years, eelgrass largely recovered from the 1930s' wasting disease, although in some areas it has not regained its previous distribution[5].

A second seagrass found in the region is widgeon grass (*Ruppia maritima*); it occurs sporadically, mainly in low-salinity, brackish and freshwater areas, marsh pools and some tidal rivers. Relatively little is known about the distribution and ecology of widgeon grass; in most areas it is much less common than eelgrass.

Throughout the western North Atlantic region, large meadows of eelgrass are found from Canada through New Jersey. Despite fluctuations in some areas due to recent episodes of wasting disease and recovery, the trend over the past 30 years has been a steady decrease in eelgrass distribution and abundance due to anthropogenic impacts. In the few areas of this region where habitat change analysis has been carried out, dramatic declines in eelgrass populations have been documented[12, 13]. Despite some laws that recognize the habitat value of eelgrass, there is no direct protection of seagrasses in Canada or the United States.

In Canada, eelgrass is found on the east coast south of the Arctic Circle, where it occupies vast intertidal and subtidal areas. Eelgrass is found in Hudson Bay, in the harbors of Newfoundland's rocky shoreline and in large meadows in the Northumberland Strait off Prince Edward Island. Eelgrass beds circumscribe much of Nova Scotia, though absent from the northern coastline of the Bay of Fundy.

In Maine, the northernmost US state on the east coast, extensive eelgrass beds occupy the full range of eelgrass habitat conditions. In northern Maine, with the highest tidal ranges in the world (more than 8 m), eelgrass occurs mostly in protected bays and harbors, as well as tidal rivers. In mid-coast Maine, eelgrass is also found on exposed coasts and around islands, for instance in Penobscot and Casco Bays. Eelgrass distribution in southern Maine ranges from sheltered areas to exposed coasts, but eelgrass occurs less frequently, due to the coastal geomorphology which is dominated by salt marshes; widgeon grass is found in many salt marsh ponds and tidepools.

New Hampshire, south of Maine, has a coastline of only 27 km, and a drowned river valley estuary consisting of the Piscataqua River and Great and Little Bays[14]. Eelgrass meadows occur at the mouth of the river in Portsmouth Harbor (see Case Study 20.1), both intertidally and to a depth of 12 m below mean sea level, in small patches along the sides of the deeper river channels, sparsely in Little Bay and in a large monospecific expanse within Great Bay. Along the rest of the New Hampshire coastline, eelgrass is found in deep meadows along the exposed coast and in smaller patches in sheltered areas and harbors. Widgeon grass occurs in salt marsh ponds and, to a limited extent, in upper Great Bay[15, 16]. Further south, the state of

Massachusetts has extensive eelgrass meadows in the physically protected areas of Cape Cod, the offshore islands and along its glacially striated southern shoreline. In Boston Harbor, north of Cape Cod, eelgrass beds are slowly recovering due to improved water clarity from installation of an offshore sewage discharge. South of Cape Cod, in Buzzards Bay and Vineyard Sound, eelgrass is disappearing due to anthropogenic impacts[12]. No known intertidal eelgrass occurs in Massachusetts or further south. Widgeon grass occurs in some salt marshes and brackish ponds.

Rhode Island is the smallest state in the United States, with a long shoreline for its size. Heavily developed Narragansett Bay, a water body dominating the state,

Case Study 20.1
PORTSMOUTH HARBOR, NEW HAMPSHIRE AND MAINE

The western North Atlantic exhibits a range of environmental conditions supporting eelgrass, from pristine to highly developed and from intertidal to deep subtidal. Cold winters produce conditions of ice scour while summer heat can desiccate intertidal eelgrass. Portsmouth Harbor, the mouth of the Great Bay Estuary on the border of Maine and New Hampshire, typifies many of these conditions. Within the harbor, eelgrass flourishes in large intertidal flats, often exposed for several hours around low tide and, at high tide, submerged by over 3 m of water. Eelgrass plants on these flats are small and thin bladed, typically with leaves less than 30 cm long. Across the channel in water about 3 m mean sea level, exceptionally long-leaved (2 m) eelgrass grows in a protected area behind a US Coast Guard pier; the bed is subject to frequent boat activity and mooring impacts.

Upstream from the Coast Guard station in the highly developed commercial harbor, some eelgrass thrives despite the nearby sewage discharge for the city of Portsmouth, because high tidal volumes deliver clear ocean water. However, adjacent to these beds, dredge spoil illegally dumped in a shallow subtidal zone buries former eelgrass habitat, while across the channel the hardened shoreline of the Portsmouth Naval Shipyard has virtually eliminated eelgrass and the possibility of its recovery. In less heavily developed areas upstream in the Piscataqua River eelgrass has been transplanted as mitigation for port expansion, replacing beds lost in the early 1980s to an outbreak of wasting disease[57]. Some of the transplants expanded into beds which continue to thrive eight years after transplanting; others died soon after transplanting due primarily to bioturbation[58]. Further up-estuary, the Piscataqua River connects to Little Bay and then Great Bay with its extensive intertidal eelgrass meadows.

At the mouth of Portsmouth Harbor and along the open New Hampshire coast, eelgrass beds thrive in the clear Gulf of Maine waters to a depth of 12 m mean sea level. Here, at depths below most human impacts, eelgrass forms a lush green expanse only rarely disturbed by lobster pots or fishing activity.

The mix of habitats in which eelgrass grows in Portsmouth Harbor is representative of many of the eelgrass habitats found throughout the western North Atlantic. North of Portsmouth Harbor, in Maine and Canada, eelgrass may grow in vast intertidal meadows in areas of extreme tidal fluctuation. South of Portsmouth Harbor, eelgrass is rarely found intertidally, but often forms shallow meadows in back barrier lagoons or salt ponds. However, the eelgrass habitats in Portsmouth Harbor capture many of the conditions affecting its distribution throughout the western North Atlantic, including large tidal variation, temperature extremes, ice, both significant human impacts and restoration efforts, and ongoing wasting disease episodes.

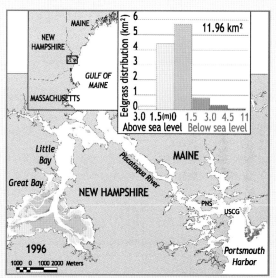

Eelgrass distribution by depth in Portsmouth Harbor, Great Bay Estuary, on the border of New Hampshire and Maine, United States.

Notes: Depths plotted as mean low water, with an average tidal range of 2.7 meters. Data from 1996; eelgrass polygons were determined from image analysis and ground-truthing of aerial photography using the C-CAP protocol[42]. USCG United States Coast Guard station; PNS Portsmouth Naval Shipyard.

has only a few remaining small eelgrass beds. More extensive eelgrass occurs in sheltered coastal ponds behind barrier beaches on the south shore of Rhode Island but losses are occurring (see Case Study 20.2). Widgeon grass is found in the salt ponds in areas of groundwater intrusion and in some freshwater ponds. Connecticut, on Long Island Sound, has eelgrass in subtidal habitats of some protected bays, with offshore beds in the shelter of Fishers Island. All eelgrass beds in Connecticut occur in the eastern third of the coastline due to poor water quality in the western part of Long Island Sound. Widgeon grass occurs in some salt marshes. New York State has eelgrass only around Long Island with no known beds on the north shore of the island, but substantial meadows in some bays and inlets of the east and south shores, including the Peconic Estuary. Widgeon grass is found in the upper brackish portions of some bays and salt marshes. New Jersey, the southernmost state in the region, has extensive eelgrass beds in the southern bays with Barnegat Bay having the best-documented populations. In New Jersey, eelgrass is predominantly found in shallow, open lagoons, while widgeon grass occurs in brackish water areas.

Within the western North Atlantic region, eelgrass restoration has been undertaken sporadically; considerable research has taken place on both transplanting and seed planting. New transplanting methods have been developed which simplify the transplanting process and increase its level of success; projects have transplanted up to 2.62 hectares of eelgrass[2]. These relatively small and expensive projects also demonstrate that preservation of seagrass is vastly preferable to, and less costly than, restoring lost habitat.

BIOGEOGRAPHY

In the western North Atlantic, eelgrass is found in both intertidal and subtidal areas, from a depth of +2 m to −12 m mean sea level[5]. Depth distribution is limited by water clarity and the large tidal range along the North Atlantic coastline. Eelgrass distribution ranges from the protected low-salinity (5 psu) waters of inner estuaries and coastal ponds to high-energy locations fully exposed to the Gulf of Maine and the North Atlantic with salinity of 36 psu. Eelgrass inhabits a range of sediment conditions from soft, highly organic muds to coarse sand and partial cobble[5]. In a comparative study of eelgrass populations from Maine to North Carolina, some general patterns of variation with latitude were found[5]. In summary, eelgrass shoot density decreases (1 275 to 339 shoots/m^2) with increasing latitude (south to north), while leaf biomass (from 106 to 249 g dry weight/m^2) and plant size (35 cm to 125 cm average leaf length) increase with increasing latitude[5]. Additionally, eelgrass leaf growth showed a significant increase from south to north over this geographic range, from a

Map 20.1
The western North Atlantic

low of 0.7 to a high of 19.1 g dry weight/m^2/day. At the southern end of its western North Atlantic range, eelgrass distribution is limited by high summer temperatures, the relatively small tidal range, and the generally low-organic, sandy sediments of the back barrier island lagoons. North of Cape Cod, eelgrass grows most commonly in estuarine environments or along the open coast where the cooler water temperatures, higher tidal ranges, and fine-grained organic sediments create conditions that support larger plants and greater biomass.

Eelgrass in the western North Atlantic provides habitat for numerous commercially important fish and shellfish species[17]. Young winter flounder (*Pseudopleuronectes americanus*) concentrate in eelgrass beds[18,19]; juvenile lobsters (*Homarus americanus*) likewise favor eelgrass habitat and have been shown to overwinter in burrows within eelgrass beds[20,21]; Atlantic cod (*Gadus morhua*) is documented as using eelgrass beds as nursery habitat in Canada[22]. Other commercially and recreationally important species that use eelgrass habitat include smelt (*Osmerus mordax*), which spends time in eelgrass as part of its migratory cycle[14], and striped bass (*Morone saxatilis*) which have been tracked moving into eelgrass beds to feed[23]. Shellfish, including bay scallops (*Argopecten irradians*) and blue mussels (*Mytilus*

edulis), have been shown to utilize eelgrass beds, sometimes for settlement of juvenile phases and sometimes as adults[24-26].

While some species associated with eelgrass habitat are in decline, such as flounder, cod and scallops, no species are officially designated as threatened. The eelgrass limpet, *Lottia alveus*, became extinct after the 1930s' wasting disease[27]. The brant goose (*Branta bernicla*), a species dependent on eelgrass as a primary food source, was abundant before the 1930s and has only partially recovered. Ducks, swans and other species of goose use eelgrass as food and are known to stop in eelgrass areas during migration.

HISTORICAL PERSPECTIVES

Major losses of eelgrass area in the western North Atlantic occurred before any documentation of distribution was accomplished. Areas such as Boston Harbor (Massachusetts) and the Providence River (Rhode Island) have experienced human modification and impact for the last 400 years; sites within these harbors which probably supported eelgrass are now filled or degraded. Historical reports and anecdotal information from fishermen, as well as early navigation charts, all indicate that eelgrass extent was previously much greater than it is today. For example, in Narragansett Bay, Rhode Island, S. Nixon has found charts dating to the 1700s showing eelgrass well up into the Providence River in the upper estuary. Today, the small amount of eelgrass in Narragansett Bay extends only two thirds of the way up the bay[28]. Quantitative studies of seagrass loss in the western North Atlantic have occurred only in the past decade or so, and in only a few locations.

The wasting disease of the 1930s almost eliminated eelgrass from much of the area reported here. This decline had major ecological impacts[4]. The human use of eelgrass wrack for insulation, bedding, stuffing and as mulch, which constituted a commercial effort in Canada and northern New England states, dropped off during the 1930s and 1940s and never revived[29,30]. Today only a few home gardeners collect eelgrass wrack for mulch.

AN ESTIMATE OF HISTORICAL LOSSES

Since the arrival of Europeans in the region, the western North Atlantic has lost eelgrass populations in virtually all areas of intense human settlement. Today, most of these areas remain devoid of eelgrass although, with improved sewage treatment and environmental controls of discharge, some industrialized areas (Boston Harbor and New Bedford Harbor, Massachusetts) are beginning to show eelgrass recovery or are now suitable for restoration. Dredging, filling, marina development, boat activity, fishing practices, hardening of the shoreline and anthropogenic nutrient and sediment discharge all continue to impact eelgrass habitat and areas where it could return. The loss of eelgrass has not been quantified in the region but certainly differs in two areas of the coast. North of Cape Cod, Massachusetts, eelgrass loss since settlement is estimated to be in the order of 20 percent, while south of Cape Cod, which is more heavily populated and industrialized, we estimate that 65 percent of eelgrass distribution has been lost.

Two locations in New England have documentation of the rapid decline of eelgrass populations resulting from anthropogenic nutrient loading by way of contaminated groundwater discharge: Waquoit Bay, Massachusetts, and Ninigret Pond, Rhode Island. In Waquoit Bay, the decline in eelgrass associated with nitrogen loading rates was documented in a space-for-time substitution of seven sub-estuaries having varying degrees of housing development[12]. The greatest eelgrass loss occurred in the sub-estuaries with most development; overall, 60 percent of the eelgrass was lost from this estuary in five years. In Ninigret Pond, eelgrass distributions were compared over a 32-year period using historical and recent maps; areal distribution of eelgrass declined by 41 percent[13] (see Case Study 20.2).

In New Jersey's Little Egg Harbor and Barnegat Bay, eelgrass beds were mapped through the 1970s and 1980s, and again in 1999[31]. Throughout the period, 20 km² of eelgrass were lost. Little Egg Harbor and the adjacent Barnegat Bay are the only two areas in New Jersey that still support eelgrass to any extent. Other areas in New Jersey which supported eelgrass historically have declined and show no recovery.

Maquoit Bay in Maine (northern Casco Bay) has been impacted by mussel dragging, a fishery practice in which a weighted steel frame and net are dragged through eelgrass beds to harvest blue mussels[32] (see Case Study 20.3). Dragging for mussels in 1999 created a 28.3-hectare bare area in the center of a large eelgrass meadow[33].

Great Bay, New Hampshire, experienced a recurrence of the wasting disease in the 1980s. Eelgrass populations went from 824 hectares in 1986 to 130 hectares in 1989. This loss, accounting for 80 percent of the eelgrass in Great Bay, was reversed by rapid recruitment from seed production and a recovery of eelgrass to 1015 hectares by 1996.

Changes in the physical environment that may result from eelgrass loss include seafloor subsidence and loss of fine particle sediments and organic matter[7,34], increase in sediment transport and decrease in sediment deposition[35], and short-term water quality degradation caused by resuspended sediment[36,37]. Biological changes may include a shift in the benthic infauna from a predominantly

Case Study 20.2
NINIGRET POND, RHODE ISLAND

Loss of eelgrass is a problem in shallow nutrient-enriched estuaries of the urban and urbanizing northeastern United States. From 1960 to 1992 there was a clear relationship between increased housing density and decreased eelgrass area in Ninigret Pond, Rhode Island, a shallow estuarine embayment behind a barrier beach[13]. With increased housing density, and corresponding increased nutrient loading via enriched groundwater which produced macroalgal blooms, eelgrass area in Ninigret Pond decreased rapidly between 1974 and 1992, primarily in shallow areas of the pond (see figures). The major loss of eelgrass in the pond occurred in shallow areas where macroalgae and groundwater enrichment had the greatest impact; eelgrass in the deeper areas showed little change.

Maps of eelgrass distribution were compared with the number of houses in the watershed, and a significant linear trend of eelgrass area loss with increased housing over time was demonstrated (see figure)[13]. Over the 32-year period examined, housing in the watershed quadrupled, while eelgrass areal distribution declined 41 percent.

In shallow estuarine systems, such as Ninigret Pond, throughout the northeastern United States[1], especially within watersheds dominated by highly permeable sand/gravel glacial outwash aquifers, groundwater is a dominant source of freshwater and associated nitrate contamination[12, 59]. There is minimal removal of nitrate as groundwater discharges from highly permeable and low-organic soils into estuarine shorelines. In Ninigret Pond, groundwater discharge entering the pond was clearly visible in thermal infrared photographs, ultimately contributing to eelgrass loss, mostly in shallow areas and due to macroalgal smothering[13]. Subsequently, eelgrass has continued to decline in Ninigret Pond.

Eelgrass distribution in Ninigret Pond, Rhode Island (United States) plotted by depth for 1974 and 1992.

Notes: Data from image analysis of aerial photography and ground-truthing[13].

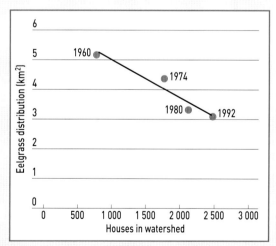

Change in eelgrass area in Ninigret Pond, Rhode Island (United States) plotted against increasing number of houses in the watershed[13].

deposit-feeding community to a suspension-feeding community[7] and a reduction in epifaunal species abundance[38]. These types of physical and biological changes reduce estuarine productivity and can prevent natural recolonization of eelgrass even when water quality becomes adequate.

Table 20.1
The area of eelgrass, *Zostera marina*, in the western North Atlantic

Location	Area (km²)	Year	Method
Maine	128.10	1992-97	C-CAP
New Hampshire	11.88	1996	C-CAP
Massachusetts	158.94	1995	C-CAP
Rhode Island	3.56	1999 and 1992	C-CAP C-CAP
Connecticut	2.56	1993-95	Diver survey
New York (data for Peconic Estuary only)	8.50	1994	Ground survey
New Jersey (data for Barnegat Bay and Little Egg Harbor only)	60.83	1999	Aerial photo
Total	374.37	1990s	

Notes: Year indicates the date of sampling. C-CAP is the US National Oceanic and Atmospheric Administration's Coastal Change Analysis Program, which includes a protocol for assessing seagrass from aerial photography[42]. The estimated area of eelgrass was obtained from comprehensive surveys within each location, except for New York and New Jersey, where more cursory information was available. No quantitative data were available for Canada.

Sources: Maine: S. Barker, Maine Department of Marine Resources; New Hampshire: F.T. Short, University of New Hampshire; Massachusetts: C. Costello, Massachusetts Department of Environmental Protection; Rhode Island: Narragansett Bay Estuary Program and Short *et al.*[13]; Connecticut: R. Rozsa, Connecticut Department of Environmental Protection; New York: Peconic Estuary Program, NY; New Jersey: Center for Remote Sensing and Spatial Analysis, Rutgers University and Lathrop *et al.*[31].

AN ESTIMATE OF PRESENT COVERAGE

Eelgrass in Canada was mapped in the early 1980s but never digitized. The primary known areas of eelgrass distribution in eastern Canada are summarized here. There is eelgrass in James Bay, Quebec, part of Hudson Bay[39]. Extensive eelgrass meadows are reported in the Northumberland Strait between New Brunswick and Prince Edward Island. Eelgrass beds are found in parts of the St Lawrence River[40]. In Nova Scotia, on the Atlantic coast, eelgrass grows in coves, tidepools and on the exposed coast[41]. There is probably more eelgrass in eastern Canada than in the US states of the western North Atlantic region combined, but no quantitative data are available for Canada.

In the United States, current distribution by state ranges from 250 hectares in Connecticut to over 15000 hectares in neighboring Massachusetts (Table 20.1). The majority of eelgrass area occurs north of Cape Cod. There are no known areal estimates for widgeon grass.

Potential seagrass habitat is difficult to measure because the depth distribution of eelgrass in most areas of the northeastern United States varies depending on water clarity. Methods have been developed for determining potential seagrass habitat[2], but have not been comprehensively applied to the region.

A DESCRIPTION OF PRESENT THREATS

Over the last decade, eelgrass populations have declined in some parts of New England and elsewhere due to pollution associated with increased human populations[12,43] and episodic recurrences of the wasting disease[44,45], as well as other human-induced and natural disturbances[46].

Seagrass is often impacted by direct damage from boating activities such as actual cutting by propellers, propeller wash and boat hulls dragging through vegetated bottom[47,48]. Other activities relating to boat operation and storage that impact eelgrass include docks which can shade the tide flat and prohibit light penetration[49,50], moorings which create holes within meadows from the swing of the anchor chain[5,48], and channel and marina dredging[51].

Certain fishing and aquaculture practices also impact eelgrass[52]. For example, harvesting mussels by trawling or dragging through *Zostera marina* meadows can eliminate areas of eelgrass or reduce shoot density and plant biomass[32] (see Case Study 20.3). Clam digging can disturb eelgrass either by direct removal or increased turbidity.

The following threats to eelgrass in the western North Atlantic are listed roughly in order of magnitude, except for the last three where the level of threat is difficult to quantify. The impact of brown tide can be severe in localized areas, occurs frequently in part of the region (Long Island) and is unknown in others[53]. The relative impacts to eelgrass health and distribution from both climate change and sea-level rise are presently unknown[54]. Wasting disease has the potential for very great impact, as seen in the 1930s, but most recent outbreaks of the wasting disease have been followed by rapid recovery.

o Point and non-point source nutrient loading: Anthropogenic inputs of nutrients from land development, sewage disposal, agriculture and the increase in impervious surfaces all contribute, resulting in overenrichment which promotes algal blooms[55]. In many places, nutrient-contaminated groundwater discharge to bays and coastal ponds is a major contributor to eelgrass loss[12,56].

o Sediment runoff: Land disturbance and deforestation produce increased loads of sediment to

coastal waters, increasing turbidity and decreasing light levels.
- o Dredging: Despite laws regulating dredge and fill, routine dredging is permitted in bays and harbors for channel maintenance, deepening of mooring fields, and improved boat access. All these activities often cause direct and indirect destruction of eelgrass.
- o Fisheries and shellfisheries harvest practices: Net dragging for fish and shellfish in parts of the region can have severe local impacts, uprooting eelgrass over large areas of the bottom. Dragging scars persist in eelgrass beds for many years.
- o Hardening of the shoreline: Creation of bulkheads and sea walls, as well as elimination of shoreline vegetation, increase sediment input to coastal waters and exacerbate sediment resuspension.
- o Filling: Historically, filling had a large impact on eelgrass, but now regulations limit fill activity in coastal waters.
- o Boating, including boat docks and moorings: Boating activities in shallow waters resuspend sediments and create propeller scars, damaging eelgrass beds. Docks shade eelgrass to the point of elimination and bed fragmentation. Moorings create holes in eelgrass beds as the long mooring chains, needed for the high tidal ranges in the region, drag across the bottom.
- o Aquaculture pens and rafts: The rapid expansion of Atlantic salmon aquaculture in Maine and Canada has led to deployment of pens within sheltered estuarine areas. High nutrient loads resulting from excess feed and fish waste create local eutrophication conditions. Blue mussel rafts shade the bottom and promote macroalgal growth that causes eelgrass loss.
- o Brown tide: Algal blooms shade and eliminate eelgrass.
- o Climate change and sea-level rise: The potential impacts of these changes are great[54].
- o Wasting disease: Historically, the wasting disease severely impacted eelgrass distribution; currently, the disease impacts populations at less severe levels. The conditions that led to the widespread disease outbreak of the 1930s are not known, but there is the potential for recurrence.

POLICY

In Canada, seagrass receives no specific legal protection. In the United States, seagrass is protected under the Clean Water Act as a "special aquatic site" and falls within Essential Fish Habitat as a "habitat area of particular concern" under the Magnuson-Stevens Fishery Conservation and Management Act. Neither of these laws provides complete or direct protection of eelgrass, but eelgrass habitat may be given special consideration in permit review processes.

There are no marine protected areas having eelgrass in the western North Atlantic region. There are five National Estuarine Research Reserve (NERR) sites

---

Case Study 20.3
MAQUOIT BAY, MAINE

Maquoit Bay, Maine (United States) is the location of a vast eelgrass bed representing one of the more extensive stands of intertidal seagrass in the western North Atlantic. The plants range from small, thin-bladed eelgrass with extensive flowering in the shallow intertidal to robust, densely growing large plants (more than 2 m) at the lower extent of the intertidal. This estuary in mid-coast Maine has a tidal range of 5 m, resulting in the twice daily exposure of much of the eelgrass; subtidal beds extend to a depth of 10 m in the clear waters of the outer bay. The bay is harvested for wild blue mussels (*Mytilus edulis*), soft-shell clams (*Mya arenaria*) and clam worms (*Nereis virens*). Fishing practices include dragging for mussels, which has destroyed up to 0.3 km² of eelgrass in a season (see photograph). Such uprooted areas in the eelgrass meadow are predicted to require 10-17 years to recover via a combination of seed recruitment and rhizome elongation[32]. In contrast, the local practice of digging for clams and clam worms with a shovel or rake results in partial disturbance to the inshore edge of the eelgrass meadow; the beds recover after two years.

Maquoit Bay, Maine (United States), showing scars in the eelgrass bed caused by mussel dragging.

Note: Below the scrape marks, the round areas of disturbance to the bed are caused by the mussel draggers discharging debris from their nets as they return for another pass.

Students sampling intertidal *Zostera marina* at a SeagrassNet monitoring site in Portsmouth Harbor, New Hampshire.

which contain eelgrass. While these reserves afford no legal or direct protection to eelgrass, they are managed for research purposes and knowledge of eelgrass distribution within these reserves has been established. Most NERR sites have programs that include public awareness and outreach education.

There is some increasing scientific, policy and public awareness and interest in seagrass within the region. A group of scientists and managers meets annually under the auspices of the US Environmental Protection Agency to discuss eelgrass and receive an update on research activities. A multimillion dollar port development project in Maine was denied a construction permit in the mid-1990s based largely on its potential impact on eelgrass habitat. The US Army Corps of Engineers, the agency primarily responsible for dredge, fill and other construction activities in coastal waters of the United States, has an increasing awareness of the importance of eelgrass habitat, and has recently undertaken mitigation for some routine mooring area dredge activities that impacted an eelgrass bed. Clearly, more specific protection of seagrass habitat is needed in the western North Atlantic.

ACKNOWLEDGMENTS

We thank the University of New Hampshire Agricultural Experiment Station for support. Thanks to Jamie Adams for GIS contributions. Eelgrass distribution coverages and data were provided by Seth Barker, Maine Department of Marine Resources; Charles Costello, Massachusetts Department of Environmental Protection; the Narragansett Bay Estuary Program, Rhode Island; Ron Rozsa, Connecticut Department of Environmental Protection; the Peconic Estuary Program, New York; Richard Lathrop, Grant F. Walton Center for Remote Sensing and Spatial Analysis, Rutgers University, New Jersey. This is Jackson Estuarine Laboratory contribution number 392 and AES scientific contribution number 2144.

AUTHORS

Frederick T. Short and Catherine A. Short, University of New Hampshire, Jackson Estuarine Laboratory, 85 Adams Point Road, Durham, NH 03824, USA. **Tel:** +1 603 862 2175. **Fax:** +1 603 862 1101. **E-mail:** fred.short@unh.edu

REFERENCES

1. Roman CT, Jaworski N, Short FT, Findlay S, Warren RS [2000]. Estuaries of the Northeastern United States: Habitat and land use signatures. *Estuaries* 23(6): 743-764.
2. Short FT, Davis RC, Kopp BS, Short CA, Burdick DM [2002 in press]. Site selection model for optimal restoration of eelgrass, *Zostera marina* L. *Marine Ecology Progress Series*.
3. Short FT, Burdick DM, Short CA, Davis RC, Morgan PA [2000]. Developing success criteria for restored eelgrass, salt marsh and mud flat habitats. *Ecological Engineering* 15: 239-252.
4. Thayer GW, Kenworthy WJ, Fonseca MS [1984]. *The Ecology of Eelgrass Meadows of the Atlantic Coast: A Community Profile*. US Fish and Wildlife Service FWS/OBS-84/24. 85 pp.
5. Short FT, Burdick DM, Wolf J, Jones GE [1993]. *Eelgrass in Estuarine Research Reserves along the East Coast, USA, Part I: Declines from Pollution and Disease and Part II: Management of Eelgrass Meadows*. Jackson Estuarine Laboratory, University of New Hampshire, Durham, NH. 107 pp.
6. Heck KL Jr, Able KW, Roman CT, Fahay MP [1995]. Composition, abundance, biomass and production of macrofauna in a New England estuary: Comparisons among eelgrass meadows and other nursery habitats. *Estuaries* 18(2): 379-389.
7. Rasmussen E [1977]. The wasting disease of eelgrass (*Zostera marina*) and its effects on environmental factors and fauna. In: McRoy CP, Helfferich C (eds) *Seagrass Ecosystems: A Scientific Perspective*. Marcel Dekker, New York. pp 1-52.
8. Muehlstein LK, Porter D, Short FT [1991]. *Labyrinthula zosterae* sp. nov., the causative agent of wasting disease of eelgrass, *Zostera marina*. *Mycologia* 83(2): 180-191.
9. Short FT, Muehlstein LK, Porter D [1987]. Eelgrass wasting disease: Cause and recurrence of a marine epidemic. *Biology Bulletin* 173: 557-562.
10. Muehlstein LK, Porter D, Short FT [1988]. *Labyrinthula* sp., a marine slime mold producing the symptoms of wasting disease in eelgrass, *Zostera marina*. *Marine Biology* 99: 465-472.
11. Milne LJ, Milne MJ [1951]. The eelgrass catastrophe. *Scientific American* 184: 52-55.
12. Short FT, Burdick DM [1996]. Quantifying eelgrass habitat loss in relation to housing development and nitrogen loading in Waquoit Bay, Massachusetts. *Estuaries* 19(3): 730-739.
13. Short FT, Burdick DM, Granger S, Nixon SW [1996]. Long-term decline in eelgrass, *Zostera marina* L., linked to increased housing development. In: Kuo J, Phillips RC, Walker DI, Kirkman H (eds) *Seagrass Biology: Proceedings of an International Workshop, Rottnest Island, Western Australia, 25-29 January 1996*. University of Western Australia, Nedlands, Western Australia. pp 291-298.
14. Short FT (ed) [1992]. *The Ecology of the Great Bay Estuary, New Hampshire and Maine: An Estuarine Profile and Bibliography*. Jackson Estuarine Laboratory, University of New Hampshire, Durham, NH. 222 pp.
15. Richardson FD [1980]. Ecology of *Ruppia maritima* L. in New Hampshire (USA) tidal marshes. *Rhodora* 82: 403-439.
16. Richardson FD [1983]. Variation, Adaptation, and Reproductive Biology in *Ruppia maritima* L. Populations from New Hampshire Coastal and Estuarine Tidal Marshes. PhD dissertation, University of New Hampshire, Durham. 147 pp.
17. Kurland JM [1994]. Seagrass habitat conservation: An increasing challenge of coastal resource management in the Gulf of Maine. In: Wells PG, Ricketts PJ (eds) *Coastal Zone Canada '94: Cooperation in the Coastal Zone: Conference Proceedings*. Vol. 3. Coastal Zone

Regional map: North America XI

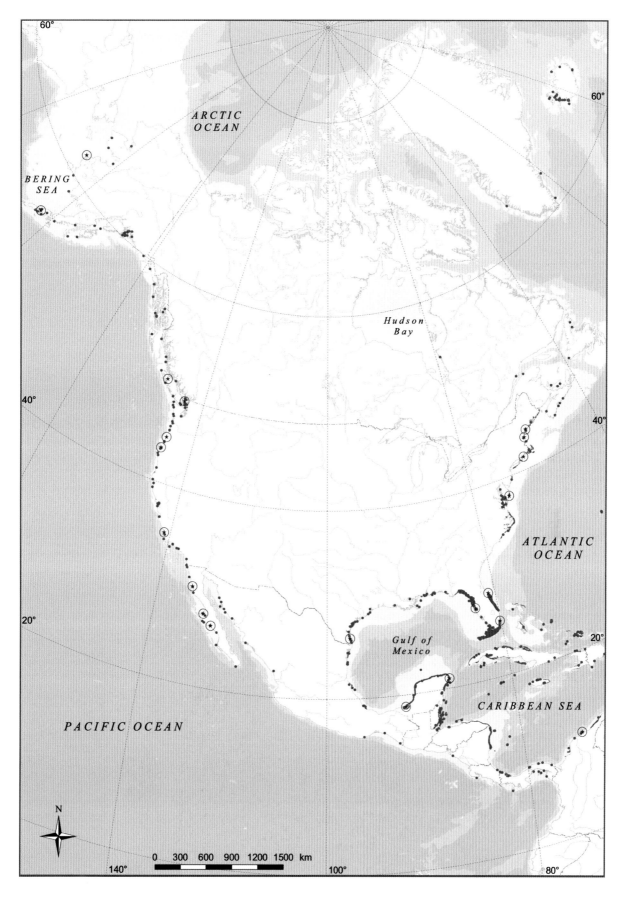

DIVERSITY OF SEAGRASS HABITATS

Intertidal *Halodule wrightii* meadow in Cabo Frio, Brazil.

Flowering shoots of *Zostera marina* in the Netherlands.

Enhalus acoroides and *Cymodocea rotundata* in Puerto Galera, Philippines.

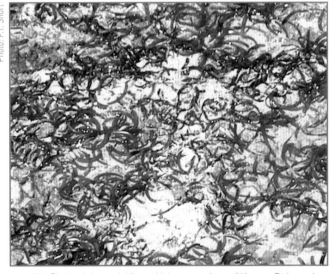

Intertidal *Thalassia hemprichii* on a high energy shore of Kosrae, Federated States of Micronesia.

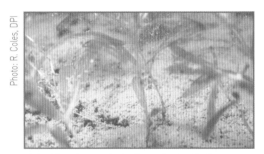

Halophila tricostata in the Great Barrier Reef, Queensland, Australia.

Halophila spinulosa in Papua New Guinea.

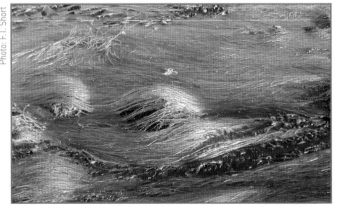

Phyllospadix torreyi, growing on solid rock with kelp in Ensenada, Mexico.

Canada Association, Bedford Institute of Oceanography, Dartmouth, Nova Scotia.
18. Armstrong MP [1995]. A Comparative Study of the Ecology of Smooth Flounder, *Pleuronectes putnami*, and Winter Flounder, *Pleuronectes americanus*, from Great Bay Estuary, New Hampshire. PhD dissertation, University of New Hampshire. 147 pp.
19. Short FT, Burdick DM, Bosworth W, Grizzle RE, Davis RC [1998]. New Hampshire Port Authority Mitigation Project Progress Report, June 1998. Prepared for the New Hampshire Port Authority and the New Hampshire Dept. of Transportation. 59 pp.
20. Karnofsky EB, Atema J, Elgin RH [1989]. Natural dynamics of population structure and habitat use of the lobster, *Homarus americanus*, in a shallow cove. *Biological Bulletin* 176: 247-256.
21. Short FT, Matso K, Hoven H, Whitten J, Burdick DM, Short CA [2001]. Lobster use of eelgrass habitat in the Piscataqua River on the New Hampshire/Maine Border, USA. *Estuaries* 24: 249-256.
22. Gotcetlas V, Fraser S, Brown JA [1997]. Use of eelgrass beds (*Zostera marina*) by juvenile Atlantic cod (*Godus morhua*). *Canadian Journal of Fisheries and Aquatic Sciences* 54: 1306-1319.
23. Normadeau Associates [1979]. Piscataqua River Ecological Studies, 1978 Monitoring Studies, Report No. 9 for Public Service Company of New Hampshire. Volume I: Physical/chemical Studies, Biological Studies. Normadeau Associates, Inc., Bedford, NH. 479 pp.
24. Short FT [1988]. Eelgrass-scallop Research in the Niantic River. Waterford-East Lyme, Connecticut Shellfish Commission, Final Report. 12 pp.
25. Newell CR, Hidu H, McAlice BJ, Podniesinski G, Short F, Kindblom L [1991]. Recruitment and commercial seed procurement of the blue mussel *Mytilus edulis* in Maine. *Journal of the World Aquaculture Society* 22: 134-152.
26. Grizzle RE, Short FT, Newell CR, Hoven H, Kindblom L [1996]. Hydrodynamically induced synchronous waving of seagrasses: "Monami" and its possible effects on larval mussel settlement. *Journal of Experimental Marine Biology and Ecology* 206: 165-177.
27. Carlton JT, Vermeij GJ, Lindberg DR, Carlton DA, Dudley EC [1991]. The first historical extinction of a marine invertebrate in an ocean basin: The demise of the eelgrass limpet *Lottia alveus*. *Biological Bulletin* 180: 72-80.
28. Nixon SW. Personal communication.
29. Wyllie-Echeverria S, Cox PA [1999]. The seagrass (*Zostera marina*, [ZOSTERACEAE]) industry of Nova Scotia (1907-1960). *Economic Botany* 53: 419-426.
30. Wyllie-Echeverria S, Arzel P, Cox PA [2000]. Seagrass conservation: Lessons from ethnobotany. *Pacific Conservation Biology* 5: 329-335.
31. Lathrop RG, Styles RM, Seitzinger SP, Bognar JA [2001]. Using GIS modeling approaches to examine the spatial distribution of seagrasses in Barnegat Bay, New Jersey. *Estuaries* 24: 904-916.
32. Neckles HA, Short FT, Barker S, Kopp B [2001]. Evaluation of Commercial Fishing Impacts to Eelgrass in New England. Abstract. Estuarine Research Federation Conference, Nov. 3-8, St Petersburg, FL.
33. Barker S, Marine Department of Marine Resources. Personal communication.
34. Christiansen C, Christoffersen H, Dalsgaard J, Nornberg P [1981]. Coastal and nearshore changes correlated with die-back in eelgrass (*Zostera marina*). *Sedimentary Geology* 28: 168-178.
35. Hine AC, Evans MW, Davis RA, Belknap DA [1987]. Depositional response to seagrass mortality along a low-energy, barrier-island coast: West-central Florida. *Journal of Sedimentary Petrology* 57(3): 431-439.
36. Duarte CM [1995]. Submerged aquatic vegetation in relation to different nutrient regimes. *Ophelia* 41: 87-112.
37. Olesen B [1996]. Regulation of light attenuation and eelgrass *Zostera marina* depth distribution in a Danish embayment. *Marine Ecology Progress Series* 134: 187-194.
38. Connolly RM [1995]. Effects of removal of seagrass canopy on assemblages of small, motile invertebrates. *Marine Ecology Progress Series* 118: 129-194.
39. Lalumiere R, Messier D, Fournier JJ, McRoy CP [1994]. Eelgrass meadows in a low arctic environment, the northeast coast of James Bay, Quebec. *Aquatic Botany* 47: 303-315.
40. Wyllie-Echeverria S. Personal communication.
41. Robertson AI, Mann KH [1984]. Disturbance by ice and life history adaptations of the seagrass, *Zostera marina*. *Marine Biology* 80: 131-142.
42. Dobson JE, Bright EA, Ferguson RL, Field DW, Wood LL, Haddad KD, Iredale H, Jensen JR [1995]. NOAA Coastal Change Analysis Program (C-CAP): Guidance for Regional Implementation. NOAA Technical Report NMFS 123. US Dept. of Commerce, Seattle, Washington. 92 pp.
43. Kemp WM, Boyton WR, Stevenson JC, Twilley RR, Means JC [1983]. The decline of submerged vascular plants in the upper Chesapeake Bay: A summary of results concerning possible causes. *Marine Technology Society Journal* 7: 78-89.
44. Short FT, Mathieson AC, Nelson JI [1986]. Recurrence of the eelgrass wasting disease at the border of New Hampshire and Maine, USA. *Marine Ecology Progress Series* 29: 89-92.
45. den Hartog C [1994]. The dieback of *Zostera marina* in the 1930s in the Wadden Sea: An eye witness account by van der Werff A. *Netherlands Journal of Aquatic Ecology* 28: 51-54.
46. Short FT, Wyllie-Echeverria S [1996]. Natural and human-induced disturbance of seagrasses. *Environmental Conservation* 23(1): 17-27.
47. Zieman J [1976]. The ecological effects of physical damage from motorboats on turtle grass beds in southern Florida. *Aquatic Botany* 2: 127-139.
48. Walker DI, Lukatelich RJ, Bastyan G, McComb AJ [1989]. Effect of boat moorings on seagrass beds near Perth, Western Australia. *Aquatic Botany* 36: 69-77.
49. Burdick DM, Short FT [1998]. Dock Design with the Environment in Mind: Minimizing Dock Impacts to Eelgrass Habitat. UNH Media Services, Durham, NH. CD-ROM.
50. Burdick DM, Short FT [1999]. The effects of boat docks on eelgrass beds in coastal waters of Massachusetts. *Environmental Management* 23: 231-240.
51. Short FT, Jones GE, Burdick DM [1991]. Seagrass decline: Problems and solutions. *Proceedings of Seventh Symposium on Coastal and Ocean Management/ASCE*. Long Beach, CA.
52. ASMFC [1999]. *Evaluating Fishing Gear Impacts to Submerged Aquatic Vegetation and Determining Mitigation Strategies*. Atlantic States Marine Fisheries Commission, Washington, DC.
53. Dennison WC, Marshall GJ, Wigand C [1989]. Effect of "brown tide" shading on eelgrass (*Zostera marina* L.) distributions. *Coastal and Estuarine Studies* 35: 675-692.
54. Short FT, Neckles H [1999]. The effects of global climate change on seagrasses. *Aquatic Botany* 63: 169-196.
55. Short FT, Burdick DM, Kaldy JE [1995]. Mesocosm experiments quantify the effects of eutrophication on eelgrass, *Zostera marina* L. *Limnology and Oceanography* 40: 740-749.
56. Deegan LA, Wright A, Ayvazian SG, Finn JT, Golden H, Merson RR, Harrison J [2002]. Nitrogen loading alters seagrass ecosystem structure and support of higher trophic levels. *Aquatic Conservation* 12(2): 193-212.
57. Burdick DM, Short FT, Wolf J [1993]. An index to assess and monitor the progression of the wasting disease in eelgrass, *Zostera marina*. *Marine Ecology Progress Series* 94: 83-90.
58. Davis R, Short FT [1997]. An improved method for transplanting eelgrass, *Zostera marina* L. *Aquatic Botany* 59: 1-16.
59. Valiela I, Costa J, Foreman K, Teal JM, Howes B, Aubrey D [1990]. Transport of groundwater-borne nutrients from wastelands and their effects on coastal waters. *Biogeochemistry* 10: 177-197.

21 The seagrasses of THE MID-ATLANTIC COAST OF THE UNITED STATES

E.W. Koch
R.J. Orth

The mid-Atlantic region of the United States includes four states: Delaware, Maryland, Virginia and North Carolina. It is characterized by numerous estuaries and barrier-island coastal lagoons with expansive salt marshes and seagrass beds in most shallow-water areas[1]. There are no rocky shores. Hard substrates are either man-made (rock jetties and riprap or wood pilings) or biogenically generated (oyster and worm reefs). Sediments are predominantly quartz sand in shallow exposed areas with finer grain sediments in deeper or well-protected areas. Marsh peat outcroppings or cohesive sediments are sometimes found in the subtidal areas adjacent to eroding marshes. Climatic variations are large with air temperatures ranging from –10°C to 40°C and water temperatures ranging from 0°C to 30°C. Tides are equal and semi-diurnal but relatively small in range (maximum of 1.3 m during spring tides).

The largest estuary in the country, the Chesapeake Bay (18 130 km^2), occurs in this area. Its watershed covers 165 760 km^2, drains from six states and is inhabited by more than 15 million people. Additionally, the estuarine system of the state of North Carolina is the third largest in the country, encompassing more than 8 000 km^2 with a watershed of more than 63 000 km^2. Other estuaries in the mid-Atlantic include the Delaware Bay and a series of barrier-island coastal lagoons.

Flowering aquatic plants are common in the estuaries of the mid-Atlantic region. They are often referred to as submersed aquatic vegetation (SAV). This term includes all flowering aquatic plants from freshwater to marine habitats. The term "seagrass" is used exclusively for species that occur in the higher salinity zones (>10 psu)[2,3]. Only three seagrass species are found in the mid-Atlantic region: *Halodule wrightii* (shoal grass), *Ruppia maritima* (widgeon grass) and *Zostera marina* (eelgrass). The northernmost area of the mid-Atlantic (Delaware estuaries and bays) is presently unvegetated. In contrast, the middle and southern areas are colonized by monospecific stands or by intermixed beds of seagrass (usually two species). The beds can vary from small and patchy to quite extensive. The largest seagrass bed in the Chesapeake Bay is composed of a mixture of *Zostera marina* and *Ruppia maritima* and covers 13.6 km^2.

Seagrass habitat provides food and refuge from predators for a wide variety of species, some of which have recreational and commercial significance. The invertebrate production in just one seagrass bed in the lower Chesapeake Bay was estimated to be 0.4 metric tons per year[4]. Seagrass beds in Chesapeake Bay are reported to be important nursery areas for the blue crab, *Callinectes sapidus*, whose commercial harvest can yield close to 45 000 metric tons in a good year. The bay scallop (*Argopecten irradians*) fishery is also closely tied to seagrass abundance because the larval stage attaches its byssal thread to seagrass leaves. The decline of seagrasses in Virginia's coastal bays in the 1930s led to the complete disappearance of the bay scallop, and loss of a substantial commercial fishery. Seagrasses have not returned to this region, nor have bay scallops. Other important local fisheries sometimes (but not always) associated with seagrasses include hard clams (*Mercenaria mercenaria*) and fish of commercial and recreational importance, e.g. striped bass (*Morone saxatilis*), spotted sea trout (*Cynoscion nebulosus*), spot (*Leiostomus xanthurus*) and gag grouper (*Mycteroperca microlepis*)[5].

BIOGEOGRAPHY

The state of North Carolina is an interesting biogeographical boundary for seagrasses in the North Atlantic. On the east coast of the United States it is the southernmost limit for the distribution of the temperate seagrass *Zostera marina* and the northernmost limit for the distribution of the tropical seagrass *Halodule wrightii*[6]. Due to their existence at

the limits of their thermal tolerance, the seagrasses found in this boundary zone are expected to show early effects of global warming in this area. *Ruppia maritima* is able to tolerate a broad range of temperatures and is found throughout the mid-Atlantic region and possibly along the coasts of South Carolina and Georgia.

Seagrasses in the mid-Atlantic region occur in wave-protected habitats. The extensive lagoon system (from Delaware to North Carolina) is delimited to the east by long barrier islands. These islands provide shelter from oceanic waves, making the lagoons ideal habitats for *Zostera marina*, *Ruppia maritima* and *Halodule wrightii*. No seagrasses (but seagrass wrack, including reproductive shoots with viable seeds) have been reported for the exposed shores of the Atlantic Ocean. The seagrasses in the mid-Atlantic region also colonize areas covering a wide range of salinities: from full-strength seawater (30-32 psu) near the mouths of the estuaries to mesohaline zones (10-20 psu) in the middle portion of the estuaries. Due to its ability to tolerate relatively low salinities, *Ruppia maritima* is usually the seagrass that extends farthest into the estuaries.

The distribution of seagrasses in the mid-Atlantic region is restricted to shallow waters because of the high suspended sediment and nutrient loadings leading to relatively turbid waters in seagrass habitats (light attenuation coefficients higher than 1 per m^2 are quite common). In relatively pristine areas (North Carolina sounds adjacent to barrier islands and Chincoteague Bay), the maximum depth to which seagrasses grow can be as great as 2 m, while in habitats associated with the mainland and eutrophic (i.e. nutrient enriched) conditions (Chesapeake Bay, North Carolina sounds near the mainland), the maximum vertical distribution only reaches depths of 0.5 to 1.0 m[7, 8]. In other areas, such as the Delaware coastal bays, seagrasses are almost completely absent due to high water turbidity.

HISTORICAL PERSPECTIVES

No record exists of the extent of the vegetation prior to the 1930s, but anecdotal evidence of historical changes in eelgrass[9, 10] suggest that seagrasses occurred in the Chesapeake Bay region in the mid- to late 1800s[11]. In the pre-colonial period (1800s), seagrasses are believed to have formed extensive beds in estuaries and lagoons in the mid-Atlantic region covering the coastal bays in their entirety. It is not known to what depths seagrasses used to grow in the estuaries, but it may have been as deep as 4 m. When *Zostera marina* beds were extensive, the seagrass was used for packing and upholstery stuffing. It was also used for insulation of buildings due to its low flammability and excellent insulating properties.

A massive decline of seagrasses in the mid-

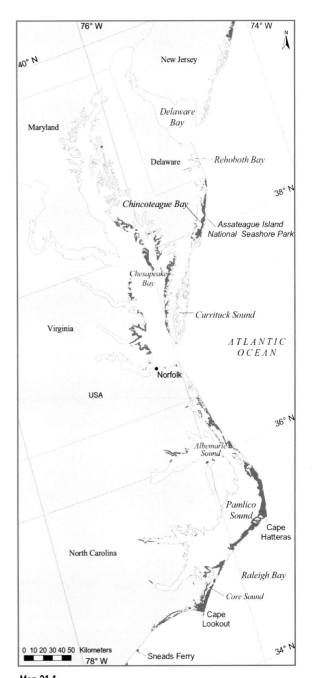

Map 21.1
The mid-Atlantic coast of the United States

Atlantic region occurred in the 1930s as *Zostera marina* was affected, and in many locations eliminated, by wasting disease[12, 13]. The loss of eelgrass was reported throughout the northern Atlantic. In some areas in the mid-Atlantic (Chesapeake Bay, Chincoteague Bay, North Carolina sounds), eelgrass beds slowly recovered. In the Delaware coastal bays (Indian River and Rehoboth Bays), recovery of eelgrass through the 1950s ended, apparently due to eutrophication. In the coastal bays of the lower eastern shore of Virginia, eelgrass was completely eliminated and never

Figure 21.1
Seagrass distribution (mainly *Zostera marina* and *Ruppia maritima* but possibly also a few hectares of other SAV species) in Chesapeake Bay

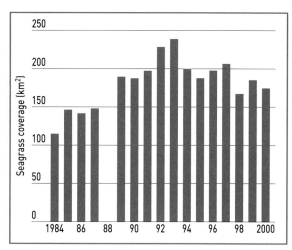

Source: Based on data from the Virginia Institute of Marine Science SAV mapping program.

Figure 21.2
Changes in seagrass (*Zostera marina* and *Halodule wrightii*) distribution in the Cape Lookout area (southern Core Sound, North Carolina) between 1985 and 1988

Note: Areas of seagrass coverage that did not change between the two years are shown in green cross-hatch; areas of gain are shown by the vertical white hatching and areas of loss are shown by the horizontal white hatching.

Source: Poster produced by the Beaufort Lab entitled *SAV Habitat in 1985 and 1988: Cape Lookout to Drum Inlet, North Carolina,* by Randolph Ferguson, Lisa Wood and Brian Pawlak.

recovered. The decline in the 1930s was complicated by a hurricane of unprecedented proportions in August 1933. There is no evidence of eelgrass ever being present in Delaware Bay.

PRESENT DISTRIBUTION

Although rare, sparse and small eelgrass beds are present in the coastal bays of Delaware (a result of restoration efforts). They are too small to map and also ephemeral in nature. There is very little seagrass in the state of Delaware.

Unprecedented changes to eelgrass populations in Chesapeake Bay occurred following Tropical Storm Agnes in June 1972. Eelgrass beds in the upper portions of Chesapeake Bay were the most influenced by the effects of the runoff (low salinities and high turbidity), which occurred during the peak growth period for eelgrass. While the distribution of seagrasses in Chesapeake Bay (Maryland and Virginia) had been partially documented in 1971 and 1974, the first baywide survey was conducted in 1978, and annual surveys began in 1984. Based on these data, seagrass distribution in Chesapeake Bay was observed to increase 63 percent between 1985 and 1993, but distribution then declined 27 percent between 1993 and 2000 (Figure 21.1). In contrast, from 1986 to 2000, seagrass distribution in the coastal bays of Maryland and Virginia increased 238 percent (see Case Study 21.1). Presently, the seagrasses in Chesapeake Bay show declines in some areas while recovering in others. There is great interannual variation, making it difficult to estimate the area of seagrass.

In North Carolina, where the seagrass habitats are dominated by shallow areas protected by extensive barrier islands, seagrass distribution has only recently been mapped. Core Sound was mapped in 1988 and inside of Cape Hatteras in 1990. The area south of Cape Lookout has not yet been mapped but it is known that no seagrasses are found south of Sneads Ferry (80 km north of the city of Wilmington)[14]. The lack of seagrasses in Albermarle Sound is believed to be the result of the high water turbidity in this area. The western portion of Pamlico Sound is also mostly unvegetated due to the long fetch and consequent high turbidity during strong wind events. Although there has not been a sustained effort to map seagrasses in North Carolina, researchers have been investigating aspects of seagrass ecology and report no noticeable changes in species composition or distribution since the 1970s[15]. One quantitative effort (Figure 21.2) confirms this. In the Core Sound area (between Drum Inlet and Cape Lookout) seagrass distribution was generally consistent between the two years in which it was mapped. In 1985 there were 7 km² of seagrass and in 1988 there were 6.6 km², only a 5.7 percent loss. There were 151 beds in 1985 and 149 in 1988. Two anthropogenic impacts on seagrasses were noted between 1985 and 1988: a clam harvesting operation dug up seagrasses, while in another area dredge spoil was deposited on a seagrass bed[16]. In North Carolina, seagrass beds have been relatively stable since the 1970s at approximately 80 km². It is not clear if seagrass beds in North Carolina also suffered the declines observed in the Chesapeake Bay before researchers began to work in these habitats in the 1970s. *Zostera marina* was affected by the wasting disease of the 1930s in North Carolina, but recovered, as in Chincoteague Bay.

PRESENT THREATS

The main threats to seagrasses in the mid-Atlantic region today are eutrophication and high turbidity from poor land-management practices. As the coastal zone continues to be developed, nutrient loads and suspended sediments in the water column tend to increase[17]. These nutrients may come from well-defined sources such as a sewage treatment plant, a pig farm or a golf course, but a large amount of nutrients also comes from non-point sources such as farmland and groundwater nutrient enrichment by septic systems. As a result of increased nutrient loading, epiphytic algae may grow directly on the seagrass leaves while blooms of phytoplankton or macroalgae may occur in the water column. These processes decrease the amount of light that reaches the seagrasses and cause their decline or death. Most water bodies in the mid-Atlantic are now phytoplankton dominated, and the few pristine lagoons are showing signs of deterioration resulting from blooms of nuisance macroalgae such as *Chaetomorpha linum* and *Ulva lactuca* (mats up to 1.5 m thick). These algal blooms have adversely impacted healthy seagrass beds (see Case Study 21.1) as well as recent eelgrass restoration efforts in the Delaware coastal bays.

Seagrass beds are vulnerable to disruption by commercial fishing practices, especially clam and scallop dredging. Hydraulic clam dredging digs deep trenches or circles into the sediments (see Case Study 21.1). If these are vegetated by seagrasses, the plants are lost and the recovery is relatively slow[18]. Clam dredging also has a negative impact on other fisheries. The trenches caused by hydraulic clamming in seagrass beds prevent crabbers from pulling their scrapes through the seagrass beds (a practice that causes relatively little damage to the plants), directly threatening their livelihood.

As coastal areas become more heavily populated, more individuals also want to enjoy water-related activities. Boat-generated waves and turbulence have a negative impact on seagrasses and their habitats[19]. There is also no doubt that propeller scars have a detrimental effect on seagrasses[20,21]. The effect is similar to that described for clam dredging although the scars are narrower. This problem is most severe in North Carolina but has also been documented in Maryland and Virginia.

Dredging and maintenance dredging of channels is a threat to seagrasses in all mid-Atlantic states. This operation increases the turbidity of the water, may bury seagrasses and may increase the nutrient concentration in the water column. Regulations in North Carolina suggest (but do not require) that damage to seagrasses be minimized during dredging activities. Maryland is currently re-evaluating its dredging regulations.

Sea-level rise has the potential to pose a threat to seagrasses in the mid-Atlantic. The vulnerability of coastal zones to sea-level rise has been classified as very high in this region, the highest risk on the east coast of the United States. Unfortunately, our understanding of how sea-level rise affects seagrasses is in its infancy. It is known that sea-level rise leads to marsh erosion[22-24] and the eroded sediments are then transported to coastal waters where seagrass beds may occur. This may lower the light available to seagrasses and may lead to their decline or loss. The loss of the seagrasses could then lead to further coastal erosion due to the loss of wave attenuation previously provided by the seagrasses.

Although a natural event, a storm can be detrimental to seagrasses. Hurricanes are quite common in the mid-Atlantic, especially in the state of North Carolina, and have shown to be detrimental to seagrasses by removing the plants, eroding the sediment, burying seagrass beds and/or increasing turbidity of the water[25]. It is expected that with global warming hurricane frequency and intensity will increase. With that, the threat to seagrasses is also expected to increase. However, little quantitative data exist on the effects of hurricanes on long-term stability of seagrass beds in this region. Hurricanes are more frequent in the fall period (September and October) and it is possible that water quality effects may be marginal as temperatures are lower and growth is generally less than in the spring.

POLICIES AND REGULATIONS

No state or federal marine parks exist in the mid-Atlantic region, but several protected islands include the adjacent waters in their jurisdiction. The national estuarine research reserves in Maryland and North Carolina include seagrass habitats, although no protection is afforded by this designation. The Assateague Island National Seashore Park protects its

adjacent seagrasses. The state of Delaware currently has no protection for seagrasses in its regulatory framework. The total area of protected seagrass beds has not been identified for the mid-Atlantic.

At the federal level, seagrasses are afforded some protection under Section 404 of the Clean Water Act (33 USC 1341-1987) and Section 10 of the Rivers and Harbors Act (33 USC 403), which regulate the discharge of dredged or fill material into US waters. Authority for administering the Clean Water Act rests with the US Environmental Protection Agency. Seagrass protection under the Act is provided by a federal permit program that is delegated to and administered by the US Army Corps of Engineers. Potential impacts on "special aquatic sites", such as seagrass beds, are considered in the permit review process. Section 10 of the Rivers and Harbors Act, also administered by the Army Corps, regulates all activities in navigable waters including dredging and placement of structures.

On a regional basis, considerable and cooperative efforts by scientists, politicians, federal and state resource managers, and the general public have developed policies and plans to protect, preserve and enhance the seagrass populations of Chesapeake Bay[26]. The foundation for the success of these management efforts has been the recognition of the habitat value of seagrasses to many fish and shellfish, and the elucidation of linkages between seagrass habitat health and water quality conditions. Because of these linkages, the distribution of seagrasses in Chesapeake Bay and its tidal tributaries is being used

---

Case Study 21.1
SEAGRASSES IN CHINCOTEAGUE BAY: A DELICATE BALANCE BETWEEN DISEASE, NUTRIENT LOADING AND FISHING GEAR IMPACTS

Chincoteague Bay is one of the most pristine water bodies in the mid-Atlantic. It is a relatively shallow coastal lagoon (average depth 1.2 m) with limited freshwater input and long residence times (flushing of 7.5 percent per day). Salinities are close to those of seawater (26-31 psu) and nutrient levels are relatively low (<10 µM total nitrogen, <4 µM phosphate[29]). The western shore of Chincoteague Bay is characterized by extensive salt marshes and isolated, small towns representing an area of low developmental pressure (less than 0.04 person per hectare). The eastern shore is located adjacent to an unpopulated (but accessible to tourists) barrier island (Assateague Island National Seashore) with an extensive dune system along the Atlantic coast and marshes along the Chincoteague Bay shoreline.

Seagrasses in Chincoteague Bay are found almost exclusively on the eastern shores. Due to its relatively shallow depth, it is believed that the entire bay used to be colonized by *Zostera marina*. In the 1930s, *Zostera marina* disappeared as a result of wasting disease after which it slowly began to recolonize the eastern shore. The recovery of the seagrasses in Chincoteague Bay has been well documented since 1986 (see figure, left). Although there was a 40 percent increase in the human population on the western shore of Chincoteague Bay between 1980 and 2000, the total nitrogen and phosphorus loadings declined between 1987 and 1998 (in some areas as much as 50 percent). This is believed to be due to the construction of sewage treatment plants and the reduction of the amount of fertilizers used on the farms west of Chincoteague Bay. As a result, phytoplankton concentration is low and light penetration relatively deep. Seagrasses flourished during this period showing a 238 percent increase in distribution between 1986 and 1999. In 1996, seagrasses even began colonizing the western shore which had remained unvegetated since the 1930s.

One of the first threats to seagrass in Chincoteague Bay since its decimation in the 1930s came from a fisheries practice[28]. In 1997, severe damage to the seagrass beds was noted and attributed to two types of hard clam fishing gear: hydraulic dredges and modified oyster dredges (see

Recovery and recent decline of seagrass (*Zostera marina* and *Ruppia maritima*) distribution in Chincoteague Bay.

as an initial measure of progress in the restoration of living resources and water quality. Restoration targets and goals have been established to link demonstrable improvements in water quality to increases in seagrass abundance[27]. The states of Maryland and Virginia each have separate regulatory agencies to oversee activities that could be injurious to seagrass populations. Both states are committed to protecting seagrass habitat while maintaining viable commercial fisheries and aquaculture operations.

Maryland State Code COMAR 4-213 specifically prohibits damage to seagrasses for any reason except for commercial fishing activities and certain specific situations such as clearing seagrasses from docks, piers and navigable waters. If seagrasses will be adversely affected, the Maryland Department of the Environment and the Maryland Department of Natural Resources are responsible for issuing a permit, which includes a plan showing the site at which the activity is proposed, a dated map of current seagrass distribution and the extent of seagrass to be removed. Maryland does prohibit one type of commercial fishing activity, hydraulic clam dredging, in specific regions of its state waters. Hydraulic clam dredging is prohibited both within a specified distance from shore, which varies by political boundaries (NRA 4-1038), and in existing seagrass beds (NR 4-1006.1), as determined by annual aerial mapping surveys.

In Virginia, permits to use state-owned submerged lands now include seagrass presence as a factor to be considered in the application process (Code 28.2-1205 (A) [6], amended in 1996). On-bottom shellfish photograph, right). The seagrass area affected by hydraulic dredging increased from 0.53 km² in 1996 to 5.08 km² in 1997, while modified oyster dredge scars increased from 10 in 1995 to 218 scars in 1997. Analysis of the recovery from both types of scarring showed that some scars require more than three years to revegetate to undisturbed levels. Once notified of these impacts, resource managers in Maryland and Virginia responded within several months to protect seagrasses through law and regulation preventing clam dredging within seagrass beds. In Virginia, the new regulation was successful in reducing scarring, but required later revisions for successful enforcement. In Maryland, however, procedural requirements to fully implement the law required additional time, during which scarring increased to 12.57 km² in 1999. This issue has demonstrated the importance of close linkages between the scientific research community, politicians, management agencies, law enforcement agencies and the public, as well as the importance of sanctuaries or protection zones to prevent damage to critical seagrass habitats.

Over the last three years, seagrasses in Chincoteague Bay have been exposed to another stress: the blooms of the nuisance macroalga *Chaetomorpha linum*, suggesting that this formerly pristine area may be experiencing eutrophication. Indeed, nutrient data shows a renewed increase in total nitrogen and phosphorus loads in 1999 and 2000. While pristine systems are dominated by seagrasses, systems in the early and late stages of eutrophication are dominated by macroalgae and/or phytoplankton, respectively[30]. The macroalgal mats observed in Chincoteague Bay over the last two years can be as thick as 1.5 m, killing the seagrasses beneath and leaving long scars visible via aerial photography. Managers are currently attempting to determine the source of the nutrients fueling these macroalgal blooms and threatening the seagrasses of Chincoteague Bay.

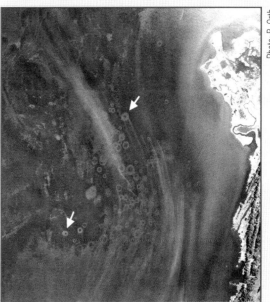

Aerial photograph taken in 1998 of a portion of Chincoteague Bay, Virginia, seagrass bed showing damage to the bed from a modified oyster dredge.

Notes: Arrows point to circular "donut-shaped" scars created by the dredge being pulled by a boat in a circular manner. The light areas in each circle represent areas that had vegetation that was uprooted and are now unvegetated. The dark spot within each circle is seagrass that was not removed. The long, light-colored streaks emanating from some of the scars are sediment plumes created by the digging activities of sting rays.

aquaculture activities requiring structures are now prohibited from being placed on existing seagrass beds (4-VAC 20 335-10, effective January 1998). In 1999, the Virginia Marine Resources Commission was directed (Code 28.2-1204.1) to develop guidelines with criteria to define existing beds and to delineate potential restoration areas. Dredging for clams (hard and soft) in Virginia is prohibited in waters less than 1.2 m where seagrasses are likely to occur. A special regulation was passed for seagrasses in the Virginia portion of Chincoteague Bay (4-VAC 20-1010) where clam and crab dredging is prohibited within 200 m of seagrass beds. Because of enforcement issues, the Virginia regulation has recently been modified (4-VAC 20-70-10 seq.) to include permanent markers with signs delineating the protected seagrass[28].

In the state of North Carolina, regulations involving seagrasses are not as strong as in Virginia and Maryland. North Carolina protects seagrass beds along underdeveloped areas. These areas are to be used mainly for education and research although some recreational activities are permitted. The dredging of channels is regulated such that seagrass beds must be avoided. Damage to seagrasses is also to be minimized when docks, piers, bulkheads, boat ramps, groins, breakwaters, culverts and bridges are constructed.

ACKNOWLEDGMENTS

Mark Finkbeiner, Ben Anderson and Brian Glazer are thanked for their help and for sharing their data. Mike Durako and Mark Fonseca contributed valuable information. Lisa Wood and Randolph Ferguson developed the seagrass vectors in Figure 21.2. Mike Naylor and several of the above listed colleagues provided comments on the report. Melissa Wood and Dave Wilcox provided technical support. This is contribution number 3613 from Horn Point Laboratory, University of Maryland Center for Environmental Science and contribution number 2491 from the Virginia Institute of Marine Science, College of William and Mary, Gloucester Point, Virginia.

AUTHORS

Evamaria W. Koch, Horn Point Laboratory, University of Maryland Center for Environmental Science, P.O. Box 775, Cambridge, MD 21613, USA. **Tel:** +1 410 221 8418. **Fax:** +1 410 221 8490. **E-mail:** koch@hpl.umces.edu

Robert J. Orth, Virginia Institute of Marine Science, School of Marine Science, College of William and Mary, Gloucester Point, VA 23062, USA.

REFERENCES

1. Orth R, Heck JKL Jr, Diaz RJ [1991]. Littoral and intertidal systems in the mid-Atlantic coast of the United States. In: Mathieson AC, Nienhuis PH (eds) *Ecosystems of the World 24: Intertidal and Littoral Ecosystems of the World*. Elsevier, Amsterdam. pp 193-214.
2. Orth R [1976]. The demise and recovery of eelgrass, *Zostera marina*, in the Chesapeake Bay, Virginia. *Aquatic Botany* 2: 141-159.
3. Moore KA, Wilcox DJ, Orth RJ [2000]. Analysis of the abundance of submersed aquatic vegetation communities in the Chesapeake Bay. *Estuaries* 23: 115-127.
4. Fredette TJ, Diaz RJ, van Montfrans J, Orth RJ [1990]. Secondary production within a seagrass bed (*Zostera marina* and *Ruppia maritima*) in lower Chesapeake Bay. *Estuaries* 13: 431-440.
5. Ross SW, Moser ML [1995]. Life history of juvenile gag, *Mycteroperca microlepis*, in North Carolina estuaries. *Bulletin of Marine Science* 56: 222-237.
6. Ferguson RL, Pawlak BT, Wood LL [1993]. Flowering of the seagrass *Halodule wrightii* in North Carolina, USA. *Aquatic Botany* 46: 91-98.
7. Dennison WC, Orth RJ, Moore KA, Stevenson JC, Carter V, Kollar S, Bergstrom P, Batiuk RA [1993]. Assessing water quality with submersed aquatic vegetation. *Bioscience* 43: 86-94.
8. Ferguson RL, Korfmacher K [1997]. Remote sensing and GIS analysis of seagrass meadows in North Carolina, USA. *Aquatic Botany* 58: 241-258.
9. Cottam C [1934]. Past periods of eelgrass scarcity. *Rhodora* 36: 261-264.
10. Cottam C [1935]. Further notes on past periods of eelgrass scarcity. *Rhodora* 37: 269-271.
11. Orth RJ, Moore KA [1984]. Distribution and abundance of submersed aquatic vegetation in Chesapeake Bay: An historical perspective. *Estuaries* 7: 531-540.
12. Renn CE [1934]. Wasting disease of *Zostera* in American waters. *Nature* 134: 416.
13. Tutin TG [1938]. The autecology of *Zostera marina* in relation to its wasting disease. *New Phytology* 37: 50-71.
14. Durako. Personal communication.
15. Fonseca MS. Personal communication.
16. Finkbeiner. Personal communication.
17. Kemp WM, Twilley RT, Stevenson JC, Boynton WR, Means JC [1983]. The decline of submerged vascular plants in Upper Chesapeake Bay: Summary of results concerning possible causes. *Marine Technology Society Journal* 17: 78-89.
18. Stephan CD, Peuser RL, Fonseca MS [2000]. *Evaluating Fishing Gear Impacts to Submerged Aquatic Vegetation and Determining Mitigation Strategies*. Atlantic States Marine Fisheries Commission Habitat Management Series 5. 35 pp.
19. Koch EW [2002]. The impact of boat-generated waves on a seagrass habitat. *Journal of Coastal Research* 37: 66-74.
20. Clark PA [1995]. Evaluation and management of propeller damage to seagrass beds in Tampa Bay, Florida. *Florida Scientist* 58: 193-196.
21. Dawes CJ, Andorfer J, Rose C, Uranowski C, Ehringer N [1997]. Regrowth of the seagrass *Thalassia testudinum* into propeller scars. *Aquatic Botany* 59: 139-155.
22. Kearney MS, Stevenson JC, Ward LG [1994]. Spatial and temporal changes in marsh vertical accretion rates at Monie Bay: Implications for sea-level rise. *Journal of Coastal Research* 10: 1010-1020.
23. Ward LG, Kearney MS, Stevenson JC [1998]. Variations in sedimentary environments and accretionary patterns in estuarine marshes undergoing rapid submergence, Chesapeake Bay. *Marine Geology* 151: 111-134.
24. Rooth JE, Stevenson JC [2000]. Sediment deposition patterns in *Phragmites australis* communities: Implications for coastal areas threatened by rising sea-level. *Wetlands Ecology and Management* 8: 173-183.

25 Fonseca MS, Kenworthy WJ, Whitfield PE [2000]. Temporal dynamics of seagrass landscapes: A preliminary comparison of chronic and extreme disturbance events. In: Pergent G, Pergent-Martini C, Buia MC, Gambi MC (eds) *Proceedings 4th International Seagrass Biology Workshop, Sept. 25-Oct. 2, 2000, Corsica, France.* pp 373-376.

26 Orth RJ, Batiuk RA, Bergstrom PW, Moore KA [2002a, in press]. A perspective on two decades of policies and regulations influencing the protection and restoration of submerged aquatic vegetation in Chesapeake Bay, USA. *Bulletin of Marine Science*.

27 Batiuk RA, Orth RJ, Moore KA, Dennison WC, Stevenson JC, Staver LW, Carter V, Rybicki NB, Hickman RE, Kollar S, Bieber S, Heasly P [1992]. Chesapeake Bay Submerged Aquatic Vegetation Habitat Requirements and Restoration Targets: A Technical Synthesis. Chesapeake Bay Program, Annapolis, MD, CBP/TRS 83/92. 248 pp.

28 Orth RJ, Fishman JR, Wilcox DW, Moore KA [2002]. Identification and management of fishing gear impacts in a recovering seagrass system in the coastal bays of the Delmarva Peninsula, USA. *Journal of Coastal Research* 37: 111-119.

29 Boynton WR, Murray L, Hagy JD, Stokes C, Kemp WM [1996]. A comparative analysis of eutrophication patterns in a temperate coastal lagoon. *Estuaries* 19: 408-421.

30 Valiela I, McClelland J, Hauxwell J, Behr PJ, Hersh D, Foreman K [1997]. Macroalgal blooms in shallow estuaries: Controls and ecophysiological and ecosystem consequences. *Limnology and Oceanography* 42: 1105-1118.

22 The seagrasses of
THE GULF OF MEXICO

C.P. Onuf
R.C. Phillips
C.A. Moncreiff
A. Raz-Guzman
J.A. Herrera-Silveira

The Gulf of Mexico is a vast basin of water, spanning 12° of latitude, from 18° to 30°N, and 17° of longitude, from 81° to 98°W. It is bisected by the Tropic of Cancer and is largely subtropical; however, along the northern edge, up to five days with freezing temperatures are probable on an annual basis. The coastal fringe is moist, with annual precipitation in excess of 1000 mm, except for southern Texas and northern Mexico. Precipitation is concentrated in the summer period, most pronounced along the coast of Mexico and least pronounced along the coast of Louisiana. Most of the Gulf of Mexico is fringed by a broad coastal plain, except for northwestern Cuba and sections of the Mexican coast near Veracruz. The inner continental shelf to a depth of 20 m is broad off the western side of the Yucatán Peninsula, along the coast of Louisiana and along the western side of Florida, extending as much as 80 km offshore to the tip of Florida. Elsewhere, the inner shelf is relatively narrow. Most of the rivers draining into the Gulf of Mexico have restricted catchments, except along the north shore, most obviously the Mississippi River, and parts of the western gulf, including the Ríos Bravo (Grande), Panuco, Grijalva and Usumacinta. Barrier islands and spits are prominent features along much of the coast, and coral reefs shelter the large expanse of water off the southern tip of Florida and off the coasts of Veracruz, Campeche, Yucatán and northwestern Cuba. Lunar spring tides are less than 1 m throughout the region.

CUBA
No recent assessment has been performed of the seagrass resources on Cuba's coast bordering the Gulf of Mexico[1]. Therefore, we must resort to an extensive report of a survey of the northwestern Cuban shelf conducted in 1972-73[2]. Fortuitously, this region from 22 to 23°N and 83 to 85°W is essentially all of the Cuban coast that borders the Gulf of Mexico. At that time, seagrasses covered 75 percent of the 2740 km² area of the northwestern Cuban shelf. Seagrasses were limited to the part of the shelf shoreward of a fringing reef. A total of four species were found at 282 stations: *Thalassia testudinum* was found to a depth of 14 m and accounted for 97.5 percent of total angiosperm biomass (190 g/m²), *Syringodium filiforme* to 16.5 m and 3.5 g/m², *Halophila engelmanni* to 14.4 m and 0.25 g/m², and *Halophila decipiens* to 24.3 m and 0.14 g/m².

UNITED STATES
Florida
In 1995, there were 9888 km² of seagrasses along the Gulf of Mexico coast of Florida[3]. This includes the seagrass beds of Monroe County off the Florida Keys but does not include the seagrasses in Card Sound of Biscayne Bay in Dade County. Nor does it include the large sparse offshore beds of *Halophila decipiens* from the Florida Keys to the Big Bend area.

The southern tip of the Florida Peninsula bordering on the Gulf of Mexico (Monroe and Collier Counties) contained 5901 km² of seagrasses. Monroe County alone, mostly in Florida Bay and the Florida Keys, contained 54.6 percent of the state's seagrasses. The middle section of the Florida Peninsula's Gulf of Mexico coast (Lee County to Pinellas County) contained 446 km² of seagrasses. The Big Bend region (Pasco County to Wakulla County) contained 3346 km² while the Florida Panhandle (Franklin to Escambia Counties) contained 195 km² of seagrasses[3].

Florida Bay and the Florida Keys are relatively shallow and, because of the tourism in the area, the seagrasses have been particularly subject to damage from boat traffic and sewage pollution from greatly expanded residential and hotel development, and marina and boat usage. As much as 17.3 percent of seagrass meadow in Monroe County had been scarred by boat propellers, and 32 percent of the 17.3 percent

had been severely scarred[3]. This scarring often leads to a loss of seagrass because of erosion and blow-outs. In addition, Florida Bay suffered a massive die-off of *Thalassia testudinum* beginning in 1987 that continued at least through 1994. Researchers have hypothesized the involvement of several factors in the initiation of the die-off, but few have been investigated adequately[4]. Beginning in 1991, algal blooms and persistent high turbidity were widespread, accounting for further deterioration of seagrass meadows, such that between 1984 and 1994 *Thalassia testudinum* biomass declined by 28 percent, *Syringodium filiforme* by 88 percent and *Halodule wrightii* by 92 percent[5].

The seagrasses in the Big Bend of Florida (31 percent of all seagrasses in Florida) have experienced relatively little impact from poor water quality problems or scarring from boat usage. This is due to their remoteness from population centers and the relatively low population density of the area. However, the area is on the brink of a huge development effort. It can only be hoped that the state and local jurisdictions will demand proper sewage disposal (not septic tanks) and will not engage in the dredge and fill activities that occurred in Tampa Bay and Sarasota Bay in the 1950s and 1960s.

Historical analysis is limited to only a few bay systems along the west coast of Florida. Charlotte Harbor lost approximately 30 percent of its seagrasses prior to the 1980s[6]. There was a further harbor-wide decline of 3.3 percent (2.43 km²) between 1988 and 1992. In 1992, the Southwest Florida Water Management District initiated a biennial mapping project to assess trends in Charlotte Harbor. Data from these studies will be used to assess the effectiveness of pollutant load reduction strategies on water quality. Between 1992 and 1994, a 4 percent (2.91 km²) increase was observed in the seagrass beds of the harbor, followed by an additional 3.6 percent increase (2.74 km²) between 1994 and 1996.

Prior to the 1980s, seagrass losses in Sarasota Bay were estimated to total approximately 30 percent. However, changes in seagrass coverage in Sarasota Bay have been dramatic since 1988. Between 1989 and 1990, nitrogen loads to the bay from wastewater treatment plants diminished by as much as 25 percent. Water transparency off the city of Sarasota and Manatee County increased from a mean Secchi disc depth of 1.1 m to 1.5 m, often deeper. Between 1988 and 1994, seagrass coverage in Manatee County increased 6.4 percent (1.42 km²) and another 7.8 percent (1.85 km²) between 1994 and 1996. In Sarasota County, seagrass coverage increased 10.1 percent (0.78 km²) between 1988 and 1994, and 22.7 percent (1.93 km²) between 1994 and 1996. Most of these increases were along the deep (>1.0 m) edges of existing seagrass beds[6]. These observed increases are believed to be directly linked to improving water quality and light penetration resulting from reductions in point-source pollutant loads.

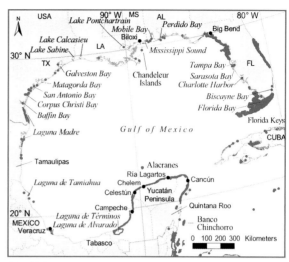

Map 22.1
The Gulf of Mexico

Alabama

Alabama has only 90 km of exposure to the Gulf of Mexico, and much of that is encompassed by Mobile Bay which receives river discharge volumes second only to the Mississippi River along the US portion of the Gulf of Mexico. Consequently, there is little opportunity for the establishment of true seagrass beds. A few small patches of *Halodule wrightii* have been reported at the south end of Mobile Bay along the western shore, and 2.5 km² are present in Perdido Bay, shared with Florida, down from 4.9 km² in 1940-41[13]. At the west end of the Alabama coast and shared with Mississippi, ephemeral beds of *Ruppia maritima* cover ca 2 km² in Grand Bay during the late spring and early summer, and *Halodule wrightii* has also been documented in this area.

Mississippi

Historically, populations of *Halodule wrightii*, *Halophila engelmanni*, *Ruppia maritima*, *Syringodium filiforme* and *Thalassia testudinum* were present and abundant along the northern shores of Mississippi's barrier islands[14, 15, 16]. Overall, Mississippi has lost most of the seagrass cover that was present in 1967-69, and only one marine species, *Halodule wrightii*, still exists in measurable quantities in Mississippi Sound. *Ruppia maritima* (widgeon grass) occurs in isolated but well-developed patches along the immediate coastline, and as an occasional component in *Halodule wrightii* beds along the barrier islands in Mississippi Sound.

Some well-established populations of *Halodule*

wrightii, Halophila engelmanni, Ruppia maritima, Syringodium filiforme and *Thalassia testudinum* exist along the western shorelines and in the small internal bayou systems of the Chandeleur Islands, in southeastern Louisiana. These islands begin 37 km due south of Biloxi, and are a likely source of vegetative propagules and possibly seeds supplementing or repopulating seagrass beds in some areas of the Mississippi coast.

Seagrass distributions from a 1967-69 Gulf of Mexico estuarine inventory[14] were used as a historical baseline, while data from a 1992 US Geological Survey aerial imagery study[17] were ground-truthed to document recent distribution patterns. Potential seagrass habitat was also identified using a 2 m critical depth limit which had been previously established in a National Park Service seagrass monitoring project[18].

Seagrasses and potential seagrass habitat in Mississippi Sound lie mainly along the northern shorelines of the offshore islands. From east to west, Petit Bois Island supported 6.8 km^2 of seagrass meadow according to the 1969 survey, down to 1.5 km^2 in 1992; Horn Island seagrass cover decreased from 22.5 km^2 to 2.2 km^2 over the same period; Dog Keys Pass beds from 8.4 km^2 to none; Ship Island from 6.2 km^2 to 1.0 km^2; and Cat Island from 2.4 km^2 to 0.7 km^2. Only on T-shaped Cat Island do seagrass beds occur in protected areas along its southwest shoreline as well as along its north side. *Ruppia maritima* occurs at two locations at opposite ends of the mainland shore: Point aux Chenes Bay at the Alabama border – 5.3 km^2 in 1969 down to 0.5 km^2 in 1992, and Buccaneer State Park, 10 km from the Louisiana border – 0.8 km^2 in 1969 down to 0.2 km^2 in 1992.

State-wide in 1969 submersed aquatic vascular plants covered an estimated 55.2 km^2 of coastal waters, mostly true seagrasses on the north sides of offshore islands. In 1992, only 8.1 km^2 of submersed vascular plants were found, 2.1 km^2 of which were from areas not included in the 1967-69 survey[14]. Almost all of the 2.1 km^2 were located in Grand Bay and are shared with Alabama, as noted in the Alabama section. Therefore, seagrasses in Mississippi suffered a decline of between 85 and 89 percent over 23 years. Information from the early survey[14] indicated that 67.6 percent of potential seagrass habitat was vegetated, in comparison to only 13.4 percent in 1992. Physical loss of seagrass habitat is assumed for areas where 1969 coverage exceeds current estimates of seagrass habitat. This total is estimated to be 19.6 percent. Since the discrimination of seagrasses from macroalgae in the 1967-69 survey was less precise than in the 1992 assessment, losses may be somewhat overestimated.

In Mississippi Sound, seagrasses appear to be threatened by the cumulative effects of both natural events and anthropogenic activities in the coastal marine environment. The primary vector for the disappearance of seagrasses is presently thought to be an overall decline in water quality. Development may be a major factor, as it often results in elevated nutrient levels, higher sediment loads, and the introduction of contaminants, which lead to a loss of water quality. Cyclic shifts in precipitation patterns that affect both salinity and turbidity, and extreme events, especially hurricanes, are also involved. Areas of seagrass habitat

> **Case Study 22.1**
> **TAMPA BAY**
>
> Based on the available habitat at that time, Tampa Bay is estimated to have supported 309.6 km^2 of seagrasses in 1879[7]. By 1981, only 57.5 km^2 remained. The most dramatic decrease occurred between 1950 and 1963, when approximately 50 percent of the total seagrass cover disappeared. During this period, Hillsborough Bay lost 94 percent of its grass beds, Old Tampa Bay lost 45 percent and Tampa Bay proper lost 35 percent[8, 9, 10].
>
> The losses up to the 1950s were due to poorly treated wastewater discharges and industrial wastes from phosphate mines, citrus canneries and other industrial sources[10], as well as extensive dredging, and dredge and fill activities that changed water circulation patterns and caused extensive turbidity in the waters[11].
>
> Recent work has shown that the trend of seagrass loss in Tampa Bay has been reversed. In 1988, 93.1 km^2 of seagrass were present (this was the first year of monitoring conducted by the Southwest Florida Water Management District)[9]. In 1990, this coverage increased to 99.3 km^2, and increased again in 1994 to 105.7 km^2. The bay-wide seagrass coverage in 1997 was estimated at 109.3 km^2 [12]. In Hillsborough Bay, seagrass increased from near zero in 1984 to about 0.57 km^2 in 1998.
>
> This expansion apparently started in response to water quality improvements from the late 1970s to the mid-1980s. These improvements followed a nearly 50 percent reduction in the early 1980s in external nitrogen loading from domestic and industrial point sources, primarily discharging to Hillsborough Bay. A slight decline in seagrass coverage occurred in Tampa Bay in the late 1990s, presumably a result of high rainfall during 1995, 1996 and the 1997-98 El Niño event, all of which increased nitrogen loading to the bay[12].

loss coincide with areas where rapid coastal erosion[19] and massive long-term movement of sand have been documented[20]. Physical loss of habitat and decreased light availability in combination with declining water quality are the most visible features that directly affect seagrass communities.

Louisiana

The coast of Louisiana features a wide band of fresh to brackish marsh, with some large lakes. Extensive beds of submersed aquatic vascular plants occur there, but the salinity is generally too low for seagrasses to thrive. Since the mid-1950s, seagrasses have been lost from Lake Pontchartrain and White, Calcasieu and Sabine Lakes and from behind the south coast barrier islands[13]. Small amounts of *Halodule wrightii* have reappeared on sand flats along the north shore, and still smaller amounts along the southeastern shore of Lake Pontchartrain in the 1990s. However, most of Louisiana's seagrasses are confined to the mixed species beds along the western shore of the Chandeleur Islands. The beds are relatively stable, at least by regional standards, only decreasing from 64.1 km^2 in 1978 to 56.6 km^2 in 1989 despite the passage of two hurricanes in that period. Apparently, the islands provide a protected shallow-water environment far enough removed from the plume of the Mississippi River and other influences of developed coastlines for seagrasses to thrive in the middle of an otherwise inhospitable shore.

Texas

Based on a recent compilation of surveys from the late 1980s and 1990s[21], the coast of Texas supports 951 km^2 of seagrass meadow. Unlike Florida, seagrasses in Texas do not occur seaward of the barrier islands and along the open coast. Rather, they are confined to the more protected waters of the bays behind the barrier islands, especially in the lagoonal segments extending away from river mouths. Seagrasses are limited in occurrence along the upper Texas coast to 16.6 km^2 in Galveston and Matagorda Bays, covering less than 1 percent of the bottom. Along the middle Texas coast, including the San Antonio, Aransas-Copano and Corpus Christi Bay systems, cover by seagrasses increases by an order of magnitude (174.8 km^2 and 12 percent of bay bottom). Along the lower Texas coast, encompassing upper and lower Laguna Madre and Baffin Bay, seagrasses define the ecosystem, covering 751.9 km^2 and 50 percent of bay bottom. In Laguna Madre proper, seagrasses carpet more than 70 percent of bay bottom.

Halodule wrightii is the dominant seagrass and *Halophila engelmanni* and *Ruppia maritima* are at least sporadically present in all bay systems along the Texas coast. *Ruppia maritima* can dominate some beds in the north. *Thalassia testudinum* is present at one location at the extreme west end of the Galveston Bay system and then is next seen 200 km to the southwest, where, within 10 km of the gulf outlet at Aransas Pass, it is the dominant species over a quarter of the vegetated bottom. The association of *Thalassia testudinum* with a natural gulf outlet is even stronger in Laguna Madre. Within 20 km of Brazos Santiago Pass at the south end of the lagoon, *Thalassia testudinum* is the dominant species over 90 percent of the seagrass meadow. Farther from the outlet, *Thalassia testudinum* is uncommon. *Syringodium filiforme* occurs only south from Aransas Bay near Aransas Pass and is uncommon in Aransas and Corpus Christi Bays. It is the dominant species over 7 percent of the vegetated bottom in upper Laguna Madre and 30 percent of vegetated bottom in lower Laguna Madre.

Precipitation, inflow of freshwater and bathymetry are the most influential environmental determinants of the gradient of increasing seagrass abundance from northeast to southwest along the Texas coast. The much higher precipitation and inflow of the upper Texas coast ensure that its estuaries receive higher loads of sediments and nutrients and greater freshening of bay waters than do estuaries of the middle or lower Texas coast. This tends to result in higher turbidity and a reduced area of suitable salinity for seagrass growth. In addition Laguna Madre is the shallowest Texas bay so that sufficient light to support seagrass growth reaches much more of the bottom than in other bays.

The Galveston Bay system supported more than 20 km^2 of submersed aquatic vegetation in 1956, but by 1987 only 4.5 km^2 remained[22,23]. Seagrasses were limited to West Bay and its tributary embayment, Christmas Bay, which is furthest removed from riverine influences. In West Bay, seagrass beds declined from 4.6 km^2 in 1956 to 1.3 km^2 in 1965 and were absent in 1987. In Christmas Bay the 5.0 km^2 of seagrasses in 1971-72 declined to 1.1 km^2 in 1987. Losses are attributed to shorefront development and lower water quality associated with the urbanization of the area, including discharges of six sewage treatment plants, two of which have been discharging since the early 1960s. In the main stem of Galveston Bay extensive beds of *Ruppia maritima* present along the western shore in 1956 were destroyed by Hurricane Carla in 1961 and have not recovered, whereas beds in Trinity Bay have changed little. Reduced water clarity and subsidence resulting from excessive groundwater withdrawal coupled with bulkheading of the western shore of Galveston Bay may be responsible for the failure of submersed vegetation to re-establish. The appearance of some patches of *Halophila engelmanni* in West Bay since 2000 may be an early indication that

Case Study 22.2
LAGUNA MADRE

Laguna Madre accounts for 75 percent of seagrass cover in the state of Texas, while making up only 20 percent of the state's embayment area. Seagrasses are the foundation of the Laguna Madre ecosystem. As a result of the high-quality nursery habitat provided by large expanses of continuous meadow, Laguna Madre supports more than 50 percent of the Texas inshore finfish catch. Together with its sister lagoon, the Laguna Madre of Tamaulipas, which lies across the delta of the Río Grande in Mexico, the area is even more important for the redhead duck, *Aythya americana*, providing wintering habitat for more than 75 percent of the world population.

As a result of the importance of Laguna Madre as a natural resource, whole system seagrass surveys have been conducted at approximately decade intervals since the mid-1960s. Although seagrasses strongly dominate the Laguna Madre ecosystem, they have been undergoing profound change. In lower Laguna Madre seagrasses covered almost the entire bottom in 1965, but between 1965 and 1974 large tracts of deep bottom went bare and, with small adjustments in configuration, have remained bare to the present day. Shifts in species composition have been even more far reaching, with *Halodule wrightii* dominant over 89 percent of the seagrass meadow in 1965 and only 41 percent in 1998-99, being replaced by other species from the south. *Syringodium filiforme* achieved maximal coverage in the 1988 survey but already was being replaced in the south by *Thalassia testudinum*. By the 1998-99 survey this displacement was much farther advanced. *Halophila engelmanni* is a transient in this system, occupying a considerable area at the outer edge in the 1988 survey but not dominant at any other time. In upper Laguna Madre, between 1967 (see figure, opposite) and 1988 (not shown), cover of *Halodule wrightii* increased from 118 to 249 km^2 and was continuous from shore to shore over the northern third of the lagoon. However, by 1998-99 *Halodule wrightii* cover had decreased to 214 km^2, with a central deep area in the north reverting to bare bottom and a large patch of *Syringodium filiforme* taking over at the north end.

From the mid-1960s to the present, for the lagoon as a whole, the area of bare bottom has increased by 10 percent. The area of *Thalassia testudinum* has increased from barely present to dominating some 11 percent of lagoon bottom. *Syringodium filiforme* increased from 7 to 17 percent, and then fell back to 15 percent, all at the expense of *Halodule wrightii* which covered 67 percent of lagoon bottom in the earliest survey, and then declined to 41 percent at present. The loss of *Halodule wrightii* may have serious consequences because it is almost the sole food source for 400 000-600 000 wintering redhead ducks.

WATERWAY'S PROMINENT ROLE
The Gulf Intracoastal Waterway plays a prominent role in most of the radical change seen in Laguna Madre's seagrass meadows over the last 35 years. The loss of seagrass cover in deep parts of lower Laguna Madre between 1965 and 1974, and its failure to revegetate since, is probably due to maintenance dredging of the Gulf Intracoastal Waterway. Intensive monitoring of the light regime before, during, and up to 15 months after maintenance dredging in 1988 documented significantly increased light attenuation to the end of the study in the region where seagrasses had been lost. Laguna Madre is a notoriously windy location and frequent episodes in which sediment from the mounds of unconsolidated, fine-textured dredge deposits is resuspended and dispersed by currents account for the propagation of dredging effects over large areas and long periods of time. Since dredging frequency is in the order of two years, the light reduction is chronic[25].

The Gulf Intracoastal Waterway is also implicated in the other big change in seagrasses in lower Laguna Madre, the displacement of *Halodule wrightii* by more euhaline (adapted narrowly to marine salinity) species moving north over time. Prior to completion of the Gulf Intracoastal Waterway in 1949, there was no permanent water connection between upper Laguna Madre and lower Laguna Madre. A 30 km reach of seldom flooded sand flats prevented this. Before 1949, salinities in excess of 60 psu, or about twice the salt content of the adjacent Gulf of Mexico, were not uncommon in lower Laguna Madre, and in the southern extremity of upper Laguna Madre salinities in excess of 100 psu were measured several times. The breaching of the sand flat barrier at the midpoint of the lagoon greatly enhanced exchange within the lagoon and between the lagoon and the Gulf of Mexico. Prevailing southeasterly winds now drive lagoon water north across Corpus Christi Bay and ultimately out into the gulf, with inflow from the gulf at the south end replenishing the system. The net result was a moderation of hypersaline conditions within the lagoon. Now, salinities seldom reach 50 psu anywhere in the lagoon.

Halodule wrightii is the only species that can tolerate salinities greater than 60 psu and is a

superior colonizer compared to *Syringodium filiforme* or *Thalassia testudinum*. Consequently, *Halodule wrightii* was probably widespread before construction of the waterway, although there are only incidental reports from a few locations for that period, and it is not surprising that according to the first systematic survey of seagrass cover in the mid-1960s it dominated the lower Laguna Madre overwhelmingly. Gradual displacement from the south by *Syringodium filiforme*, in its turn later displaced by *Thalassia testudinum*, is consistent with what we know of the relative salinity tolerances of the species, and their colonizing and competitive abilities under conditions of moderate salinity. The euhaline species had been confined to the immediate vicinity of the natural gulf outlet at the extreme south end of the lagoon until moderation of the salinity regime. Thereafter, life history and competitive characteristics of the plants set the time course of change[26].

THE UPPER LAGUNA MADRE

Presumably, upper Laguna Madre has experienced the same system shift as lower Laguna Madre, yet its trajectory of seagrass change has been very different. *Halodule wrightii* increased rather than decreased through 1988 and *Syringodium filiforme* was not evident until 1998-99. Changes in upper Laguna Madre seem to lag those in lower Laguna Madre by 20 or 30 years. Almost certainly, the reason for the lag is the extreme hypersalinity of the southern section of upper Laguna Madre before completion of the waterway. Even *Halodule wrightii* cannot tolerate 100 psu salinity and must have been absent from most of upper Laguna Madre, except close to Corpus Christi Bay. The much greater distance to source populations probably accounts for expansion of *Halodule wrightii* meadows through 1988 in the upper Laguna Madre, whereas it was already at a maximum in the lower Laguna Madre by 1965. Similarly, the possible sources of *Syringodium filiforme* for the colonization of upper Laguna Madre are the lower Laguna Madre or across Corpus Christi Bay, near the closest gulf outlet to the north. Not surprisingly, establishment in upper Laguna Madre was long delayed compared to lower Laguna Madre, where it was present from the outset.

The last large historical change is the loss of vegetation from deep parts of upper Laguna Madre between 1988 and the present. Through 1990, Laguna Madre was renowned for its crystal clear water. However, in June 1990, a phytoplankton bloom was first noted that was dense and long-lived enough to earn it its own name, the "Texas brown tide". The bloom varied in intensity but was continually present from 1990 to 1997 and has flared up sporadically since. Light intensity at 1 m depth was reduced by half over large areas, and gradually *Halodule wrightii* died back in deep areas. Although a suite of factors played a role in the initiation and unprecedented persistence of the brown tide, nutrients regenerated from the gradual die-back of the seagrass meadow were almost certainly involved in sustaining the bloom, until steady state was reached between seagrass distribution and the brown tide-influenced light regime. A disturbing aspect of this perturbation is that as yet there is little sign of recovery. Apparently because of the loss of seagrass cover, the bottom is much more prone to sediment resuspension. Because new recruits have no reserves to tide them over episodes of low light, establishment has not occurred[27].

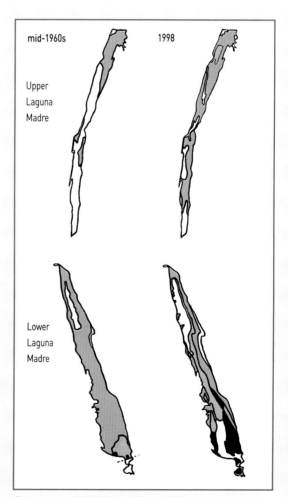

Seagrass cover in the Laguna Madre of Texas

Notes: *Halodule wrightii* (green), *Syringodium filiforme* (grey), *Thalassia testudinum* (black), bare/no seagrass (white).

management efforts are resulting in improved water quality in the bay.

Little net change in seagrass cover is evident along the middle Texas coast[21]. The dredging of navigation channels, boating activities and nutrient enrichment from non-point sources are the suspected causes of a loss of 3.3 km² of *Thalassia* near Aransas Pass, while in the same general area, subsidence has led to inundation of previously emergent flats and colonization of 8.7 km² by *Halodule wrightii*. Thus, in the absence of bulkheading in this part of Corpus Christi, the effect of subsidence on submersed vegetation is the opposite of what it was with bulkheading in Galveston Bay. A growing management concern along this part of the coast is that large areas of shallow seagrass meadow close to population centers show moderate to heavy propeller scarring[24].

MEXICO

The southwestern coast of the Gulf of Mexico has approximately 15 estuarine systems. The largest include Laguna Madre (2 000 km²) in the state of Tamaulipas, Laguna de Tamiahua (880 km²) and Laguna de Alvarado (118.3 km²) in the state of Veracruz, and Laguna de Términos (1 700 km²) in the state of Campeche. The study of seagrasses and their distribution in Mexico dates from the 1950s. In the southwestern Gulf of Mexico there are five genera of seagrasses: *Thalassia*, *Syringodium*, *Halodule*, *Halophila* and *Ruppia*, of which *Thalassia* has the widest distribution. It is found from Tamaulipas in the north to Quintana Roo in the south as well as in various reef systems[28].

Tamaulipas

The seagrass species reported for this coast include *Thalassia testudinum*, *Syringodium filiforme*, *Halodule wrightii* and *Halophila engelmanni*. Laguna Madre has four inlets that change position over time. As a result of its location in a semi-arid area, its restricted communication with the sea and a minimum river runoff from the San Fernando river, this lagoon is hypersaline, with marine conditions restricted to the areas of tidal influence near the inlets. The western margin has extensive beds of macroalgae, while *Halodule wrightii* is established along both the eastern and western margins of the lagoon and covers 357.4 km² or 18 percent of the lagoon area. Associated macrofauna includes 76 species of mollusks, 42 of crustaceans and 105 of fish[29].

Veracruz

The system of coral reefs in front of the port of Veracruz boasts five seagrass species: *Halodule wrightii*, *Thalassia testudinum*, *Syringodium filiforme*, *Halophila engelmanni* and *Halophila decipiens*. *Halodule wrightii* is found in the shallower areas where it tolerates changes in temperature and salinity, and *Halophila decipiens* is found in the deeper parts down to 10 m. Other species associated with the *Thalassia testudinum* here include green, brown and red macroalgae, foraminifers, sponges, anemones, corals, polychaetes, mollusks, crustaceans, sea urchins, sea stars and fish[30]. First identified in 1977[28] and continuing largely unchanged to the present, the main environmental problems in this area are caused by local fisheries as well as by the great loads of sediment, fertilizers, insecticides and herbicides that are transported along the coastal rivers to the coral reefs and seagrass beds, only to be resuspended during winter storm seasons.

The waters of Laguna de Tamiahua are predominantly euhaline because the lagoon has only two small sea inlets at the north and south ends and freshwater inflow is limited to several very small creeks. The western margin has extensive beds of macroalgae while *Halodule wrightii* is established along the eastern margin of the lagoon and covers 106.2 km² or 12 percent of the lagoon area. Associated macrofauna includes 67 species of mollusks, 32 of crustaceans and 129 of fish[29].

By contrast Laguna de Alvarado has two sea inlets, one of which is a narrow channel, and four rivers, one of which carries a great volume of water. Consequently, the lagoon is oligo-mesohaline with salinities below 18 psu throughout the greater part of the year. *Ruppia maritima* covers 3.2 km² or 3 percent of the lagoon area. This is the only species of seagrass in Alvarado where it forms dense beds along the inner margin of the sand barrier. Associated macrofauna includes 49 species of mollusks, 26 of crustaceans and 106 of fish[29].

Tabasco

Very little is known about seagrasses in the state of Tabasco. Only *Halodule wrightii* and *Ruppia maritima* have been recorded for the coastal lagoons with a cover of 8.1 km². The scarcity of seagrasses along the coast is the result of the large sediment load and high turbidity delivered by the Grijalva-Usumacinta river system and transported west along the coast.

Yucatán Peninsula

The Yucatán Peninsula is located in the southeastern Gulf of Mexico, includes the states of Campeche, Yucatán and Quintana Roo, and encompasses approximately 1 800 km of coastline, including islands. The karstic nature of the peninsula is responsible for non-point groundwater discharges through springs directly into coastal lagoons, the broad continental

Regional map: The Caribbean

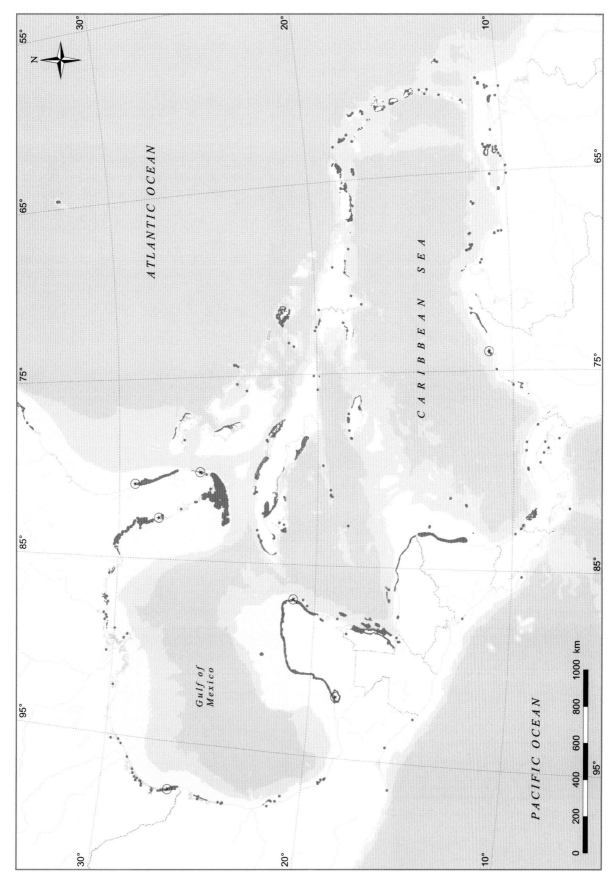

XIV Regional map: South America

shelf and the open sea. Based on satellite images and field observations during 2000-01, there are 5 911 km² of seagrasses[31] in the coastal lagoons, the coastal sea and the coral reef lagoons of Chinchorro and Alacranes.

Seagrasses along the coasts of the peninsula include the six species present in the southwestern Gulf of Mexico. The most widespread species are *Thalassia testudinum*, *Syringodium filiforme* and *Halodule wrightii*, with this last species also found in the hypersaline Ría Lagartos, Yucatán. *Halophila engelmanni* and *Halophila decipiens* are found in small areas while *Ruppia maritima* is found in the shallow brackish coastal lagoons of Términos, Celestún, Chelem, Dzilám, Ría Lagartos, Nichupté, Ascensión and Chetumal[32]. Beds of *Thalassia testudinum* and *Syringodium filiforme* have also been sighted around the reefs of Alacranes and Cayo Arcas, Campeche.

Thalassia testudinum is dominant in open waters, especially in the coral reef lagoons. The maximum total biomass and shoot density recorded are 2 000 g/m² and 1 222 shoots/m² [33,34]. *Halodule wrightii* is the most widely distributed species in the peninsula. It is found in mixed and monospecific stands, in shallow waters (<1 m), around freshwater springs and at salinities between 20 and 57 psu. The maximum total biomass and shoot density recorded are 700 g/m² and 14 872 shoots/m² [35,36]. *Syringodium filiforme* has been observed mainly in open waters mixed with *Halodule wrightii* and *Thalassia testudinum*, dominating in regions where strong currents are observed. The maximum total

---

Case Study 22.3
LAGUNA DE TÉRMINOS

Laguna de Términos is the best studied coastal lagoon in the Mexican Gulf of Mexico. It is located at the transition area between a western terrigenous region and an eastern calcareous region and is also characterized by a marked north-south salinity gradient, established by the tides that enter through two sea inlets and the freshwater provided by three rivers[41]. The seagrasses *Thalassia testudinum*, *Syringodium filiforme* and *Halodule wrightii* cover 496.4 km² or 29 percent of the lagoon area.

Thalassia testudinum covers extensive areas along the northern, eastern and southeastern margins where salinity is high and the water is relatively clear, creating an ideal environment for this tropical, polyeuhaline species[42]. *Syringodium filiforme* is restricted to the northeastern region where salinity is high and the sediment is biogenic, sandy and calcareous. This species favors calcareous substrates and is found forming dense seagrass beds along the eastern Caribbean coast of Mexico. *Halodule wrightii* is found along the northern and western margins of the lagoon, the first with high salinity, clear water and sandy substrates, and the second with low salinity, turbid water and muddy substrates, where the great Grijalva-Usumacinta river system drains into the lagoon. Its distribution shows it is a tropical euryhaline, pioneering and adaptable species that tolerates a variety of environmental characteristics. *Ruppia maritima* has also been observed in the low-salinity areas of the southwest of the lagoon.

In this lagoon, all seagrass beds, but mainly those of *Thalassia testudinum*, play an important part as nursery, feeding and protection areas for the larvae and juveniles of commercially important species, such as the shrimp *Litopenaeus setiferus*, *Farfantepenaeus aztecus* and *Farfantepenaeus duorarum*, and the fish *Caranx hippos*, *Lutjanus analis*, *Bagre marinus*, *Centropomus undecimalis* and *Archosargus rhomboidalis*, that migrate through the lagoon during their life cycles and constitute offshore fisheries of great economic value, among which the shrimp fishery is particularly noteworthy.

Associated macrofauna include 174 species of mollusks, 60 of crustaceans and 214 of fish[29], establishing Laguna de Términos as the most species rich (448) of Mexico's four large lagoons along the Gulf of Mexico, supporting almost twice as many species as the other lagoons (181 species in Laguna de Alvarado, 228 species in Laguna de Tamiahua and 223 species in Laguna Madre). The environmental heterogeneity and complexity of Laguna de Términos arise from the presence and distribution of the four seagrass species, macroalgae, mangrove forests and the marked salinity gradient, all of which favor the recruitment of a great number of stenohaline and euryhaline estuarine species, and some marine species. These establish communities with different trophic structures[43] subdividing the lagoon according to their preferences for different habitat types[44].

No long-term data on seagrass coverage have been recorded for Laguna de Términos. However, the extent of the seagrass beds, as well as shoot density, along the inner margin of the barrier island were markedly reduced when Hurricane Roxanne passed over the lagoon twice in October 1996. Recovery occurred within three years.

biomass and shoot density recorded are 1000 g/m² and 7140 shoots/m²[36]. *Halophila decipiens* and *Halophila engelmanni* can be observed in small patches mixed with other seagrasses and macroalgae in Yucatán and Quintana Roo, mainly on shallow (<1-5 m) sandy bottoms. *Ruppia maritima* has been observed growing in mixed stands with *Halodule wrightii* in polyhaline (18 psu) waters, and in monospecific stands in mesohaline (5-18 psu) waters with a biomass of 1000 g/m².

The seagrasses of the peninsula have been affected by trawling and eutrophication in Campeche, by trawling, tourism, eutrophication and port development in Yucatán, and by tourism and eutrophication in Quintana Roo, all suggesting that water quality is a major concern for the seagrasses in this region.

Historical analysis is limited. Observations from 1985 to 2001 in Progreso, Yucatán, show a loss of 95 percent of seagrass cover and a replacement by green macroalgae (*Caulerpa* spp.), with eutrophication again the main cause[37]. However not all decline has proved permanent. Following the construction of port infrastructure in the north of Yucatán the hydrology of the Chelem Lagoon changed and the seagrass community was negatively impacted. Four years later hydrological restoration took place and the seagrass community recovered[38]. Natural events such as hurricanes have significant effects on seagrass communities. However, these have shown an important recovery capacity. For example, during Hurricane Gilbert in 1988, *Halodule wrightii* beds in the coastal lagoon of Celestún, Yucatán, lost 93 percent of their area. Within three years, they had recovered to their initial condition[39]. A similar pattern of recovery was observed for *Thalassia testudinum* in Cancún[40].

CONCLUSION

The Gulf of Mexico is a globally important seagrass area. Extensive beds cover about 19000 km² from Cuba and the southern tip of Florida through Texas to the Yucatán Peninsula of Mexico. In many places impacts appear to be low and seagrass beds healthy; however, some recent perturbations are disturbing because of their magnitude and our inadequate understanding of causation. Most worrisome are the persistent and recurring algal blooms, high turbidity and changes in species composition afflicting seagrass meadows of Florida Bay on the east side and Laguna Madre on the west side of the Gulf of Mexico.

Ray on low-density *Syringodium filiforme*, Mexican Caribbean.

AUTHORS

Christopher P. Onuf, US Geological Survey, National Wetlands Research Center, Campus Box 339, 6300 Ocean Drive, Corpus Christi, Texas, USA. **Tel:** +1 361 985 6266. **Fax:** +1 361 985 6268. **E-mail:** chris_onuf@usgs.gov

Ronald C. Phillips, 3100 South Kinney Road #77, Tucson, Arizona 85713, USA.

Cynthia A. Moncreiff, Gulf Coast Research Lab, P.O. Box 7000, 703 East Beach Drive, Ocean Springs, Mississippi, USA.

Andrea Raz-Guzman, Instituto de Investigaciones sobre los Recursos Naturales, Universidad Michoacana de San Nicolas de Hidalgo, Avenida San Juanito s/n, San Juanito Itzícuaro, Morelia 58330, Michoacán, Mexico.

Jorge A. Herrera-Silveira, CINVESTAV-IPN, Merida, Carr. Ant. Progreso km 6, Merida 97310, Yucatán, Mexico.

REFERENCES

1. Beatriz Martínez, Instituto de Oceanología, Habana, Cuba. Personal communication.
2. Buesa RJ [1975]. Population biomass and metabolic rates of marine angiosperms on the northwestern Cuban shelf. *Aquatic Botany* 1: 11-23.
3. Sargent FJ, Leary TJ, Crewz DW, Kruer CR [1995]. Scarring of Florida's Seagrasses: Assessment and Management Options. FMRI Technical Report TR-1. Florida Marine Research Institute, St Petersburg, FL.
4. Fourqurean JW, Robblee MB [1999]. Florida Bay: A history of recent ecological changes. *Estuaries* 22: 345-357.
5. Hall MO, Durako MJ, Fourqurean JW, Zieman JC [1999]. Decadal changes in seagrass distribution and abundance in Florida Bay. *Estuaries* 22: 445-459.
6. Kurz RCX, Tomasko DA, Burdick D, Ries TF, Patterson SK, Finck R [2000]. Recent trends in seagrass distributions in southwest Florida coastal waters. In: Bortone SA (ed) *Seagrasses: Monitoring, Ecology, Physiology and Management*. CRC Press, Boca Raton, FL. pp 157-166.
7. Lewis RR III, Durako MJ, Moffler MD, Phillips RC [1985]. Seagrass meadows of Tampa Bay: A review. In: Treat SF, Simon JL, Lewis RR III, Whitman RL (eds) *Proceedings, Tampa Bay Area Scientific Information Symposium*. Burgess Pub. Co., Minneapolis, MN.

8. Johansson JOR, Ries TF [1997]. Seagrass in Tampa Bay: Historic trends and future expectations. In: Treat SF, Clark PA (eds) *Proceedings, Tampa Bay Scientific Information Symposium 3*. Tampa, FL. pp 139-150.
9. Lewis RR III, Clark PA, Fehring WK, Greening HS, Johansson RO, Paul RT [1998]. The rehabilitation of the Tampa Bay estuary, Florida, USA, as an example of successful integrated coastal management. *Marine Pollution Bulletin* 37: 468-473.
10. Johansson JOR, Lewis RR III [1991]. Recent improvements in water quality and biological indicators in Hillsborough Bay, Florida. In: Treat SF, Clark PA (eds) *Proceedings, Tampa Bay Scientific Information Symposium 2*. Tampa, FL. pp 1199-1215.
11. Lewis RR III, Estevez ED [1988]. The Ecology of Tampa Bay, Florida: An Estuarine Profile. US Fish and Wildlife Service Biological Report 85(7.18).
12. Johansson JOR, Greening H [2000]. Seagrass restoration in Tampa Bay: A resource-based approach to estuarine management. In: Bortone SA (ed) *Seagrasses: Monitoring, Ecology, Physiology and Management*. CRC Press, Boca Raton, FL. pp 279-293.
13. Handley LR [1995]. Seagrass distributions in the northern Gulf of Mexico. In: LaRoe ET, Farris GS, Puckett CE, Doran PD, Mac MJ (eds) Our Living Resources: A Report to the Nation on the Distribution, Abundance, and Health of U.S. Plants, Animals, and Ecosystems. US Department of the Interior, National Biological Service, Washington, DC. pp 273-275.
14. Eleuterius L N [1973]. The distribution of certain submerged plants in Mississippi Sound and adjacent waters. In: Christmas JY (ed) *Cooperative Gulf of Mexico Estuarine Inventory and Study, Mississippi, Phase IV: Biology*. State of Mississippi, Gulf Coast Research Laboratory, Ocean Springs, MS. pp 191-197.
15. Eleuterius LN, Miller GJ [1976]. Observations on seagrasses and seaweeds in Mississippi Sound since Hurricane Camille. *Journal of the Mississippi Academy of Sciences* 21: 58-63.
16. Sullivan MJ [1979]. Epiphytic diatoms of three seagrass species in Mississippi Sound. *Bulletin of Marine Science* 29: 459-464.
17. US Geological Survey [1992]. Draft maps: Isle au Pitre, LA-MS, Cat Island, MS-LA, Ship Island, Dog Keys Pass, MS, Horn Island West, MS, Horn Island East, MS, Petit Bois Island, MS, Grand Bay SW, MS-AL, Kreole, MS-AL. Scale 1:24 000. National Wetlands Research Center, Lafayette, LA.
18. Heck KL, Sullivan MJ, Zande JM, Moncreiff CA [1996]. An Ecological Analysis of Seagrass Meadows of the Gulf Islands National Seashore, Years One, Two, and Three: Seasonal assessment and inventory, interaction studies and continuing assessment/inventory, August 1996. Final Report to the National Park Service.
19. Oivanki SM [1994]. Belle Fontaine, Jackson County, Mississippi: Human history, geology, and shoreline erosion. *Mississippi Department of Environmental Quality, Office of Geology, Bulletin No. 130*.
20. Otvos EG [1981]. Barrier island formation through nearshore aggradation: Stratigraphic and field evidence. *Marine Geology* 43: 195-243.
21. Pulich W Jr [1999]. Introduction. In: Seagrass Conservation Plan for Texas. Texas Parks and Wildlife Department, Resource Protection Division, Austin, TX. pp 14-29.
22. Pulich W Jr, White WA [1991]. Decline of submerged vegetation in the Galveston Bay system: Chronology and relationships to physical processes. *Journal of Coastal Research* 7: 1125-1138.
23. Adair SE, Moore JL, Onuf CP [1994]. Distribution and status of submerged vegetation in estuaries of the upper Texas coast. *Wetlands* 14: 110-121.
24. Dunton KH, Schonberg SV [2002]. Assessment of propeller scarring in seagrass beds of the south Texas coast. *Journal of Coastal Research* Special Issue No. 37: 100-110.
25. Onuf CP [1994]. Seagrasses, dredging and light in Laguna Madre, Texas, USA. *Estuarine Coastal and Shelf Science* 39: 75-91.
26. Quammen ML, Onuf CP [1993]. Laguna Madre: Seagrass changes continue decades after salinity reduction. *Estuaries* 16: 303-311.
27. Onuf CP [2000]. Seagrass responses to and recovery (?) from seven years of brown tide. *Pacific Conservation Biology* 5: 306-313.
28. Lot A [1977]. General status of research on seagrass ecosystems in Mexico. In: McRoy CP, Helfferich C (eds) *Seagrass Ecosystems. A Scientific Perspective*. Marcel Dekker Inc, NY. pp 233-245.
29. Raz-Guzman A, Reguero M, Huidobro L, Corona A [2001]. Estuarine community composition in Mexican Gulf of Mexico coastal lagoons. *16th Biennial Conference of the Estuarine Research Federation, Conference Abstracts*. p 114.
30. Lot A [1971]. Estudios sobre fanerógamas marinas en las cercanías de Veracruz, Ver. *Anales del Instituto de Biologia Seria Botanica UNAM* 42: 1-48.
31. http://www.dumac.org.mx
32. Espinoza J [1996]. Distribution of seagrasses in the Yucatan Peninsula, Mexico. *Bulletin of Marine Science* 59: 449-454.
33. Gallegos ME, Merino M, Marbá N, Duarte CM [1993]. Biomass and dynamics of *Thalassia testudinum* in the Mexican Caribbean: Elucidating rhizome growth. *Marine Ecology Progress Series* 95: 185-192.
34. Van Tussenbroek BI, Hermus K, Tahey T [1996]. Biomass and growth of the turtle grass *Thalassia testudinum* (Banks and Konig) in a shallow tropical lagoon system in relation to tourist development. *Caribbean Journal of Science* 32: 357-364.
35. Herrera-Silveira JA [1994]. Phytoplankton productivity and submerged macrophyte biomass variation in a tropical coastal lagoon with groundwater discharge. *Vie Milieu* 44: 257-266.
36. Gallegos ME, Merino M, Rodríguez A, Marbá N, Duarte CM [1994]. Growth patterns and demography of pioneer Caribbean seagrasses *Halodule wrightii* and *Syringodium filiforme*. *Marine Ecology Progress Series* 109: 99-104.
37. Herrera-Silveira JA, Troccoli, GL, Aranda CN, Alvarez GC, Trejo J [2000]. Evaluación de la calidad ambiental de la zona costera de Progreso: Hidrología y clorofila-a. Informe Técnico CINVESTAV-LPP, p 50.
38. Herrera-Silveira JA, Ramírez-Ramírez J, Gómez N, Zaldivar A [2000]. Seagrass bed recovery after hydrological restoration in a coastal lagoon with groundwater discharges in the north of Yucatán. In: Bortone S (ed) *Seagrass: Monitoring Ecology, Physiology and Management*. CRC Press, Boca Raton, FL. pp 123-135.
39. Herrera-Silveira JA [1993]. Ecología de los productores primarios en la laguna de Celestún, México. Patrones de variación espacial y temporal. Doctoral thesis, University of Barcelona.
40. Marbá N, Gallegos ME, Merino M, Duarte CM [1994]. Vertical growth of *Thalassia testudinum*: Seasonal and interannual variability. *Aquatic Botany* 47: 1-11.
41. Raz-Guzman A, de la Lanza G [1991]. Evaluation of photosynthetic pathways of vegetation, and of sources of sedimentary organic matter through $\delta^{13}C$ in Terminos Lagoon, Campeche, Mexico. *Anales del Instituto de Biologia Seria Botanica UNAM* 62: 39-63.
42. Phillips RC [1980]. Role of seagrasses in estuarine systems. In: Fore PL, Peterson RD (eds) *Proceedings Gulf of Mexico Coastal Ecosystems Workshop*. US Fish and Wildlife Service, Albuquerque, New Mexico, FWS/OBS-80/30. pp 67-96.
43. Raz-Guzman A, Sánchez AJ [1996]. Trophic structure related to seagrass habitat complexity. In: Kuo J, Phillips RC, Walker DI, Kirkman H (eds) *Seagrass Biology. Proceedings of an International Workshop. Rottnest Island, Western Australia*. University of Western Australia, Nedlands, Western Australia. pp 241-248.
44. Sánchez AJ, Raz-Guzman A [1997]. Distribution patterns of tropical estuarine brachyuran crabs in the Gulf of Mexico. *Journal of Crustacean Biology* 17: 609-620.

23 The seagrasses of THE CARIBBEAN

J.C. Creed
R.C. Phillips
B.I. Van Tussenbroek

The Caribbean region includes the following sub-regions[1]: western Caribbean (Mexico, Belize, Guatemala and Honduras), southern Caribbean (Nicaragua, Costa Rica and Panama, and South American Colombia), the Lesser Antilles (Venezuela, and the islands Aruba, Curaçao, Bonaire, Trinidad and Tobago, Barbados, Grenada, St Vincent and the Grenadines, St Lucia, Martinique, Dominica, Guadeloupe, Antigua and Barbuda, Montserrat, St Kitts and Nevis, St Martin (Sint Maarten), St Eustarius, Saba, Anguilla, the British Virgin Islands, Turks and Caicos Islands, and the US Virgin Islands) and the Greater Antilles (Puerto Rico, Hispaniola, Jamaica, the Cayman Islands and Cuba). In the wider Caribbean context, here we also consider the Guyanas (Guyana, Suriname and French Guiana), the Bahamas and the east coast of Florida.

The coastlines of the mainlands and islands under consideration stretch from about 26°N to 4°S and from 88°W to 52°W, and are influenced by the Gulf of Mexico, the Caribbean Sea and the Atlantic Ocean. They stretch from the Tropic of Cancer (only the northern Bahamian islands and Florida are subtropical) to just north of the equator.

In a review of seagrass ecosystems and resources of Latin America in 1992, Phillips[1] stated that "the most basic research is needed in almost every place". During the subsequent decade, the number of published reports and surveys on the biology and ecology of the seagrasses of Central and South America has increased. The seagrass communities of some of the islands of the Bahamas, and the Greater and Lesser Antilles, have also been quite well studied, but both quantitative and qualitative information on the status of seagrasses is highly variable, reflecting the large number of countries and territories which make up the region and their individual political and economic approaches to research, as well as exploitation and protection of coastal marine resources. Substantial research is still needed in order to provide a comprehensive assessment of Caribbean seagrasses.

Seagrasses are found throughout the Caribbean. They grow in the reef lagoons between the beaches and coral reefs or form extensive meadows in more protected bays and estuaries. Seven seagrass species are recognized. Turtle grass, *Thalassia testudinum*, is the most abundant seagrass in the region, but is not known south of Venezuela. Plants are erect, leaves generally varying from 5 to 15 mm wide and from 10 to 50 cm long, but can reach up to 1 m. This seagrass forms dense rhizome mats below the sediment creating extensive meadows on shallow sand or mud substrates from the lower intertidal to a maximum 10-12 m depth, but has also been reported below 20 m. Manatee grass, *Syringodium filiforme*, has a similar geographical distribution, with cylindrical, narrow leaves which form canopies up to 45 cm high. It usually grows intermixed with *Thalassia testudinum*, but can grow in monospecific areas, beds or patches from the upper sublittoral down to more than 20 m.

Shoal grass, *Halodule wrightii*, is found throughout the wider Caribbean region. It has small, supple, grass-like leaves varying in width between 2 and 5 mm and in length between 4 and 10 cm, but sometimes reaching more than 50 cm in length. It is found growing on sand and mud from the intertidal down to 5 m. *Ruppia maritima*, widgeon grass, is also found throughout the Caribbean and, like *Halodule wrightii*, has small grass-like leaves. It is a shallow-water species found in the brackish waters of bays and estuaries between 0 and 2.5 m deep. The three sea vine species belonging to the genus *Halophila* – *Halophila baillonii*, *Halophila engelmanni* and *Halophila decipiens* – are small and delicate. Their leaves are paddle shaped, are less "grass-like" than the other species and lack a basal sheath. *Halophila decipiens* is found in deep water (to 30 m), while *Halophila*

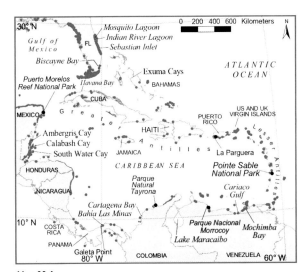

Map 23.1
The Caribbean

engelmanni is found only down to 5 m and is restricted to the Bahamas, Florida, the Greater Antilles and the western Caribbean. *Halophila baillonii* is only found in the Lesser Antilles[2].

ECOSYSTEM DESCRIPTION

The coastlines of the Caribbean are characterized by three ecosystems: seagrasses, coral reefs and mangroves, with numerous linkages and trophic interactions existing between these ecosystems[3]. Seagrasses are considered to be open systems, exporting leaves and other components of primary production in the form of organic material to other habitats. At Galeta Point, Caribbean Panama, a 0.01-km² seagrass bed has been estimated to export 37-294 kg/month of *Thalassia testudinum* leaves, and 3-171 and 3-74 kg/month, respectively, of the associated macroalgae *Laurencia* and *Acanthophora* spp. Seagrass and its associated algal material can even end up as offshore foodfalls, which have been shown to be a significant pathway by which energy enters the deep sea[4]. Seagrasses, by stabilizing sediments in their extensive systems of roots and rhizomes, prevent abrasion and burial by sediments of the adjacent corals during storms.

Migratory movements of various animals such as fish, spiny lobsters, prawns and sea urchins enhance the links between the seagrasses, reefs and mangroves. These migratory movements can occur on a daily basis (e.g. foraging in the seagrass beds during the day and sheltering from predation in the reefs during the night) or seasonally, when juvenile stages of species migrate from mangroves or seagrasses to the reefs when reaching adulthood[3].

Thalassia testudinum typically dominates seagrass vegetation in the reef lagoons where it often coexists with *Syringodium filiforme*, *Halodule wrightii* and calcareous rhizophytic green algae belonging to the order Caulerpales, amongst which *Halimeda* spp. are the most conspicuous members, and play an important role in production of sediments. Calcium carbonate sand production of *Halimeda* spp. in seagrass beds can exceed 2 kg/m²/year[5]. Other non-calcified rhizophytic algae of the same order such as *Caulerpa* spp. and *Avrainvillea* spp. are also found in these beds. Drifting or free-floating masses of algae (e.g. *Laurencia* spp., *Lobophora* sp., *Euchema* sp., *Hypnea* sp. and *Acanthophora* sp.) may be abundant locally. *Thalassia testudinum*-dominated communities in reef environments are usually preceded on the beach side by a small fringe of *Halodule wrightii*. In deeper waters, other seagrasses such as *Syringodium filiforme* and *Halophila decipiens* replace *Thalassia testudinum*.

In more protected areas, such as estuarine environments, or zones influenced by mangroves, and depending on prevailing salinity, nutrient conditions, light and sediment conditions, seagrass vegetation consists of virtually monospecific beds or alternating monospecific patches of *Thalassia testudinum* and *Halodule wrightii* with rhizophytic *Caulerpa* spp. and loosely attached green algae, many of which belong to the order Dasycladales (e.g. *Batophora* sp., *Acetabularia* spp.). Drifting mats, when occurring, are typically formed of filamentous red and green algae in these areas. *Ruppia maritima* is found in brackish waters in bays and estuaries, sometimes with *Halodule wrightii*. *Halodule wrightii* forms monospecific stands in lagoons with high salinity fluctuations. *Halophila* spp. grow in finer sands and sediments, forming monospecific or mixed species beds with the above-mentioned seagrasses. *Halophila* spp. require less light than the other seagrass species, and can be found in very deep waters or in very shallow areas with turbid conditions.

Seagrasses are colonized by calcareous and filamentous epiphytes. In classic models of Caribbean seagrass succession, rooted vegetation starts with rhizophytic algae followed by *Halodule wrightii* (and sometimes *Syringodium filiforme*) which are considered to be pioneer species, with a *Thalassia testudinum*-dominated vegetation as climax[6, 7, 8].

An enormous diversity of fauna is associated with the Caribbean seagrasses. Groups that contribute most to the richness of seagrass systems in the Caribbean are fish, echinoderms, decapods, gastropods, mollusks and sponges. Foraminifera, polychaetes, oligochaetes, nematodes, coelenterates, amphipods, isopods, hydrozoans and bryozoans are important mesofaunal groups. The Caribbean seagrasses have distinct demersal fish assemblages. Fish assemblages in the seagrasses vary depending on their affinity with mangrove and coral reef communities. At Martinique, French West Indies, 65 species of fishes belonging to 28 families were collected in *Thalassia testudinum* beds. In Belize barrier reef lagoons, the fish community is dominated numerically and in biomass by grunts, apogonids and tetraodontiforms. Most fishes are either juveniles of species that occur as adults on the reef, or are small species that reside in the lagoon. In Panama and the Virgin Islands, juvenile snappers, scorpion fishes, grunts and goatfishes in seagrass meadows feed predominantly on decapod crustaceans and other fishes.

Macroinvertebrate diversity in seagrass beds is high; for example, in Venezuela, 127 macroinvertebrate species are associated with the seagrass beds at Mochimba Bay[8]. Throughout the Caribbean, sea urchins such as *Lytechinus variegatus* and *Tripneustes ventricosus* are also major herbivores of seagrass blades, consuming 0.155 (Jamaica) and 1.4 (St Croix, US Virgin Islands) g dry weight/individual/day, respectively, of *Thalassia testudinum*. *Diadema antillarum* is another important herbivorous urchin in some places. Leaves of seagrasses are used as a substratum by invertebrates such as hydroids and sponges. Foraminiforans play a large role in the production of calcareous sediments.

Two potentially important mesofaunal top-down regulators in *Thalassia testudinum* beds are the isopod *Limnoria simulata*, which bores into the live tissue of *Thalassia testudinum*[9], and the small green snail *Smaragdia viridens*, which grazes on the chloroplast-containing epidermis. Macroalgae in seagrass beds can provide food and shelter for associated fauna. The inconspicuous epiphytes are a major food source to members of the mesofauna. The seagrass-associated alga *Batophora* sp. is a preferred source of food for the queen conch, and *Laurencia* spp. are the

---

Case Study 23.1
FLORIDA'S EAST COAST

The seagrasses of Florida's east coast, from Mosquito Lagoon (29°N) in the north to lower Biscayne Bay (25°N) south of Miami, occur in the shallow lagoonal coastal river systems typical of this area. (The seagrasses of the Florida Keys are covered in Chapter 22.) Approximately 2800 km² of seagrass habitat is found along the east coast of Florida, much of it in the Indian River Lagoon, 250 km long and encompassing population centers and many canal-side and marina-oriented housing developments. Estimates of seagrass decline include a loss of about 30 percent from the Indian River and, since 1950, a 43 percent loss from the northern and urbanized section of Biscayne Bay. However, Indian River seagrasses showed some increase during the years 1994-98[22].

In the Indian River, dredge spoil islands vegetated with mangroves and Australian pines dot the intercoastal waterway; bridges and supported highways cross the river to provide access to the coastal beaches. The east coast of Florida is subject to hurricanes. Manatee live in the Indian River, eating seagrass; many manatee are killed or scarred by boat propellers. In the cooler winter months, manatees cluster in the warm water at the outlets of electric plants, deep boat basins and other large facilities.

Seven species of seagrass are found along Florida's east coast, usually in mixed beds rather than pure stands: *Thalassia testudinum*, *Syringodium filiforme*, *Halodule wrightii*, *Halophila decipiens*, *Halophila engelmanni*, *Ruppia maritima* and *Halophila johnsonii*. The first six are also found throughout the Caribbean. Similar to its growth habit in other areas, *Thalassia testudinum* is a climax species, slow growing, long lived and requiring high light levels. *Thalassia testudinum* is more abundant in the south, reaching its northern limit in the mid-Indian River Lagoon; beds often persist for decades. Shoal grass, *Halodule wrightii*, is the most abundant seagrass in the main part of the Indian River Lagoon, an early colonizer, and grows in both shallow and mid-depths (to 2 m). *Syringodium filiforme* is known as "manatee grass" and is habitat and food for the manatee and the second-most commonly occurring species in the Indian River Lagoon, although in Mosquito Lagoon, a sub-estuary of the Indian River Lagoon, *Ruppia maritima* is more common than *Syringodium filiforme*. Elsewhere in the Indian River, *Ruppia maritima* is common and grows in both salt and fresh environments. The *Halophila* species in Florida include the cosmopolitan *Halophila decipiens* and *Halophila engelmanni*, and the rare and fragile *Halophila johnsonii*, a seagrass found only along the southeast coast of Florida between Sebastian Inlet (27°51'N) and Virginia Key (25°45'N), primarily in the Indian River.

Halophila johnsonii was given threatened species status in 1998 by the National Marine

principal settling substrate for the recruitment of the spiny lobster.

Seagrass beds have been recognized as productive fishery areas in the Caribbean[10]. Soft-bottom demersal fisheries exploit sciaenids, mullets, snappers, groupers, grunts, sharks, penaeid shrimp, loliginid squid and octopods over seagrass beds. Other resources are free-living macroalgae *Eucheuma* and *Hydropuntia*, which are collected and used to make seamoss drinks in the Lesser Antilles and seamoss porridge in Belize. The queen conch and the spiny lobster are major fisheries resources. The queen conch *Strombus gigas* is associated with *Thalassia testudinum* beds and is now seriously threatened by overfishing. The spiny lobster *Panulirus argus* is also a very important resource fished in the seagrass beds and on the nearby reefs. Though its abundance has decreased due to overfishing, present restrictions in fisheries aim to protect this resource in various countries. In Belize, spiny lobster and queen conch contribute most of the total value of exported seafood, estimated at US$10.4 million in 1995. Spiny lobster fishing alone is worth over US$23 million per year in Nicaragua.

The Caribbean is exposed to three types of natural hazard: hurricanes, volcanic eruptions and earthquakes. Hurricane activity may result in loss of seagrass vegetation because of sediment erosion or sediment deposition on the seagrass beds. In 1989 Hurricane Hugo, with squalls exceeding 160 knots, was the most violent of the century to pass over Guadeloupe and Puerto Rico[11]. It had a more destructive impact on *Syringodium filiforme* beds than on *Thalassia testudinum*. Hugo eroded nearshore areas, and tens of square kilometers of highly productive seagrass meadows were destroyed by the formation of large sediment "blow-outs", holes in otherwise continuous seagrass meadows. Such

Fisheries Service under the Federal Endangered Species Act. It has one of the most limited geographical distributions of any seagrass in the world, although it is currently under genetic study to investigate the possibility that it may be an introduced species[23]. Beds of *Halophila johnsonii* are highly transient and the plants are quick growing, with individual plants reaching mature size in about two weeks and beds often persisting no longer than a few months. Beds are often discontinuous and patchy. Neither male flowers of *Halophila johnsonii* nor seeds or seedlings have been found, and it is speculated that the plant's populations are maintained by vegetative growth alone[24]. The St Johns River Water Management District has been monitoring seagrass in the Indian River since 1994. The monitoring program has documented large changes in the distribution of some seagrasses, with *Halophila johnsonii* and other *Halophila* species increasing more than 500 percent in four years at some locations.

Grouper, snapper, sea trout and flounder use seagrass habitat as nursery on the Florida east coast. Bay scallops, shrimp and blue crabs also depend on seagrasses. Controls on dredge and fill activities set standards for turbidity, water color and other physical parameters. There are guidelines to protect manatee habitat and to preserve indigenous life forms, including seagrasses, on State of Florida submerged lands. Removal or destruction of seagrasses in state parks is forbidden, and some areas limit boats with engines in order to decrease propeller scarring of the grass beds. Educational efforts aimed at sports fishers and boaters are attempting to decrease human impacts on seagrasses, though the impacts from land-based human development are probably of greater concern.

Halodule wrightii with a male flower, Florida.

A "blow-out" in the Turks and Caicos Islands.

blow-outs can migrate and expand, taking many years to recover[6].

However, in many cases, either hurricanes had no visible effects at all[8, 12] or recovery was relatively fast, which was the case for *Thalassia testudinum* in a Mexican Caribbean reef lagoon in 1988 after the passage of Hurricane Gilbert, a class 5 hurricane with the lowest atmospheric pressure (888 mb) ever reported in the area[13]. In Costa Rica, the Limón earthquake of 1991 resulted in a 0.5 m uplift of a lagoon and *Thalassia testudinum* completely overgrew it, although the following year there was an equivalent reduction in seagrass area[7]. Perez and Galindo[14] reported mass defoliation of *Thalassia testudinum* in Parque Nacional Morrocoy, Venezuela, due to hyposalinity after torrential rains, but recovery was fast (several months) as the apical-shoot meristems did not die and shortly after the event formed leaves again. These reports indicate that overall, under natural conditions, Caribbean seagrass beds seem to be fairly resistant or resilient to major natural disturbances.

PRESENT DISTRIBUTION

The only data available on area coverage of seagrasses are from specific studies in which seagrasses have been mapped at a very local scale. The following estimates are compiled from various sources: Mexico 500 km^2; Belize 1 500 km^2; Guatemala (one site) 20 km^2; Nicaragua (Great Corn Island) 2.4 km^2; Costa Rica (Parque Nacional Cahuita) 0.2 km^2; Venezuela (Cariaco Gulf) 500 km^2; Curaçao 8 km^2; Bonaire 2 km^2; Tobago 0.64 km^2; Martinique 41.4 km^2; Guadeloupe 82.2 km^2; Antigua (Seatons Harbour) 1 km^2; Puerto Rico (La Parguera, Guayanilla Bay) 27.68 km^2; Jamaica (Discovery Bay) 0.5 km^2; Cuba (Cayo Coco, Sabana-Camagüey Archipelago) 75 km^2; Grand Cayman 25 km^2.

More extensive mapping may have been carried out but area estimates not published; further mapping and distribution studies in the Caribbean are badly needed. Distribution data are required in order to assess the real or potential threats to seagrass and losses. Determining the distribution of Caribbean seagrasses is a problem because mapping of coastal marine resources using remote-sensing techniques has only recently been refined to be able to differentiate seagrass communities, and most countries in the Caribbean have neither the infrastructure nor the funding to execute projects of this nature.

HISTORICAL PERSPECTIVES

There are few historical reports on permanent losses of seagrass beds in the Caribbean. Barbados and Carriacou in the Grenadine Island chain lost seagrasses between 1969 and 1994. In Trinidad and Tobago seagrass beds once found in Scotland Bay, Grand Fond Bay (Monos Island), Five Islands, Cocorite (near the mouth of the Diego Martin River) and Speyside in Tobago have disappeared.

In the past, grazing by the green turtles and manatees must have had an enormous impact on the seagrass beds. The common names "turtle grass" and "manatee grass" are a testament to the importance of seagrasses in the diet of the green turtle *Chelonia mydas* and the West Indian manatee *Trichechus manatus*. The manatee and the green turtle are threatened throughout the Caribbean because of overfishing and, in the case of the turtle, the collection of eggs for food. Although both animals have received considerable conservation status, they are still hunted. We can only speculate about their past impact on the seagrass communities and how these communities have changed since populations of these large herbivores have diminished or disappeared.

PRESENT THREATS

The vast expanses of the seagrass beds in the Caribbean, together with their relatively high resistance or resilience to major natural disturbances, may give a false sense of security and lead to the perception that they are immune to human impacts. Socio-economically, the Caribbean region includes high proportions of urban human populations and rapidly expanding agricultural and tourist-industrial frontiers. The population growth of the Caribbean over the last 20 years has been estimated to be 58 percent[15], which has led to increasing pressure on the adjacent coastlines and their seagrasses. Additionally, an estimated 12 million tourists per year visit the Caribbean region. The number of tourists who visited the Belize Barrier Reef Complex in 1994 was 128 000, generating an estimated US$75 million.

On local scales, seagrasses are being destroyed or removed by the construction of coastal developments associated with tourism or other coastal

activities. Tourist developments are accompanied by the construction of harbors and docks, channel dredging and recreational moorings. In Venezuela, houses were constructed over seagrass beds[16]. At La Parguera, Puerto Rico, increased traffic of ships and recreational vessels are causes of anchor damage, littering, trampling, propeller scarring, fuel impacts, detrimental shading of the seagrasses by marinas and piers, and damage of the beds by dredging. Seagrass beds in front of hotel beaches are often removed, for example in Pointe Sable National Park on St Lucia, in Venezuela and in Mexico. At other sites, seagrasses have been removed to make way for salt production and mariculture; seagrass beds were used for the cultivation of seamoss (*Gracilaria* spp.) in St Lucia. In Guatemala 95 km² were lost between 1965 and 1984 to shrimp and salt production. In St Lucia, dynamite fishing has destroyed seagrass areas. Damage of seagrasses through illegal sand mining from beaches is widespread, particularly in the smaller islands. Sand mining suspends sediments and alters local hydrodynamics. The seagrasses at Ambergris Cay, Belize, have been damaged by dredging (sand extraction and deposition), and Belize has suffered coastal erosion because of sand (and seagrass) removal.

Pollution from land-based sources varies from country to country. The greatest threats are from eutrophication (sewage and agricultural fertilizers), hydrocarbons, pesticides and other toxic wastes[17]. Eutrophication is characterized according to type of effluent discharge, being diffused through freshwater surface runoff (for example, rivers), distinct point sources (effluents from sewage treatment plants) or multiple point sources (such as submarine springs connected to the aquifers that are contaminated by land-based human activities). The detrimental effects of diffuse river loads are exacerbated by erosion of watersheds caused by deforestation, urbanization and agricultural activities. Throughout the Caribbean, the impact of moderate eutrophication on seagrass ecosystems results in changes in community structure or species composition and higher productivity of epiphytes on seagrass leaves. As nutrient loads increase, epiphytes or drifting algal masses become more abundant, resulting in a decline in seagrass shoot density, leaf area and biomass.

Rapidly increasing development throughout the Caribbean will result in an ever-increasing load of wastewater nutrients into coastal marine environments; such loading has already been particularly damaging to seagrasses at Curaçao, at Antigua and Barbuda and in Jamaica. Oil is drilled in the southern region of the Caribbean with Venezuela and Trinidad and Tobago being the principal producing countries[17]. In 1986, 8 million liters of crude oil spilled onto

---

Case Study 23.2
PARQUE NATURAL TAYRONA, BAHÍA DE CHENGUE, COLOMBIA

The small bay (3.3 km²) of Bahía de Chengue is situated on the Caribbean coast of Colombia in the Parque Natural Tayrona[9]. The bay contains sedimentary beaches, rocky shores, small lagoons and small rivers. Mangroves, coral reefs and seagrass beds consisting mainly of *Thalassia testudinum* occur in the bay. Four other seagrass species (*Syringodium filiforme*, *Halodule wrightii*, *Halophila baillonii* and *Halophila decipiens*) are also found in the bay. *Syringodium filiforme* can form monospecific patches.

Corals of the genera *Manicina*, *Siderastrea*, *Millepora*, *Diaporia*, *Porites* and *Cladocora* grow within the *Thalassia testudinum* beds. Sea urchins and seaweeds, such as *Halimeda opuntia*, are common within the beds.

Biological diversity is considered to be high at Parque Natural Tayrona because of the range of habitats. For example 372 fish species have been reported for this area. *Thalassia testudinum* biomass has been estimated as between 631 and 1831 g dry weight/m² with green leaves representing less than 10 percent of that weight. Productivity has been estimated as 1.71-5.36 g dry weight/m²/day. Twenty-six shrimp species have been identified within the *Thalassia testudinum* beds.

The importance of the site lies in its proximity to Santa Marta and the Instituto de Investigaciones Marinas y Costeras (INVEMAR) and because it is relatively well preserved. Only one family lives on the bay and there is no road, so tourists are rare. There is fishing and small-scale salt mining in the bay, and fishermen sometimes use dynamite to fish. Gill nets and beach seines are extended across the seagrass beds.

Bahía de Chengue is a CARICOMP site and is thus regularly monitored. Such programs usually attract future studies to the regions where they are initiated. This seagrass site is representative of the region, although it is relatively protected from human impacts and receives special attention from scientists.

Case Study 23.3
PUERTO MORELOS REEF NATIONAL PARK

Puerto Morelos Reef National Park (Parque Nacional Arrecifes Puerto Morelos) is situated at 21°00'00" and 20°48'33"N, and 86°46'39"W and has an area of ca 90 km². It extends along the northern part of an extensive barrier-fringing reef complex that runs from Belize to the Yucatán Strait (Mexico), the second largest barrier-fringing reef complex in the world. In this park, three seagrass species and 264 macroalgal species have been reported, together with 669 species of marine invertebrate and vertebrate fauna[25]. The dominant ecosystems are coral reefs, seagrass beds and the inland mangroves which are separated from the marine environment by sand berms 2-3 m high. During periods of exceptionally heavy rains, overflow of mangrove wetlands exports brackish tannin-colored waters into the lagoon. The Yucatán limestone is extremely karstic, and rainwater rapidly infiltrates into the aquifer, resulting in the absence of surface drainage or rivers. Rainfall varies between 1.1 and 1.3 m per year, and the water passes through an immense network of underground caves and channels to vent into marine coastal areas through submarine springs (*ojos del agua*) and fissures. Thus, the lagoon environment is principally governed by marine conditions. The water in the lagoon is oligotrophic: low mean nitrite (0.06 µM), nitrate (13.9 µM) and phosphate (0.46 µM) concentrations were recorded during 1982 and 1983. Salinity varies little throughout the year, generally fluctuating between 35.8 and 36.2 psu. Surface water temperature varies seasonally, from ca 26°C in the winter (extreme minimum of 12.5°C) to 31°C in the summer (extreme maximum 34.5°C). The vegetation in the lagoon largely consists of *Thalassia testudinum*, accompanied by *Syringodium filiforme* (occasionally *Halodule wrightii*) and rhizophytic and calcareous algae growing on coarse carbonate sand. *Halodule wrightii* forms very narrow fringing zones near the beach. During 1990-91, total biomass of *Thalassia testudinum* in Puerto Morelos reef lagoon attained annual mean values of 573 g dry weight/m² in a back-reef station, 774 g dry weight/m² in a coastal fringe area and 811 g dry weight/m² in a lush bed in a mid-lagoon station: leaf biomass constituted between 4.8 and 8.6 percent of total biomass[26]. Total biomass of *Syringodium filiforme* or *Halodule wrightii* is usually small when growing intermixed with *Thalassia testudinum* (between ca 20 and 250 g dry weight/m²), but can attain high values (>500 g dry weight/m²) in small monospecific patches or fringes[27].

The Mexican Caribbean coast of the Yucatán Peninsula has undergone immense growth over the last four decades. It is now one of the premier destinations for resort tourism within the Caribbean. Amongst the major attractions are the crystal-clear seas, the white-sand beaches and the reef ecosystems. The reefs of Puerto Morelos received the status of National Park through presidential declaration in February 1998. The effects of increased population pressure throughout the region have been substantial. Puerto Morelos has changed from a tiny fishing village to a rapidly growing community of approximately 3 000 residents and 2 500 hotel rooms[25]. Several significant potential sources of nutrients occur in the region: hotels, intensive farming, rubbish disposal and residences.

Although the Puerto Morelos Reef National Park lagoon is relatively pristine, the increasing pressure of human development is starting to be noticeable. The village of Puerto Morelos is not yet equipped with a central sewer system, and wastes are discharged into septic tanks or directly, without any treatment, into holes in the ground. These land sources of nutrients can enter the water table and flow through to the reef lagoon kilometers away. Reefs in the coastal seas thrive under low natural nutrient concentrations (also the reason why the waters are so clear), which implies that any increase in nutrient input into these areas may cause drastic changes in the coastal ecosystems in the near future.

Thalassia testudinum shoot with a female flower.

seagrasses at Bahía Las Minas, Panama[18]. *Thalassia testudinum* initially suffered blade damage and browning but eventually recovered, except for a 20-90 cm wide shoreward margin where the seagrass died off. *Syringodium filiforme*, which proved to be more sensitive, still had lower biomass two to three years after the event. The density of seagrass infauna was reduced by a factor of three at oiled sites. Additionally, effluent from bauxite mining has been reported to damage seagrass beds in Jamaica, Suriname, Guyana, the Dominican Republic and Haiti[17].

Often different factors of human impact act synergistically and together they are responsible for severe seagrass loss. For example, overfishing of wrasses and triggerfish off the coast of Haiti and the US Virgin Islands caused an explosion in sea urchins which then destroyed the seagrass beds by overgrazing. In Jamaica, urban and industrial pollution, dredging of canals, landfilling, bauxite mining, oil spills, channelization, urban runoff, urban sewage, construction of river bulkheads and docks, artificial beach nourishment, thermal effluents and cement tailings all degrade seagrass ecosystems[19]. Other seagrass beds near industrial areas are also highly impacted, such as those at Lake Maracaibo (Venezuela), the "El Mamonal" industrial complex (Cartagena Bay, Colombia), the west coast of Trinidad and Havana Bay in Cuba[17]. The replanting of seagrasses has successfully mitigated a fraction of these impacts.

As seagrasses actively form and maintain extensive subtidal flat structures in the Caribbean, there is concern about the effects of global warming and sea-level rise on seagrasses. Models of global climate change predict considerable changes for the coastal environments of the Caribbean, including rising sea level, increasing water temperature and more frequent hurricanes. Seagrasses should be able to maintain vertical rates of habitat accretion in pace with predicted rises in sea level until at least the middle of this century, and a rise in sea level is not expected to seriously affect the predominant species unless a general deterioration of the habitat occurs[20].

POLICY, REGULATION, PROTECTION

In the Caribbean, two major intergovernmental efforts to protect the environment can be singled out: the Caribbean Environment Programme of the United Nations Environment Programme (CEP-UNEP) and the Meso American Barrier Reef Project (Sistemas Arrecifales Mesoamericanos) which forms part of the Corredor Biológico Mesoamericano (CBM). Both include countries of the Caribbean which have recognized the desirability of managing marine coastal areas; seagrasses are included, but not singled out for

Sand dunes adjacent to *Halodule wrightii* and *Thalassia testudinum* beds, Florida, USA.

specific environmental legislation. Most countries are formulating or have formulated marine management system plans. Systems of marine protected areas vary from country to country, but most include seagrasses. Of the 31 fully managed marine protected areas of the Caribbean, 24 (74 percent) include seagrasses[21].

The CARICOMP network (Caribbean Coastal Marine Productivity network) was set up to monitor coral reefs, mangroves and seagrasses[7]. In this network, associated marine laboratories and conservation units work together, using standardized techniques to measure the target ecosystems. CARICOMP was established in 1985, the associated network in 1990 and operation of the network was initiated at the end of 1992. The stated aims of CARICOMP are "to determine the dominant influences on coastal productivity, to monitor for ecosystem change, and ultimately to discriminate human disturbance from long-term natural variation in coastal systems over the range of their distribution". To these ends, CARICOMP is coordinated through a Data Management Centre at the University of the West Indies in Jamaica. Seagrass parameters such as biomass, areal productivity and turnover, shoot density, leaf width and length, and leaf area indices are measured twice yearly. CARICOMP involves 27 institutions in 17 countries where seagrasses are being, or will be, monitored[7].

ACKNOWLEDGMENTS

We would like to acknowledge the help and information provided by Phanor Montoya-Maya, Eduardo Klein, Daisy Pérez and Ricardo Bitter-

Soto. The first author was supported by fellowships from CNPq and UERJ (Prociência) during the preparation of this chapter. The third author is grateful to Tracy Blanchon for critically reviewing the manuscript.

AUTHORS

Joel C. Creed, Laboratório de Ecologia Marinha Bêntica, Departamento de Ecologia, Instituto de Biologia Roberto Alcântara Gomes, Universidade do Estado do Rio de Janeiro – UERJ, PHLC Sala 220, Rua São Francisco Xavier 524, CEP 20559-900, Rio de Janeiro RJ, Brazil. **Tel:** +55 (0)21 2587 7328. **Fax:** +55 (0)21 2587 7655. **E-mail:** jcreed@openlink.com.br

Ron C. Phillips, 3100 So. Kinney Road #77, Tucson, Arizona 85713, USA.

Brigitta I. Van Tussenbroek, Unidad Acdemica Puerto Morelos, Instituto de Ciencias del Mar y Limnología, Universidad Nacional Autónoma de México, Apdo. Postal 1152, Cancún, 77500, Quintana Roo, Mexico.

REFERENCES

1. Phillips RC [1992]. The seagrass ecosystem and resources in Latin America. In: Seeliger U (ed) *Coastal Plant Communities of Latin America*. Academic Press, San Diego, California. pp 107-121.
2. Littler DS, Littler MS [2000]. *Caribbean Reef Plants*. Offshore Graphics, Washington.
3. UNESCO [1983]. *Coral Reefs, Seagrass Beds and Mangroves: Their Interaction in the Coastal Zones of the Caribbean*. UNESCO Reports in Marine Science 23.
4. Kilar JA, Norris JN [1988]. Composition, export, and import of drift vegetation on a tropical, plant-dominated, fringing-reef platform (Caribbean, Panama). *Coral Reefs* 7: 93-103.
5. Freile D, Hillis L [1997]. Carbonate productivity by *Halimeda incrassata* in a land proximal lagoon, Pici Feo, San Blas, Panama. *Proceedings of 8th International Coral Reef Symposium* 1: 767-772.
6. Patriquin DG [1975]. "Migration" of blowouts in seagrass beds at Barbados and Carriacou, West Indies, and its ecological and geological implications. *Aquatic Botany* 1: 163-189.
7. UNESCO [1998]. *CARICOMP – Caribbean Coral Reef, Seagrass and Mangrove Sites*. Coastal Region and Small Island Papers 3.
8. Williams SL [1990]. Experimental studies of Caribbean seagrass development. *Ecological Monographs* 60: 449-469.
9. Van Tussenbroek BI, Brearley A [1998]. Isopod burrowing in leaves of turtle grass, *Thalassia testudinum*, in a Mexican Caribbean reef lagoon. *Marine and Freshwater Research* 49: 525-531.
10. Sturm MG de L [1991]. The living resources of the Caribbean Sea and adjacent regions. *Caribbean Marine Studies* 2: 18-44.
11. Rodriguez RW, Webb RMT, Bush DM [1994]. Another look at the impact of Hurricane Hugo on the shelf and coastal resources of Puerto Rico, USA. *Journal of Coastal Research* 10: 278-296.
12. Glynn PW, Almodovar LR, Gonzalez JG [1964]. Effects of hurricane Edith on marine life in La Parguera, Puerto Rico. *Caribbean Journal of Science* 4: 335-345.
13. Van Tussenbroek BI [1994]. The impact of Hurricane Gilbert on the vegetative development of *Thalassia testudinum* in Puerto Morelos reef lagoon, Mexico: A retrospective study. *Botanica Marina* 37: 421-428.
14. Perez D, Galindo L [2000]. Effects of hyposalinity in *Thalassia testudinum* (Hydrocharitaceae) from Parque Nacional Morrocoy, Venezuela. *Revista Biologia Tropical* 48: (Supplement) 243-249.
15. UNEP [1997]. *Global Environment Outlook*. Oxford University Press, New York.
16. Vera B [1992]. Seagrasses of the Venezuelan coast: Distribution and community components. In: Seeliger U (ed) *Coastal Plant Communities of Latin America*. Academic Press, San Diego, California. pp 135-140.
17. UNEP [1994]. Regional Overview of Land-based Sources of Pollution in the Wider Caribbean Region. CEP Technical Report No. 33.
18. Marshall MJ, Batista V, Matias D [1993]. Effects of the 1986 Bahia Las Minas, Panama, oil spill on plants and animals in seagrass communities. In: Keller BD, Jackson JBC (eds) Long-term Assessment of the Oil Spill at Bahia Las Minas, Panama Synthesis Report. US Department of the Interior Minerals Management Service OCS Study MMS 93-0048.
19. Thorhaug A, Miller B, Jupp B, Bookers F [1985]. Effects of a variety of impacts on seagrass restoration in Jamaica. *Marine Pollution Bulletin* 16: 355-360.
20. UNEP [1993]. Ecosystem and Socioeconomic Response to Future Climatic Conditions in the Marine and Coastal Regions of the Caribbean Sea, Gulf of Mexico, Bahamas, and the Northeast Coast of South America. CEP Technical Report No. 22.
21. Kelleher G, Bleakley C, Wells S [1995]. *A Global Representative System of Marine Protected Areas*. Vol. 4. Environment Department, The World Bank, Washington.
22. Morris LJ, Virnstein RW, Miller JD, Hall LM [1999]. Monitoring seagrass changes in Indian River Lagoon, Florida using fixed transects. In: Bortone SA (ed) *Seagrasses: Monitoring, Ecology, Physiology, and Management*. CRC Press, Boca Raton. pp 167-176.
23. Waycott, M. Personal communication.
24. Heidelbaugh WS, Hall LM, Kenworthy WJ, Whitfield P, Virnstein RW, Morris LJ, Hanisak MD [1999]. Reciprocal transplanting of the threatened seagrass *Halophila johnsonii* (Johnson's seagrass) in the Indian River Lagoon, Florida. In: Bortone SA (ed) *Seagrasses: Monitoring, Ecology, Physiology, and Management*. CRC Press, Boca Raton. pp 197-210.
25. INE (Instituto Nacional de Ecología, Comunidad de Puerto Morelos, Quintana Roo) [2000]. Programa de manejo del Parque Nacional Arrecife de Puerto Morelos. Instituto Nacional de Ecología, México, DF. 222 pp.
26. Van Tussenbroek BI [1998]. Above- and below-ground biomass and production of *Thalassia testudinum* in a tropical reef lagoon. *Aquatic Botany* 61: 69-82.
27. Gallegos ME, Merino M, Rodríguez A, Marbà N, Duarte CM [1994]. Growth patterns and demography of pioneer Caribbean seagrasses *Halodule wrightii* and *Syringodium filiforme*. *Marine Ecology Progress Series* 109: 99-104.

24 The seagrasses of SOUTH AMERICA: BRAZIL, ARGENTINA AND CHILE

J.C. Creed

The geographical region considered includes the coastlines and islands of the following countries: Brazil, Uruguay, Argentina, Chile, Peru and Ecuador. These coastlines stretch from about 5°N to 57°S and from 82°W to 34°W; thus they are influenced by both the Pacific and Atlantic Oceans.

Phillips[1] stated in 1992 in a review of seagrass ecosystems and resources of Latin America that "the most basic research is needed in almost every place". During the subsequent decade, the number of published reports of surveys, biology and ecology of the seagrasses of South America has increased somewhat. However, despite the fact that some progress has been made with respect to Phillips's comment, it must be recognized that much important information remains unavailable as gray literature or, as is often the case, no information has been collected.

ECOSYSTEM DESCRIPTION

Although South America's seagrasses are the subject of some taxonomic debate, at least six seagrass species have been reported for the region; nearly all are restricted to the Atlantic coast. In the southwest Atlantic, seagrasses are common but rarely form very extensive meadows. Remarkably, the only seagrasses known on the Pacific coast of South America are a couple of small populations of *Heterozostera tasmanica* (now *Zostera tasmanica*) in northern Chile at Coquimbo[2]. Intriguingly, this species is otherwise known only from Australia; it has been suggested that these are remnants of formerly widely distributed Chilean populations[1]. No seagrasses are known for Peru or Ecuador.

Shoalgrass (*Halodule wrightii*) and sea vines (*Halophila decipiens*) have a tropical-subtropical distribution in the southwest Atlantic, stretching from the Caribbean to the Brazilian states of Paraná and Rio de Janeiro, respectively. Surprisingly, *Syringodium filiforme* and *Thalassia testudinum*, which one might also expect to find, are restricted to the Caribbean and are not found in the southwest Atlantic. Consequently, in the northeast of Brazil *Halodule wrightii* is able to form large monospecific beds, such as those found at Itamaracá Island.

In Brazil, *Halodule wrightii* is the seagrass that most frequently occurs and has the widest distribution. *Halophila baillonii*, which is also found widely in the Caribbean, has been reported twice (in 1888 and in the 1980s, though not since) at Itamaracá Island, in the northeast of Brazil. Large extents of the Brazilian coastline have no recorded seagrasses, either because seagrasses have not yet been found or because they are not present. The continental shelf is sometimes as narrow as 15 km, and for this and other reasons extensive reefs have not developed in Brazil. Consequently, along most of the coast *Halodule wrightii* is restricted to estuaries, bays and other protected ecosystems.

Where the continental platform widens, such as in the Abrolhos region of Brazil, the formation of extensive reefs allows the establishment of seagrasses, which thrive in these protected areas. Here, beds of *Halodule wrightii* (2-7 m) and *Halophila decipiens* (5-22 m) become more common. *Halophila decipiens* reaches its southernmost limit in the Atlantic Ocean in Guanabara Bay at Rio de Janeiro, right under the famous Sugarloaf Mountain. Because it can occupy deeper waters, *Halophila decipiens* may be of greater ecological importance than has previously been thought, but its distribution is still not very well known. As these species reach the southeast of Brazil they form smaller and more isolated populations.

Halodule emarginata is a species endemic to Brazil, and forms small populations from northeast to southeast Brazil. However, there is no agreement as to whether *Halodule emarginata* should be maintained as a distinct species from *Halodule wrightii*. Leaf-tip characteristics are used to distinguish the species, which rarely flower or fruit.

Ruppia maritima is found sporadically from Brazil down to Argentina, where it forms the southernmost populations of seagrass in the world, at the Magellan Straits[3]. Such records reflect the species' wide latitudinal distribution and tolerance to variable environmental conditions, as it can be found growing in coastal lagoons and estuaries with salinities from 0 to 39 psu. At the Patos Estuarine Lagoon in southern Brazil, a large (about 120 km^2) area of *Ruppia maritima* dominates the benthos and local primary productivity.

An unattached leaf of what was reported as *Zostera* has been found at Montevideo, Uruguay. Phillips[1] commented that the leaf tip resembled that of *Heterozostera tasmanica* but that it was unlikely that it came from so far away as Chile.

The dearth of studies dealing with seagrasses in South America can be exemplified by the Brazilian experience, although Brazil has the most studies available. After a pioneering study of the biota associated with *Halodule wrightii* by Kempf[4], few studies were carried out until the 1980s. From 1980 until the present, an average of 22 additional companion species a year have been reported associated with the Brazilian seagrasses and the trend suggests that this rate will continue[5], demonstrating that seagrass habitats are attracting research effort (Figure 24.1). However only one or two research papers relating to seagrasses of the region are published each year.

Case Study 24.1
ITAMARACÁ ISLAND, NORTHEAST BRAZIL

The regional importance of the populations of seagrasses at Itamaracá Island, Pernambuco State (7°45'S, 34°50'W) has been recognized for some time[4, 9, 13]. At Itamaracá there are large expanses of *Halodule wrightii* on the eastern side of the island in shallow-water flats protected from the open sea by reefs. Locally, *Halodule* is known as *capim agulha* which means needlegrass. In 1967, *Halodule* stretched approximately 1.2 km seaward in a mapped portion 1.2 km wide (1.4 km^2), although the total area is greater but unknown[6]. *Halophila baillonii*, only found at Itamaracá in Brazil, is known from collections in the Santa Cruz Channel made in 1888 and in the 1980s, and has not been found since[9].

ECOLOGICAL INTERACTIONS
There are ecological interactions between the seagrasses and the local mangrove, estuarine and reef systems. The fauna is taxonomically diverse: over 100 macrofaunal and 46 epiphyte floral taxa have been identified in or on *Halodule* at Itamaracá[6]. Amongst the *Halodule* beds at Itamaracá are clams, shrimp, lobsters, stone and blue crabs, and common and ballyhoo halfbeaks, all of which are fished recreationally and commercially. The region's seagrass beds are feeding grounds for the West Indian manatee. About 12 metric tons a year of *Halodule wrightii* are collected for fodder for captive-reared manatees at the Centro de Pesquisas do Peixe-boi Marinho (National Manatee Research Centre) run by the Instituto Brasileiro do Meio-Ambiente e Recursos Naturais Renováveis IBAMA, which is located on the island.

The area of Itamaracá is one of the most productive fisheries in the state of Pernambuco, and the seagrasses contribute as a nursery and foraging area. Local inhabitants have reported regression in the areal coverage of *Halodule* at Itamaracá. Furthermore, the local fishermen report that a reduction in the area of *Halodule* has resulted in a drop in fisheries production, especially of prawn and halfbeaks. It has been suggested that coastal development, bad landuse practices, landfill, pollution and an increase in tourism are responsible. Local researchers have recommended that environmental education programs be implemented to help preserve and manage the ecosystem.

Three basic research needs have been highlighted at Itamaracá Island:
o   the realization of a survey of the distribution of the seagrass meadows and their associated flora and fauna;
o   the identification of impacting anthropogenic agents;
o   the development of research programs which identify the regional ecological importance of seagrasses[6].

The Itamaracá area stands out because:
o   there is a co-occurrence of seagrass species;
o   *Halodule wrightii*, a known pioneer in the Caribbean seagrass succession, here is the climax dominant;
o   the seagrass beds are of commercial importance and may be suffering die-back due to human activities.

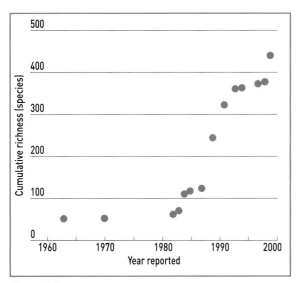

Figure 24.1
Cumulative number of companion species to the Brazilian seagrasses reported since 1960[20]

Map 24.1
South America

BIOGEOGRAPHY

South American seagrasses are often found near to, or closely trophically linked with, other marine and coastal ecosystems and habitats, and this juxtaposition results in heightened diversity. For example, recent studies which have compared the macrofauna associated with *Halodule wrightii* and *Halophila decipiens* beds with nearby areas devoid of vegetation have shown that total density, richness and the diversity of the infauna is enhanced by the presence of seagrasses. Coral reefs extend from the Caribbean to Abrolhos, Brazil, and mangroves to Santa Catarina State, Brazil, where they are replaced by salt marshes and mud flats.

Halodule is associated with shallow habitats without much freshwater input, such as reefs, algal beds, coastal lagoons, rocky shores, sand beaches and unvegetated soft-bottom areas and nearby mangroves without too much salinity fluctuation. *Halophila* is associated with deeper reefs, algal and marl beds, and deeper soft-bottom vegetated areas. *Ruppia maritima* can be found in low-salinity (coastal lagoon, estuary, fishpond, mangrove, salt marsh and soft-bottom unvegetated) and high-salinity (coastal lagoon, salt pond, soft-bottom unvegetated) habitats.

Seagrass beds in South America are known to be important habitat for a wide variety of plants and animals. About 540 taxa (to genus or species level) of organisms associated with the Brazilian seagrasses have been compiled by Creed[5]. The groups that contributed most to species diversity are polychaetes, fish, amphipods, decapods, foraminifera, gastropod and bivalve mollusks, macroalgae and diatoms.

Two threatened species which feed directly on seagrasses from the Caribbean to Brazil are the green turtle *Chelonia mydas*[5] and the West Indian manatee *Trichechus manatus*[6]. Both have benefited from specific conservation action sponsored privately and by the Brazilian environmental agency IBAMA (green turtles by the Projeto TAMAR and manatees by the Projeto Peixe-Boi Marinho). The black-necked swan *Cygnus melancoryphus* and the red-gartered coot *Fucila armillata* also feed directly on *Ruppia maritima* in southern Brazil and Argentina but are not endangered[7]. Recently, the semi-aquatic capybara *Hydrochaeris hydrochaeris*, which is the world's largest rodent, was observed feeding on *Ruppia maritima* near Rio de Janeiro. Amongst the fauna which use seagrasses as a habitat are two corals, *Meandrina brasiliensis* and *Siderastrea stellata*. They grow unattached in *Halodule wrightii* beds and are sold as souvenirs locally.

While the seagrasses of South America contribute to coastal protection and local productivity, and thus fisheries, there is hardly any information available about the value of seagrasses to the local economy. Economically important fish species such as the bluewing searobin (*Prionotus punctatus*), whitemouth croaker (*Micropogonias furnieri*) and mullet (*Mugil platanus*) are found and fished in Brazilian seagrass beds. Local fisheries exploit commercially important crustaceans such as blue

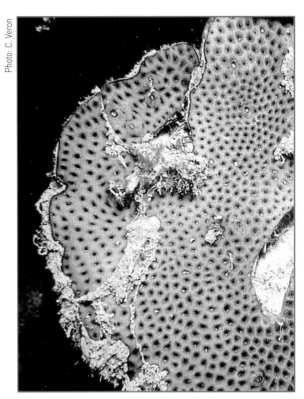

Siderastrea stellata, a coral that occurs in a wide range of shallow habitats and is common in seagrass beds.

Ruppia maritima is found sporadically from Brazil down to Argentina.

crabs (*Callinectes sapidus*), stone crab (*Menippe nodifrons*), lobster (*Panulirus argus* and *Panulirus laevicauda*) and shrimp (*Penaeus brasiliensis* and *Penaeus paulensis*), all associated with seagrass beds. Other shellfish which are commercially collected from seagrass beds are clams (*Anomalocardia brasiliana, Tagelus plebeius, Tivela mactroides*), volutes (*Voluta ebraea*), rockshells (*Thais haemastoma*), oysters (*Ostrea puelchana*) and cockles (*Trachycardium muricatum*)[5]. In Chile, the Chilean scallop (*Argopecten purpuratus*) preferentially settles in *Zostera tasmanica* beds[8].

HISTORICAL PERSPECTIVES

As no detailed seagrass mapping work has been carried out in South America, there is little anecdotal or factual information about changes in seagrass distribution and abundance. It is thought that the *Zostera tasmanica* beds in Chile are historically remnant populations of much larger meadows[1]. Researchers from the Universidade Federal Rural of Pernambuco are currently mapping *Halodule wrightii* beds at Itamaracá Island, Brazil. This should allow some measure of loss over the last 40 years to be estimated, as the area was partially mapped before, in the 1960s. For now, the only quantified seagrass loss is of *Halodule wrightii* beds at Rio de Janeiro. These were listed by Oliveira *et al.*[9] and were revisited ten years later[10]. Seagrass was no longer found at 16 percent of these sites. Losses are not due to direct use of seagrasses, as the only known use of *Halodule wrightii* is to feed captive-reared manatees.

PRESENT DISTRIBUTION

Estimating the real area of seagrass cover in South America is at present almost impossible because of the dearth of studies. The little data available allow only an educated "best guess". Brazil probably has about 200 km^2 of seagrasses, Chile 2 km^2 and Argentina about 1 km^2.

PRESENT THREATS

As Phillips[1] pointed out, all seagrasses found from Mexico to southern Brazil are species characteristic of the western tropical Atlantic Ocean, so management problems concerning them should be relatively similar. In fact, some general observations of threats to marine coastal ecosystems are pertinent. The population of Latin America continues to grow, from 179 million to 481 million between 1950 and 1995[11]. The concentration of population growth in urban areas and marginal agricultural lands is the main factor responsible for pressures exerted by human population on the environment[12]. In South America, this pressure is concentrated in the coastal cities. The continent has several large urban centers: São Paulo, Brazil, population in 2000 27.9 million (second largest city in the world); Buenos Aires, Argentina, 11.4 million (12th); Rio de Janeiro, Brazil, 10.2 million (16th)[12]. By 2020, over 80 percent of the population of South America is expected to live in urban areas[11]. Such concentrations put incredible stresses on the coastal marine environment. Human activity affects the environment in three major ways: landuse and landcover change, the extraction and depletion of natural resources, and the production of wastes.

It is necessary at least to know the distribution of seagrasses in order to assess potential or real threats to them. This is a problem when considering the seagrasses of South America. Direct reports of impacts on seagrasses on the continent are few. However, pollution by heavy metals from sporadic mining and metalworking activities, by polychlorinated biphenyl congeners and organochlorine compounds, and by nutrients from agricultural runoff and sewage discharge have all been reported. Effects of physical damage by anchors and trampling on seagrass and associated macroalgae have also been identified. Loss of water area, because of sediments produced after erosion due to deforestation, infilling for construction and dredging activities, has also reduced the area occupied by South America's seagrasses. *Ruppia*

Case Study 24.2
ABROLHOS BANK, BAHIA STATE, NORTHEAST BRAZIL

The Abrolhos Bank is formed by a widening of the continental shelf at the extreme south of the state of Bahia (18°S, 38°W). The area consists of an inner line of reef banks, a small archipelago of five islands with embryonic fringing reefs, and outer reef banks[14]. The Abrolhos Marine National Park protects the archipelago, surrounding waters and an inshore reef system. Peculiar to the area are reef columns called *chapeirões* which typically extend from a depth of about 20 m to the surface. Nearshore areas are subjected to higher turbidity because of local river inputs but offshore areas are characterized by less turbid waters.

Despite the considerable research interest invested in the national park, until recently[15] seagrasses were overlooked and not reported. In fact, *Halodule wrightii* and especially *Halophila decipiens* are more common than previously believed. *Halodule wrightii* is found in shallow sandy areas interspersed with coastal reefs and around the Abrolhos Archipelago, while *Halophila decipiens* is found down to at least 22 m. The suspicion that *Halophila decipiens* may be very abundant on the Abrolhos Bank was confirmed in a recent rapid assessment protocol of biodiversity carried out in the region[16]. Of the 45 reef edge/soft-bottom sites selected, *Halophila* was present at 18 (40 percent). Although no total area quantification was made, these sites were distributed over a study area of about 6000 km², so the potential importance of *Halophila decipiens* in the region, especially in terms of primary productivity, could be enormous. Very little is known about the biology or ecology of *Halophila decipiens* in Brazil.

The Abrolhos Bank has important reef-based and open-sea fisheries. In Abrolhos, there are trophic interactions between seagrass beds and reefs across distinct grazing halos. Large vertebrates, such as sea chub, parrotfish and surgeonfish (seagrass stomach contents: *Kyphosus* spp. 12 percent; *Acanthurus chirurgus* 8 percent; *Sparisoma* and *Scarus* 0.5-5 percent[17]) and green turtles (*Chelonia mydas* which has been observed taking 32 bites of *Halodule wrightii* per minute *in situ*) heavily graze the seagrass and 56 associated seaweed species. Predatory fish of commercial importance hunt over the seagrass beds and juvenile yellowtailed snapper and angelfish live in the seagrass. The Abrolhos Archipelago receives about 900 tourist boat visits per year and is an important ecotourist destination[15]. Despite being protected within the national park, the seagrass beds have lost 0.5 percent per year because of anchor damage, showing reduced seagrass density and a change in the community structure which can take more than a year to recover after a single impact[15]. Buoys were installed recently which should alleviate the problem, but despite its desirability no transplantation has been carried out to mitigate the losses suffered so far.

Redonda Island, Abrolhos Archipelago, Brazil.

Photo: J.C. Creed

maritima has suffered from reduced freshwater inputs because of rice irrigation, population growth and lock construction.

Acknowledging that there are intrinsic human-related pressures on the coastal zone, and that these have been quantified for the South American continent, the superimposition of known threat potential onto the known seagrass distribution can provide a measure of the threat to South American seagrasses. Using these criteria, 100 percent of Chilean and Argentine, and 40 percent of Brazilian seagrasses are "highly threatened". Thirty-six percent of Brazil's seagrasses are "moderately threatened" and 24 percent are in "low threat" areas.

POLICY, REGULATION, PROTECTION

Seagrasses are not specifically protected by legislation in Brazil, Chile or Argentina but are covered by resource

Case Study 24.3
RUPPIA MARITIMA IN THE PATOS LAGOON SYSTEM

The *Ruppia maritima* meadows in the Patos Lagoon, near the city of Rio Grande, Rio Grande do Sul (32°S, 52°W) have received more research attention than any other seagrass system in South America. The Patos Lagoon consists of shallow bays with mean annual water depths of 20-70 cm. The *Ruppia maritima* meadows occupy an area of about 120 km^2 of the estuary system. Considerable scientific knowledge has been accumulated about these seagrass beds during the last 25 years, and has recently been summarized[7,18]. *Ruppia* forms extensive beds in these shallow marginal bays with both annual and perennial populations, depending on local environmental factors, interspecific (algal) shading and epiphyte fouling[19]. When compared with other seagrass habitats worldwide, light attenuation in the waters of the Patos Lagoon is relatively high, and consequently *Ruppia* is restricted to relatively shallow water[7]. This imposes seasonal influences on primary productivity; net annual primary productivity of *Ruppia maritima* has been estimated as 39.6-43.2 g carbon/m^2/year[7].

The *Ruppia* meadows interface with sandy shorelines, unvegetated tidal flats and salt marsh habitats. The large areas of *Ruppia* meadows serve as complex habitats for a local fishery by providing substrate, refuge, nursery and feeding grounds. Associated drift algae can also be locally abundant alternative habitats. Pink shrimp (ca 2800 metric tons landed annually in the region) and the blue crab (ca 1400 tons), found foraging in the seagrass, are important local artisanal fishery resources. Whitemouth croaker (ca 7500 tons) and mullet (ca 2300 tons) also use the *Ruppia* beds as nursery or foraging grounds. The stout razorclam (*Tagelus plebius*) is another commercially important species. Predators such as the bottlenose dolphin are common (31-100 individuals within the lagoon system) in the Patos Lagoon and feed principally on whitemouthed croaker which is found in the *Ruppia* beds. Eleven percent of the water area has been lost since the 1700s and this includes substantial areas of *Ruppia* meadow; many anthropogenically filled areas were previously inhabited by seagrasses. Areas of preservation, conservation and development have been proposed for the region. *Ruppia* beds would be partially protected under such a proposal. However, "management efforts of the Patos Lagoon estuary are hampered by technical and legal problems"[7]. The *Ruppia* beds continue to be studied as part of the Brazilian Long Term Ecological Research Program (PELD). Locally *Ruppia maritima* is called *lixo-capim* which means weedgrass.

Ruppia maritima.

management and conservation legislation. Brazil has comprehensive environmental legislation. The Federal Constitution of 1988 dedicates a chapter exclusively to the environment. The federal government has amongst other items the responsibility to "preserve and restore essential ecological processes and promote the ecological management of species and ecosystems ... preserve the diversity and integrity of genetic resources ... protect the flora and fauna". The coastal zone is recognized as a national resource by this constitution. In 1998, Congress approved the Environmental Crimes Law. It regulates crimes against the natural environment. Although algal beds, coral reefs and mollusk beds are specifically mentioned, seagrasses are not.

Brazil has a complex management system based on the creation of conservation units at federal, state and municipal government levels. Approximately 290 conservation units are recognized in the coastal zone, which represent about 21 million hectares that have specific legislation. Of these, seagrasses are found in both Marine National Parks (Abrolhos and Fernando de Noronha) as well as numerous ecological stations, state parks, biological reserves and environmental protected areas.

ACKNOWLEDGMENTS

I would like to acknowledge the help and information provided by the following colleagues: Alejandro Bortolus (from Argentina), Carlos Eduardo Leite Ferreira, Eduardo Texeira da Silva, Erminda da Conceição Guerreiro Couto, Karine Magalhães, Marcia Abreu de Oliveira Figueiredo and Ulrich Seeliger (from Brazil), and Evamaria W. Koch and Ronald Phillips (from the USA). The author was supported by fellowships from CNPq and UERJ (Prociência) during the preparation of this chapter.

AUTHOR

Joel C. Creed, Laboratório de Ecologia Marinha Bêntica, Departamento de Ecologia, Instituto de Biologia Roberto Alcântara Gomes, Universidade do Estado do Rio de Janeiro – UERJ, PHLC Sala 220, Rua São Francisco Xavier 524, CEP 20559-900, Rio de Janeiro RJ, Brazil. **Tel:** +55 (0)21 2587 7328. **Fax:** +55 (0)21 2587 7655. **E-mail:** jcreed@openlink.com.br

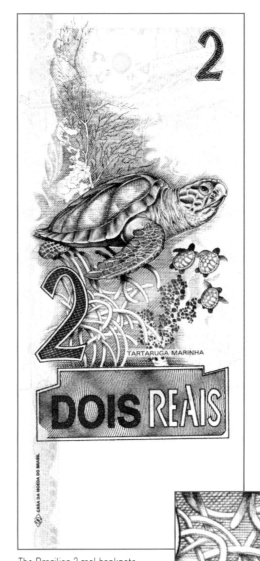

The Brazilian 2 real banknote depicts a turtle, corals, algae and seagrass. A closer look (inset), however, reveals the artist's mistake – the seagrass is probably *Thalassia*, a genus that is not found in Brazil.

REFERENCES

1. Phillips RC [1992]. The seagrass ecosystem and resources in Latin America. In: Seeliger U (ed) *Coastal Plant Communities of Latin America*. Academic Press, San Diego, California. pp 107-121.
2. Gonzalez SA, Edding ME [1990]. Extension of the range of *Heterozostera tasmanica* (Martens ex Aschers) den Hartog in Chile. *Aquatic Botany* 38: 391-395.
3. Short FT, Coles RG (eds) [2001]. *Global Seagrass Research Methods*. Elsevier Science, Amsterdam.
4. Kempf M [1967/9]. Nota preliminar sobre os fundos costeiros da região de Itamaracá (Norte do Estado de Pernambuco, Brasil). *Trabalhos do Instituto Oceanografico da Universidade do Recife* 9/11: 95-110.
5. Creed JC [2000]. The biodiversity of Brazil's seagrass and seagrass habitats: A first analysis. *Biologia Marina Mediterranea* 7: 207-210.
6. Barros HM, Eskinazi-Leça E, Macedo SJ, Lima T (eds) [2000]. Gerenciamento participativo de estuarios e manguezais. Universitaria da UFPE, Recife.
7. Seeliger U, Odebrecht C, Castello JP (eds) [1997]. *Subtropical Convergence Environments: The Coast and Sea of the Southwestern Atlantic*. Springer-Verlag, Berlin.
8. Aguilar M, Stotz WB [2000]. Settlement of juvenile scallops *Argopecten purpuratus* (Lamarck, 1819) in the subtidal zone at Puerto Aldea, Tongoy Bay, Chile. *Journal of Shellfish Research* 19: 749-755.

9. Oliveira EC de, Pirani JR, Giulietti AM [1983]. The Brazilian seagrasses. *Aquatic Botany* 16: 251-267.
10. Creed JC. Personal observations.
11. UNEP [1997]. *Global Environment Outlook*. Oxford University Press, New York.
12. UNEP [1994]. *Regional Overview of Land-based Sources of Pollution in the Wider Caribbean Region*. CEP Technical Report No. 33.
13. den Hartog C [1972]. The seagrasses of Brazil. *Acta Botanica Neerlandica* 21: 512-516.
14. Leao ZMAN, Kikuchi RKP [2001]. The Abrolhos reefs of Brazil. In: Seeliger U, Kjerfve B (eds) *Coastal Marine Ecosystems of Latin America*. Springer-Verlag, Berlin. pp 83-96.
15. Creed JC, Amado Filho GM [1999]. Disturbance and recovery of the macroflora of a seagrass (*Halodule wrightii* Ascherson) meadow in the Abrolhos Marine National Park, Brazil: An experimental evaluation of anchor damage. *Journal of Experimental Marine Biology and Ecology* 235: 285-306.
16. Figueiredo MAO [in press]. Diversity of macrophytes in the Abrolhos Bank, Brazil. Conservation International, Washington, DC.
17. Ferreira CEL. Unpublished data.
18. Seeliger U [2001]. The Patos Lagoon Estuary, Brazil. In: Seeliger U, Kjerfve B (eds) *Coastal Marine Ecosystems of Latin America*. Springer-Verlag, Berlin. pp 167-184.
19. Costa CSB, Seeliger U [1989]. Vertical distribution and resource allocation of *Ruppia maritima* L. in a southern Brazilian estuary. *Aquatic Botany* 33: 123-129.
20. Creed JC. Unpublished data.

Appendix 1: Seagrass species, by country or territory

SPECIES	Algeria	Angola	Anguilla	Antigua and Barbuda	Australia	Azerbaijan	Bahamas	Bahrain	Bangladesh	Barbados	Belize	Bermuda	Brazil	British Indian Ocean Territory	Brunei	Bulgaria	Cambodia	Canada	Cayman Islands	Chile	China	Colombia	Comoros	Costa Rica	Croatia	Cuba	Cyprus	Denmark	Dominican Republic
Amphibolis antarctica					✓																								
Amphibolis griffithii					✓																								
Cymodocea angustata					✓																								
Cymodocea nodosa	✓															✓									✓		✓		
Cymodocea rotundata					✓																		✓						
Cymodocea serrulata					✓																			✓					
Enhalus acoroides					✓									✓	✓														
Halodule beaudettei							✓	✓																					
Halodule bermudensis												✓																	
Halodule emarginata													✓																
Halodule pinifolia					✓																								
Halodule uninervis					✓			✓	✓					✓									✓	✓					
Halodule wrightii		✓		✓			✓			✓	✓	✓										✓	✓	✓		✓			✓
Halophila australis					✓																								
Halophila baillonii											✓											✓							
Halophila beccarii																						✓							
Halophila capricorni					✓																								
Halophila decipiens					✓					✓	✓	✓										✓	✓	✓		✓			✓
Halophila engelmanni																								✓					
Halophila hawaiiana																													
Halophila johnsonii																													
Halophila minor					✓																								
Halophila ovalis					✓			✓						✓								✓	✓						
Halophila ovata					✓																		✓						
Halophila spinulosa					✓									✓															
Halophila stipulacea							✓																✓				✓		
Halophila tricostata					✓																								
Phyllospadix iwatensis																					✓								
Phyllospadix japonicus																													
Phyllospadix scouleri																					✓								
Phyllospadix serrulatus																					✓								
Phyllospadix torreyi																					✓								
Posidonia angustifolia					✓																								
Posidonia australis					✓																								
Posidonia coriacea					✓																								
Posidonia denhartogii					✓																								
Posidonia kirkmanii					✓																								
Posidonia oceanica	✓																								✓		✓		
Posidonia ostenfeldii					✓																								
Posidonia robertsoniae †					0																								
Posidonia sinuosa					✓																								
Syringodium filiforme		✓	✓				✓			✓	✓	✓							✓			✓		✓		✓			✓
Syringodium isoetifolium					✓																		✓	✓					✓
Thalassia hemprichii					✓									✓									✓						
Thalassia testudinum		✓	✓				✓			✓	✓	✓							✓			✓		✓		✓			✓
Thalassodendron ciliatum					✓									✓									✓						
Thalassodendron pachyrhizum					✓																								
Zostera asiatica																					✓								
Zostera caespitosa																													
Zostera capensis																													
Zostera capricorni					✓																								
Zostera caulescens																													
Zostera japonica																		✓			✓								
Zostera marina	✓															✓		✓			✓							✓	
*Zostera mucronata**					0																								
*Zostera muelleri**					0																								
Zostera noltii	✓						✓									✓									✓		✓		
*Zostera novazelandica**																													
Zostera tasmanica					✓																✓								
TOTAL	3	2	2	3	29	1	3	3	1	5	4	3	4	2	4	3	2	5	2	1	12	5	6	4	3	5	3	2	5

Notes: ✓ indicates presence of a species; 0 indicates a species name no longer used.
† *Posidonia robertsoniae* is conspecific with *Posidonia coriacea*. * Species that are now considered to be conspecific with *Zostera capricorni*.
Not including *Ruppia* spp. *Heterozostera tasmanica* is now designated *Zostera tasmanica*.

SPECIES	Egypt	Eritrea	Estonia	Fiji	Finland	France	France – Guadaloupe	France – Martinique	France – New Caledonia	French Polynesia	Germany	Greece	Greenland	Grenada	Guadeloupe	Guatemala	Guinea Bissau	Haiti	Honduras	Iceland	India	Indonesia	Ireland	Israel	Italy	Jamaica	Japan	Jordan	Kazakhstan
Amphibolis antarctica																													
Amphibolis griffithii																													
Cymodocea angustata																													
Cymodocea nodosa	✓					✓						✓												✓	✓				
Cymodocea rotundata	✓								✓												✓	✓					✓		
Cymodocea serrulata	✓								✓												✓	✓					✓		
Enhalus acoroides									✓												✓	✓					✓		
Halodule beaudettei																													
Halodule bermudensis																													
Halodule emarginata																													
Halodule pinifolia				✓					✓												✓	✓					✓		
Halodule uninervis	✓			✓					✓												✓	✓		✓			✓	✓	
Halodule wrightii														✓	✓	✓	✓	✓	✓							✓			
Halophila australis																													
Halophila baillonii															✓														
Halophila beccarii																					✓								
Halophila capricorni									✓																				
Halophila decipiens	✓								✓	✓					✓			✓				✓		✓		✓			
Halophila engelmanni																													
Halophila hawaiiana																													
Halophila johnsonii																													
Halophila minor				✓					✓												✓						✓		
Halophila ovalis	✓	✓		✓					✓	✓											✓	✓		✓			✓	✓	
Halophila ovata	✓			✓																	✓								
Halophila spinulosa																					✓	✓							
Halophila stipulacea	✓	✓										✓									✓	✓		✓	✓			✓	
Halophila tricostata																													
Phyllospadix iwatensis																											✓		
Phyllospadix japonicus																											✓		
Phyllospadix scouleri																													
Phyllospadix serrulatus																													
Phyllospadix torreyi																													
Posidonia angustifolia																													
Posidonia australis																													
Posidonia coriacea																													
Posidonia denhartogii																													
Posidonia kirkmanii																													
Posidonia oceanica	✓					✓						✓												✓	✓				
Posidonia ostenfeldii																													
Posidonia robertsoniae [†]																													
Posidonia sinuosa																													
Syringodium filiforme														✓	✓											✓			
Syringodium isoetifolium	✓			✓					✓												✓	✓					✓		
Thalassia hemprichii	✓	✓							✓												✓	✓					✓		
Thalassia testudinum							✓	✓							✓	✓		✓	✓							✓			
Thalassodendron ciliatum	✓																				✓	✓						✓	
Thalassodendron pachyrhizum																													
Zostera asiatica																											✓		
Zostera caespitosa																											✓		
Zostera capensis																													
Zostera capricorni																													
Zostera caulescens																											✓		
Zostera japonica																											✓		
Zostera marina			✓		✓	✓					✓	✓	✓							✓			✓		✓		✓		
Zostera mucronata*																													
Zostera muelleri*																													
Zostera noltii						✓					✓	✓													✓				✓
Zostera novazelandica*																													
Zostera tasmanica																													
TOTAL	12	3	1	6	1	4	1	1	11	2	2	5	1	2	5	2	1	3	2	1	14	12	1	6	5	4	16	4	1

Notes: ✓ indicates presence of a species; 0 indicates a species name no longer used.
[†] Posidonia robertsoniae is conspecific with Posidonia coriacea. * Species that are now considered to be conspecific with Zostera capricorni. Not including Ruppia spp. Heterozostera tasmanica is now designated Zostera tasmanica.

Appendix 1

SPECIES	Kenya	Kiribati	Korea, DPR	Korea, Rep.	Kuwait	Latvia	Lebanon	Libyan Arab Jamahiriya	Lithuania	Madagascar	Malaysia	Maldives	Marshall Islands	Martinique	Mauritania	Mauritius	Mayotte	Mexico	Micronesia	Morocco	Mozambique	Myanmar	Netherlands	Netherlands Antilles	New Zealand	Nicaragua	Norway	Oman	Palau	Panama
Amphibolis antarctica																														
Amphibolis griffithii																														
Cymodocea angustata																														
Cymodocea nodosa							✓	✓							✓					✓										
Cymodocea rotundata	✓									✓	✓	✓				✓	✓		✓		✓								✓	
Cymodocea serrulata	✓									✓	✓					✓	✓		✓		✓								✓	
Enhalus acoroides	✓									✓	✓						✓		✓		✓	✓							✓	
Halodule beaudettei											✓																			
Halodule bermudensis																														
Halodule emarginata																														
Halodule pinifolia										✓									✓										✓	
Halodule uninervis	✓			✓						✓	✓					✓			✓		✓	✓							✓ ✓	
Halodule wrightii	✓										✓			✓				✓		✓				✓		✓				✓
Halophila australis																														
Halophila baillonii																								✓						✓
Halophila beccarii											✓										✓									
Halophila capricorni																														
Halophila decipiens											✓					✓	✓				✓			✓					✓	
Halophila engelmanni																	✓													
Halophila hawaiiana																														
Halophila johnsonii																														
Halophila minor	✓										✓						✓		✓										✓	
Halophila ovalis	✓			✓						✓	✓					✓			✓		✓	✓							✓ ✓	
Halophila ovata																														
Halophila spinulosa											✓																			
Halophila stipulacea	✓			✓						✓							✓				✓									
Halophila tricostata																														
Phyllospadix iwatensis			✓	✓																										
Phyllospadix japonicus				✓																										
Phyllospadix scouleri																		✓												
Phyllospadix serrulatus																														
Phyllospadix torreyi																		✓												
Posidonia angustifolia																														
Posidonia australis																														
Posidonia coriacea																														
Posidonia denhartogii																														
Posidonia kirkmanii																														
Posidonia oceanica								✓												✓										
Posidonia ostenfeldii																														
Posidonia robertsoniae†																														
Posidonia sinuosa																														
Syringodium filiforme														✓				✓						✓		✓				✓
Syringodium isoetifolium	✓									✓	✓	✓				✓			✓		✓								✓ ✓	
Thalassia hemprichii	✓	✓								✓	✓	✓	✓			✓			✓		✓								✓	
Thalassia testudinum														✓				✓						✓		✓				✓
Thalassodendron ciliatum	✓									✓	✓					✓ ✓	✓		✓										✓ ✓	
Thalassodendron pachyrhizum																														
Zostera asiatica			✓	✓																										
Zostera caespitosa			✓	✓																										
Zostera capensis	✓									✓											✓									
Zostera capricorni																									✓					
Zostera caulescens				✓																										
Zostera japonica			✓	✓																										
Zostera marina			✓	✓		✓			✓	✓										✓			✓				✓			
*Zostera mucronata**																														
*Zostera muelleri**																														
Zostera noltii															✓					✓			✓					✓		
*Zostera novazelandica**																									0					
Zostera tasmanica																														
TOTAL	12	1	5	7	2	1	2	3	1	12	12	3	2	2	3	7	3	8	10	4	12	5	2	5	1	3	2	4	10	5

Notes: ✓ indicates presence of a species; 0 indicates a species name no longer used.
† *Posidonia robertsoniae* is conspecific with *Posidonia coriacea*. * Species that are now considered to be conspecific with *Zostera capricorni*.
Not including *Ruppia* spp. *Heterozostera tasmanica* is now designated *Zostera tasmanica*.

SPECIES	Papua New Guinea	Philippines	Poland	Portugal	Puerto Rico	Qatar	Romania	Russian Federation	Samoa	São Tomé and Principe	Saudi Arabia	Senegal	Seychelles	Sierra Leone	Singapore	Solomon Islands	Somalia	South Africa	Spain	Sri Lanka	St Kitts and Nevis	St Lucia	St Vincent & the Grenadines	Sudan	Sweden
Amphibolis antarctica																									
Amphibolis griffithii																									
Cymodocea angustata																									
Cymodocea nodosa				✓							✓	✓							✓						
Cymodocea rotundata	✓	✓									✓	✓	✓												
Cymodocea serrulata	✓	✓									✓	✓	✓							✓				✓	
Enhalus acoroides	✓	✓									✓	✓	✓							✓					
Halodule beaudettei																									
Halodule bermudensis																									
Halodule emarginata																									
Halodule pinifolia	✓	✓											✓												
Halodule uninervis	✓	✓			✓						✓	✓	✓				✓			✓				✓	
Halodule wrightii					✓						✓														
Halophila australis																									
Halophila baillonii					✓																				
Halophila beccarii		✓											✓		✓										
Halophila capricorni																									
Halophila decipiens	✓	✓			✓						✓	✓								✓	✓	✓	✓		
Halophila engelmanni					✓																				
Halophila hawaiiana																									
Halophila johnsonii																									
Halophila minor	✓	✓											✓		✓										
Halophila ovalis	✓	✓			✓				✓		✓	✓	✓				✓	✓		✓				✓	
Halophila ovata	✓	✓							✓																
Halophila spinulosa	✓												✓												
Halophila stipulacea					✓						✓													✓	
Halophila tricostata																									
Phyllospadix iwatensis								✓																	
Phyllospadix japonicus																									
Phyllospadix scouleri																									
Phyllospadix serrulatus																									
Phyllospadix torreyi																									
Posidonia angustifolia																									
Posidonia australis																									
Posidonia coriacea																									
Posidonia denhartogii																									
Posidonia kirkmanii																									
Posidonia oceanica																			✓						
Posidonia ostenfeldii																									
Posidonia robertsoniae †																									
Posidonia sinuosa																									
Syringodium filiforme					✓																		✓		
Syringodium isoetifolium	✓	✓							✓		✓	✓	✓		✓					✓				✓	
Thalassia hemprichii	✓	✓									✓	✓	✓		✓	✓								✓	
Thalassia testudinum					✓																		✓		
Thalassodendron ciliatum	✓	✓									✓	✓					✓	✓	✓					✓	
Thalassodendron pachyrhizum																									
Zostera asiatica								✓																	
Zostera caespitosa																									
Zostera capensis																		✓							
Zostera capricorni																									
Zostera caulescens								✓																	
Zostera japonica								✓																	
Zostera marina			✓	✓	✓	✓													✓						✓
Zostera mucronata*																									
Zostera muelleri*																									
Zostera noltii				✓		✓	✓												✓						✓
Zostera novazelandica*																									
Zostera tasmanica																									
TOTAL	12	14	1	3	6	3	2		6	3		1	10	1	9	1	11		3	5	3	5	7		2	1		1	7	2

Notes: ✓ indicates presence of a species; 0 indicates a species name no longer used.
† *Posidonia robertsoniae* is conspecific with *Posidonia coriacea*. * Species that are now considered to be conspecific with *Zostera capricorni*.
Not including *Ruppia* spp. *Heterozostera tasmanica* is now designated *Zostera tasmanica*.

Appendix 1

SPECIES	Syrian Arab Republic	Tanzania	Thailand	Tonga	Trinidad and Tobago	Tunisia	Turkey	Turkmenistan	Turks and Caicos Islands	Ukraine	United Arab Emirates	USA – total	USA – Pacific Islands	US Virgin Islands	UK	Vanuatu	Venezuela	Viet Nam	Western Samoa	Yemen
Amphibolis antarctica																				
Amphibolis griffithii																				
Cymodocea angustata																				
Cymodocea nodosa	✓					✓	✓													
Cymodocea rotundata		✓	✓								✓		✓			✓				✓
Cymodocea serrulata		✓	✓								✓		✓			✓		✓	✓	✓
Enhalus acoroides		✓	✓								✓		✓			✓		✓		✓
Halodule beaudettei																				
Halodule bermudensis																				
Halodule emarginata																				
Halodule pinifolia			✓										✓	✓		✓		✓		
Halodule uninervis		✓	✓	✓							✓	✓	✓			✓		✓		✓
Halodule wrightii			✓		✓				✓			✓		✓			✓			
Halophila australis																				
Halophila baillonii												✓					✓			
Halophila beccarii			✓															✓		
Halophila capricorni																				
Halophila decipiens			✓		✓							✓	✓	✓		✓	✓			
Halophila engelmanni												✓						✓		
Halophila hawaiiana												✓	✓							
Halophila johnsonii												✓								
Halophila minor		✓	✓									✓	✓			✓		✓	✓	
Halophila ovalis		✓	✓	✓								✓	✓	✓		✓		✓	✓	✓
Halophila ovata																				
Halophila spinulosa																				
Halophila stipulacea	✓	✓					✓					✓								✓
Halophila tricostata																				
Phyllospadix iwatensis																				
Phyllospadix japonicus																				
Phyllospadix scouleri												✓								
Phyllospadix serrulatus												✓								
Phyllospadix torreyi												✓								
Posidonia angustifolia																				
Posidonia australis																				
Posidonia coriacea																				
Posidonia denhartogii																				
Posidonia kirkmanii																				
Posidonia oceanica	✓					✓	✓													
Posidonia ostenfeldii																				
Posidonia robertsoniae [†]																				
Posidonia sinuosa																				
Syringodium filiforme					✓				✓			✓		✓			✓			
Syringodium isoetifolium		✓	✓	✓								✓	✓			✓		✓	✓	✓
Thalassia hemprichii		✓	✓									✓	✓			✓		✓		✓
Thalassia testudinum					✓							✓		✓			✓			
Thalassodendron ciliatum		✓														✓		✓		✓
Thalassodendron pachyrhizum																				
Zostera asiatica												✓								
Zostera caespitosa																				
Zostera capensis		✓																		
Zostera capricorni																	✓			
Zostera caulescens																				
Zostera japonica												✓						✓		
Zostera marina					✓	✓			✓			✓			✓					
Zostera mucronata*																				
Zostera muelleri*																				
Zostera noltii	✓				✓	✓	✓	✓							✓					
Zostera novazelandica*																				
Zostera tasmanica																				
TOTAL	4	12	11	3	4	4	5	1	2	2	3	23	11	4	2	12	6	11	4	9

Notes: ✓ indicates presence of a species; 0 indicates a species name no longer used.
[†] *Posidonia robertsoniae* is conspecific with *Posidonia coriacea*. * Species that are now considered to be conspecific with *Zostera capricorni*.
Not including *Ruppia* spp. *Heterozostera tasmanica* is now designated *Zostera tasmanica*.

Appendix 2: Marine protected areas known to include seagrass beds, by country or territory

Few of these sites are managed directly to support seagrass protection, and in many cases they do not protect the most important areas of seagrass in a region. Total protected area is given in hectares but this is not indicative of the area of seagrass.

Summary of IUCN management categories – more detailed information at: http://www.unep-wcmc.org/protected_areas/categories/ index.html
- Ia: Strict Nature Reserve: protected area managed mainly for science
- Ib: Wilderness Area: protected area managed mainly for wilderness protection
- II: National Park: protected area managed mainly for ecosystem protection and recreation
- III: Natural Monument: protected area managed mainly for conservation of specific natural features
- IV: Habitat/Species Management Area: protected area managed mainly for conservation through management intervention
- V: Protected Landscape/Seascape: protected area managed mainly for landscape/seascape conservation and for recreation
- VI: Managed Resource Protected Area: protected area managed mainly for the sustainable use of natural ecosystems
- u/a Unavailable

IUCN management category does not always equate with management effectiveness.

Country	Area name	Designate	Size (ha)	IUCN cat.	Year
Anguilla	Crocus Bay	Marine Park	–	u/a	
	Sombrero Island	Marine Park	–	u/a	
Antigua and Barbuda	Cades Bay	Marine Reserve	–	u/a	1999
Australia	Ashmore Reef	National Nature Reserve	58 300	Ia	1983
	Corner Inlet	Marine and Coastal Park	18 000	VI	1986
	Great Barrier Reef	Commonwealth Marine Park	34 480 000	VI	1979
	Hinchinbrook Island	National Park	39 900	II	1989
	Marmion	Marine Park	9 500	VI	1987
	Ningaloo	Marine Park	225 564	VI	1987
	Ningaloo Reef	Commonwealth Marine Park	232 600	u/a	
	Rowley Shoals	Marine Park	23 250	VI	1990
	Shark Bay	Marine Park	748 735	VI	1990
	Shoalwater Islands	Marine Park	6 545	VI	1990
	Wilsons Promontory	National Park	49 000	II	1898
Bahamas	Union Creek	Managed Nature Reserve	1 813	Ia	1965
Bahrain	Hawar Islands	Other area	–	u/a	
Belize	Half Moon Cay	National Monument	3 925	III	1982
	Hol Chan	Marine Reserve	411	IV	1987
	Port Honduras	Marine Reserve	84 700	IV	2000
	South Water Cay	Marine Reserve	29 800	IV	1996
Brazil	Abrolhos	Marine National Park	91 300	II	1983
	Fernando de Noronha	Marine National Park	11 270	II	1988
	Saltinho	State Forest Reserve 2	10	u/a	1986
British Indian Ocean Territory	Diego Garcia	Restricted Area	–	V	1994
Cambodia	Ream	National Park	15 000	II	1993
Canada	Race Rocks	Ecological Reserve	220	Ia	1980
Cayman Islands	Little Sound (Grand Cayman)	Environmental Zone	1 731	Ib	1986
	North Sound (Grand Cayman)	Replenishment Zone	3 310	IV	1986
	South Sound (Grand Cayman)	Replenishment Zone	317	IV	1986
	Spott Bay (Cayman Brac)	Replenishment Zone	33	IV	1986
China	Shan Kou	Nature Reserve	8 000	V	1990

Country	Area name	Designate	Size (ha)	IUCN cat.	Year
Colombia	Corales del Rosario y de San Bernardo	Natural National Park	120 000	II	1977
	Old Providence McBean Lagoon	Natural National Park	995	II	1996
	Tayrona	Natural National Park	15 000	II	1964
Costa Rica	Cahuita	National Park	14 022	II	1970
	Gandoca–Manzanillo	National Wildlife Refuge	9 449	IV	1985
Croatia	Briuni	National Park	4 660	V	1983
Cuba	Punta Francé+D225s – Punta Pederales	Parque Nacional Marino	17 424	II	1985
Cyprus	Lara–Toxeftra	Marine Reserve	650	IV	1989
Dominica	Cabrits	National Park	531	II	1986
Dominican Republic	Del Este	National Park	80 800	II	1975
	Jaragua	National Park	137 400	II	1983
	Los Haitises	National Park	154 300	II	1976
	Montecristi	National Park	130 950	II	1983
France	Cote Bleue	Marine Park	3 070	VI	1982
	Golfe du Morbihan	Nature Reserve (by Decree)	1 500	u/a	
	Scandola	Nature Reserve (by Decree)	1 669	IV	1975
French Polynesia	Scilly (Manuae)	Territorial Reserve	11 300	IV	1992
Germany	Strelasund Sound/Greifswald Lagoon/Isle Greifswald	Wetland Zone of National Importance	–	V	1980
	Wismar Bight/Salzhaff area	Wetland Zone of National Importance	–	V	1980
Guadeloupe	Grand Cul de Sac Marin	Nature Reserve	3 736	IV	1987
Guam	Guam	Territorial Seashore Park	6 135	VI	1978
Guatemala	Punta de Manabique/ Bahía La Graciosa	Wildlife Refuge	38 400	u/a	
Honduras	Guanaja	Marine Reserve	28 000	u/a	
	Jeanette Kawas	National Park	78 162	II	1988
	Punta Izopo	Wildlife Refuge	11 200	IV	1992
India	Gulf of Kutch	Marine National Park	16 289	II	1980
	Gulf of Kutch	Marine Sanctuary	29 303	IV	1980
	Gulf of Mannar	Marine National Park	623	II	1986
	Gulf of Mannar	Biosphere Reserve (National)	1 050 000	VI	1989
	Wandur	Marine National Park	28 150	II	1983
Indonesia	Arakan Wowontulap	Nature Reserve	13 800	Ia	1986
	Bali Barat	National Park	77 727	II	1982
	Kepulauan Karimata	Nature Reserve	77 000	Ia	1985
	Kepulauan Togian	Nature Reserve	100 000	u/a	1989
	Pulau Bokor	Nature Reserve	15	Ia	1921
	Pulau Rambut	Nature Reserve	18	Ia	1939
	Ujung Kulon	National Park	122 956	II	1992
Israel	Elat Coral	Reserve	50	IV	
Italy	Archipelago Toscano	Zona di Tutela Biologica Marina (Italy)	–	IV	1982
	Cinque Terre	Zona di Tutela Biologica Marina (Italy)	–	IV	1982
	Golfo di Portofino	Zona di Tutela Biologica Marina (Italy)	–	IV	1982
	Miramare	Zona di Tutela Biologica Marina (Italy)	27	IV	1986
	Portofino	Regional/Provincial Nature Park	4 660	u/a	
Jamaica	Discovery Bay	Marine Park	–	u/a	
	Montego Bay	Marine Park	1 530	II	1991
	Negril	Marine Park	–	u/a	1998

Country	Area name	Designate	Size (ha)	IUCN cat.	Year
Jamaica	Negril Bay/Bloody Bay–Hanover FIS	Fisheries Sanctuary	–	u/a	
	Ocho Rios	Protected Area	–	V	1966
	Palisadoes–Port Royal Cays	National Park	100	u/a	
Kenya	Kiunga	Marine National Reserve	25 000	VI	1979
	Malindi	Marine National Park	630	II	1968
	Malindi–Watamu	Marine National Reserve	17 700	VI	1968
	Mpunguti	Marine National Reserve	1 100	VI	1978
	Watamu	Marine National Park	12 500	II	1968
Korea, Republic of	Nakdong River Mouth	Natural Ecological System Preservation Area	3 421	IV	1989
Madagascar	Grand Recif	Marine National Park	–	u/a	
	Mananara Marine	National Park	1 000	II	1989
Malaysia	Pulau Besar	Marine Park	8 414	II	
	Pulau Perhentian Besar	Marine Park	9 121	II	1999
	Pulau Perhentian Kecil	Marine Park	8 107	II	
	Pulau Redang	Marine Park	12 750	II	1999
	Pulau Sibu	Marine Park	4 260	II	
	Pulau Sipadan	Marine Reserve	710	u/a	
	Pulau Tengah	Marine Park	5 149	II	
	Pulau Tiga	Park	15 864	II	1978
	Pulau Tinggi	Marine Park	10 180	II	
	Pulau Tioman	Marine Park	25 115	II	
	Talang-Satang	National Park	19 414	u/a	
	Tunku Abdul Rahman	Park	4 929	II	1974
	Turtle Islands Heritage	Protected Area	136 844	u/a	1996
Martinique	Caravelle	Nature Reserve	422	IV	1976
Mauritania	Banc d'Arguin	National Park	1 173 000	II	1976
Mauritius	Baie de l'Arsenal Marine National Park	Marine National Park	100	u/a	
	Balaclava	Marine Park	–	II	1997
	Flacq	Fishing Reserve	600	IV	1983
	Port Louis	Fishing Reserve	500	IV	1983
	Trou d'Eau Douce Fir	Fishing Reserve	700	IV	1983
Mexico	Arrecifes de Puerto Morelos	National Park	10 828	II	1998
	Banco Chinchorro	Biosphere Reserve (National)	144 360	VI	1996
	El Vizcaíno	Biosphere Reserve (National)	2 546 790	VI	1988
	La Blanquilla	Other area	66 868	IV	1975
	Ría Lagartos	Other area	47 840	u/a	1979
	Sistema Arrecifal Veracruzano	National Marine Park	52 239	II	1992
Monaco	Larvotto	Marine Reserve	50	IV	1976
Mozambique	Bazaruto	National Park	15 000	II	1971
	Ilhas da Inhaca e dos Portugueses	Faunal Reserve	2 000	IV	1965
	Maputo	Game Reserve	90 000	IV	1969
	Marromeu	Game Reserve	1 000 000	IV	1969
	Nacala-Mossuril	Marine National Park	–	Ib	
	Pomene	Game Reserve	10 000	IV	1972
	Primeira and Segunda Islands	National Park	–	u/a	
	Zambezi	Wildlife Utilization Area	1 000 000	VI	1981

Appendix 2

Country	Area name	Designate	Size (ha)	IUCN cat.	Year
Netherlands Antilles	Bonaire	Marine Park	2 600	u/a	1979
	Saba	Marine Park	820	u/a	1987
Nicaragua	Cayos Miskitos	Marine Reserve	50 000	Ia	1991
Palau	Ngerukewid Islands	Designation unknown	1 200	III	1956
Panama	Comarca Kuna Yala (San Blas)	Indigenous Commarc	320 000	u/a	1938
Papua New Guinea	Kamiali	Wildlife Management Area	47 413	VI	1996
	Lou Island	Wildlife Management Area	–	u/a	
	Maza (I)	Wildlife Management Area	184 230	VI	1978
	Motupore Island	Wildlife Management Area	–	u/a	
	Nanuk Island	Provincial Park	12	IV	1973
	Talele Islands	Provincial Park	40	IV	1973
Philippines	St Paul Subterranean River	National Park	5 753	II	1971
	Tubbataha Reefs National Marine Park	Marine Park	33 200	u/a	1988
Puerto Rico	Boqueron	Wildlife Refuge (Refugio de Vida Silvestre)	237	IV	1964
	Cayos de la Cordillera	Nature Reserve	88	IV	1980
	Estuarina Nacional Bahia Jobos	Hunting Reserve	1 133	IV	1981
	Isla Caja de Muerto	Nature Reserve	188	IV	1988
	Jobos Bay	National Estuarine Research Reserve	1 168	IV	1981
	La Parguera	Nature Reserve	4 973	IV	1979
Réunion	Cap la Houssaye–Ravine Trois Bassins	Fishing Reserve	–	VI	1978
	Iles Glorieuses	Nature Reserve	–	IV	1975
	Ilot d'Europa	Nature Reserve	–	IV	1975
	Pointe de Bretagne–Pointe de l'Etang Sale	Fishing Reserve	–	VI	1978
	Ravine Trois Bassins–Pointe de Bretagne	Fishing Reserve	–	VI	1978
Russian Federation	Astrakhansky	Zapovednik	66 816	Ia	1919
	Dalnevostochny Morskoy	Zapovednik	64 316	Ia	1978
	Kedrovaya Pad	Zapovednik	17 900	Ia	1925
St Lucia	Maria Islands	Nature Reserve	12	IV	1982
	Pigeon Island	Other area	20	III	1978
	Soufriere	Marine Management Area	–	VI	1994
St Vincent and the Grenadines	Tobago Cays	Marine Reserve	3 885	IV	1987
Saudi Arabia	Dawhat Ad Dafi, Dawhat Al-Musallamiyah & Coral Islands	Other area	210 000	u/a	
	Farasan Islands	Protected Area	69 600	Ia	1989
Seychelles	Aldabra	Special Nature Reserve	35 000	Ia	1981
	Port Launay	Marine National Park	158	II	1979
	St Anne	Marine National Park	1 423	II	1973
Singapore	Southern Islands	Marine Nature Area	980	u/a	1996
Slovenia	Strunjan	Landscape Park	192	V	1990
South Africa	Agulhas	National Park	–	u/a	
	Cape Peninsula	National Park	–	u/a	
	Greater St Lucia	Wetland Park	258 686	II	1895
	Knysna	Other area	15 000	u/a	
Spain	Doñana	National Park (State Network)	50 720	II	1969
	Illa de Tabarca	Marine Nature Reserve (Spain)	1 463	IV	1986
	Illes Medes	Submarine Nature Reserve	418	IV	1983

Country	Area name	Designate	Size (ha)	IUCN cat.	Year
Tanzania	Bongoyo Island	Marine Reserve	–	II	1975
	Chumbe Island Coral Park (CHICOP)	Marine Sanctuary	30	II	1994
	Fungu Yasini	Marine Reserve	–	II	1975
	Mafia Island	Marine Park	82 200	VI	1995
	Maziwi Island	Marine Reserve	–	II	1981
	Mbudya	Marine Reserve	–	II	1975
	Menai Bay	Conservation Area	47 000	VI	1997
	Misali Island	Conservation Area	2 158	VI	1998
	Mnemba	Conservation Area	15	VI	1997
	Pangavini	Marine Reserve	–	II	1975
Thailand	Haad Chao Mai	National Park	23 086	II	1981
	Mu Ko Libong	Non-hunting Area	44 749	III	1979
Tonga	Fanga'uta and Fanga Kakau Lagoons	Marine Reserve	2 835	VI	1974
	Pangaimotu Reef	Reserve	49	IV	1979
Trinidad and Tobago	Buccoo Reef	Nature Reserve	650	Ia	1973
Tunisia	Ichkeul	National Park	12 600	II	1980
Turks and Caicos Islands	West Caicos Marine	National Park	397	IV	1992
Ukraine	Arabats'kiy	State Zakaznik	600	u/a	
	Karadagskiy	Nature Zapovednik (Ukraine)	2 874	Ia	1979
	Karkinits'ka zatoka	State Zakaznik	27 646	u/a	
	Kazantypskyi	Nature Zapovednik (Ukraine)	450	u/a	
	Molochniy liman	State Zakaznik	1 900	u/a	
	Mys Martiyan	Nature Zapovednik (Ukraine)	240	Ia	1973
United Kingdom	Helford River	Voluntary Reserve (UK)	–	u/a	1987
	Isles of Scilly	Area of Outstanding Natural Beauty (UK)	1 600	V	1976
	Skomer	National Nature Reserve (UK)	307	IV	1959
	Skomer	Marine Nature Reserve (UK)	1 500	IV	1990
United States	Acadia	National Park	15 590	II	1919
	Apalachicola	National Estuarine Research Reserve	99 630	IV	1979
	Assateague Island	National Seashore	16 038	V	1965
	Bahia Honda	State Park	212	V	1961
	Biscayne	National Park	72 900	II	1980
	Breton	National Wildlife Refuge	3 661	IV	1904
	Cape Cod	National Seashore	18 018	V	1961
	Channel Islands	National Park	100 987	II	1980
	Channel Islands	National Marine Sanctuary	428 466	IV	1980
	Chesapeake Bay (MD)	National Estuarine Research Reserve	2 374	IV	1981
	Chesapeake Bay (VA)	National Estuarine Research Reserve	1 796	IV	1991
	Dry Tortugas	National Park	26 203	II	1992
	Everglades	National Park	606 688	II	1947
	Fire Island	National Seashore	7 834	V	1964
	Florida Keys	Wilderness (Fish and Wildlife Service)	2 508	Ib	1975
	Galveston Island	State Park	786	Ia	
	Grand Bay	National Estuarine Research Reserve	7 452	IV	1999
	Great Bay	National Estuarine Research Reserve	2 138	IV	1989
	Gulf Islands (Florida)	National Seashore	54 928	V	1971
	Hawaiian Islands (8 sites)	National Wildlife Refuge	102 960	Ia	1945
	Izembek	National Wildlife Refuge	122 660	IV	1960

Country	Area name	Designate	Size (ha)	IUCN cat.	Year
United States (continued)	John Pennekamp Coral Reef	State Park	22 684	V	1959
	Merritt Island	National Wildlife Refuge	55 953	IV	1963
	Narragansett Bay	National Estuarine Research Reserve	1 286	IV	1980
	Padilla Bay	National Estuarine Research Reserve	4 455	IV	1980
	Pinellas	National Wildlife Refuge	159	IV	1956
	Rookery Bay	National Estuarine Research Reserve	5 062	IV	1978
	South Slough	National Estuarine Research Reserve	1 903	IV	1974
	St Marks	National Wildlife Refuge	26 467	IV	1931
	Waquoit Bay	National Estuarine Research Reserve	1 013	IV	1988
	Wells	National Estuarine Research Reserve	648	IV	1984
United States minor outlying island	Baker Island	National Wildlife Refuge	12 843	Ia	1974
Venezuela	Archipiélago Los Roques	National Park	221 120	II	1972
	Ci	Wildlife Refuge	25 723	u/a	
	Cuare	Wildlife Refuge	11 825	IV	1972
	Laguna de la Restinga	National Park	18 862	II	1974
	Laguna de Tacarigua	National Park	39 100	II	1974
	Médanos de Coro	National Park	91 280	II	1974
	Mochima	National Park	94 935	II	1973
	Morrocoy	National Park	32 090	II	1974
	San Esteban	National Park	43 500	II	1987
Viet Nam	Con Dao	National Park	15 043	II	1982
Virgin Islands (British)	Little Jost Van Dyke	Natural Monument	450	u/a	
	Norman Island	National Park	390	u/a	
	North Sound	National Park	3 800	u/a	
	The Dogs	Protected Area	2 000	u/a	
	Wreck of the Rhone	Marine Park	324	III	1980
Virgin Islands (US)	Green Cay	National Wildlife Refuge	6	IV	1977
	Salt River Submarine Canyon NHS	Protected Area	1 000	u/a	
	Sandy Point	National Wildlife Refuge	134	IV	
	St James	Marine Reserve and Wildlife Sanctuary	–	u/a	1994
	Virgin Islands	National Park	5 308	II	1956

Appendix 3: Species range maps

The range maps have been created to establish a guide to where individual seagrass species might be expected to occur. Range boundaries were drawn to encompass all points where there was sufficient documentation to determine the occurrence of a seagrass species in a location. It is possible that seagrass species occur beyond the ranges shown since not all areas have been adequately surveyed. In some instances, isolated observations of species occurrence were marked as distinct entries. Range maps were not prepared for *Ruppia* species as the existing data were deemed insufficient. The species range maps update earlier work by den Hartog[1] and by Phillips and Meñez[2]. Estimates of species range area are given with each map.

Morphological differences in seagrass plants have led to some confusion regarding species designation, and several species are now being revised based on genetic and morphometric research. Two recent species revisions have been made which combine identified conspecific species (see *Halophila ovalis* and *Zostera capricorni*, below); revised species range maps have been included. The species range maps for species formerly accepted have also been included to make a connection with the literature documenting these species distributions. Several other species designations remain a matter of debate and are currently under genetic and morphometric investigation. Closely linked species that may actually be conspecific have often been grouped together in a complex while additional research is undertaken. These species do not have adjusted range maps, but descriptions of the *Posidonia ostenfeldii* complex and the *Halodule* spp. complexes are included below.

Halophila ovalis revision

A group of *Halophila* species found around the world have recently received detailed genetic evaluation. Four species, *Halophila johnsonii* Eiseman from the east coast of Florida, *Halophila hawaiiana* Doty and Stone from Hawaii, *Halophila ovata* Gaudichaud from the Indo-Pacific and *Halophila minor* (Zollinger) den Hartog, were all determined to be morphological variations of, and therefore conspecific with, *Halophila ovalis* (R. Brown) Hooker f.[3]. Range maps are presented for each former species individually as well as for the newly redefined *Halophila ovalis*.

Zostera capricorni revision

Zostera novazelandica Setchell and *Zostera capricorni* Ascherson in New Zealand are conspecific, based on detailed genetic and morphometric analysis[4]. From the same analysis, the Australian species *Zostera muelleri* Irmisch ex. Aschers. and *Zostera mucronata* den Hartog are also considered to be conspecific with *Zostera capricorni*. We have included *Zostera novazelandica* and *Zostera capricorni* on the same range map. *Zostera muelleri* and *Zostera mucronata* are displayed separately. Range maps are presented for former species individually as well as for the newly redefined *Zostera capricorni*[4].

Posidonia ostenfeldii complex

In the case of the *Posidonia ostenfeldii* complex, new genetic testing suggests that some of the species are conspecific. The complex consists of five species: *Posidonia ostenfeldii* den Hartog, *Posidonia denhartogii* Kuo & Cambridge, *Posidonia robertsoniae* Kuo & Cambridge, *Posidonia coriacea* Cambridge & Kuo and *Posidonia kirkmanii* Kuo & Cambridge[5].

However, published results of genetic testing and analysis of morphological characteristics shared by species within the complex have shown that *Posidonia coriacea* and *Posidonia robertsoniae* are not separate species[6]; they are treated as *Posidonia coriacea* here. More genetic research is needed in order to re-evaluate and define the complex as a whole[6]. Without publication of conclusive information, we have continued to view the species separately but realize that the complex is likely to contain conspecific species.

Halodule spp. complexes

The most recent genetic analysis indicates that two major groups of *Halodule* exist[7]. The Atlantic *Halodule* spp. complex consists of *Halodule wrightii* Ascherson, *Halodule beaudettei* (den Hartog) den Hartog and *Halodule bermudensis* den Hartog. The three species are distinguished by leaf tip morphology. However, some researchers have suggested that leaf tip morphology changes under different environmental conditions[8]. Further genetic studies and/or common garden experiments are needed. The Indo-Pacific *Halodule* species, consisting of *Halodule uninervis* (Forsskål) Ascherson and *Halodule pinifolia* (Miki) den Hartog, are suggested to be conspecific in unpublished genetic studies. These results indicate that *Halodule uninervis* is the only *Halodule* species in the Pacific and Indian Oceans, except in eastern Africa and India where *Halodule wrightii* is also found. Without conclusive published information, we present the *Halodule* species separately but realize that both complexes are likely to contain conspecific species.

References

1. den Hartog C [1970]. *The Sea-Grasses of the World*. North Holland Publishing Co., Amsterdam. 275 pp.
2. Phillips RC, Meñez EG [1988]. *Seagrasses*. Smithsonian Contributions to the Marine Sciences 34. Smithsonian Institution Press, Washington DC.
3. Waycott M, Freshwater DW, York RA, Calladine A, Kenworthy, WJ [2002]. Evolutionary trends in the seagrass genus *Halophila* (Thouars): insights from molecular phylogeny. *Bulletin of Marine Science*.
4. Les DH, Moody ML, Jacobs SWL, Bayer RJ [2002]. Systematics of seagrasses (Zosteraceae) in Australia and New Zealand. *J Sys Botany* 27: 468-484.
5. Kuo J, Cambridge ML [1984]. A taxonomic study of the *Posidonia ostenfeldii* complex (Posidoniaceae) with description of four new Australian seagrasses. *Aquatic Botany* 20: 267-295.
6. Campey ML, Waycott M, Kendrick GA [2000]. Re-evaluating species boundaries among members of the *Posidonia ostenfeldii* species complex (Posidoniaceae) – morphological and genetic variation. *Aquatic Botany* 66: 41-56.
7. Waycott M. Personal communication.
8. McMillan C, Williams SC, Escobar L, Zapata O [1981]. Isozyymes, secondary compounds and experimental cultures of Australian seagrasses in *Halophila*, *Halodule*, *Amphibolis*, and *Posidonia*. *Aust J Bot* 29: 247-260.

In the maps below, * indicates species designations that are a matter of debate and currently under genetic and morphometric investigation.

Family Hydrocharitaceae
(3 genera)

Genus *Enhalus* L.C. Richard (1 species)

Dioecious robust perennial with creeping, coarse unbranched or sparsely monopodially branched rhizomes with short internodes. Fleshy thick roots are unbranched. Male inflorescence has a short stalk and a 2-bladed spathe surrounding many flowers that break off and release pollen to float on the surface of the water. There are 3 sepals, 3 petals and 3 stamens. Female inflorescence with a long stalk and a 2-bladed spathe containing 1 large flower. There are 3 sepals and 3 petals. The ovary is rostrate, composed of 6 carpels and 6 styles, and is forked from the base. The stalk of the female flower coils and contracts after anthesis. Fruits are fleshy. Leaves are distichously arranged and sheathed at the base; persistent fibrous strands from previously decayed leaves enclose the stem. Leaf apex rounded.

Enhalus acoroides

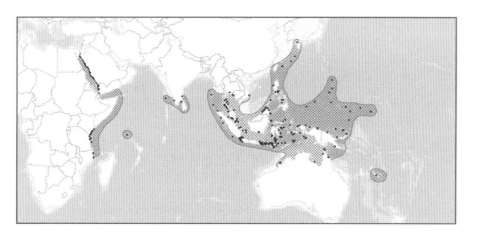

MAP 1: *Enhalus acoroides* (L.f.) Royle (Hydrocharitaceae)
Shaded area = 5 005 000 km^2, actual species distribution is much less

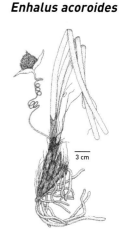

Enhalus acoroides

Genus *Halophila* Thouars (14 species)

Monoecious and dioecious small, fragile plants with long internodes on rhizomes each bearing 2 scales and a lateral shoot. Each node has 1 unbranched root. Single inflorescence covered by 2 spathal bracts. Male flowers have short stalks, 3 tepals, 3 stamens, sessile anthers and pollen grains in ellipsoid chains. Female flowers are sessile, have 3-6 styles, and ellipsoid to globular fruit that hold numerous globular seeds. Leaves are either distichously arranged along upright shoot, in pairs on long petioles or as pseudo-whorls at top of lateral shoots. Leaves have ovate, elliptic, lanceolate or linear blades with 1 mid-vein and intramarginal veins linked by cross-veins. Leaf margins smooth or serrate, with leaf surface smooth or hairy.

Halophila australis, Halophila baillonii, Halophila beccarii, Halophila capricorni, Halophila decipiens, Halophila engelmanni, Halophila hawaiiana, Halophila johnsonii, Halophila minor, Halophila ovalis, Halophila ovata, Halophila spinulosa, Halophila stipulacea, Halophila tricostata

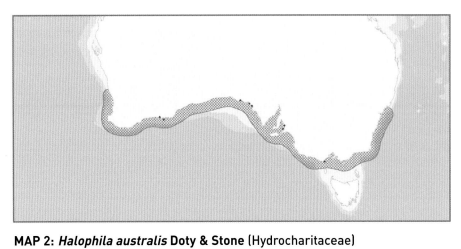
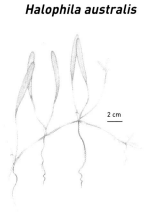

Halophila australis

MAP 2: *Halophila australis* **Doty & Stone** (Hydrocharitaceae)
Shaded area = 424 000 km², actual species distribution is much less

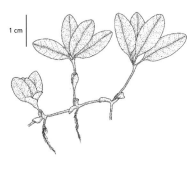

Halophila baillonii

MAP 3: *Halophila baillonii* **Ascherson** (Hydrocharitaceae)
Shaded area = 139 000 km², actual species distribution is much less

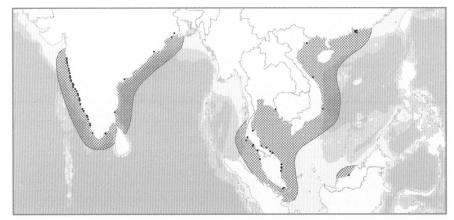
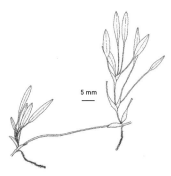

Halophila beccarii

MAP 4: *Halophila beccarii* **Ascherson** (Hydrocharitaceae)
Shaded area = 1 511 000 km², actual species distribution is much less

Appendix 3   265

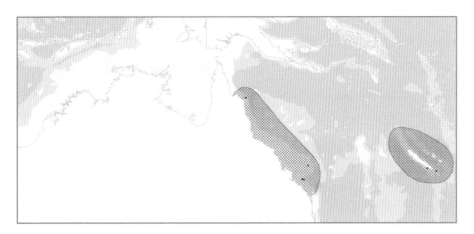

Halophila capricorni

MAP 5: *Halophila capricorni* **Larkum** (Hydrocharitaceae)
Shaded area = 269 000 km², actual species distribution is much less

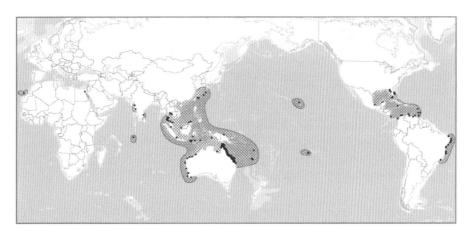
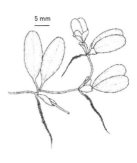

Halophila decipiens

MAP 6: *Halophila decipiens* **Ostenfeld** (Hydrocharitaceae)
Shaded area = 7 702 000 km², actual species distribution is much less

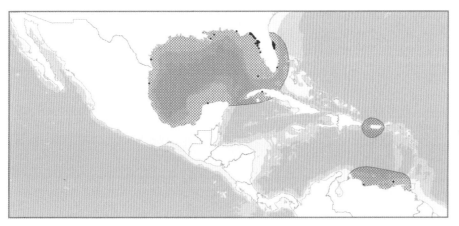
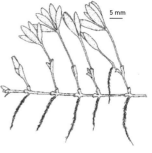

Halophila engelmanni

MAP 7: *Halophila engelmanni* **Ascherson** (Hydrocharitaceae)
Shaded area = 725 000 km², actual species distribution is much less

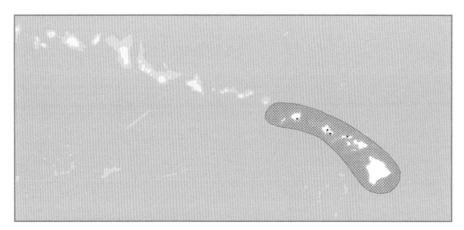
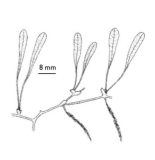

MAP 8: *Halophila hawaiiana* Doty & Stone (Hydrocharitaceae)
Shaded area = 7 000 km², actual species distribution is much less
Note: *Halophila hawaiiana* is now conspecific with *Halophila ovalis*[3].

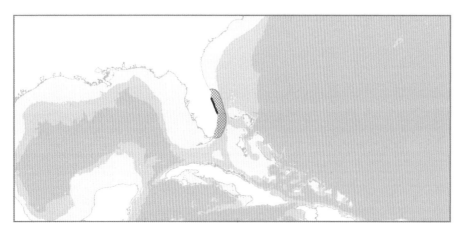
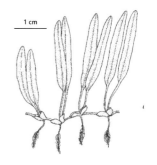

MAP 9: *Halophila johnsonii* Eiseman (Hydrocharitaceae)
Shaded area = 12 000 km², actual species distribution is much less
Note: *Halophila johnsonii* is now conspecific with *Halophila ovalis*[3].

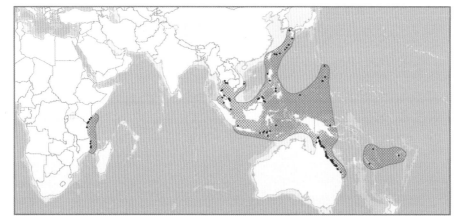
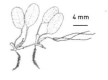

MAP 10: *Halophila minor* (Zollinger) den Hartog (Hydrocharitaceae)
Shaded area = 3 761 000 km², actual species distribution is much less
Note: *Halophila minor* is now conspecific with *Halophila ovalis*[3].

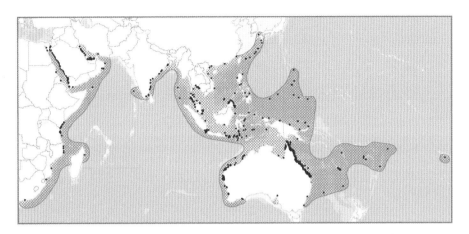
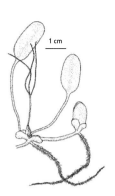

Halophila ovalis

MAP 11: *Halophila ovalis* (R. Brown) Hooker f. (Hydrocharitaceae)
Shaded area = 7 614 000 km², actual species distribution is much less
Note: Map represents the range of the former *Halophila ovalis*, before its current revision. See Map 12.

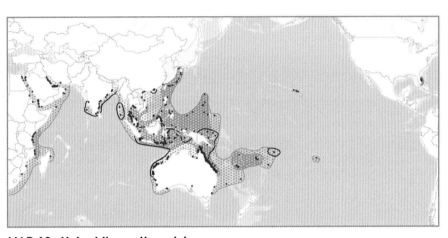

Halophila ovalis revision

MAP 12: *Halophila ovalis* revision
Shaded area = 7 633 000 km², actual species distribution is much less
Note: *Halophila johnsonii*, *Halophila hawaiiana*, *Halophila minor* and *Halophila ovata* are now considered to be conspecific with *Halophila ovalis*[3]. The range map presented here is for the newly redefined *Halophila ovalis*.

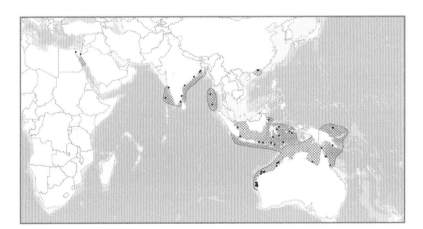
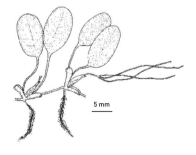

Halophila ovata

MAP 13: *Halophila ovata* Gaudichaud (Hydrocharitaceae)
Shaded area = 3 186 000 km², actual species distribution is much less
Note: *Halophila ovata* is now conspecific with *Halophila ovalis*[3].

268 WORLD ATLAS OF SEAGRASSES

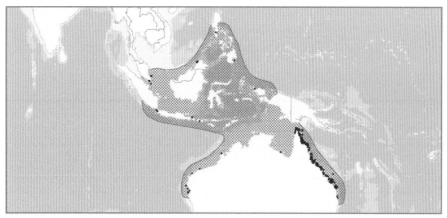
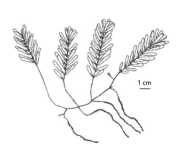

Halophila spinulosa

MAP 14: *Halophila spinulosa* (R. Brown) Ascherson (Hydrocharitaceae)
Shaded area = 3 796 000 km^2, actual species distribution is much less

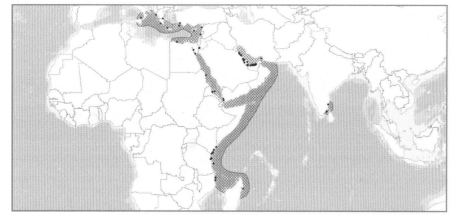
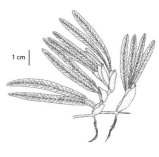

Halophila stipulacea

MAP 15: *Halophila stipulacea* (Forsskål) Ascherson (Hydrocharitaceae)
Shaded area = 924 000 km^2, actual species distribution is much less

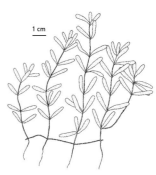

Halophila tricostata

MAP 16: *Halophila tricostata* Greenway (Hydrocharitaceae)
Shaded area = 415 000 km^2, actual species distribution is much less

Genus *Thalassia* Banks ex König (2 species)

Dioecious perennial with creeping rhizomes that have many small internodes and 1 scale leaf at each node. At intervals there are 1 or more unbranched roots and a short erect stem with 2-6 leaves. Male flowers have a short stalk, 3 perianth sections, 3-12 light yellow stamens and globular pollen grains linked into chains. Female flowers also have 3 perianth segments with 6-8 styles each split into 2 lengthy stigmata. Fruit is prickly and spherical with a fleshy pericarp that splits into non-uniform valves, releasing pear-shaped seeds with membranous testa. Seeds germinate immediately. The linear leaf blade, sometimes slightly bowed, has 9-17 longitudinal veins and a round, finely serrulated apex. Tannin cells are present but stomata are not.

Thalassia hemprichii, Thalassia testudinum

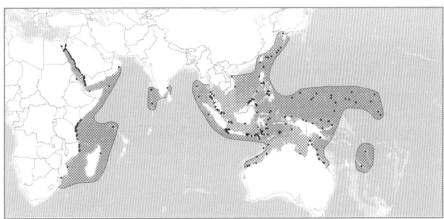

Thalassia hemprichii

MAP 17: *Thalassia hemprichii* **(Ehrenberg) Ascherson** (Hydrocharitaceae)
Shaded area = 6 094 000 km^2, actual species distribution is much less

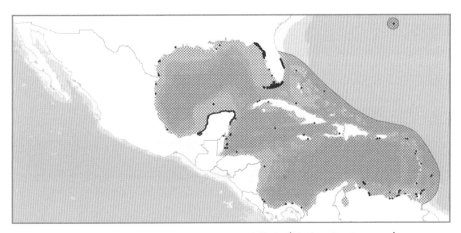

Thalassia testudinum

MAP 18: *Thalassia testudinum* **Banks ex König** (Hydrocharitaceae)
Shaded area = 1 165 000 km^2, actual species distribution is much less

Family Cymodoceaceae
(5 genera)

Genus *Amphibolis* C. Agardh (2 species)

Dioecious perennial with woody sympodially branched rhizomes. 1-2 wiry but abundantly branched roots at each node. Nodes may have long, thin abundantly branched stiff stems with crown leaves on each branch. The singular, terminal flowers are enclosed by several leaves. Male flowers have 2 anthers connected at the same height to a short stalk. Female flowers are sessile with 2 free ovaries, each with a short style spilt into 3 long stigmata with pericarpic lobes at each ovary base. Seedlings are viviparous and have comb-shaped structures extending from the pericarpic lobes that act as anchors. Leaf sheaths shed leaving circular scar on the erect stems. The linear leaf blade has 8-21 longitudinal veins and a bidentate apex.
Amphibolis antarctica, Amphibolis griffithii

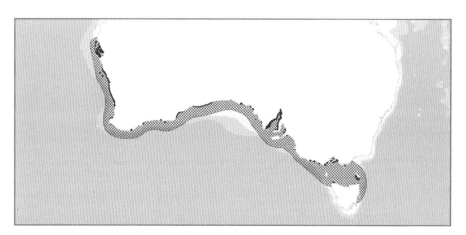
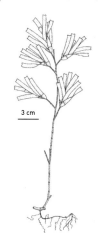

Amphibolis antarctica

MAP 19: *Amphibolis antarctica* (Labill.) Sonder et Ascherson (Cymodoceaceae)
Shaded area = 535 000 km², actual species distribution is much less

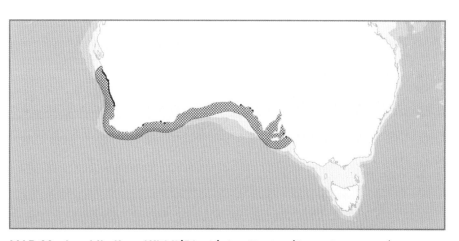
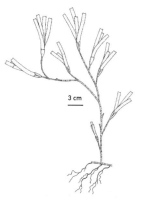

Amphibolis griffithii

MAP 20: *Amphibolis griffithii* (Black) den Hartog (Cymodoceaceae)
Shaded area = 330 000 km², actual species distribution is much less

Genus *Cymodocea* König (4 species)

Dioecious perennial with creeping herbaceous monopodially branched rhizomes. 1 to several branched roots with a short stiff stem bearing 2-7 leaves found at each node. Stalked male flower has 2 anthers connected at same height on stalk. Female flowers are sessile with 2 free ovaries each with a short style that splits into 2 long stigmata. Fruit are semi-circular to elliptical shaped with a solid pericarp. The linear leaf blade has 7-17 longitudinal veins and smooth margins. Apex rounded, sometimes notched or serrate. Leaf sheaths shed leaving circular scar on erect stems.

Cymodocea angustata, Cymodocea nodosa, Cymodocea rotundata, Cymodocea serrulata

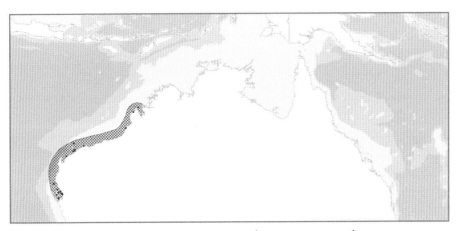

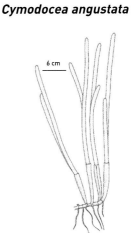

Cymodocea angustata

MAP 21: *Cymodocea angustata* **Ostenfeld** (Cymodoceaceae)
Shaded area = 160 000 km^2, actual species distribution is much less

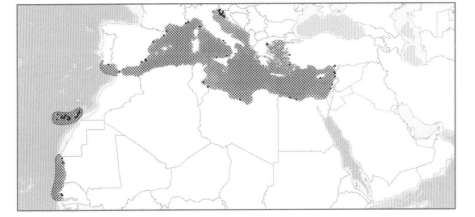

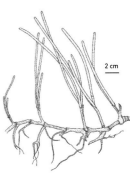

Cymodocea nodosa

MAP 22: *Cymodocea nodosa* **(Ucria) Ascherson** (Cymodoceaceae)
Shaded area = 610 000 km^2, actual species distribution is much less

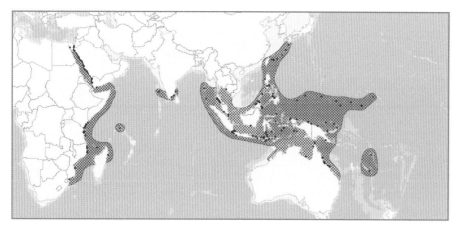
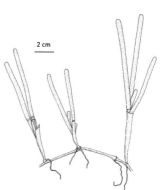

Cymodocea rotundata

MAP 23: *Cymodocea rotundata* **Ehrenberg & Hemprich ex Ascherson (Cymodoceaceae)**
Shaded area = 5 323 000 km^2, actual species distribution is much less

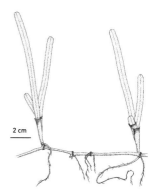

Cymodocea serrulata

MAP 24: *Cymodocea serrulata* **(R. Brown) Ascherson (Cymodoceaceae)**
Shaded area = 5 578 000 km^2, actual species distribution is much less

Genus *Halodule* Endlinger (6 species)

Dioecious perennial with creeping herbaceous monopodially branched rhizomes. 1 or more unbranched roots and a short erect stem with 1-4 leaves found at each node. The singular, terminal flowers are enclosed by a leaf. Stalked male flower has 2 anthers connected at same height on stalk. Female flowers have 2 free ovaries each with a long, continuous and undivided style. Fruit has stony, solid pericarp. Leaf sheaths shed leaving circular scar on the stems. The linear leaf blade has 3 longitudinal veins and a variable apex shape. The genus has 2 species in the Pacific and 4 species in the Atlantic, all largely distinguished by leaf tip morphology.

Halodule beaudettei, Halodule bermudensis, Halodule emarginata, Halodule pinifolia, Halodule uninervis, Halodule wrightii

Appendix 3

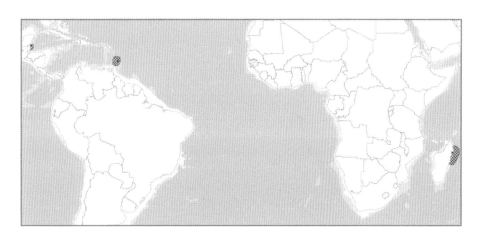

MAP 25: *Halodule beaudettei** (den Hartog) den Hartog (Cymodoceaceae)
Shaded area = 74 000 km², actual species distribution is much less

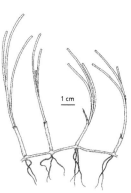

Halodule beaudettei*

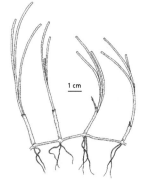

Halodule bermudensis*

MAP 26: *Halodule bermudensis** den Hartog (Cymodoceaceae)
Shaded area = 1 000 km², actual species distribution is much less

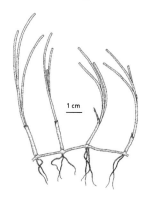

Halodule emarginata*

MAP 27: *Halodule emarginata** den Hartog (Cymodoceaceae)
Shaded area = 141 000 km², actual species distribution is much less

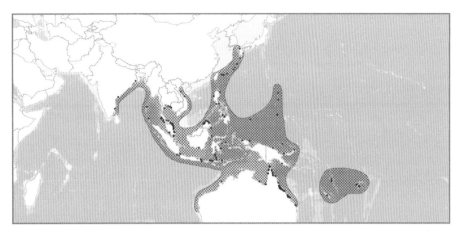
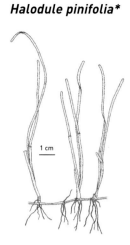

MAP 28: *Halodule pinifolia** (Miki) den Hartog (Cymodoceaceae)
Shaded area = 5 580 000 km², actual species distribution is much less

Halodule pinifolia*

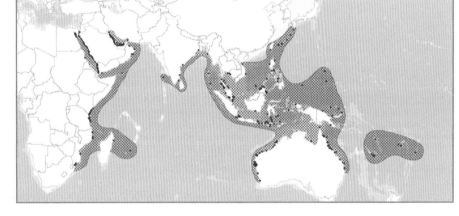

MAP 29: *Halodule uninervis** (Forsskål) Ascherson (Cymodoceaceae)
Shaded area = 6 734 000 km², actual species distribution is much less

Halodule uninervis*

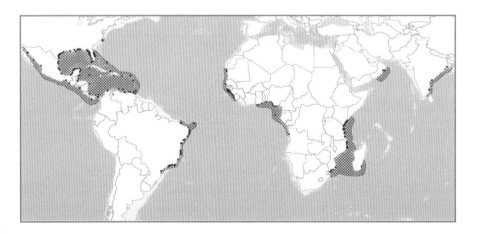
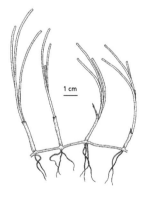

MAP 30: *Halodule wrightii* Ascherson (Cymodoceaceae)
Shaded area = 2 625 000 km², actual species distribution is much less

Halodule wrightii

Genus *Syringodium* Kützing (2 species)

Dioecious perennial with creeping herbaceous monopodially or sympodially branched rhizomes. Rhizomes with 1-4 little branched roots and an erect shoot bearing 2-3 round leaves at each node. Inflorescence cymose, flowers are encompassed by a reduced leaf. Stalked male flower has 2 anthers connected at same height on stalk. Female flowers have 2 free ovaries each with a short style and 2 short stigmata. Fruit has stony, solid pericarp. Leaf sheaths shed leaving circular scar on the rigid stems. Round leaf blades tapering to the tip.
Syringodium filiforme, Syringodium isoetifolium

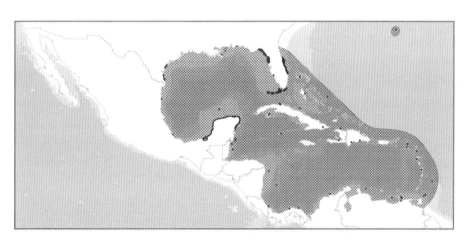
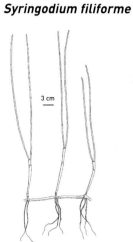

MAP 31: *Syringodium filiforme* **Kützing** (Cymodoceaceae)
Shaded area = 1 174 000 km^2, actual species distribution is much less

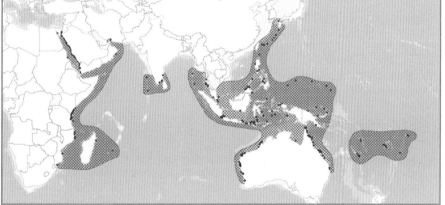
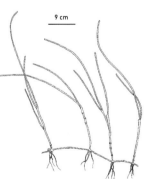

MAP 32: *Syringodium isoetifolium* **(Ascherson) Dandy** (Cymodoceaceae)
Shaded area = 5 919 000 km^2, actual species distribution is much less

Genus *Thalassodendron* den Hartog (2 species)

Dioecious perennial with woody sympodially branched rhizomes. 1 or more robust, woody, little branched roots occur at the nodes preceding the erect stem-bearing nodes. From every fourth node there are long, wiry infrequently branched stems each bearing crown leaves. Single flowers grow at the end of the stem and are enclosed by several bracts. Male flowers with 2 anthers connected at the same height to the stalk. Female flowers are sessile with 2 free ovaries, each with a short style split into 2 stigmata. Seedlings are viviparous. Leaf sheaths shed leaving circular scar on the erect stems. The linear leaf blade has 13-27 longitudinal veins and the margin and rounded apex are finely denticulate.

Thalassodendron ciliatum, Thalassodendron pachyrhizum

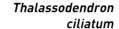

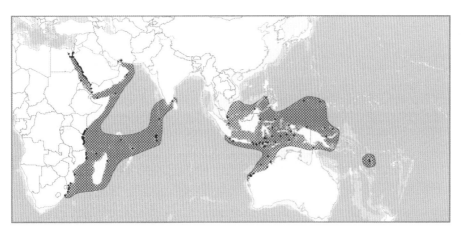

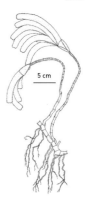

MAP 33: *Thalassodendron ciliatum* (Forsskål) den Hartog (Cymodoceaceae)
Shaded area = 4 087 000 km^2, actual species distribution is much less

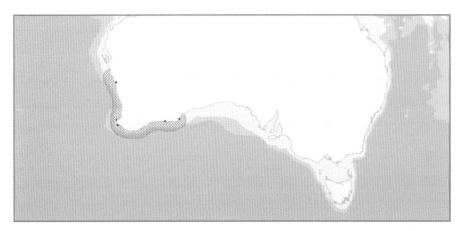

MAP 34: *Thalassodendron pachyrhizum* den Hartog (Cymodoceaceae)
Shaded area = 114 000 km^2, actual species distribution is much less

Family
Posidoniaceae
(1 genus)

Genus *Posidonia* König (8 species)

Monoecious perennial with creeping, monopodially branching rhizomes with 1-2 branched or unbranched roots and a shoot. Inflorescence racemose with many barbs. Flowers are hermaphroditic, have 3 stamens but no perianth. Stigma are disc-shaped with inconsistent lobe shapes. Stone fruit with fleshy pericarp. Pericarp splits to release oblong seeds with membranous testa. Leaves are distichous, ligulate and auriculate with a distinct sheath and blade. The leaf sheath is persistent and frequently breaks into fibrous strands covering rhizome internodes. The leaf blade is either flat and biconvex or terete and linear with 5-21 longitudinal veins and apex obtuse or truncate.

Posidonia angustifolia, Posidonia australis, Posidonia coriacea (includes conspecific *Posidonia robertsoniae*), *Posidonia denhartogii, Posidonia kirkmanii, Posidonia oceanica, Posidonia ostenfeldii, Posidonia sinuosa*

MAP 35: *Posidonia angustifolia* **Cambridge & Kuo** (Posidoniaceae)
Shaded area = 284 000 km^2, actual species distribution is much less

Posidonia angustifolia

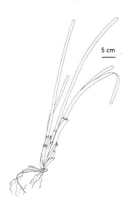

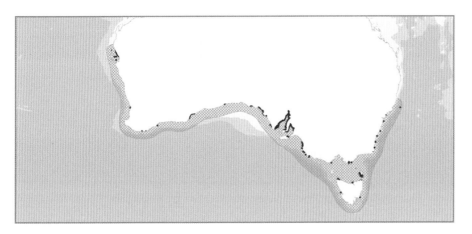

MAP 36: *Posidonia australis* **Hooker f.** (Posidoniaceae)
Shaded area = 600 000 km^2, actual species distribution is much less

Posidonia australis

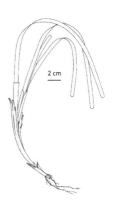

WORLD ATLAS OF SEAGRASSES

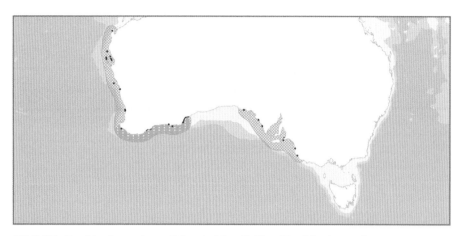
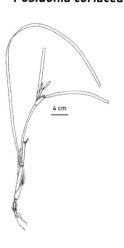

*Posidonia coriacea**

MAP 37: *Posidonia coriacea* Cambridge & Kuo (Posidoniaceae)
Shaded area = 324 000 km^2, actual species distribution is much less
Note: *Posidonia robertsoniae* has been combined with *Posidonia coriacea* as recent research[6] suggests they are conspecific.

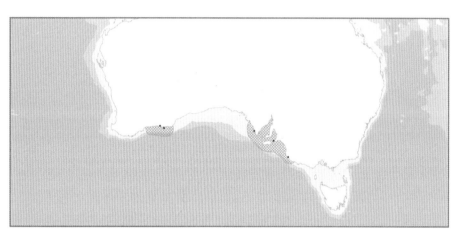
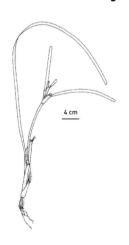

*Posidonia denhartogii**

MAP 38: *Posidonia denhartogii* Kuo & Cambridge (Posidoniaceae)
Shaded area = 137 000 km^2, actual species distribution is much less

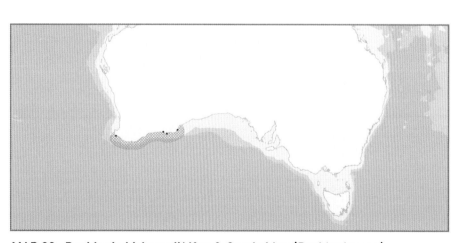
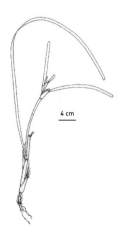

*Posidonia kirkmanii**

MAP 39: *Posidonia kirkmanii* Kuo & Cambridge (Posidoniaceae)
Shaded area = 66 000 km^2, actual species distribution is much less

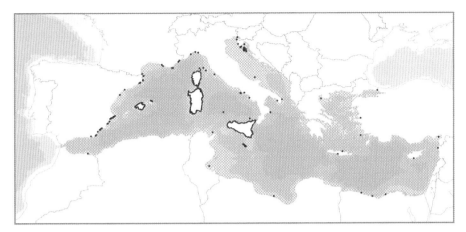
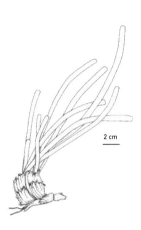

Posidonia oceanica

MAP 40: *Posidonia oceanica* **(L.) Delile** (Posidoniaceae)
Shaded area = 533 000 km², actual species distribution is much less

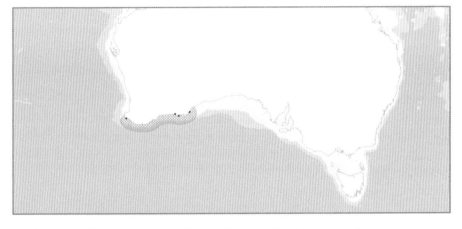
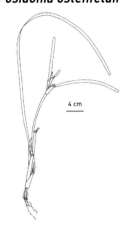

Posidonia ostenfeldii*

MAP 41: *Posidonia ostenfeldii** **den Hartog** (Posidoniaceae)
Shaded area = 66 000 km², actual species distribution is much less

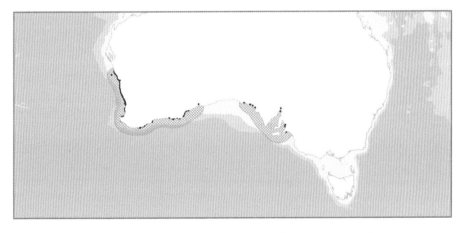
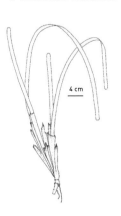

Posidonia sinuosa

MAP 42: *Posidonia sinuosa* **Cambridge & Kuo** (Posidoniaceae)
Shaded area = 266 000 km², actual species distribution is much less

WORLD ATLAS OF SEAGRASSES

Family Zosteraceae (2 genera)

Genus *Zostera* L. (9 species)

Monoecious perennial (sometimes annual) that has creeping herbaceous rhizomes with 1 to several roots and 1 shoot with 2-6 leaves. Tannin cells absent; stomata absent. The inflorescence shoots show sympodial branching; the spathe is stalked and bears alternate male and female flowers in 2 rows on the spadix without perianth. The retinacula can be present or absent. Pollination hydrophilous. The fruit is ovoid to ellipsoid. The leaf blade is linear, flattened, with 3-11 longitudinal veins. The apex shape is variable and the sheath can be open or closed. *Zostera asiatica, Zostera caespitosa, Zostera capensis, Zostera capricorni* (includes the conspecific *Zostera mucronata, Zostera muelleri* and *Zostera novazelandica*), *Zostera caulescens, Zostera japonica, Zostera marina, Zostera noltii, Zostera tasmanica* (formerly *Heterozostera tasmanica*)

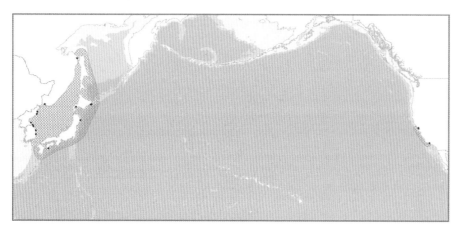

Zostera asiatica

MAP 43: *Zostera asiatica* Miki (Zosteraceae)
Shaded area = 1 311 000 km², actual species distribution is much less

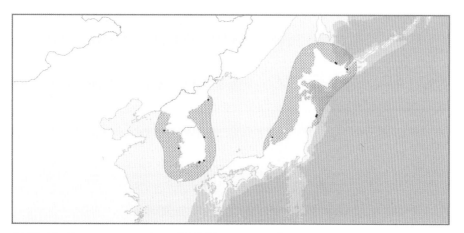

Zostera caespitosa

MAP 44: *Zostera caespitosa* Miki (Zosteraceae)
Shaded area = 445 000 km², actual species distribution is much less

Appendix 3

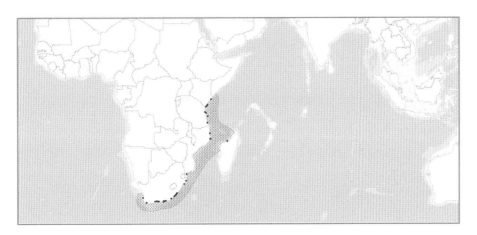

Zostera capensis

MAP 45: *Zostera capensis* Setchell (Zosteraceae)
Shaded area = 363 000 km², actual species distribution is much less

Zostera capricorni

MAP 46: *Zostera capricorni* Ascherson (Zosteraceae)
Shaded area = 543 000 km², actual species distribution is much less
Notes: *Zostera novazelandica* Setchell is now considered to be conspecific with *Zostera capricorni*.. The two show complete overlap of occurrence in New Zealand. *Zostera muelleri* and *Zostera mucronata* are also now considered to be conspecific with *Zostera capricorni*[4].

Zostera capricorni revision

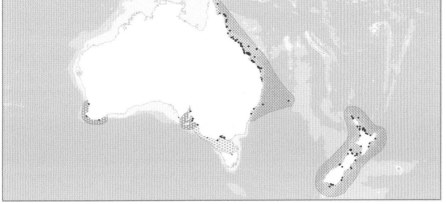

MAP 47: *Zostera capricorni* revision
Shaded area = 773 000 km², actual species distribution is much less
Note: *Zostera mucronata*, *Zostera muelleri* and *Zostera novazelandica* are now considered to be conspecific with *Zostera capricorni*[4]. The range map presented here is for the newly redefined *Zostera capricorni*.

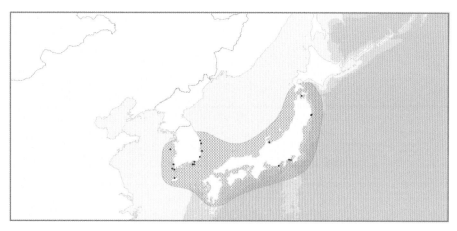
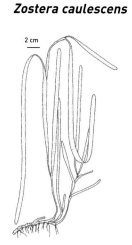

Zostera caulescens

MAP 48: *Zostera caulescens* Miki (Zosteraceae)
Shaded area = 442 000 km², actual species distribution is much less

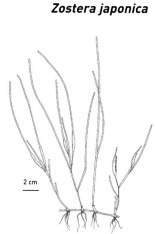

Zostera japonica

MAP 49: *Zostera japonica* Aschers. & Graebner (Zosteraceae)
Shaded area = 2 819 000 km², actual species distribution is much less

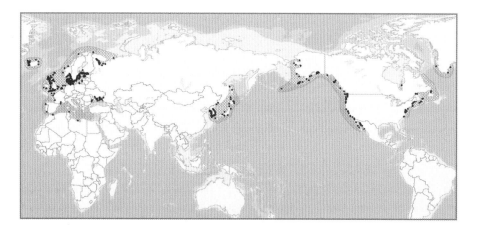

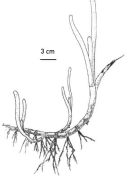

Zostera marina

MAP 50: *Zostera marina* Linnaeus (Zosteraceae)
Shaded area = 5 738 000 km², actual species distribution is much less

Appendix 3 283

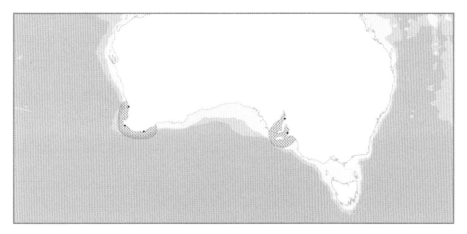
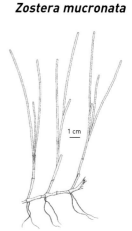

Zostera mucronata

MAP 51: *Zostera mucronata* den Hartog (Zosteraceae)
Shaded area = 116 000 km², actual species distribution is much less
Note: *Zostera mucronata* is now considered to be conspecific with *Zostera capricorni* [4].

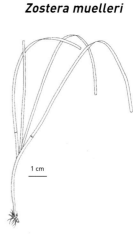

Zostera muelleri

MAP 52: *Zostera muelleri* Irmisch ex Aschers. (Zosteraceae)
Shaded area = 144 000 km², actual species distribution is much less
Note: *Zostera muelleri* is now considered to be conspecific with *Zostera capricorni* [4].

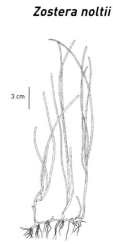

Zostera noltii

MAP 53: *Zostera noltii* Hornemann (Zosteraceae)
Shaded area = 1 571 000 km², actual species distribution is much less

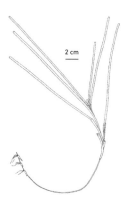

Zostera tasmanica

MAP 54: *Zostera tasmanica* (Martens ex Aschers.) den Hartog (Zosteraceae)
Shaded area = 479 000 km², actual species distribution is much less
Note: Formerly *Heterozostera tasmanica*[4].

Genus *Phyllospadix* Hooker (5 species)

Dioecious perennial with creeping herbaceous rhizomes bearing 2 to several short unbranched roots and 1 leaf at each node. The spathe is stalked and bears alternate male and female flowers in 2 rows on the spadix. The retinacula is present. Fruit are crescent-shaped and have lateral arms with hard bristles. The leaf blade is linear, flattened, subterete, sometimes leathery, sometimes rolled, with 3-7 longitudinal veins. The apex shape is variable and the sheath is open.

Phyllospadix iwatensis, Phyllospadix japonicus, Phyllospadix scouleri, Phyllospadix serrulatus, Phyllospadix torreyi

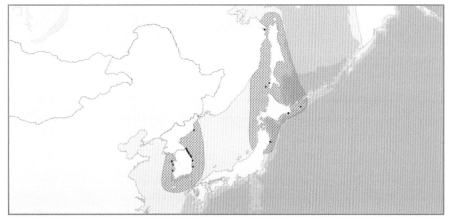

Phyllospadix iwatensis

MAP 55: *Phyllospadix iwatensis* Makino (Zosteraceae)
Shaded area = 722 000 km², actual species distribution is much less

Appendix 3   285

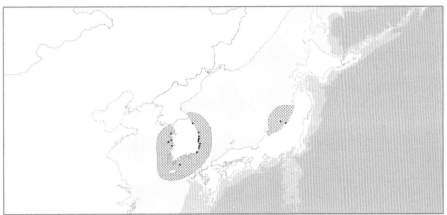
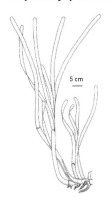

Phyllospadix japonicus

MAP 56: *Phyllospadix japonicus* **Makino** (Zosteraceae)
Shaded area = 248 000 km^2, actual species distribution is much less

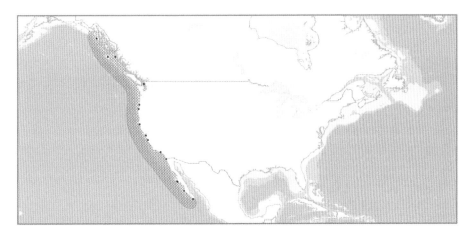

Phyllospadix scouleri

MAP 57: *Phyllospadix scouleri* **Hooker** (Zosteraceae)
Shaded area = 263 000 km^2, actual species distribution is much less

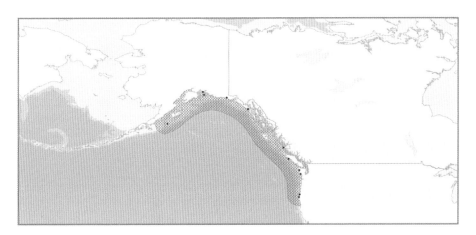
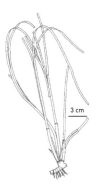

Phyllospadix serrulatus

MAP 58: *Phyllospadix serrulatus* **Ruprecht ex Aschers.** (Zosteraceae)
Shaded area = 363 000 km^2, actual species distribution is much less

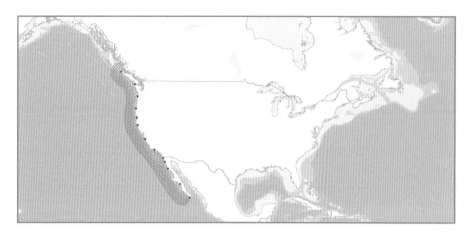
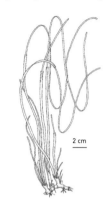

Phyllospadix torreyi

MAP 59: *Phyllospadix torreyi* S. Watson (Zosteraceae)
Shaded area = 181 000 km², actual species distribution is much less

Family Ruppiaceae (1 genus)

Genus *Ruppia* (4 marine species)

Dioecious annual or perennial with monopodially branched rhizomes and 1-2 unbranched roots per node. Root hairs abundant. Inflorescence a spike of 1-2 flowers on opposite faces of the axis, enclosed at first in the inflated sheath. Peduncle short, stout, erect or elongating greatly before anthesis to a fine thread raising the flowers to the water surface and becoming tightly spirally coiled, retracting the developing fruits. Pollination either on or below the water surface. Fruit is a fleshy drupe on a long stalk. Leaves alternate (except the 2 immediately below the flower which are sub-opposite), sheath open, edges overlapping. Blade narrow-linear to filiform, more or less concavo-convex with a large air canal either side of an inconspicuous median vein. Tannin cells present in most tissues.

Ruppia cirrhosa, Ruppia maritima, Ruppia megacarpa, Ruppia tuberosa

Range maps were not prepared for *Ruppia* species as the existing data were deemed insufficient.

Acknowledgement

We are grateful to the following sources for the information on genera taxonomy given in this appendix:

Kuo J, den Hartog C [2001]. Seagrass taxonomy and identification key. In: Short FT, Coles RG (eds) *Global Seagrass Research Methods*. Elsevier Science, Amsterdam. pp 31-58.

Womersley HBS [1984]. *The Marine Benthic Flora of Southern Australia*. Part 1. DJ Woolman, Government Printer, South Australia. 329 pp.

den Hartog C [1970]. *The Sea-Grasses of the World*. North Holland Publishing Co., Amsterdam. 275 pp.

THE GLOBAL SEAGRASS WORKSHOP

The Global Seagrass Workshop was organized and convened by UNEP-WCMC, with considerable assistance from the World Seagrass Association, in St Petersburg, Florida, USA on 9 November 2001. Twenty-three delegates (see photograph) from 15 countries prepared discussion papers for their areas of expertise, and a further five papers were received from people unable to attend (see map).

A preliminary study prepared by UNEP-WCMC, including seagrass distribution and diversity maps based on an extensive literature search, framed much of the discussion at the workshop. Delegates debated the map results and marked corrections. A standard species list was agreed upon and information on economic value, uses and threats, associated species and management interventions was shared. The workshop ended with a discussion of global priorities for seagrass research and policy and a commitment by all delegates to contribute a regional chapter to this *World Atlas*.

Delegates at the Global Seagrass Workshop. From left, front row: Caroline Ochieng, Mark Spalding, Fred Short, Michelle Taylor, Hitoshi Iizumi. Second row: Evamaria Koch, Chatcharee Supanwanid, Graeme Inglis, Joel Creed, Nataliya Milchakova, Salomão Bandeira. Third row: Paul Erftemeijer, Rob Coles, Tanaji Japtap, Miguel Fortes, Diana Walker, Hugh Kirkman, Jorge Herrera-Silveira, Japar Sidik Bujang. Back row: Kun-Seop Lee, Ron Phillips, Andrea Raz-Guzman, Sandy Wyllie-Echeverria.

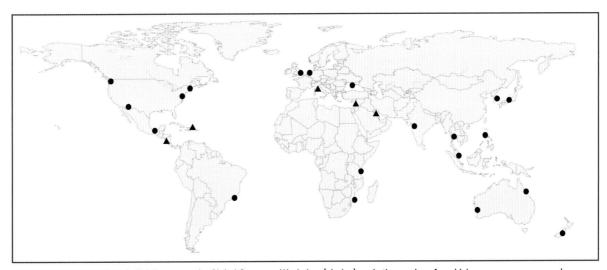

Map showing the location of all delegates at the Global Seagrass Workshop (circles), and other regions for which papers were prepared (triangles).

Index to
THE WORLD ATLAS OF SEAGRASSES

Page references in **bold** refer to figures in the text; those in *italics* refer to tables or boxed material

A

Abrolhos Bank, Brazil 243, *247*
Abu Dhabi Emirate 74, 78, 80
Acanthopagrus schlegeli 196
Acanthophora 235
adaptations of seagrasses 5
Adriatic Sea *50*, *52*
Aegean Sea 65, **67**, 68, 71, 72
Africa *see* East Africa; South Africa; West Africa
agriculture
   as threat to seagrasses 99, 130
   uses of seagrasses 61, 62, 188, 197
aircraft runways *164-5*, 166
Akkeshi, Hokkaido 187, *189*
Al Iskandarîya (Alexandria) 66-7
Alabama 225
Åland Islands, Finland **29**, 30
Alaska 199-204
Albermarle Sound, North Carolina 218
Alexandria 66-7
Alfacs Bay, Spain *54*
algae *see* epiphytic algae; filamentous algae; macroalgae; phytoplankton
Algeria *251*
alien species 49-50, 114, 203-4, 205
   macroalgae 55, 68, 69, 71, *155*
   seagrasses 114, 203-4
Amami Islands, Japan 186
Amblygobius albimaculatus 96
Ambon Bay, Indonesia *179*
Ammonia 175
Amphibolis antarctica 113, *117*, 122, *251*-5, **270**
Amphibolis griffithii *251*-5, **270**
Amphiroa fragilissima 155
Anambas Island, Indonesia *178*
anchor damage 43, 204, *247*
Andaman and Nicobar Islands 101, 102-3, *104*, 105

Andaman Sea 144-50
Angola *251*
Anguilla *251*, *256*
Anguilla anguilla 33
animal fodder 61, 62
annual populations 121
antibiotics 114
antifouling compounds 114
Antigua and Barbuda 238, *251*, *256*
Antilles, Netherlands *253*, *259*
Aplysia punctata 40
Apseudes chilkensis 175
Aqaba (Elat), Gulf of **67**, 68-9, 70-1, 74
aquaculture
   fish 114, 154, *158*, 213
   seaweeds 87-8, **176**, 197
   shellfish 43, 154, 197, 205, 213
Arabian Gulf
   biogeography 74-8
   policy and management 80
   seagrass distribution 78
   threats and seagrass losses 78, 80
   turtles and dugong 79
Arabian Sea 75
Aral Sea 59, **61**, 63
Aransas Bay, Texas 227
Arenicola marina 44
Argentina *14*, 246
Argopecten irradians 209-10, 216
Argopecten purpuratus 246
ark shell 97
Aru Islands, Indonesia *179*
associated biota 10-12
   Brazil 244, 245
   Caribbean 235-6
   East Africa 83, 85
   endangered/threatened *12*, 17, 40, 183
   India 101-2, **105**, *107*
   Indonesia 174-80
   major taxonomic groups *11*
   Mediterranean 51
   migratory movements 235
   New Zealand 134-5, *140*
   north western Atlantic 209-10
   Philippines and Viet Nam 183
   Scandinavia 27, 28, 30
   western Europe 40

Asterina pancerii 51
Atlantic Ocean *see* mid-Atlantic region; North Atlantic
Australia
   associated species 11
   endemic species *12*, *110*, *117*
   Northern Territory 111
   policies and protection 23
   seagrass species *251*
   southern coast 9, 122, 128
   see also Eastern Australia; Western Australia
Austrovenus stutchburyi 135
Avon-Heathcote Estuary, New Zealand *135*, *137*, 138, 139-40
Aythya americana 228
Azerbaijan 62, *251*
Azov Sea 59, 61-2

B

Bagamoyo, Tanzania 87
Baguala Bay, Indonesia *179*
Bahamas *251*, *256*
Bahía de Chengue, Colombia 239
Bahrain 74, *75*, 78, 80, *251*, *256*
   causeway 6, *76*
Baja California 199, 201
Balearic Islands 53
Baltic Sea 28-30, 33, 34, 35
Bangka, Indonesia *178*
Bangladesh *251*
banknote, Brazilian **249**
"banquette" 53
Banten Bay, Indonesia 175-6, *178*, 180
Barbados *251*
Barbatia fusca 97
Barnegat Bay, New Jersey 209, 210
bass, striped 209
Batophora 235-6
bay scallop 209-10, 216
beach cast material 86, *88-9*
beach seines 148-9
Belitung, Indonesia *178*
Belize *14*, 236, 237, 239, *251*, *256*

Benoa Bay, Indonesia *178*
Bermuda *251*
biodiversity
   centers of 9, 10, 185
   in seagrass ecosystems 10-12, 17
   seagrasses 9, **22**
biogeographic patterns 9-10
biomass 16
   Baltic Sea **28**
   below-ground 172
   Gulf of Mexico 231-2
   India 104
   Indonesia 172-3
   Japan 187
   Zostera 196
birds 40, 141, *189*, 199, *200*, 210, *228*, 245
Biscay, Bay of *45*
bivalves
   digging/dragging for 98, *99*, 157, *158*, 210, 212, *213*
   dredging for 43-4, 219, *220-1*
   East Africa 83, 85
   endemic *140*
   Malaysia 157, *158*
   New Zealand 135, *140*
   protected 51
   western Europe 40, 43-4
   western north Atlantic 209-10, 212, *213*
black brant *200*
Black Sea 59-61
"blow-out" areas 18, 237-8
boating 18
   Brazil *247*
   East Africa 86-7
   mid-Atlantic region 219
   North America 204, 212, 213
   western Europe 42, 43
   see also shipping
Bonaire *14*, 238
Boston Harbor 208
Botany Bay 123
bottlenose dolphin *248*
Boundary Bay, Canada 201-2
Branta bernicla (brant goose) 40, 210
Branta bernicla nigricans (black brant) *200*
Brazil
   Abrolhos Bank 243, *247*
   biogeography 245-6
   ecosystem description 243-4
   Itamaracá Island 243, *244*
   Patos Estuarine Lagoon 244, *248*
   policy and protection 248-9, *256*
   seagrass losses and coverage *14*, 246
   seagrass species *251*
Brest, Bay of 40
British Columbia 199-204
British Indian Ocean Territory *251*, *256*
Brunei *251*
Buenos Aires 246
Bulgaria *251*
"bullata" ecophene 66

C

Calcarina calcar 175
Callinectes sapidus 216, 245-6, *248*
Callophyllis rhynchocarpa 197
Calotomus carolinus 85-6
Calotomus spinidens 96, 97
Cambodia *251*, *256*
Canada
   Atlantic coast 207, 212
   Hudson Bay 207, 212
   marine protected areas *256*
   Pacific coast 199-204
   seagrass species *251*
Cape Cod 208, 210
Cape Lookout, North Carolina 218
capybara 245
carbon cycle 17, *84*, 85
carbon dioxide levels 85
Caribbean 9, 234-41
   ecosystem description 235-8
   historical perspectives 238
   policy and protection 241
   species and coverage 234-5, 238
   threats to seagrasses 238-9, 241
   see also named islands
Caribbean Coastal Marine Productivity (CARICOMP) network *239*, 241
CARICOMP network *239*, 241
Carpentaria, Gulf of 119, 120, 125
Caspian Sea **61**, 62-3
Caulerpa 232, 235
   eastern Mediterranean 68, 69, 71
   Malaysia *155*, *156*
   western Mediterranean *52-3*, *54*, 55
Caulerpa cactoides 126
Caulerpa prolifera 66, 68, 71, 155
Caulerpa racemosa 55, 68, 69, 71, 155
Caulerpa scalpelliformis 68
Caulerpa serrulata 70
Caulerpa taxifolia 55
causeway development
   Bahrain 6, *76*
   Kosrae *164-5*, 166, 167
Cayman Islands *251*, *256*
Ceramium 60, 61, 62
Cerithium tenellum 175
Chaetomorpha linum 219, *221*
Champia sp. 197
Chandeleur Islands, Louisiana 226, 227
Chara 62
Charlotte Harbor, Florida 225
Charophyceae 63
Cheilio enermis 177
Chelonia mydas see green turtle
Chesapeake Bay, US 216, 218, 220-1
Chicoreus ramosus 98
chicoric acid 55
Chile 10, *14*, 243, 246, *251*
China 185, *251*
Chinocoteague Bay *220*

Christchurch, New Zealand 137, 138
Christmas Bay, Texas 227
Chrysochromulina 31
Chwaka Bay, Tanzania 85, 86
Cladophora 60, 61
Cladophora glomerata 33
clam worms 213
clams 246
   dredging for 43-4, 219, *220-1*
   hand digging 43, 212, *213*
   soft-shell *213*
Clean Water Act (US) 220
climate change 167, 191, 241
   potential response of seagrasses 17, 85
coastal development
   Arabian region *76*, 78, 80
   Caribbean 238-9, 241
   Japan 188-9
   Malaysia 157
   Mediterranean 55
   northeastern United States *211*
   Pacific islands 161
   South America 246-7
   Western Australia 112
   western Pacific islands 165, 166-7
   see also land reclamation
coastal management
   East Africa 89-90
   rapid assessment technique *77*
coastal protection 17, 86-7, *88-9*, *116-17*, 147
Cockburn Sound, Western Australia 111, 112, 113
cockles 135, 246
cod, Atlantic 209
Colombia *239*, 241, *251*, *257*
Comoros 95, *251*
Connecticut, US 209, 212
Coorong Lakes, Australia 130
coot, red-gartered 245
coral reefs **2**
   East Africa 82, *84*
   Eastern Australia 120-1
   "halo zone" 71
   Malaysia 1, 153, *155*, 157-8
   Red Sea 69-70
   South America 245, **246**
   Thailand 145-6
   western Pacific 161-3
corals 245, **246**
Core Sound, North Carolina 218
Corpus Christi Bay, Texas 227, *229*
Corsica *51*, *53*, 55
Costa Rica 238, *251*, *257*
crabs 40, 83, *156*, 196-7
   blue 216, 245-6, *248*
   hermit 101
   New Zealand endemic 135
   Thailand 148
   threatened species 12
croaker, whitemouth 248
Croatia *251*, *257*
crustaceans 40, 97-8, *156*
   Caribbean 236-7
   Indonesia 175

seed consumption *190*
South America 245-6
see also named groups and species
Cuba 224, 238, 241, *251, 257*
Curaçao *14*, 238
curio goods 85, 98
Curonian Spit, Lithuania 30, 33
Cuyo Island *183*
Cygnus atratus 141
Cygnus cygnus 189
Cygnus melancoryphus 245
Cymodocea angustata 251-5, **271**
Cymodocea nodosa
   by country/territory *251-5*
   eastern Mediterranean 65-8
   range map **271**
   western Mediterranean 48, *49, 50, 52-3, 54*
Cymodocea rotundata
   by country/territory *251-5*
   Eastern Australia *128*
   India *104*, **105**, *106-7*
   Indonesia 171, *172, 173-4, 178-9*
   Japan 187
   Malaysia *156*, 157
   Mozambique 95, *96, 99*
   photosynthetic studies 85
   range map **272**
   Red Sea 69-70, *71*
   Thailand 144, *146, 147*, 148
   western Pacific *168, 169*
Cymodocea serrulata
   by country/territory *251-5*
   India *104*, **105**
   Indonesia 171, *172, 173-4, 178-9*
   Japan 187
   Malaysia 153, *155*
   Mozambique 95, *96, 99*
   range map **272**
   Red Sea 68-9
   Thailand 144, 145, *146, 147*
   western Pacific 163
Cymodoceaceae 6
Cypraea tigris 85
Cyprus **67**, 68, 71, *251, 257*

D

Dahlak Archipelago 69
Dar es Salaam 87
Dasycladales 235
David and Lucile Packard Foundation *168*
Deception Bay, Queensland 123
decline, global 20
deepwater seagrasses 69, 70, 120-1, 123, *124-5*, 130
definitions 5-7
deforestation 138-9
Delaware 216-22, *220-1*
Denmark
   policies and protection 34-5
   seagrass distribution and losses 27-8, 31-2, **33**

seagrass species *251*
uses of seagrasses 30
depth distribution
   Australia 120-1, 123, *124-5*, 130
   Denmark 27, 32, **33**
   Japan 186-7
   Malaysia 154
   Red Sea 69, 70
   and water quality 27, 32, **33**, 112-13, 154
   western Pacific 162
   Zostera marina 27, 32, **33**, 232
Derawan Islands, Indonesia *178*
detritus
   direct uses 55, 197
   ecological value 40
   see also beach cast
Diadema antillarum 236
Diadema savignyi 86
Diadema setosum 86, 98
Diani-Chale Lagoon, Kenya 86
diatoms 101
Dick, Chief Adam *202*
die-back *see* wasting disease
Diogenes 175
direct habitat maps 7
distribution maps
   calculating global areas 13-16
   development 3, 7-8
   geographic regions 9-10
   limitations 13
   seagrass habitat 7, 13
   seagrass species 8-10, 262, **263-86**
dolphin, bottlenose 248
Dominica 257
Dominican Republic 241, *251, 257*
Donuzlav Salt Lake 60
dragging, net 213
dredging
   Arabian Gulf *76*, 78, 80
   for bivalves 43-4, 219, *220-1*
   Malaysia 157
   mid-Atlantic US 219
duck, redhead 228
dugong (*Dugong dugon*) *2*, 183
   Arabian Gulf **78**, *79*
   captivity 180
   East Africa 83, 88
   Eastern Australia 119
   feeding *147*, **149**
   India 102, 105
   Indonesia 180
   Japan 189
   Malaysia 154, *156*
   Thailand 145, *147*, 148-9
   western Pacific 164

E

earthquakes 238
East Africa 82-90
   biogeography 82-5
   policies and protection 88-90, *258, 260*
   seagrass coverage 87-8

seagrass productivity and value 85-7
threats to seagrasses 86-8
Eastern Australia 119-31
   biogeography 120-2
   physical characteristics 119
   policies and protection 23, 130-1, *256*
   seagrass coverage *14*, 125-8
   seagrass losses 122-5, *126-7*
   threats to seagrasses 129-30
   uses of seagrasses 122, 128-9
echinoderms 97-8, 176
   see also named groups and species
Echinometra mathaei 86
Echinothrix diadema 86
ecological value *15*, 16-17, 24, 129
   carbon sequestration 17
   coastal protection/sediment stabilization 17, 86-7, *88-9, 116-17*, 147
   nutrient cycling 40, 84, 85
economic value 17-18
ecosystems
   adjacent to seagrasses 84
   seagrass 7, 10-13
   see also associated biota
ecotypes 66, 194-5
eel 33
eelgrass *see Zostera marina*
Egretta garzetta 158
Egypt 65, 66-7, **67**, *252*
El Dab'a 66
El Niño Southern Oscillation (ENSO) 205, *226*
El Suweis (Suez) **67**, 68-9, 70-1
Elat, Gulf of **67**, 68-9, 70-1, 74
Elpidium 175
emperor fish
   pink ear 95
   variegated 96
endangered species
   associated biota 17, 40, 183
   seagrasses 12-13, 188, *236-7*
endemic species 12, *110, 117*, 140
Enhalus acoroides **13**
   by country/territory *251-5*
   India 103, *104*
   Indonesia 171-80, **181**
   Japan 187
   Malaysia 154, *156*
   Mozambique 95, *96, 99*
   Philippines 183
   range map **263**
   Red Sea 68, 70
   seeds **149**, 166-7
   Thailand 144, 145, *146, 147*
   transplantation **176**
   uses 87, 122, 154, 166-7
   western Pacific 162, 166-7, *168-9*
ENSO (El Niño Southern Oscillation) 205, *226*
Enteromorpha 60, 61, 62, 87
environmental impact assessment *6, 76, 77*

epibenthos *124-5*
epifauna 27, 28, 101, *190*
Epinephelus malabaricus 147
epiphytic algae 10-11
   Black Sea 60
   Caribbean 235, 236-7
   India 101
   Korea 197
   Malaysia *156*
   Western Australia 113
   western Europe 40
Eritrea 69, *252*
Esox lucius 33
Estonia **29**, 34, *252*
Eucheuma 237
Eucheuma spinosa 87-8
Euro-Asian seas
   seagrass coverage *14*
   see also Aral Sea; Azov Sea; Black Sea; Caspian Sea
Europe, western
   historical perspectives and losses 40-3
   policies and protection 44-5
   seagrass coverage *14*, 43
   threats to seagrasses 43-4
   uses and value of seagrasses 38
   see also Mediterranean Sea; Scandinavia
European Union (EU)
   Habitats Directive 35, 44, 46, 52
   Water Framework Directive 35
eutrophication 3
   Black Sea 60
   Caribbean 239
   Denmark 32, **33**, 34, 35
   East Africa 87
   Gulf of Mexico 232
   Mediterranean 55
   Western Australia 112-13
   western Europe 42, 44
evolutionary origin 10

F

Farewell Spit, New Zealand 141
Faure Sill, Western Australia *116-17*
ferry terminals 204, **205**
Fiji 162, 166, 167, *169*, *252*
filamentous algae 33, 60, 61
Finland **29**, 30, 35, *252*
First Nations people 199
fish farming 154, *158*, 213
fish habitat areas (FHAs) 131
fish species
   associated with seagrasses 11, *12*
   Caribbean 235-6, 237
   Indonesia 176-7, 180
   Korea 196
   Mozambique 95-6
   threatened *12*
   see also named fishes
fisheries
   Arabian Gulf 77-8
   Brazil 245-6
   Caspian Sea 63
   East Africa 83, 85-6, 88
   Eastern Australia 128-9
   Korea 196-7
   Malaysia 154, *156*, *158*
   mid-Atlantic region 216
   Mozambique 95-8
   New Zealand 135
   north western Atlantic 209-10
   shrimps 77-8, 128-9, *231*
   South America 245-6
   Thailand 147-9
   value of seagrasses 17
   western Europe 40
   western Pacific 163-4
fishing methods
   invertebrates 97-8, **99**, 212, *213*, 219, *220-1*
   mechanized 148-9
   seine net 85-6
   traps 96-7
   Western Australia 115
Flores Sea 173-4
Florida 11
   east coast *14*, *236-7*
   Gulf of Mexico 224-5
flounder, winter 209
flowering 187-8, 201
flowers
   female **9**, **240**
   male **189**, **237**
foraminifera 175
forestry 138-9
France
   Atlantic coast 39, 40, 43, 44, *45*
   Caribbean islands *252*
   Mediterranean coast *51*, 55, 56
   protected areas *257*
   seagrass species *252*
fruiting 121, **149**, 187-8
Fucila armillata 245
Funakoshi Bay, Honshu *190*
fungi 101

G

Gadus morhua 209
gains, seagrasses *128-9*
Galeta Point, Panama 235
Galveston Bay, Texas 227, 230
gastropods 28, 44, 83, 85, 155, *156*, 175-6, 236, 237
Gazi Bay, Kenya 82, *84*, 85, 86
genetic testing 262
Geographe Bay, Western Australia 109-10, 111, 112
geographical information systems (GIS) 7, 166
geographical regions 9-10
Germany 28, **29**, 30, 33, *252*, *257*
Gerres oyena 83, 96
Gerupuk Bay, Indonesia 173, 175, **177**, *178*
Gilimanuk Bay, Indonesia *178*
Gippsland Lakes, Australia 123
Glénan Archipelago *45*
global habitat distribution 13
global seagrass area 13-16
Global Seagrass Workshop 2, 287
global warming *see* climate change
goatfish, dash-dot 97
goby, tailspot 96
Gracilaria coronopifolia 155, *156*
Grand Cayman *14*, 238
grazing
   birds 28, 40, 141, 189, 199, *200*, 210, 245
   dugong 147
   fish 83, 85-6, 183
   sea turtles *247*
   sea urchins 86, *236*, 241
   snails 28, 44
Great Barrier Reef 120-1, 126-7
   deepwater seagrasses *124-5*
   Green Island seagrass meadows *128-9*
   Marine Park 23, 131
Great Bay, New Hampshire 207, *208*, 210
Greece 66, **67**, 68, *252*
Green Island seagrass meadows *128-9*
green turtle 102
   Arabian Gulf *79*
   Caribbean 238
   East Africa 83, 88-9
   Eastern Australia 119
   Pacific 164
   South America 245, *247*
Greenland *252*
Grenada 52
groundwater 85, *211*
groupers *12*, **176**
   Malabar 147
Guadeloupe *14*, 238, *252*, *257*
Guam 162, *257*
Guatemala *252*, *257*
Guinea Bissau *252*
Gulf of Arabia *see* Arabian Gulf
Gulf of Carpentaria 119, 120, 125
Gulf of Kutch 105
Gulf of Mannar 102, *104*, **105**
Gulf of Mexico *14*, 224-32
Gulf of Thailand 144-50
Gulf War oil spill 75-7
Guyana 241

H

Haad Chao Mai National Park, Thailand 144-5, *146*, 147, 149
habitat distribution, global 13
habitat maps 7, 13
Habitats Directive (EU) 44, 46
Haiti 241, *252*
Halimeda 235, *239*
Halodule spp.
   leaf morphology 82-3
   taxonomy 243-4, 262, 272
Halodule beaudettei 251-5, 262, **273**

Halodule bermudensis 251-5, 262, **273**
Halodule emarginata 243-4, 251-5, **273**
Halodule pinifolia
  by country/territory 251-5
  India *104*
  Indonesia 171, *172*, *178-9*
  Malaysia 153, 155-7
  range map **274**
  Thailand 144, *146*, *147*
Halodule uninervis
  Arabian region 74, 75-8, **80**
  by country/territory 251-5
  Eastern Australia 128
  genetic studies 262
  India *104*, **105**
  Indonesia 171, 172, 173, *178-9*
  Malaysia 154-5, *156*, 157
  Mozambique 95, *96*, *99*
  range map **274**
  Red Sea 69-71
  shoot density 173
  Thailand 144, *146*, *147*
  western Pacific 166
Halodule wrightii
  Brazil 246
  by country/territory *251-5*
  Caribbean 234, 235, *236*, **237**, *239*, *240*
  East Africa 82-3, 87
  Gulf of Mexico 225, 227, *228-9*, 230, 231-2
  India *104*
  male flower **237**
  mid-Atlantic region 216-17
  Mozambique 95, *96*, *99*
  northeast Pacific 200, 203
  northernmost limit 216-17
  range map **274**
  salinity tolerance *228-9*, 231-2
  South America 243, *244*, 245, *247*
  taxonomy 82-3, 262
Halophila spp. 121, 235, 263
Halophila australis 251-5, **264**
Halophila baillonii
  Brazil 243, *244*
  by country/territory 251-5
  Caribbean 234-5
  range map **264**
Halophila beccarii 171
  by country/territory 251-5
  India 102, 103-4
  Malaysia 153, 156-7
  range map **264**
  Thailand 144, *146*
Halophila capricorni **9**, 251-5, **265**
Halophila decipiens
  by country/territory 251-5
  Caribbean 234, 235
  Eastern Australia *125*
  Gulf of Mexico 224, 230, 231, 232
  India *104*, 105
  Indonesia 171, *178-9*
  Malaysia 153, 154, *155*
  range map **265**

  South America 243, 245
  southernmost limit 243
  Thailand 144, 145, *146*, 148
Halophila engelmanni
  by country/territory *251-5*
  Caribbean 234-5
  Gulf of Mexico 224, 225-6, 227-31
  range map **265**
Halophila hawaiiana 251-5, **266**
Halophila johnsonii 236-7, 251-5, **266**
Halophila minor 262
  by country/territory *251-5*
  Indonesia *178-9*
  Mozambique *99*
  range map **266**
  Thailand 144, *146*
Halophila ovalis **2**
  Arabian region 74, 75, 77, 78
  by country/territory *251-5*
  dugong grazing *147*
  Eastern Australia 122, *125*, 128
  India *104*
  Indonesia 171, *172-3*, *178-9*
  Japan 186
  Malaysia 153, 154-6
  Mozambique 95, *96*, *99*
  photosynthetic studies 85
  range maps **267**
  Red Sea 69, 70, 71
  salinity tolerance 74
  taxonomy 262
  Thailand 144, 145, *147*, 148
  var. *ramamurtiana* 104
  western Pacific 162-3, **167**
Halophila ovata
  by country/territory *251-5*
  India *104*
  range map **267**
  Red Sea 69
Halophila spinulosa
  Australia *124-5*
  by country/territory *251-5*
  Indonesia 171, *177-8*
  Philippines 183
  range map **268**
Halophila stipulacea
  Arabian region 74-6, 78
  by country/territory *251-5*
  East Africa 87
  eastern Mediterranean 65-6, 68
  India *104*
  Mozambique 95, *99*
  range map **268**
  Red Sea 68-72
  salinity tolerance 74
  western Mediterranean 48, 49-50, 52-3
Halophila tricostata
  by country/territory *251-5*
  Eastern Australia 121, 124, *125*
  range map **268**
Hawaiian Islands 166
heavy metal pollution 33, 87
Helsinki Convention (HELCOM) 35
herring, Pacific 199

Hervey Bay, Queensland 121, 122, 123, *125*
Heterozostera tasmanica see *Zostera tasmanica*
Hippocampus 40, 51
Holothuria atra 148, 164
Holothuria scabra 85, *96-7*, 148, 176
Holothuroidea 85, *96-7*, 98, 148, 154, 163-4, 176
Homarus americanus 209
Honduras 252, 257
hotspots, biodiversity 9, 10, 185
Hudson Bay 207, 212
human food 87, 122, 136, 148, 166-7, 199, 204
Hurricane Carla 227
Hurricane Gilbert 238
Hurricane Hugo 237-8
Hurricane Roxanne 231
hurricanes 218, 219, 227, *231*, 237-8
hydraulic dredges 219, *220-1*
Hydrobia 28, 44
Hydrocharis hydrochaeris 245
Hydrocharitaceae *6*
Hydropuntia 237
hydrothermal vents 72

I

Iceland 27, 252
India
  associated biota 101-2, **105**
  biogeography 102-3
  Kadmat Island *106-7*
  policies and protection 105, 108, 257
  seagrass coverage *14*, *102*
  seagrass species 252
  threats to seagrasses 104-5
Indian River, Florida *236*
Indonesia 257
  associated biota 174-7, 179-80
  historical perspective 180
  policy and protection 180-1, 257
  seagrass coverage *14*, **178-9**, 180
  seagrass species and ecology 171-4, 252
Inhaca Island, Mozambique *96-7*, *99*, 100
integrated coastal zone management 89-90
international agreements 35
introduced species *see* alien species
invertebrates
  India *107*
  New Zealand 134-5, *140*
  southeastern Africa 97-8, **99**
  see also named species and groups
Ionian Sea 66, **67**
Iran 74, *75*
Iraq *75*
Ireland 38-9, 252
  Northern 43
Israel 67-8, *70*, 252, 257

Italy *50*, 51, *52, 252, 257*
Itamaracá Island, Brazil 243, *244*
Iwate Prefecture, Japan *190*
Izembek Lagoon, Alaska 199, *200*

J

Jamaica 238, 241, *252, 257-8*
Japan 185-6
   biogeography 186-8
   historical losses 188
   losses of seagrasses 189
   seagrass coverage *14*, 188
   seagrass species *252*
   threats to seagrasses 188-9
   uses of seagrasses 188
Jervis Bay, New South Wales 11
Jordan 69, *252*
Jubail Marine Wildlife Sanctuary, Saudi Arabia 78

K

Kadmat Island, India *106-7*
Kaduk Island, Korea 197
Karimata Island, Indonesia *178*
Karkinitsky Gulf 60
Kattegat Strait, Denmark 34, 35
Kazakhstan *252*
Kenya
   biogeography 82-3, 85
   Gazi Bay 82, *84*, 85, 86
   policies and protection 88-90, *258*
   seagrass coverage 87-8
   seagrass productivity and value 85-7
   seagrass species *253*
   threats to seagrasses 86-8
Kepulauan Seribu reefs 175
Kerch Strait 60
Kiel Bight 28
Kimberley coast, Western Australia 110
Kiribati *253*
Ko Samui, Thailand 145-6
Ko Talibong, Thailand 145, *146*, 149
Korea, Republic of 185
   biogeography 193-7
   historical losses 197
   policy and protection 198, *258*
   seagrass distribution and coverage *14*, 193, 197
   seagrass research *196*
   seagrass species *253*
   threats to seagrasses 198
   uses of seagrasses 197
Kos 68
Kosrae, Micronesia *14*, *164-5*, 166, *168*, **169**
Kotania Bay, Indonesia 175, *176*, *178-9*
Krasnovodsky Bay 62-3
Kung Krabane Bay, Thailand 145, *146*
Kuta Bay, Indonesia *177, 178*, 180
Kutch, Gulf of 105
Kuwait 74, *75, 253*
Kuwait Action Plan 80
Kwakwaka'wakw Nation *202*
Kwangyang Bay, Korea 197
Kwazulu-Natal, South Africa 94
Kylinia 60

L

Labyrinthula zosterae 18, 38, 138, 207
Laguna de Alvarado, Mexico 230, *231*
Laguna de Tamiahua, Mexico 230, *231*
Laguna de Términos, Mexico 230, *231*
Laguna Madre, Texas 227, *228-9*
Laguna Ojo de Liebre, Baja California 199
Lakshadweep Islands 102, 104, *106-7*
Lampung Bay, Indonesia *178*
Lamu Archipelago, Kenya 83, 87
land reclamation
   Arabian Gulf *76*, 78, 80
   East Africa 88
   Japan 189
   Korea 198
   Malaysia 155-7
   Philippines 183-4
   see also coastal development
latitude 13, 209
Latvia **29**, 34, *253*
Laurencia 62, 235, 236-7
Law of the Sea Treaty 167
Lebanon **67**, 68, *253*
Lelu Island, Micronesia *164-5*
Lepidochelys olevacea 102
Lepilaena 6
Leptoscarus vaigiensis 95, 97
Lethrinus lentjan 95
Lethrinus variegatus 96
Libyan Arab Jamahiriya *253*
light availability
   Australia 111-13
   Scandinavia 27, 32, **33**
   western Europe 44
   see also eutrophication; sediment loading
light requirements 111-12, 235
Limassol, Bay of 65
Limnoria simulata 236
Limón earthquake 238
limpet
   eelgrass *140*, 210
   New Zealand *140*
Lithuania **29**, 30, 33, *253*
Little Egg Harbor, New Jersey 210
lobster 209
   spiny 203, 237
Lombok, Indonesia 171, *172*, 173, 174-6, *177*, 180
Lomentaria hakodatensis 197
Long Island, US 209, 212
losses, global 20
Lottia alveus *140*, 210

Louisiana, US 226, 227
Lutjanus fulviflamma 97

M

macroalgae
   alien species 55, 68, 69, 71, *155*
   Baltic Sea 34
   Caribbean 235, 236-7
   culture 87-8, **176**, 197
   East Africa *84*, 87
   Eastern Australia *126-7*
   Euro-Asian enclosed seas 60, 63
   Gulf of Mexico 232
   India 102
   Indonesia 174
   Japan *188*
   Malaysia *155-6*
   Mediterranean *52, 53*, 55
   mid-Atlantic region 219, *221*
   Mozambique 93
   Scandinavia 33, 34
   Western Australia 113
   western Europe 40, 44
Macrophthalamus hirtipes 135
Madagascar 94, 100, *253, 258*
maerl bed **46**
Magellan Straits 244
Maia squinado 40
Maine, US 207-10, *212, 213*
Makassar Strait *179*
Malaka Strait *179*
Malaysia
   biogeography 153-4
   ecosystem description 152-3
   historical perspectives 152, 154
   macroalgae community *155-6*
   policies and protection 152, 158-9, *258*
   seagrass species *253*
   seagrass losses and present coverage *14*, 154-7
   Tanjung Adang Laut shoal 154, *155*
   threats to seagrasses 157-8
Maldives *253*
Maluku Island, central *179*
manatee 12, *236*, 238, 245
Manatee County, Florida 225
manatee grass *see Syringodium filiforme*
mangroves 82, *84*, 85, 110, 245
Manila Bay *183*
Mannar, Gulf of 102, *104*, **105**
Maori people 135-6
maps
   calculating global areas 13-16
   data sources and methods 3, 7-8
   deepwater seagrasses *124-5*
   limitations 13
   seagrass habitat 7, 13
   species ranges 3, 262, **263-86**
Maputo Bay, Mozambique *96-7*, 99
Maquoit Bay, Maine 210, *213*
marema fish traps 96-7
mariculture *see* aquaculture

marine protected areas
  Australia 115-17, 131, *256*
  East Africa 88, *258, 260*
  global *19*, 20, 23, *256-61*
  Malaysia 152, 158-9, *258*
  New Zealand 141
  western Pacific 167
Marsa Matrûh Harbor, Egypt 66
Marseille-Cortiou region, France *51*
Marshall Islands 166, *253*
Martinique *14*, 236, 238, *252, 253, 258*
Maryland, US 218, 221
Massachusetts, US 207-8, 210, 212
Mauritania *253, 258*
Mauritius **94**, 95, 99, 100, *253, 258*
Mayotte *253*
Meandrina brasiliensis 245
Mecklenburger Bight 28
Medes Islands, Spain *54*
Mediterranean Sea, eastern 65-8, 71-2
Mediterranean Sea, western 9, 48-56
  associated species 51
  productivity and biomass 52-3, 55
  seagrass coverage *14*, 51-2
  species distribution 48-51, *52-3*
  threats to seagrasses 55-6
meiofauna
  India *107*
  Indonesia 174-5
Mexico
  Caribbean coast 238, *240*
  estimated coverage *14*
  Gulf of Mexico coast 230-2
  Pacific coast 199, 201
  protected areas *258*
  seagrass species *253*
Mexico, Gulf of *14*, 224-32
Micronesia 161
  biogeography 162
  historical perspectives 164-6
  Kosrae Island *14*, 164-5, 166, *168*, **169**
  policy and protection 167, 170
  seagrass species *253*
  SeagrassNet *168-9*
  uses and threats to seagrasses 166-7
mid-Atlantic region
  biogeography 216-17
  historical perspectives 217-18
  policy and protection 219-22
  seagrass distribution 218-19
  threats to seagrasses 219, *220-1*
Miliolina 175
mining 55, 149, 155-7, 239, 241
Mississippi 225-7
mojarra, blacktip 96
mollusk shell middens 165
mollusks
  digging/dragging for 98, **99**, 157, *158*, 210, 212, *213*
  dredging for 219, *220-1*
  East Africa 83, 85
  Indonesia 175-6
  New Zealand 135, *140*

western Pacific 163
see also named species and groups
Mombasa Marine Park, Kenya *88-9*
Monaco *258*
monitoring, global *168-9*, 184
Monroe County, Florida 224-5
Montepuez Bay, Mozambique **23**, 95, 96, 97, 98
Moreton Bay, Queensland 119, 123
Morocco *253*
Morone saxatilis 209
morphological characteristics 5
  Halodule spp. 82-3
  Halophila stipulacea 66
  and nutrients 82-3
  Phyllospadix spp. 195
  Zostera spp. 190, 194-5
moshiogusa 188
Mozambique **23**, **94**
  biogeography 93
  fisheries 95-8
  protected areas *258*
  seagrass losses and coverage *14*, 98-100
  seagrass species *253*
  threats to seagrasses 97
MPAs *see* marine protected areas
mtimbi 86
mullet 248
Mullus surmuletus 40
murex 98
Musculista senhousia 205
mussels 205
  blue 28, 34, 209-10, *213*
  culture 43, 213
  harvesting 210, 212, *213*
Mya arenaria 213
Myanmar *253*
Mytilus edulis 28, 34, 209-10, *213*

N

names, local, for seagrasses 135-6, 188
Nantuna Island, Indonesia *178*
Narragansett Bay, Rhode Island 208, 210
natural hazards 18, 218, 219, 227, *231*, 237-8, 238
net dragging 213
Netherlands 39, *41*, 43, 44, *253*
Netherlands Antilles *253*, *259*
New Caledonia *252*
New England, US 210, 212
New Hampshire, US 207-8, *208*, 210, *212*, **214**
New Ireland, Papua New Guinea *169*
New Jersey, US 207, 209, *212*
New South Wales 121, 127
New York State, US 11, 209, *212*
New Zealand
  ecosystem description 134-5
  endemic species *140*
  estimated coverage *14*

historical changes in distribution 135-8, **139**, 141
  policies and protection 139-41
  seagrass species and distribution 134, *135*, 136-7, *253*
  threats to seagrasses 138-9
  use of seagrasses 136-7
New Zealand Fisheries Act (1996) 140-1
Newfoundland 207
Nicaragua 237, 238, *253*, *259*
Nicobar Islands 101, 102-3, *104*, 105
Ninigret Pond, Rhode Island 210, *211*
nomenclature 5-7
non-governmental organizations 150, 167, 170
North America
  mid-Atlantic coast 216-22
  Pacific coast 199-204
North Atlantic, western 9
  biogeography 209-10
  ecosystem description 207-9
  historical perspectives 210
  policy and protection 213-14
  seagrass coverage 212
  seagrass losses 210-11
  threats 212-13
North Carolina, US 216, 216-17, 218-20, 222
northeast Pacific 9, 199-205
  biogeography 200-4
  historical perspectives 203-4
  policy and protection 205
  seagrass coverage 204
  threats to seagrasses 204-5
Northern Ireland 43
Northumberland Strait, Canada 207, 212
Norway 27, 30-1, *253*
Notoacmea helmsi (scapha) 140
Nova Scotia 207, 212
nutrient cycling 17, *84*, 85
nutrient loading 1
  Gulf of Mexico 232
  Indonesia 171-2
  and leaf morphology 82-3
  mid-Atlantic coast 219, *220*
  New Zealand 139-40
  northeastern US *211*
  Red Sea 71-2
  Tampa Bay *226*
  Western Australia 112-13
  western North Atlantic *211*, 212
  see also eutrophication; sewage pollution
Nyali-Shanzu-Bamburi Lagoon, Kenya 86, *88-9*

O

ocean currents 191
Odontodactylus scyllarus 175
oil pollution
  Arabian Gulf 75-7, 78, 80
  Caribbean 239-40

East Africa 87
Europe 44, 50
Malaysia 155
northeast Pacific 205
Okinawa Island, Japan 189
Oman 75, *253*
Öresund region 28, 31, 34
Osmerus mordax 209
Otsuchi Bay, Honshu *190*
overwater structures 34, 204, **205**
oysters 97
    culture 43, 154, 197, 205
    dredging *220-1*

P

Pacific flyway *200*
Pacific Ocean *see* northeast Pacific; western Pacific islands
Pagrus auratus 135
Pagurus spp. 175
Palau 162, 164, 166, 167, *169*, *253*, *259*
Pamlico Sound, North Carolina 218
Panama 235, 236, 241, *253*, *259*
Panulirus argus 237, 246
Panulirus interruptus 203
Papua New Guinea 161, 165, 166
    biogeography 162
    policies and protection 167, *259*
    seagrass monitoring *169*
    seagrass species *254*
Pari Island, Indonesia 176-7
Parque Nacional Arrecifes Puerto Morelos *240*
Parque Natural Tayrona, Colombia *239*
parrotfish 83, **85**, 86, 95, 96-7
Parupeneus barberinus 97
Patos Estuarine Lagoon, Brazil 244, *248*
Pecten novazelandiae 135
Pengkalan Nangka, Malaysia *158*
Perth City, Australia 112, 115
Philippines *14*, 183-4, *254*, *259*
phosphorus levels 172, 221
photosynthetic studies 85
Phyllospadix spp. 203, 284
Phyllospadix iwatensis
    by country/territory *251-5*
    Japan 186
    Korea 193, 194, 197-8
    morphological features 195
    range map **284**
Phyllospadix japonicus
    by country/territory *251-5*
    Japan 186
    Korea 193, 194, *197*, 198
    morphological features 195
    range map **285**
Phyllospadix scouleri
    by country/territory *251-5*
    northeast Pacific 199, 202-3
    range map **285**
    uses 199

Phyllospadix serrulatus
    by country/territory *251-5*
    northeast Pacific 202, 203
    range map **285**
Phyllospadix torreyi
    by country/territory *251-5*
    northeast Pacific 199, 202, 203
    range map **286**
    uses 199, **205**
phytoplankton 44, 102, 113
Picnic Cove, Shaw Island *203*
pike 33
Pilbara coast, Western Australia 110
Pinctada nigra 97
Pinna muricata **3**, 97
Pinna nobilis 51
pinna shell **3**, 51, 97
Pleuroploca trapezium 98
Pohnpei, Micronesia *168*
Poland 33, *254*
pollution
    Caribbean 239, 241
    heavy metals 33, 87
    Mediterranean 71-2
    Red Sea 71-2
    sewage *51*, *128-9*, 161
    South America 247
    thermal 72
    toxic chemicals 113-14
    Western Australia 112, 113-14, 115
    see also eutrophication; nutrient loading; oil pollution
polychaetes 43, *107*, 114
Polysiphonia 60, 61, 62
Polysiphonia japonica 197
Port Phillip Bay, Australia 123
Portsmouth Harbor, Maine 207, *208*, 214
Portugal 39, 43, *254*
Portunus pelagicus 83, 148, *156*
Posidonia angustifolia 251-5, **277**
Posidonia australis **114**
    by country/territory *251-5*
    Eastern Australia 122
    range map **277**
    Western Australia *113*, *117*
Posidonia coriacea 251-5, 262, **278**
Posidonia denhartogii 251-5, 262, **278**
Posidonia kirkmanii 251-5, 262, **278**
Posidonia oceanica
    by country/territory *251-5*
    eastern Mediterranean 65-7, 68
    range map **279**
    seaweed competition 55
    uses 55
    western Mediterranean 48, *49*, *50*, *52-4*, 55-6
Posidonia ostenfeldii 251-5, 262, **279**
Posidonia robertsoniae 251-5, 262
Posidonia sinuosa *113*, 123, 251-5, **279**
Posidoniaceae *6*
Potamogeton 6, 29
Potamogeton pectinatus 61, 63
Princess Charlotte Bay 121

productivity *15*, 16-17
    East Africa 85
    Indonesian seagrasses 171-4
    measurement **180**
    Mediterranean 52-3, 55
    and nutrient availability 171-2
    Thalassodendron ciliatum 84, 85
    western Europe 40
    Zostera 196
protection of seagrasses 20, 23
Protoreaster nodosus 176
Pseudopleuronectes americanus 209
Pseudosquilla ciliata 175
Pteragogus flagellifera 97
Puerto Galera, Philippines 184
Puerto Morelos Reef National Park *240*
Puerto Rico 238, 239, *254*, *259*
Puget Sound 204, 205
push seines 148-9
Pyrene versicolor 175

Q

Qatar 74, **75**, *254*
queen conch 236, 237
Queensland, Australia 119, 122, 123, *125*
    seagrass protection 130-1
Queensland Fisheries Act 130-1
Quirimba Islands, Mozambique 95, 98, *99*, 100

R

rabbitfish 97, 183
Ramsar Convention 141
rapid assessment technique *77*
ray **232**
razorclam, stout *248*
Red Data Book species 188
Red Sea **67**
    pollution 71-2
    seagrass distribution 66, 68-71, 77
Redonda Island, Brazil *27*
reefs, artificial 197
remote sensing 8, 86
reproduction 162, 187-8
research
    Korea *196*
    Mediterranean 48
    photosynthesis 85
    South America 244
Resource Management Act (1991), New Zealand 140
restoration of seagrasses 23
    north western Atlantic *208*, 209
    Wadden Sea *41*
restricted range species 12-13
Réunion *259*
Rhode Island, US 208-9, 210, *211*, *212*
Rhodes (Rodos) 65
Rias Coast, Honshu *190*
rimurāhia 136

Rio de Janeiro 246
roach 33
Rodos (Rhodes) 65
Romania *254*
Rotaliina 175
Rufiji Delta, Tanzania 87
Ruppia cirrhosa
   Caspian Sea 62
   eastern Mediterranean 67
   western Europe 38, 39
Ruppia maritima
   Caribbean 234, 235, *236*
   Gulf of Mexico 225-6, 227, 230, 231, 232
   India *104*
   Indonesia 171, 180
   Korea 193, *194*, *197*, 198
   Malaysia 154
   mid-Atlantic region 216-17
   Mozambique *99*
   northeast Pacific 199, *200*, 203
   South America 245, **246**, 247-8
   Thailand 144, *146*
   var. *longpipes* 203
   var. *maritima* 203
   western Europe 38
   western North Atlantic 207, 208, 209
Ruppiaceae, taxonomy 6-7
Russian Federation 62, *254*, *259*
Rutilus rutilus 33
Ryukyu Islands, Japan 185, 186, 187, 188, 189

S

Sabah, Malaysia 153, 154, 157-8
Sabella spallanzani 114
SACs (special areas of conservation) 46
sailing *see* boating
St Kitts and Nevis *254*
St Lawrence River 212
St Lucia 239, *254*, *259*
St Vincent and the Grenadines *254*, *259*
Saleh Bay, Indonesia *179*
salinity
   Arabian Gulf 74
   Aral Sea 63
   Baltic Sea 28
   Caspian Sea 62
   Gulf of Mexico *228-9*, 231-2
   mid-Atlantic region 217
salinity tolerance
   Halodule wrightii *228-9*, 231-2
   Halophila spp. 74
   Zostera noltii 62, 63
Salish people *203*
Samoa 162, *254*
San Francisco Bay 204
San Juan Archipelago *203*
sand mining 155-7, 239
São Paulo 246
São Tomé and Principe *254*

Sarasota Bay, Florida 225
Sarawak 153
Sardinia 52, **55**
Sargassum spp. *155*
Sargassum muticum 44
Saudi Arabia
   Arabian Gulf 74-80
   protected areas *259*
   Red Sea coast **67**, 69-70
   seagrass species *254*
Saudi Arabia-Bahrain causeway 76
scallop
   bay 209-10, 216
   New Zealand 135
scallops, Chilean 246
Scandinavia
   historical and present coverage *14*, 30-4
   policy and protection 34-5
   seagrass species distribution 27-30
   threats to seagrasses 34
   uses of seagrasses 30
Scotland 38, 43
sea cucumbers 85, *96-7*, 98, 148, 154, 163-4, 176
sea goose (black brant) *200*
sea horses *12*, **13**, 40, 51
sea rabbit 40
sea snake, banded **166**
sea stars 51, 176
sea turtles
   Arabian Gulf *79*
   Caribbean 238
   East Africa 83, 88-9
   Eastern Australia 119
   endangered species 183
   grazing *247*
   India 102
   Pacific 164
   South America 245, *247*
sea urchins 40, 51, *96-7*, 98, 176
   grazing and competition 86, *236*, 241
sea-level rise 167, 219, 241
Seagrass-Watch *168*, 184
SeagrassNet 168-9, 184
seaweeds
   alien species 55, 68, 69, 71, *155*
   farming 87-8, **176**, 197
   see also macroalgae
Sebastes inermis 196
Secchi depth 27, 32, **33**
Seychelles *259*
sediment loading 1
   East Africa 87
   Eastern Australia 123
   New Zealand 138-9
   North America 210-11
   Western Australia 113
sediment stabilization 17, *116-17*
seeds
   consumption *190*
   Enhalus acoroides **149**, 166-7
   human uses of 148, 166-7
   production 121, 187-8

seine net fishing 95-6, 98, 148-9
Senegal *254*
Seri Indians 199
Sermata Islands, Indonesia *179*
Serranidae 147-8
sewage pollution
   Green Island, Australia *128-9*
   western Mediterranean *51*
   western Pacific islands 161
Seychelles **94**, 95, 99-100, *254*, *259*
Shark Bay, Western Australia 109-10, 115, *116-17*
Shaw Island, Washington *203*
shellfish
   Brazil 246
   culture 43, 154, 197, 205, 213
   harvesting 3, 97-8, **99**, 210, 212, *213*, 219, *220-1*
   Mozambique 97-8
   western Europe 40, 43-4
   western North Atlantic 209-10, 212, 213
   see also named groups and species of shellfish
shipping 44, 130, 204, **205**
   see also boating; oil pollution
shoal grass *see Halodule wrightii*
shoot density 173, 231-2
shoot height 194-5
shrimp fisheries 77-8, 128-9, *231*, 246
Sicily *52*
Siderastrea stellata 245, **246**
Sierra Leone *254*
Siganus canaliculatus 183
Siganus sutor 95, 96
Sinai 71
Singapore *254*, *259*
Skagerrak Strait, Denmark 34, 35
slime molds 18, 38, 138, 207
Slovenia *259*
Smaragdia virdens 236
smelt 209
snails
   mud 44
   small green 236
snake, banded sea **166**
snapper
   blackspot 97
   New Zealand 135
Solomon Islands 163, 164, *254*
Somalia *254*
South Africa 10
   biogeography 93-4
   present coverage 100
   protected areas *259*
   seagrass species *254*
South America
   biogeography 245-6
   ecosystem description 243-4, **245**
   historical perspectives 246
   policy and protection 248-9
   research data 244
   seagrass species and distribution 246-8, *254*
South Sea 193

Spain 39, 43, 51, *53-4*, 55
   marine protected areas *259*
   seagrass species *53*, *254*
special areas of conservation (SACs) 46
Spermonde Archipelago, Indonesia 171-2, 174
Sri Lanka *254*
Stethojulis strigiventer 96
stomatopods 175
storms 120, 123, 167, 218, 219
Strombus gibberulus 85
Strombus gigas 237
Strombus trapezium 85
Sudan *254*
Suez, Gulf of **67**, 68-9, 70-1
Sulawesi, Indonesia 171-2, 174, *178*
surfgrasses see *Phyllospadix* spp.
swan
   black 141
   black-necked 245
   whooper *189*
Sweden
   east coast 28-30, *31*
   policies and protection 35
   seagrass species *254*
   west coast 27-8, *31*
Syngnathoides biaculeatus 177
Syria 68, *255*
Syringodium filiforme
   by country/territory *251-5*
   Caribbean 234, 235, *236*, 239, *240*, 241
   Gulf of Mexico 224-31, **232**
   range map **275**
Syringodium isoetifolium **2**
   by country/territory *251-5*
   Eastern Australia *128-9*
   India *104*, **105**
   Indonesia *172-4*, *178-9*
   Malaysia *155*, *156*
   Mozambique and southeast Africa 95, *96*, *99*
   range map **275**
   Red Sea 69-71
   southeastern Africa 95, *96*, 100
   Thailand 144, 145, *146*, *147*
   western Pacific 162, 163, **166**

T

Tabasco state, Mexico 230
Tagelus plebius 248
Taka Bone Rate, Indonesia 174, *178*
Talibong Island, Thailand 145, *146*, 149
Tamaulipas 230
Tampa Bay, Florida *226*
Tanjung Adang Laut, Malaysia 154, *155*
Tanzania
   biogeography 82-3, 85
   policies and protection 88-90, *260*
   seagrass coverage 87-8
   seagrass productivity and value 85-7
   seagrass species *255*
   threats to seagrasses 86-8
Tarut Bay, Saudi Arabia 74-5, *75-6*, 77
Tasmania **121**, 122, 127-8
taxonomy 5-7, *262*, *263*, *269*, *270*, *271*, *272*, *275*, *276*, *277*, *280*, *284*, *286*
TBT (tributyltin) 114
Te Angiangi Marine Reserve, New Zealand 141
tectonic activity 205, 238
Teluk Kemang, Malaysia *155*
teripang trade 176
Texas 227-30
Thailand
   biogeography and seagrass species 144-6, *255*
   dugong 145, *147*, 148-9
   estimated coverage *14*
   historical perspectives and losses 146-8
   policy and protection 149-50, *260*
   threats to seagrasses 148-9
   uses and value of seagrasses 146-8
Thalassia *269*
Thalassia hemprichii **2**
   biomass and shoot density 173
   by country/territory *251-5*
   Eastern Australia *128*
   India *104*, **105**, *106-7*
   Indonesia 171, *172-4*, *177-9*
   Philippines 183
   range map *269*
   Red Sea 69-71
   southeast Africa 93, 94, 95, *96*, *99*
   Thailand 144, 145, *146*
   western Pacific 162-3, *168*, *169*
Thalassia testudinum
   by country/territory *251-5*
   Caribbean 234-41
   female flower **240**
   grazing *236*
   Gulf of Mexico 224-32
   range map *269*
Thalassodendron ciliatum
   by country/territory *251-5*
   East Africa 82, *84*
   Indonesia 171, *172-3*, 174, *177-8*
   Mozambique and southeast Africa 93, 94-5, *96*, *99*
   Pacific 162, 183
   Philippines 183
   productivity *84*, 85
   range map **276**
   Red Sea 68-71
   southern Africa 94, 95
Thalassodendron pachyrhizum *251-5*, **276**
thermal pollution 72
threatened species
   associated biota 17, 40, 183
   seagrasses 12-13, 188, *236-7*
threats to seagrasses 1, 7, 18-20
   see also named threats and under named countries and regions

Tobago *14*, 238
Tokyo Bay 188
Tonga 162, *255*, *260*
Torres Strait 120, 125-6
tourism
   Caribbean 238-9
   East Africa 86-7
toxic chemicals 113-14
   Western Australia 113-14
Tozeuma spp. 175
transplantation of seagrasses 23, **176**, *208*, 209
transport infrastructure *164-5*, 166
trap fisheries 96-7
trawling *52*, *53*, 87, 115
tributyltin (TBT) 114
Trichechus manatus *236*, 238, 245
Trinidad and Tobago 238, 239, 241, *255*, *260*
Tripneustes gratilla 86, 98, 176
Trochus niloticus 163, 175
trochus shell 163, 175
Tropical Storm Agnes 218
tropics, stresses to seagrasses 162-3
Tudor Mangrove Creek, Kenya 83
tulip shells 98
Tunisia 55, *255*, *260*
Turkey 60, 68, *255*
Turkmenistan *255*
Turks and Caicos Islands *255*, *260*
turtle grass see *Thalassia testudinum*
turtles see sea turtles

U

Ukraine *255*, *260*
Ulva 87, 102
Ungwana Bay, Kenya 88
United Arab Emirates 74, *75*, 77, 78, 80, *255*
United Kingdom
   policy and protection 44, 46, *260*
   seagrass species and coverage 38, 43, *255*
United States
   Gulf of Mexico 224-30
   marine protected areas *260-1*
   mid-Atlantic coast *14*, 216-22
   north Atlantic coast *14*, 207-14
   Pacific coast *14*, 199-204
   seagrass protection 213-14, 219-22
   seagrass species *255*
United States (US) Army Corps of Engineers 214, 220
urchins see sea urchins
Uruguay 244
uses of seagrasses *16*
   agricultural 61, 62, 188, 197
   ancient Egypt 65
   Australia 122, 128-9
   Black Sea 61
   East Africa 86

human consumption 87, 122, 136, 148, 166-7, 199, 204
Japan 188
Korea 197
Malaysia 154
medicinal 148
Mediterranean region 55
New Zealand 136-7
North America 199, 204, **205**
Scandinavia 30
seeds 148, 166-7
Thailand 146-7, 148
United States 217
western North Atlantic 210
western Pacific Islands 166-7
see also value of seagrasses

V

value of seagrasses *15*, 16-18
 to fisheries see fisheries
 see also ecological value; economic value; uses of seagrasses
Vancouver Island 201, *202*
Vanuatu 162, 164, 166, *255*
Vaucheria dichotoma 63
Venezuela 238, 239-40, *255*, *261*
Venice Lagoon *50*
Veracruz 230
Victoria, Australia 121-2, 123, 127
Viet Nam *14*, 183, 184, *255*, *261*
Virgin Islands 236, 241, *255*, *261*
Virginia, US 216, 217-18, 221-2

W

Wadden Sea 39, *41*, 43, 44
Wakatobi Island *179*
Wales 43
Waquoit Bay, Massachusetts 210
wasting disease 18
 Black Sea 59-60
 mid-Atlantic region 217-18, 219, *220*
 New Zealand 138, 139
 Scandinavia 31, **32**
 western Europe 38, *41*, 42
 western North Atlantic 207, 210, 212, 213
water quality see eutrophication; nutrient loading; pollution; sediment loading
West Africa 10
West Timor *179*
Western Australia
 biogeography 110-11
 ecosystem description 109-10
 mechanisms of seagrass decline 111-14
 policy and protection 115-16
 seagrass coverage *14*, 114
 Shark Bay *116-17*
 threats to seagrasses 114-15

western Pacific islands 161, *252*
 biogeography 161-4
 historical perspectives and seagrass losses 164-6
 policy and protection 167, 170
 present seagrass coverage 166
 seagrass monitoring *168-9*
 threats to seagrasses 162-3, 167
 uses of seagrasses 166-7
Western Samoa 255
Westernport Bay, Australia 122, 123, *126-7*
Whanganui (Westhaven) Inlet, New Zealand 134, *137*, 141
widgeon grass see *Ruppia maritima*
women 97, 98
World Heritage Sites
 Great Barrier Reef 23, 120-1, *124-5*, *126-7*, 131
 Shark Bay 109-10, 115, *116-17*
World Seagrass Distribution Map 3, 7-8, 13, 14-16, **21**
wrasse
 flagfin 97
 three-ribbon 96

Y

Yad Fon Association 150
Yamada Bay, Japan *190*
Yamdena Islands, Indonesia *179*
Yap, Micronesia 163, 166
Yellow Sea 193
Yemen *75*, *255*
Yucatán Peninsula 230-2, *240*

Z

Zannichellia palustris 33
Zanzibar 83, 85, 87-8
Zeuxo sp. *190*
Zostera
 flowering and fruiting 187-8
 morphological variation 194-5
 phylogenetic studies *196*
 taxonomy 280
 uses 61, 62
Zostera angustifolia 38
Zostera asiatica
 by country/territory *251-5*
 Japan 186, 187, 188
 Korea 193, 194, 195, 197-8
 northeast Pacific 200, 203
 range map **280**
Zostera caespitosa
 by country/territory *251-5*
 Japan 186, 187, 188
 Korea 194, 195, *197*, 198
 morphology 195
 range map **280**
Zostera capensis 93-4, *96-7*, 98, **99**, 100, *251-5*, **281**
Zostera capricorni
 Australia 122

New Zealand 134-7, 141
 range *251-5*, **281**
 taxonomic revision 262
Zostera caulescens
 by country/territory *251-5*
 Japan 187, 189, *190*
 Korea 194, 195, 196, 197
 morphological features 195
 range map **282**
Zostera japonica
 by country/territory *251-5*
 Japan 185-6, 187
 Korea 194, 195, 196, *197*, 198
 morphological features 195
 northeast Pacific 201, 202, 203-4, 205
 range map **282**
 Viet Nam 184
Zostera marina
 by country/territory *251-5*
 ecological value 199
 Euro-Asian seas 59-63
 flowering and fruiting 187-8, **189**, 201
 Japan 185, 187-8, 189
 Korea 194-5, 196, 197
 Mediterranean 48-9, *50*, *52-3*
 mid-Atlantic coast 216-22
 morphological variation 194-5
 northeast Pacific 199, 200-1, 204
 northwest Atlantic 207-14
 productivity and biomass 40
 range map **282**
 Scandinavia 27-34
 southernmost limit 216-17
 transplantation *208*
 uses 30, 61, *202*, 217
 variation with latitude 209
 western Europe 38-40, *41*, 43
Zostera mucronata *251-5*, **283**
Zostera muelleri *251-5*, **283**
Zostera noltii
 by country/territory *251-5*
 eastern Mediterranean 65, 66, 68
 Euro-Asian seas 59-63
 protection 44
 range map **283**
 salinity tolerance 62, 63
 Scandinavia 27, 28-9
 western Europe 38-44
 western Mediterranean 49, *52-3*, 54
Zostera novazelandica 134, *251-5*, 262
Zostera tasmanica
 by country/territory *251-5*
 Chile 243, 246
 Eastern Australia 121-2
 range map **284**
Zosteraceae 6